MESOPOTAMIAN MATHEMATICS, 2100–1600 BC

Technical Constants in Bureaucracy and Education

OXFORD EDITIONS OF CUNEIFORM TEXTS

VOL. XIV

MESOPOTAMIAN MATHEMATICS, 2100–1600 BC

Technical Constants in Bureaucracy and Education

BY

ELEANOR ROBSON

CLARENDON PRESS · OXFORD

OXFORD
UNIVERSITY PRESS

Great Clarendon Street, Oxford OX2 6DP

Oxford University Press is a department of the University of Oxford.
It furthers the University's objective of excellence in research, scholarship,
and education by publishing worldwide in

Oxford New York

Athens Auckland Bangkok Bogotá Buenos Aires Cape Town
Chennai Dar es Salaam Delhi Florence Hong Kong Istanbul Karachi
Kolkata Kuala Lumpur Madrid Melbourne Mexico City Mumbai Nairobi
Paris São Paulo Shanghai Singapore Taipei Tokyo Toronto Warsaw

with associated companies in Berlin Ibadan

Oxford is a registered trade mark of Oxford University Press
in the UK and in certain other countries

Published in the United States
by Oxford University Press Inc., New York

ISBN 0-19-815246-9

3 5 7 9 10 8 6 4

Printed in Great Britain
on acid-free paper by
Bookcraft (Bath) Short Run Books
Midsomer Morton

PREFACE

Mathematics was an integral part of Mesopotamian scribal culture: indeed, writing was invented towards the end of the fourth millennium BC for the express purpose of recording numerical information. Throughout the third millennium, counting and measuring systems were gradually revised in response to the demands of large-scale state bureaucracies. This development led in the end to the sexagesimal, or base 60, place value system (from which the modern system of counting hours, minutes and seconds is ultimately derived). At the same time, mathematical methods for managing labour, land and other resources evolved, as did techniques for training scribes in those methods. By the beginning of the second millennium — the Old Babylonian (hereafter OB) period — scribal education was producing what is now conventionally called Babylonian mathematics: the earliest known body of 'pure' mathematics in history. There has been much research over the last seventy years into the computational techniques, conceptualization and terminology of OB maths, but little exploration of its place in society. Yet it was one of the key elements of scribal training in the early second millennium and is thus pivotal to our understanding of the educational practices and intellectual culture of ancient Mesopotamia.

The aim of this book is twofold: to make a detailed, technical study of a small group of texts conventionally known as 'coefficient lists'; and to use the results of that study to explore the *Sitz im Leben* of mathematics in third and second millennium Mesopotamia. The introduction examines the history of the study of coefficient lists and identifies them as one of six genres of OB mathematical texts. There follows a discussion of the characteristics and function of coefficients, in both lists and mathematical problems. The main body of the work is a mathematical and philological discussion of all known OB coefficients, grouped according to their subject matter. The names and mathematical functions of many of them are established for the first time, which in turn leads to improvements in the understanding of many individual problem texts, and several large topics of OB mathematics. In particular, labour calculations for carrying building materials and the sophisticated geometry of grain-piles are elucidated.

Finally, two aspects of the wider historical and cultural context of OB are examined, the first of which is the clear origin of many coefficients — and much of the more practically orientated mathematics — in accounting and quantity surveying practices of the late third millennium. During that time a sophisticated accounting procedure was developed for some sectors of the state economy, whereby the output of work gangs could be monitored and compared to a theoretical maximum productivity. This over-bureaucratic system did not survive the end of the empire, and records of production from the following period reveal a truer picture of variable efficiency. Nevertheless, the rates of output deducible from the OB documents appear to cluster around the earlier ideals. A few of those work quotas appear in contemporary school practice texts, and many more seem to be precursors to the OB quantity surveying coefficients.

The other half of the historical exploration uses the coefficient lists and other material to examine some aspects of mathematics education in early Mesopotamia. Many of the characteristic features of OB mathematics were already discernible in the mid-third millennium, as others have convincingly demonstrated. Further, each of the text types known from OB was already in use, at least in prototype form. The fact that many of the OB coefficients have unequivocal antecedents suggests that the tail end of the third millennium saw the diffusion of mathematics across the Near East, as centrally trained numerate bureaucrats were dispersed across the empire. Later, without a formal network for the dissemination of scribal knowledge, local traditions and variants grew up, ranging from the use of technical terms and the grammatical forms in which they were couched, to mathematical refinements of problems and their solutions, and the adoption of new values for coefficients. Nevertheless, the shared character of this mathematics — its subject matter and style, and the format of the documents in which it was written — persisted.

This book began as a D.Phil. thesis in Oriental Studies, funded by the British Academy from October 1992 to September 1995, which was submitted to the University of Oxford in 1995. My examiners, Jens Høyrup and Farouk Al-Rawi, made many valuable comments and suggestions, which I have tried as far as possible to incorporate here. I am particularly indebted to Christopher Walker of the British Museum, who has so generously made available to me much hitherto unpublished material as well as his preliminary studies of some of those texts. Marie-Christine Ludwig, Aaron Shaffer and Cornelia Wunsch also freely gave me access to, and copies of, mathematical material in the British Museum tablets they were studying. The hand-copies of tablets in the appendices are all published with the permission of the Trustees of the British Museum and the Babylonian Section of the University Museum, Philadelphia.

Anne Kilmer kindly provided me with a copy of her unpublished manuscript (Kilmer 1985). Niek Veldhuis generously gave me access to his doctoral thesis (Veldhuis 1997) and much other interesting material. Marcel Sigrist allowed me to cite mathematical tablets from Sigrist (fc.). John Brinkman, Paul-Alain Beaulieu, and Tina Breckwoldt all took time to collate difficult and badly preserved texts for me at Chicago and at Yale. Steve Tinney and family were wonderfully welcoming and hospitable during my stay in Philadelphia. Brian MacGregor, John Taylor and the staff of the Griffith Institute Library, Oxford have been unfailingly helpful, as has John Moffett of the Needham Research Institute, Cambridge. I must also mention the British Society for the History of Mathematics, whose members have so often provided stimulating and positive feedback on my work. But it is a particular pleasure to thank David Fowler of the University of Warwick, who introduced me to Mesopotamian mathematics as an undergraduate and has been egging me on ever since.

In and around Oxford I have had advice, support and encouragement from many people, including, but not exclusively, Linda Bregstein, David Brown, Martin Keighery, Derek Kennet, Ezra Marcus, Roger Moorey, Michael Roaf, St John Simpson, Luke Treadwell, Andrew Wilson, and Sir Christopher Zeeman; the various habituées and residents of 77 Bullingdon Road in the summer of 1995; Anne Knowland and all my colleagues at Oriental Institute, especially those who allowed me to use their offices at one time or another; Hilary O'Shea and Enid Barker, my indefatigable editors at OUP; and most of all, Stephanie Dalley, the series editor, for being an unfailingly stimulating and sympathetic teacher, colleague, and friend.

I owe Jeremy Black for uncountably many things; it is impossible to express my gratitude adequately. Equally, my immeasurable thanks go to Annie Robson, June Barrow-Green, Katie Judge, Katja Goebs, Jaimie Lovell, Paula McMahon, and Janet Morgan, without whom life would be infinitely duller. Finally, many apologies to everyone else who deserves to be mentioned here but whom I have inadvertently overlooked. Needless to say, there is no one to blame but me for the errors, omissions, inconsistencies, and other failings of this book.

E.R.

Oxford, August 1997

CONTENTS

LIST OF FIGURES

LIST OF TABLES

ABBREVIATIONS AND CONVENTIONS

COEFFICIENT LISTS

This system of abbreviations is based on the sigla used in Kilmer (1960: 284). Transliterations of the texts can be found in §A.1.

Table 0.1: The twelve coefficient lists

Text	*Museum number*	*Provenance*[1]	*Date*	*Place of publication*	*Friberg*[2]
A	A 3553	unknown	OB	Kilmer 1960: text A[3]	Ka
B	CBS 10996	Nippur	MB/NB[4]	Kilmer 1960: text B[5]	Kb
C	IM 52916	Shaduppum	OB	Goetze 1951	G
D	TMS 3	Susa	late OB	*TMS*: text III[6]	BR
E	YBC 7243	Larsa?	OB	*MCT*: text Ue[7]	NSe
F	YBC 5022	Larsa?	OB	*MCT*: text Ud[8]	NSd
G	IM 49949	Sippar Amnanum	early OB	Edzard 1971: no. 236	E
H	IM 52301	Shaduppum	OB	Baqir 1950*b*	B
J	BM 47860	Babylon?	LB	Livingstone 1986: pl. 1[9]	L
K	BM 37096	Babylon?	LB	Here §A.1	FARa
L	BM 36776	Babylon?	LB	Here §A.1	FARb
M	U 17214r	Ur	OB	Figulla and Martin 1953: no. 881[10]	FM

OTHER ABBREVIATIONS[11]

2 N-T	tablets from the second season of excavations at Nippur by the Oriental Institute of the University of Chicago
A	Asiatic collection of the Oriental Institute of the University of Chicago
admin.	administration, administrative
AHw	= von Soden 1959–81
Akk.	Akkadian language
AO	Antiquités Orientales, held by the Louvre, Paris. With the suffix S: from Susa; with the suffix T: from Girsu (modern Tello)
approx.	approximate, approximately
ARM 6	= Kupper 1954
ARM 9	= Birot 1960
Ash	Ashmolean Museum, Oxford
AUAM	Andrews University Archaeological Museum, Berrien Springs, MI

[1] See §§2.1–2.
[2] That is, the set of abbreviations used by Friberg (1996*a*: 3–5).
[3] With collations by John Brinkman and the author, 1995 and 1997.
[4] See §2.2.1.
[5] With collations by the author, 1997.
[6] It was not possible to have this tablet collated.
[7] With collations from published photographs by the author, 1997.
[8] With collations from published photographs by the author, 1997.
[9] With collations by the author, 1995.
[10] First identified as a coefficient list by Friberg (1996*a*: 4).
[11] Bibliographical abbreviations are used to cite tablets only when the museum number is unknown to me. For places of publication of individual tablets, see in general the Index of Texts.

BM	British Museum, Department of Western Asiatic Antiquities
Böhl	Sammlung de Liagre Böhl, Leiden
CAD	= Oppenheim *et al.* 1956–
CB	Collection Béghin, Versailles (private)
CBS	Catalogue of the Babylonian Section, University Museum, University of Pennsylvania, Philadelphia
CMAA	California Museum of Ancient Art
CUA	Catholic University of America
Db_2	tablets from the Iraq Museum's second season of excavation at Tell Dhiba'i
Deir-ez- Zor	Deir-ez-Zor Museum, Syria
DP	= Allotte de la Fuÿe 1915
Duke	Duke University, Durham, North Carolina
DV 5	= Nikolskii 1915
ED	Early Dynastic period (*c*.3000–2350 BC)
ED III	Early Dynastic period, phase III (*c*.2500–2350 BC)
Erm	Hermitage Museum, Saint Petersburg
esp.	especially
fc.	forthcoming
fig.	figure
FLP	Free Library of Philadelphia
GAG	= von Soden 1952*b*
GN	geographical name
Haddad	tablets from the Iraq Museum's excavations at Me Turan (modern Tell Haddad)
HS	Hilprecht Sammlung, Jena
HSM	Harvard Semitic Museum
IM	Iraq Museum, Baghdad
Isin	First Dynasty of Isin (*c*.2017–1794 BC)
Ist Ni	Istanbul Museum, tablets from Nippur
Ist O	Istanbul Museum, tablets from Kish
Ist P	Istanbul Museum, tablets from Puzrish-Dagan
Ist S	Istanbul Museum, tablets from Sippar
Ist T	Istanbul Museum, tablets from Girsu
Ist Um	Istanbul Museum, tablets from Umma
K	Kouyunjik collection, British Museum Department of Western Asiatic Antiquities
Kass.	Kassite period (*c*.1600–1150 BC)
Larsa	Larsa Dynasty (*c*.1932–1763 BC)
LB	Late Babylonian period (538–0 BC)
M	tablets from the French excavations at Mari
M (10)	University of Michigan Collection, Ann Arbor
MA	Middle Assyrian period (*c*.1400–885 BC)
MAH	Musée d'Art et d'Histoire, Geneva
MB	Middle Babylonian period (*c*.1150–626 BC)
MCT	= Neugebauer and Sachs 1945
MDP 27	= van der Meer 1935
ME	Middle Elamite period (*c*.1350–1120)
metrol.	metrology, metrological
mill.	millennium
MKT	= Neugebauer 1935–7
MLC	Morgan Library Collection, held at Yale University
MM	Metropolitan Museum of Art, New York
N	Nippur Collection of the Babylonian Section, University Museum, University of Pennsylvania, Philadelphia
n.	note(s)
NA	Neo-Assyrian period (883–609 BC)
NB	Neo-Babylonian period (625–539 BC)
NBC	Nies Babylonian Collection, held at Yale University
NCBT	Newell Collection of Babylonian Tablets, held at Yale University
Noor	Noor Collection, held in the New Year Public Library, New York
O	Stovall Museum of Science and History, University of Oklahoma, Norman, Oklahoma
OA	Old Assyrian period (*c*.2000–1750 BC)
OAkk.	Old Akkadian period (*c*.2350–2150 BC)
OB	Old Babylonian period (*c*.2000–1600 BC)
pers. comm.	personal communication
pl.	plate
Platt	private collection, New York
Plimpton	Plimpton Library, Columbia University, New York
PN	personal name
PRAK	= de Genouillac 1924–5
PSD	= Sjöberg *et al.* 1984–
PTS	Princeton Theological Seminary
Rylands	John Rylands Library, University of Manchester
Sm	Smith collection, British Museum Department of Western Asiatic Antiquities
SP	Sumerian Proverbs Collection (see Alster 1997).
Str	Bibliothèque Nationale et Universitaire de Strasbourg
Sum.	Sumerian language
Syr	University of Syracuse, New York
TM	tablets excavated at Ebla (Tell Mardikh)
TMB	= Thureau-Dangin 1938
TMS	= Bruins and Rutten 1961
TSŠ	= Jestin 1937
U	tablets from the excavations at Ur
UET 3	= Legrain 1937–47
UET 5	= Figulla and Martin 1953
UET 6/2	= Gadd and Kramer 1966
UET 6/3	= Shaffer forthcoming
UET 7	= Gurney 1973
UM	Babylonian Section, University Museum, University of Pennsylvania, Philadelphia
Ur III	Third Dynasty of Ur (*c*.2150–2000 BC)
Uruk	Uruk period (*c*.3400–3000 BC), esp. phases III–IV (*c*.3200–3000 BC)
VAT	Vorderasiatische Abteilung, Tontafeln, Staatliche Museen, Berlin
W	tablets from the German excavations at Uruk (modern Warka).
Walker	Walker Arts Center, Minneapolis
YBC	Yale Babylonian Collection, Yale University

CONVENTIONS USED

All chapters and sections within chapters are numbered; cross-references to them take the form §1.2.3. Appendices are referred to as §A.1. Figures and tables are all labelled, and numbered consecutively by chapter, so that, for instance, Figure 3.4 is the fourth figure in §3 and Table 5.6 the sixth table in §5.

Cuneiform texts are referred to by museum number; for place of publication see Nemet-Nejat (1993: 251–90), with additions and corrections in the Index of Texts. Lower case roman numerals in brackets after a text number indicate the problem number on the tablet; upper case roman numerals without brackets refer to the columns of a tablet.

Number notation within transliterations and translations follows Friberg (1987–90: 534). Numbers are represented without any indication of absolute value (i.e., without 'sexagesimal points') in the transliteration. A semicolon marks the 'sexagesimal point' in the translation and commentary. A space is left between sexagesimal places, and where necessary decimal equivalents are given in the commentary.

Metrological relationships, unless otherwise indicated, are as given in Powell (1987–90); approximate metric equivalents of measurements are indicated in brackets where appropriate. Somewhat confusingly, the sar is the Old Babylonian metrological unit for both area and volume, as well as for brick quantity. In this work they are distinguished as follows:

> sar_a = area-sar: 1 nindan × 1 nindan (approx. 36 m^2);
> sar_v = volume-sar: 1 nindan × 1 nindan × 1 cubit (approx. 18 m^3);
> sar_b = brick-sar: 12 00 (720) bricks.

British mathematical terminology is used throughout so that, for instance, the word 'trapezium' describes a quadrilateral with one set of parallel sides while 'trapezoid' refers to a quadrilateral with no sides parallel.

In transliterations of cuneiform text,

<... > enclose signs which were unintentionally omitted by the scribe and have been restored by modern scholars;

« ... » enclose signs unintentionally added by the scribe;

[...] enclose signs which are missing because of damage to the tablet, and which have been restored by modern scholars;

⸢ ... ⸣ enclose signs which are partially damaged.

Syllabically written Akkadian is represented by *italic* typeface; logograms by SMALL CAPITALS. Texts entirely in Sumerian are in bold lettering. Because the font used to typeset the book has no capital g̃, uppercase and lowercase g̃ have been represented by g throughout.

In English translations, [...] enclose complete words or passages whose restoration is conjectural; uncertain translations are set in *italic* type.

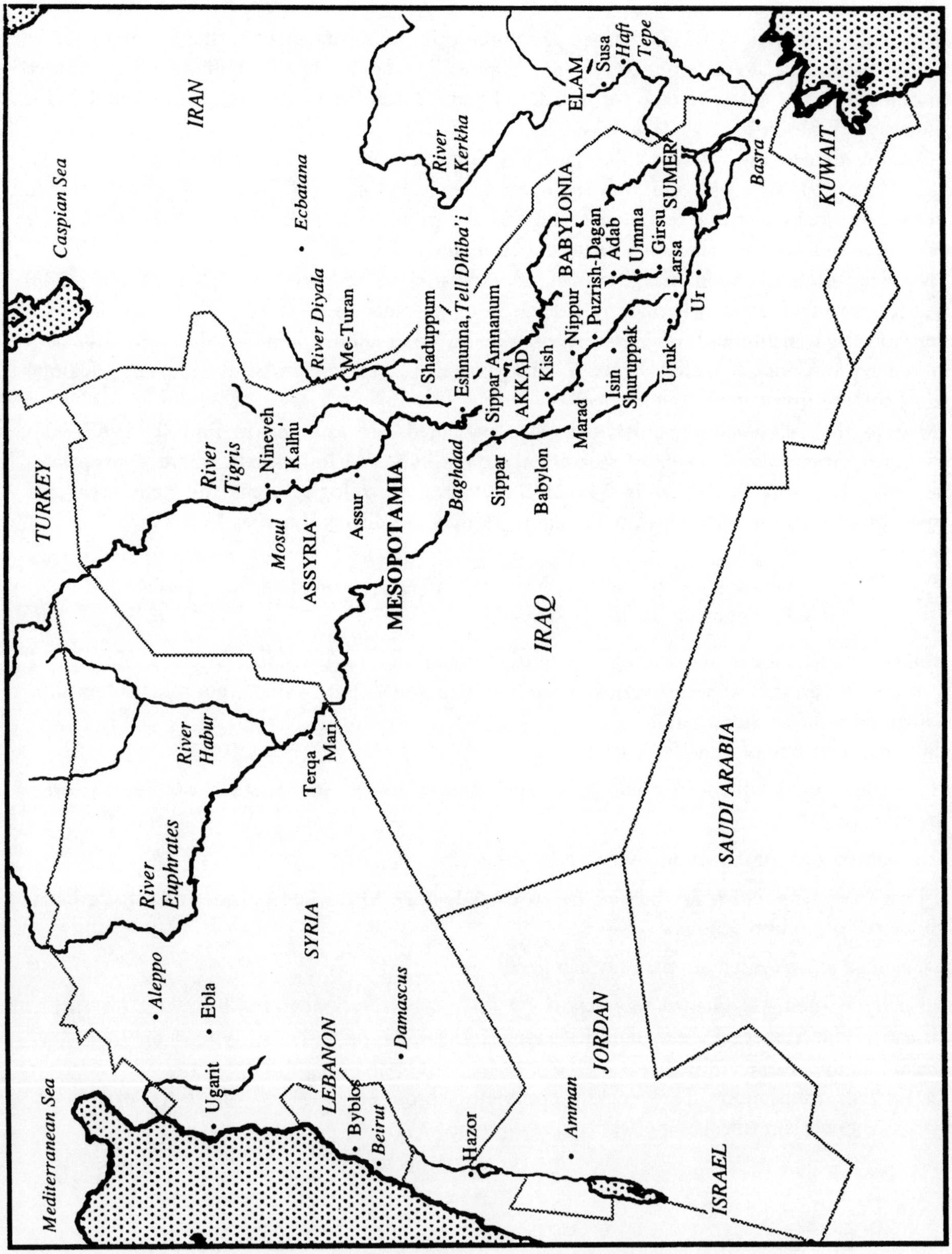

Figure 0.1: Findspots of mathematical tablets in the ancient Near East, and other sites mentioned in the book

• Ancient cities — ANCIENT REGIONS

• *Modern cities, rivers, and site names* — *MODERN STATES*

1. INTRODUCTION

Mathematics was one of the key elements of scribal training in early second millennium Mesopotamia and is thus pivotal to our understanding of its educational practices and intellectual culture as well as its administrative procedures. Many thousands of cuneiform tablets, from dozens of sites in modern-day Iraq, Iran, Syria, Lebanon, and Israel, have revealed a level of sophistication and abstraction which has also earned the ancient Near East a prominent role in the history of mathematics. This introduction briefly outlines the history of the study of Mesopotamian mathematics in general, and of coefficient lists in particular. After summarizing the aims, extent, and limits of the present work it concludes with a generic analysis of Old Babylonian mathematical texts.

1.1 MATHEMATICS WITHIN ASSYRIOLOGY

1.1.1 A brief history

Høyrup (1996*a*) divides the study of Mesopotamian mathematics into three historical phases.[1] The decade 1930–40 is characterized as 'the heroic era' (1996*a*: 1), in which Neugebauer and Thureau-Dangin pioneered, and made enormous advances in, the study of 'Babylonian' mathematics, culminating in the publication of *MKT* and *TMB* respectively. Naturally, at this stage the emphasis was on the mathematical content and terminology of the texts, with occasional references to Greek sources (e.g. Neugebauer 1931–2: 97). Too little was known to allow mathematics to be fitted into the broader picture of Mesopotamian civilization:

> Man darf ... nicht vergessen, daß wir über die ganze Stellung der babylonischen Mathematik im Rahmen der Gesamtkultur praktisch noch gar nichts wissen. (Neugebauer 1934: 204)

Tablets were either unprovenanced or without exact findspots (*MKT*: III 76). Although the mathematical texts were known to be school exercises (*MCT*: 1; *MKT*: III 83), no 'eduba' texts, which portray mathematics as an integral part of OB scribal education, were published until after both Neugebauer and Thureau-Dangin had stopped working on OB mathematics.[2] No link could be made between OB mathematics and the practices of contemporary administrative texts, as these rarely included explicit calculations. The coefficient lists, arguably the 'nexus between mathematical computation and administrative and engineering reality' (Høyrup 1994: 24), had not yet been discovered.

The second phase, from 1940 to the mid-1970s, is named by Høyrup the 'triumph of translations', during which it was 'business as usual' (1996*a*: 11, 14). The methodology which had been established before the war continued unchallenged by Bruins, Baqir and their

[1] Preceded by an 'initial struggle' from the mid-1850s to understand the various number systems in cuneiform texts (Friberg 1982: xiii). The first problem texts, BM 85194 and BM 85210, were published in hand-copy at the turn of the century (King 1900); OB mathematical problems — VAT 6598 (xviii), (xxi) — were first successfully interpreted by Weidner (1916); see now §A.4.

[2] See Sjöberg (1976) and bibliography there.

contemporaries. With few exceptions,[3] all those now working on the subject were from mathematical, not Assyriological, backgrounds, and interest in the mathematical content remained dominant over philology and history. Even as new finds were being made in the ancient kingdoms of Elam and Eshnunna, it was still the 'algebraic' and geometrical techniques of 'Babylonian' mathematics that found their way, modernized and distorted,[4] into general textbooks (e.g. Beckmann 1971: 19–20; Boyer 1989: 29–50; Eves 1983: 22–44; Kline 1972: 3–14). The study of 'pure' mathematics was seen as more significant from a modern European viewpoint,[5] but crucially it was easier, being less dependent on linguistic and historical-cultural knowledge, as this popular book on the early history of mathematics makes plain:

> We shall not discuss [OB] Applied mathematics such as compound interest, problems related to excavations and constructions, to alloys, etc. The technical terms which are involved are difficult to understand; moreover, these matters are of little importance for the history of mathematics. (van der Waerden 1954: 80)

There was still little concern with wider issues, although they were by then accessible to solution. In particular, modern knowledge of mathematics in the OB scribal schools might have been vastly different had Gadd and Kramer fully published the exercise tablets from OB Ur (1966), around forty of which contain uncopied calculations.[6] Gordon (1959) in his study of Sumerian proverbs published copies or photographs of over a dozen tablets containing metrological or mathematical material as well as the proverbs themselves,[7] yet did not refer to this fact in the text of the book. It now turns out that many of the Nippur school tablets contain a similar mixture of subjects (Falkowitz 1983–4; Veldhuis 1997). A close reading of the chapters on table texts in *MKT* shows that Neugebauer too was aware that mixed-subject school texts existed (e.g. *MKT*: I 48, 50). Material such as this could have yielded vital information on the context in which mathematics was taught, and perhaps even techniques of instruction. The vast Ur III administrative archives were starting to be published in quantity, but perhaps because of their very size, no large-scale study of the mathematical knowledge that can be inferred from them has yet been attempted.[8]

At this time Vaiman was a lone, and largely ignored, voice stating the importance of 'applied' mathematics and setting a far-sighted and still relevant agenda for its study:

> (1) Vor allem gilt es, den Umfang der Kentnisse auf dem Gebiete der angewandten und also der theoretischen Mathematik zu ergründen.
>
> (2) Ferner müssen wir feststellen, wann die oder jene Grundsätze der angewandten Mathematik auftauchten. Die angewandte Mathematik erreichte ihren Höhepunkt zur Zeit der III. Ur-Dynastie. Später haben wir es schon mit der Entwicklung der theoretischen Mathematik zu tun, die sich allerdings in der altbabylonischen Zeit entfaltete.
>
> (3) Wir müssen feststellen, inwiefern von der Mathematik abgeleitete Begriffe neue mathematische Vorstellungen förderten, die ihrerseits kaum mit praktischen Anforderungen der Gesellschaft unmittelbar zu

[3] Baqir (1950*a*, *b*; 1951), Goetze (1945; 1946; 1951), and Sachs (1947; 1952), who restricted themselves to editions of texts, and von Soden (1952*a*; 1964; 1977), whose interests were primarily philological.

[4] For instance, Plimpton 322, possibly the most unrepresentative of OB mathematical texts, was, and still is, given disproportionately wide — and modernizing — coverage (e.g. Joseph (1992: 115–18); van der Waerden (1983: 2–5). But more sensitive treatments of its contents are now beginning to appear: see, for instance, Burton (1995: 70–7); Deakin (1996*a*, *b*).

[5] Both as a precursor to Greek (and therefore modern European) mathematics and as the first manifestation of 'mathematicians', with all that implies about the nature of the mathematical texts themselves.

[6] See §§1.3.3 and 10.4; hand-copies and transliterations in §A.5.

[7] See §10.4 for a discussion of the phenomenon of mixed-subject school texts.

[8] But see now Englund (1988; 1990; 1991); Nissen *et al.* (1990); Snell (1982) on various Ur III quantitative administrative practices.

tun hatten. Wir müssen also klarlegen, inwiefern das Entwicklungsniveau der theoretischen Mathematik von dem der angewandten Mathematik abhängig war. [...]

(4) Schließlich ist die wirtschaftliche und technische Basis zu ergründen, von der die Entwicklung der sumerisch-babylonischen angewandten Mathematik bedingt war. Diese Mathematik, wie heute ersichtlich, war eher eine Mathematik der 'Buchhalter' und 'Wirtschaftler', als der 'Techniker' und 'Ingenieure'. Vor allem heißt es, nicht nur die mathematischen Texte schlechthin, sondern auch den mathematischen Inhalt der wirtschaftlichen Urkunden systematisch zu erforschen. (Vaiman 1960: 1–2)[9]

The revitalization began at the end of the 1960s, as Assyriology started to re-engage with mathematics, with a 'fresh start from sources through new approaches' (Høyrup 1996*a*: 17). Powell's thesis on Sumerian numeration and metrology (1971) was an early symptom of an increasing interest in pre-OB mathematics, metrology and numeration. Subjects of enquiry ranged from prehistoric accounting systems (Schmandt-Besserat 1977; 1978; 1979) and proto-Sumerian and proto-Elamite metrology (Friberg 1978) to Sargonic and Presargonic metro-mathematical exercises (Edzard 1969; Høyrup 1982; Powell 1976*a*, *b*) as well as three texts of mathematical interest from third millennium Ebla (Archi 1989; Friberg 1986). By 1982 Friberg could state with confidence that:

the pre-Babylonian origin of several central themes in 'Babylonian mathematics' can hardly any longer be put in doubt. It is only the theory of quadratic equations, with its many algebraic and geometric applications, that cannot yet with certainty be traced back to a date before the early OB period. (Friberg 1982: vii–viii)

1.1.2 Current thinking

Høyrup (1996*a*: 22) has hailed the 1980s as a new era in the study of Mesopotamian mathematics:

The development over the last fifteen years does express a trend, *viz.* away from the attitude that the mathematical knowledge of the Babylonians should be studied as *a step in the ladder leading to*, and hence from the *perspective of* modern mathematics, and toward the position of a multi-dimensional anthropology where mathematics is primarily studied in relation to historical context.

Indeed, the spheres of academic interest have continued to widen, both chronologically[10] and geographically,[11] and Høyrup in particular has done much to try and bring a broader perspective to the subject (e.g. 1990*a*, *b*; 1993*a*, *b*; 1994). Yet a recent major survey of Mesopotamian mathematics devoted only three of its fifty pages to 'practical mathematics' (Friberg 1987–90: 568–70),[12] concentrating instead on the 'algebraic' and geometrical techniques employed, and a potentially seminal work on practical mathematics in Mesopotamia went no further than stating that mathematical terminology was a 'reflection of everyday life' (Nemet-Nejat 1993; cf. Robson 1995*b*). It appears that Høyrup's ideal has not yet been fully realized.

[9] In short, Vaiman's aims are to establish: the extent of Mesopotamian 'applied' mathematical knowledge; the relative chronology of its development; the extent to which the 'theoretical' mathematical advances were dependent on this; and the economic and technical background which conditioned its growth.

[10] e.g. LB texts: Britton (1991–3), Friberg (1993); Friberg, Hunger and Al-Rawi (1990), von Weiher (1993); prehistory: Nissen, Damerow, and Englund (1990); Schmandt-Besserat (1992). Also Damerow and Englund (1987); Whiting (1984).

[11] e.g. Mari: Charpin (1993); Soubeyran (1984); Terqa: Rouault and Masetti-Rouault (1993: no. 298); Tell Haddad: Al-Rawi and Roaf (1984); Haft Tepe: Herrero and Glassner (1996: no. 289).

[12] But see now Friberg (1996*a*, *b*).

1.2 THE STUDY OF COEFFICIENTS

1.2.1 Previous work on coefficients

The very first mathematical texts to be published, BM 85194 and BM 85210 (King 1900), contained coefficients: parameters not given in the statement of a problem, but which were essential to their solution (see §1.3.6). By the time *MKT* and *TMB* were published in the late 1930s a further nine problem texts were known to use these factors,[13] some of which appeared to serve little purpose, and whose numerical values could seem arbitrary.[14] It was at first difficult to interpret the ancient word used to describe these numbers, owing to the ambiguous way in which it was written:

> Wenn ŠI-DU als *šiddu* ['length'] zu fassen ist, [...] keineswegs sicher ist. Möglicherweise handelt es sich aber um igi-gub 'Koeffizient'. (*MKT*: I 371 n. 5)

The phonetic complements in BM 85196 and AO 677: II 13 soon enabled the correct reading and meaning to be established as IGI.GUB/*igigubbûm* — 'fraction fixe' or 'fester Bruchteil' (Thureau-Dangin 1935: 6; *MKT*: II 42, 51). This allowed the word to be related to other terminology such as IGI and IGI.BI 'mutual reciprocals' and IGI.TE.EN 'proportion' (*TMB*: XII).[15]

However, it was not until *MCT* in 1945 that the first coefficient lists (here texts E and F) were published. These consisted of:

> lists of numbers (to which are added short explanations) which occur in the mathematical texts. We find here, e.g., coefficients referring to bricks, work assignment, etc. — in short, just those parameters which must be known by anyone dealing with various types of mathematical texts. (*MCT*: 38)

Several more problem texts which used coefficients were also included in *MCT*.[16] Over the next twenty-five years seven more lists were published (texts A–D, G–H, M in this work), along with pertinent problem texts from Susa and Shaduppum.[17] The only major study of coefficients as a group went no further than sorting them into various categories, based on the orders within the lists themselves. It was 'meant as a text presentation, rather than a mathematical or lexical contribution' (Kilmer 1960: 274) and as such contains little analysis of the material, although an attempt was made to include in that catalogue coefficients from problem texts as well.[18]

Since then, piecemeal work has been done on individual coefficients, especially those related to geometry (Friberg 1987–90: §5.4; Muroi 1992*b*; Vaiman 1959; 1962; 1963) and to a lesser extent bricks and building (e.g. Friberg 1996*a*; Powell 1982*a*; 1988; von Soden 1977). The opportunities provided by the publication of the set of problem texts in Haddad 104 (Al-Rawi and Roaf 1984), which contains an exceptionally large number of coefficients, have been largely ignored,[19] as has the intriguing LB coefficient list text J (Livingstone 1986).

[13] AO 6770, BM 85196, BM 85200, Str 366, VAT 6598, VAT 8522, VAT 8523, YBC 4669, YBC 4673.

[14] e.g. see Neugebauer's comments on the coefficient 2;15 in VAT 6598 (MKT: I 283) (cf. §4.5).

[15] See §5.6.3 for a discussion of the mathematical meaning of this word.

[16] YBC tablets 7284, 7289, 7997, 8588, 8600 and 9874, and MLC 1354.

[17] *TMS* tablets 14, 20, 21 and 25 (Susa); IM tablets 53961, 54011, 54478 and 54538 (Shaduppum). Also Böhl 1821 and Erm 15073 (both unprovenanced).

[18] Only 9 out of around 100 attestations then published were cited: see §2.3. The phenomenon of implicit coefficients (see §2.3.1) was not addressed.

[19] This tablet was discussed in 1985 at the Third Berlin Workshop on Concept Development on Babylonian Mathematics, but the results were never published (Høyrup, pers. comm. 1995).

More recently, Kilmer has updated, but not published, her initial coefficient catalogue (Kilmer 1985) and Nemet-Nejat has included coefficients in her overview of 'practical' mathematics (1993: esp. 19–21, 278–9). The coefficients which do not appear in problem texts — the large majority of them — are not discussed there but many are cited in the 'glossary of realia' (1993: 149–241). In short, there has been no comprehensive survey of this important genre of mathematical texts.[20]

1.2.2 The word 'coefficient'

The Sumerian word IGI.GUB, Akkadian *igigubbûm*, has been given several different translations over the last sixty years. Thureau-Dangin referred to an 'inverse fixe' (*TMB*: XII), Neugebauer to a 'Koeffizient' or 'fester Bruchteil' (*MKT*: II 51) and later 'coefficient' (*MCT*: 132). Kilmer used the phrase 'key number' (1960), Bruins and Rutten 'constante fixe' (*TMS*: 30). More recently Høyrup has used 'constant factor' (1990*a*: 47) while Friberg prefers 'constant' (1987–90: 544; 1996*a*). Neugebauer and Sachs' word 'coefficient' (*MCT*: 132) has remained the most commonly used translation for IGI.GUB/*igigubbûm* (e.g. Al-Rawi and Roaf 1984: 181; Edzard 1970: 9; Nemet-Nejat 1993: 19–21; Powell 1988: 165).

All of these translations have their own merits and demerits: it is almost impossible to find a satisfactory translation for any Old Babylonian mathematical word in modern English, as the concepts behind them are often so different from our own. Friberg and Høyrup have each tried to overcome this difficulty in recent years by inventing a new mathematical vocabulary.[21] The exercise, although well intentioned (Friberg 1987–90: 552), has been to a great extent self-defeating as one then has to translate a jargon-filled and visually unattractive 'conform translation' into standard English to understand it.[22] Conventional translations are used throughout this book which are as sensitive as possible but retain modern English syntax and vocabulary, while the concepts behind the words are analysed in commentary afterwards if necessary.[23] Likewise, IGI.GUB/*igigubbûm* is always translated 'coefficient' here, with no connotation of its modern mathematical sense, because it is both simple and widely accepted.

1.2.3 Extent, limits and aims of this study

This work focuses on the eight published coefficient lists which are known to date from the Old Babylonian period, namely lists A, C–H, and M (see Table 0.1). Its aims are:

- to set out formally the defining characteristics of mathematical texts and their six sub-types (§1.3);
- to make a detailed study and comparison of the coefficient lists as a genre of mathematical texts (§2);
- to identify and explain as many as possible of the individual coefficients (§§3–8);

and to use this information to briefly look at some broader issues:

[20] Two further collections of problem texts which use coefficients, BM 96954 + BM 102366 + SÉ 93 and BM 96957 + VAT 6598, are presented in §§A.3–4 of this study.

[21] So that for instance *mitḫartum* becomes a 'confrontation' (Høyrup 1990*a*: 63) or an 'equalside' (Friberg *et al.* 1990: 549–51) in translation, rather than simply a 'square'.

[22] e.g. a-šà *ši-ta mi-it-ḫa-ra-ti-ia ak-mur-ma* 21 15 (BM 13901: II 11) becomes 'The surfaces *of my two confrontations I have accumulated*: 21° 15′ ' (Høyrup 1990*a*: 278; his typesetting), instead of 'I added the areas of my two squares and 21;15 (came up)' (my translation).

[23] Compare Black (1980: 156): 'Most literal translations that have appeared in editions of cuneiform texts serve practically no purpose whatsoever. To the philologist who can understand the original they offer no help [...] and to the scholar who is not a philologist and to the layman they are useless, since they are rarely comprehensible in the language in which they purport to be written.'

- how mathematics was taught to Ur III and OB scribes (§10);
- how this school mathematics was then used by scribes in their working lives (§9);
- the regional linguistic and mathematical characteristics of OB mathematics (§10).

The study draws on several types of source material, as well as the coefficient lists themselves:

- published OB cuneiform mathematical texts, particularly problem texts;[24]
- previously unpublished OB mathematical texts (§§A.1–5);
- other genres of cuneiform texts, from school exercises and administrative documents to house and field plans;
- archaeological evidence, such as standardized baked bricks (§A.6) and excavation plans.

The four post-OB coefficient lists, texts B, J, K, and L, will not be dealt with in detail. There is a brief discussion of them in §2.2 with new transliterations in §A.1. Only those coefficients which have a direct bearing on the OB texts will be mentioned in other parts of the work. There are three reasons for their omission:

- texts B and J both include substantial amounts of non-mathematical material (a section on music in B; mystical-numerological passages in J) and a full analysis would involve substantial digression from the main topic;
- texts K and L are small fragments of presumably much larger lists, about which little can be said further than the comments made in this study;
- there are few other mathematical documents currently available from the late periods, which means that it is not yet possible to make any substantial advance for post-OB in any of the five aims listed above.

Only those OB mathematical texts that directly concern the coefficients are mentioned; no LB mathematics is discussed.

This book attempts to go some small way towards meeting three of the objectives set out by Vaiman nearly forty years ago: to establish the extent of Mesopotamian mathematical knowledge; the relative chronology of its development; and the economic and technical background which conditioned its growth (Vaiman 1960: 1–2; see §1.1.1). It is hoped that it will make a useful contribution to both the history of mathematics and Assyriology — taking the study of Mesopotamian mathematics beyond simple descriptions of 'pure' mathematical procedures, and showing mathematics to be not just a 'reflection of everyday life' but an integral part of it for the ancient scribal students and administrators.

1.3 OLD BABYLONIAN MATHEMATICAL TEXTS

The Old Babylonian period is a recognized historical era, which begins in 2004 BC with the fall of Ibbi-Sin, the last ruler of the Third Dynasty of Ur, and ends in 1595 with the death of Samsu-ditana of the First Dynasty of Babylon (Brinkman 1977: 336). During the OB period many city states fought for ascendancy until by 1755 Hammurabi had conquered and united southern Mesopotamia as far north as Eshnunna and Mari (Postgate 1992: 49). Archaeological context (e.g. Baqir 1951) and palaeography (e.g. Goetze 1945) show that the vast majority of extant mathematical texts are OB in date. Finds from the kingdoms of Mari, Eshnunna and Elam have shown that the traditional term 'Babylonian mathematics' is geographically too

[24] See §1.3 for a definition of the six types of Old Babylonian mathematical texts.

narrow, and research on the early precursors to this mathematics suggests that 'Babylonian' is also a linguistic misnomer (Powell 1976*a*: 434). However, the dialectal and chronological label 'Old Babylonian' is clearly defined, makes no assumptions about the geographical origin of the tablets and can be applied to a text using palaeographic criteria.

The mathematical texts throughout this study are Old Babylonian unless otherwise indicated. Any dating more accurate than this is generally impossible because:

- few mathematical tablets are provenanced or from well-recorded findspots within a site;
- mathematical tablets are not dated and contain little prosopographical evidence;
- although it has been assumed that the proportion of phonetic Akkadian in a mathematical text is an index to the age of that text,[25] this phenomenon has not yet been separated from purely geographical considerations.

Wherever possible the provenance of the mathematical tablets discussed has been indicated in the text; all known provenances are given in the first index at the back of this work.

Although the term 'Old Babylonian' has a generally accepted meaning, the phrase 'mathematical text' does not. Clearly not all cuneiform texts with numbers in them are mathematical, and mathematical texts do not necessarily contain many numbers.[26] A distinction must be made between texts which are of mathematical interest[27] and texts which are primarily *about* mathematics, i.e. which have been written for the purpose of communicating or recording a mathematical technique or aiding a mathematical procedure to be carried out.[28] Moreover, they use a specific system of number notation which is significantly different from everyday use (*MCT*: 2). Mathematical texts comprise six clearly definable categories: table texts and problem texts, both of which are relatively well known (e.g. *MCT*: 1; *TMB*: IX; Ritter 1990: 100), and four smaller groups: coefficient lists, 'catalogues', 'model documents', and students' rough work.[29]

1.3.1 Problem texts

Problem texts are our primary source of information on Mesopotamian mathematics. Over 160 tablets containing problems have been published,[30] twelve of which are post-OB.[31] As the name

[25] e.g. Høyrup (1990*a*: 58 n. 83) states that IM 52301 'is rather late and contains several [...] deviations from normal usage.' On the contrary: this tablet is from Level II of Tell Harmal, ancient Shaduppum (Baqir 1950*b*: 130), and must therefore date to before 1755, when Hammurabi conquered the kingdom of Eshnunna and Shaduppum was destroyed (Baqir 1946: 25). The 'deviations from normal usage' are thus geographical, not chronological, ones.

[26] e.g. IM 52301 (iii) has only the numeral 4, written twice.

[27] i.e. whose contents are of interest to modern historians of mathematics, numeration or metrology, such as M 6829, M 6861, and M 15192 + M 15236 from OB Mari, of which 'il est certes question de chiffres, mais dont il n'est pas sûr que l'usage ait été à proprement parler mathématique' (Soubeyran 1984: 35).

[28] Nemet-Nejat fails to make this distinction, including in her catalogue of mathematical texts (1993: 103–7) several economic documents that contain numerical information, such as the Mari texts referred to in the previous note (1993: 99–101), and 'Const 236' (= Ist Ni 236), an economic text dated to Shu-Sin year 3 (and thus Ur III, not Old Akkadian (1993: 130)). On the other hand, YBC 9819 (Ur III), which has been excluded, is a school text written for the purpose of teaching or learning mathematics (Dunham 1982: 41; Robson 1996*a*). See §§4.5, 10.1.

[29] Friberg (1987–90: 548) also uses the term 'algorithm text', but it is unclear what he means by this, as all OB descriptions of mathematical procedures are in some sense algorithmic (cf. Knuth 1972; 1976).

[30] Nemet-Nejat (1993: 21) counts 180 tablets, but many in her list (1993: 108–12) are not problem texts, and one has been excluded. To be deleted are: non-mathematical documents: AOT 305, 'Const 236' = Ist Ni 236, K 8705, K 11097, M 6829, M 6861, M 15192 + M 15236, *UET* 3 1386; calculations and diagrams: 'Ist 04360' = Ist O 4360, M 8613, MAH 16055, *TMS* 1, *TMS* 2, YBC 7273, YBC 7289, YBC 7290, YBC 11120, YBC 11126, 'no museum number'= MCT: 6–10; 'catalogues': BM 80209, IM 52672, *TMS* 5, *TMS* 6; table texts: CUA 34, M 10, Plimpton 322; 'model document': AOT 304; duplicate: '*UMB* 16/2' = 2 N-T 30. Add BM 96954 + BM 102366 + SÉ 93 (§A.3): total 151 OB tablets containing problem texts. (The 'catalogue' IM 52672 is apparently on the same tablet

suggests, they set out a mathematical problem to be solved. Texts which do not do this are not problem texts.[32] Problem texts may also provide an answer, method of solution[33] and illustration for a problem, but do not as a rule show how the individual arithmetical operations were performed.[34] A single tablet may contain any number of problems from one to several hundred,[35] which may or may not be thematically related to each other, and may conclude with a colophon indicating the number of problems on the tablet or the position of that tablet in a series (*MCT*: 1; *MKT*: I 384–7).[36]

Individual problems are posed in the form of a question in the first person singular; indeed, problem texts from the Eshnunna region are introduced by the phrase: 'If someone asks you, saying ...'.[37] All the information needed to solve the problem is stated in this question, except the coefficient.[38] The method of solution is given as a series of instructions in the imperative or in the second person singular, introduced by the phrase 'you, in your working' or simply 'you'.[39] The solution proceeds linearly like an algorithm[40] or a culinary recipe, in which the result of one step is used in the calculation of the next, or put aside until needed again. The instructions conclude with the phrase 'this is the method' or 'the method'.[41] The illustration, if there is one, is drawn above or to the left of the statement of the problem.

The texts are written in grammatically correct, though condensed, Akkadian with full sentences. In no sense do logograms function 'as mathematical symbols' (*pace* Nemet-Nejat 1993: 15 n. 16).[42] Their frequency and use are paralleled in other genres of 'scientific' texts such as extispicy and astronomy. In such highly structured formats with limited and predictable vocabulary, logograms serve as a kind of shorthand, adding to the ease and speed of writing. Yet at the same time they exhibit the esoteric nature of the text: because some cuneiform signs take on values and meanings specific to the genre, only the educated and initiated can comprehend them.[43]

as 'two problems' (Bruins 1954: 61); these were not published and so are excluded from the count.) 'IM 54536.' appears to be a typographical error for the missing IM 54346.

[31] AO 6484, BM 34568, BM 34724, BM 34800, BM 64696, HS 245, MM 86.11.404, Sm 162, Sm 1113, VAT 7848, W 23291, W 23291-X.

[32] Thus Plimpton 322, YBC 7289, YBC 7290, YBC 7302, YBC 11120 and YBC 11126, all presented in the 'problem texts' chapter of MCT, are not problem texts. The former is a table text, the others calculations and diagrams (see §§1.3.3, 1.3.5).

[33] Those which do are referred to as 'procedure texts' by Høyrup (1990*a*: 27).

[34] For exceptions — i.e. problem texts associated with calculations — see §1.3.3.

[35] See Nemet-Nejat (1993: 21, 113–42) for a tablet-by-tablet analysis of the structure of the 180 OB and LB texts she classifies as mathematical problems (but see note 31 above).

[36] Neugebauer at first distinguished between *Serientexte*, which bore serial numbers, and *Einzeltexte*, which did not (e.g. *MKT*: III), but later abandoned this classification (*MCT*: 37).

[37] *šumma kīam išâlka umma šūma*; e.g. IM 54478, IM 53953, IM 54538 etc.

[38] Presumably because it was a fixed (GUB) coefficient, intrinsic to the type of problem to be solved, not the individual case at hand. Giving a coefficient in the numerical information would be tantamount to asking, for instance: 'Find the area of a right-angled triangle whose length is 8 and height is 6, and whose area is equal to half the length times the height.'

[39] *atta* (*ina epēšika*), e.g. AO 6770, VAT 8512 *passim*; or ZA.E (KÍD.DA.ZU.DÈ), e.g. BM 85194, Str 366 *passim*.

[40] However, the algorithm analogy does not strictly hold: for instance, there is (as yet) no unequivocal evidence for iterative procedures in OB mathematics. On the subject of iterative procedures in square root approximations see most recently Deakin (1996*a*, *b*); Fowler and Robson (fc.).

[41] (*kīam*) *nēpešum* e.g. VAT 6469, BM 85194 *passim*.

[42] Høyrup (1990*a*: 331–2) distinguishes three sorts of Sumerographic writing in mathematical texts, namely a restricted number of terms which are always written in Sumerian (but which may give rise to Akkadian loanwords, or be provided with Akkadian glosses); pseudo-Sumerian writings, or logograms; Sumerograms used as possible alternatives for an Akkadian writing, but used ideographically.

[43] George (1993: 64) refers to 'the learned scribe, who seeks at once to show off his scholarship and to render his work inaccessible to those of lesser learning'.

The subject matter of the problem texts has been well documented. Topics range from abstract 'naïve-geometrical algebra' to those which at least use a practical *pretext* to set the problem (see Friberg 1987–90; Høyrup 1990*a*; Nemet-Nejat 1993). Problems were constructed from answers known beforehand and thus must have been compiled for the purpose of teaching mathematics, not as 'practitioners' notebooks' (Høyrup 1990*a*: 330 n. 153). Further evidence for the role of problem texts in teaching mathematics in the OB period is explored in §10.5.

1.3.2 'Catalogues'

Six Old Babylonian 'catalogues' of mathematical problems have been published.[44] They list the first lines of many problems, typically on squares or circles, in a very condensed form.[45] The first 53 lines of IM 52916 have quasi-duplicates in IM 52695 + IM 52304, after which the texts diverge; similarly *TMS* 6 contains the same types of entries as two sections of *TMS* 5, including several values not preserved in the latter. Two of the texts from Shaduppum also contain details of geometrical and more practical problems, on agricultural labour rates, and commercial arithmetic (Goetze 1951). They differ from the very condensed series of problems[46] in that the latter consist of many one-line 'variants' to be substituted into a few 'main examples' whose problems are stated in full (*MCT*: 116; Friberg 1987–90: 557). The 'catalogues', on the other hand, give only the first lines of the problems, and variants on those lines repeat the whole phrase. The entries may or may not include the questions to be answered, but the answers themselves are never given.[47] The numerical information given varies systematically, while the problem to be solved remains the same. Conversely, in series of problem texts it is often the numerical information which remains constant while the question varies.

The function of these texts has never been discussed.[48] Charpin (1986: 454) distinguishes between the inventory, a simple descriptive enumeration of the tablets in a collection, and the catalogue, an ordered list of texts which obeys a classification principle, but these mathematical texts are clearly neither of those things. The problems listed on IM 52916 are interspersed with entries from a coefficient list, and IM 52672 also contains two problem texts: this excludes the lists being inventories of tablets registered by their first lines. All six texts give several numerical variants for most of the problems they list. This makes it equally unlikely that they are text catalogues, as no near-duplicates of problem texts are known which vary only in their numerical values. Similarly, the lists cannot be outlines for a series of problems to be written on one tablet, as collections of problems on a single theme use the same numerical values throughout; it is the parameter to be found which changes from problem to problem.

The key to this question may lie in the scribbled numbers written on the reverse of BM 80209. Friberg (1981: 62) has demonstrated that they were written in the course of determining appropriate values for the last set of problems in the list.[49] The purpose of the problem texts was to teach appropriate methods for solving mathematical problems. The numerical values of the instructions in the text were purely illustrative and were chosen to

[44] BM 80209 (Sippar), IM 52672, IM 52916, IM 52695 + IM 52304 (all Shaduppum), *TMS* 5, *TMS* 6 (both Susa). The text of IM 52916, because it includes coefficient lists, is transliterated in §A.1.

[45] Typically expressed in the indicative first person preterite, or the imperative. Some parts of IM 52916, however, are expressed in the accusative infinitive; see transliteration in §A.1.

[46] e.g. A 24194, A 24195, YBC 4696, YBC 4711.

[47] e.g. BM 80209: lines I 1–8 include the question; I 9–II 13 do not.

[48] Friberg (1981) does not raise the issue in his analysis of BM 80209. Goetze (1951: 129), in his edition of the badly preserved IM 52916, simply suggests that the beginning 'may have run something like this: "I became a mathematician; here are the things I had to learn".'

[49] *TMS* 5 and 6 were also 'redigés d'une façon moins bien soignée et souvent raturés ou surchargés' (*TMS*: 51).

produce simple answers and avoid tricky arithmetical procedures. It may be that these lists of first lines provided the teacher with numerically simple variants on a particular set of problems so that the students could each be given individual numerical practice in a procedure.

If so, then some support for this hypothesis can be adduced from two sets of multiplications from Ur (§A.5.1.4). The product of three numbers (which are different on each tablet) is multiplied by 6 40 in one group;[50] in the other it is divided successively by 10 and 30.[51] We might understand them as the calculation of volumes (where the length, width and height are different in each case), which are multiplied by specific coefficients.[52] (See further §10.4.)

1.3.3 Students' rough work

The sources in this third category have been almost completely ignored by modern scholars, and have often been taken for indecipherable jottings. They may simply be scribbled numbers, perhaps with vertical or horizontal rulings, or both, between them. Some include, or consist solely of, geometrical diagrams. Some are found associated with problem texts, although they can be on otherwise blank tablets, or on the back of non-mathematical school texts, in particular lenticular tablets. These latter are almost certainly the arithmetical calculations done while solving the sorts of questions posed on problem texts. There is never any explanatory material on them,[53] and the layout is often far from neat. Many show signs of repeated erasures and re-use.

Many calculations and diagrams, from several different sites, have been published, but often more by accident than design. It has been beyond the scope of this study to examine the tablets themselves to determine their contents, but those identified from published copies and descriptions, other than the fifty-three presented for the first time in §A.5,[54] are listed in Table 1.1.[55]

On a closely related group of texts, listed in Table 1.2, the problem and answer are briefly stated, usually in a little box in one corner of the tablet. This may be to remind the student of the parameters of the problem they are solving.

Several mathematical problem texts and one 'catalogue' also show the remains of calculations on them. On some tablets the layout of the calculations is explicit and an integral part of the text: on others the numbers are written in the blank space at the end of the text and have been partially erased. The texts are given in Table 1.3.

Calculations are are rarely found on administrative texts, presumably because scribal students were taught to keep their rough work on a separate tablet from the documents for which the calculations were intended.[56]

50 *UET* 6/2 nos. 218, 247, 269, and 290, *UET* 6/3 nos. 374 and 452; possibly also *UET* 6/2 250, if 7 12 is a copying error for 6 40 (the calculation is correct for 6 40).

51 *UET* 6/2 nos. 233, 236, 254, 293, and 298.

52 Suitable candidates for these coefficients might be, firstly, the storage coefficient discussed in §7.2, and, secondly, the common labouring rate of §§6.2–6.5 and the number of days in the administrative month (not a coefficient *per se*; cf. Englund 1988).

53 Occasionally there is an isolated line or phrase from the method of solution; e.g. *TMS* 2 (ii) (§3.6), YBC 7284 (§4.3).

54 Now add: AUAM 73.2833 — squares, obverse and reverse; AUAM 73.2834 — area of a triangle (§3.3); AUAM 73.2841 — area of a circle (§3.1); and AUAM 73.2584 — square. All round tablets, except 224 which is square; provenances unknown (Sigrist fc.).

55 See §I.1 for places of publication, from which the comments in quotes are taken.

56 Some exceptions: YBC 1793, an Ur III silver account; BM 30504, a Late Babylonian field survey (unpublished, courtesy Dr Cornelia Wunsch); and the jottings on administrative texts from OB Mari — in decimal notation! — discussed by Soubeyran (1984: 33–4) and Durand (1987).

Table 1.1: Published OB calculations and diagrams

Museum no.	*Subject*	*Provenance*	*Comments*
2 N-T 174	calculation	Nippur	Square tablet.
CBS 1215	reciprocals	Sippar?	'The Technique'; cf. §A.5.1.
CBS 3551	squares	Nippur	No copy published.
CBS 7356	fragment	Nippur	'Many isolated numbers.' No copy published.
Deir-ez-Zor 2961	boat	Terqa	Published in photograph only; see further §7.2
FLP 1283	reciprocals	Uruk	Round tablet. Proverb on obverse.
HS 231	calculations	Nippur?	Square tablet. Reverse blank? (Not copied.)
HS 232	squares	Nippur?	Rectangular tablet. Reverse blank? (Not copied.)
Ist O 3865	unclear	Kish	'Exercice de calculs, chiffres.' No copy published.
Ist O 3912	unclear	Kish	'Exercice de calculs, chiffres.' No copy published.
Ist O 3920	unclear	Kish	'Exercice de calculs, chiffres.' No copy published.
Ist O 3930	unclear	Kish	'Exercice de calculs, chiffres.' No copy published.
Ist O 3932	unclear	Kish	'Exercice de calculs, chiffres.' No copy published.
Ist O 4001	unclear	Kish	'Exercice de calculs, chiffres.' No copy published.
Ist O 4052	unclear	Kish	'Exercice de calculs, chiffres.' No copy published.
Ist O 4093	unclear	Kish	'Exercice de calculs, chiffres.' No copy published.
Ist O 4306	unclear	Kish	'Exercice de calculs, chiffres.' No copy published.
Ist O 4360	geometry	Kish	Diagrams: remains of sixteen trapezoids with numbers.
Ist O 4387	unclear	Kish	'Exercice de calculs, chiffres.'
Ist S 269	geometry	Sippar	Diagrams: four trapezoids on obverse.
Ist S 428	reciprocals	Sippar	Round tablet.
Ist S 441	unclear	Sippar	Round tablet; obverse model letter (Kraus 1959–62: 18).
M 288	quantity surveying	Mari	Diagram on round tablet. Photo published; see §6.5.
MAH 16055	geometry		Obverse and reverse each show five drawings of triangles.
MDP 27, 60	unclear	Susa	Cf. *MKT*: III 50.
MDP 27, 291	unclear	Susa	Cf. *MKT*: III 50.
MLC 651	'the technique'		Cf. Sachs (1947).
N 3891	'the technique'		Cf. Sachs (1947).
PTS 247	squares		No copy published.
TMS 1	geometry	Susa	Triangles. Fragmentary numbers on reverse.
TMS 2	geometry	Susa	Hexagon on obverse; heptagon on reverse (see §3.6).
TMS 4	unclear	Susa	Multiplication table on obverse.
UM 29-15-481	unclear	Nippur	No copy published.
VAT 5457	reciprocals		Square tablet. Reverse blank? (Not published.)
VAT 7002	tabular calculation		Cf. §4.5.
W 16743k	unclear	Uruk	Numbers on a scrap of lenticular tablet.
W 16743ay	reciprocals?	Uruk	Square, misshapen tablet. Reverse blank (not copied).
YBC 1839	'the technique'		Cf. Sachs (1947).
YBC 4704	reciprocals		No copy published.
YBC 7234	tabular calculation		Rectangular tablet; no copy published.
YBC 7235	tabular calculation		Rectangular tablet; no copy published.
YBC 7289	geometry		Diagram. Round tablet; reverse unpublished. See §3.4.
YBC 7290	geometry		Diagram. Rectangular tablet.
YBC 7294	squares		No copy published.
YBC 7302	geometry		Diagram.
YBC 7353	tabular calculation		Rectangular tablet; no copy. Multiplications on reverse.
YBC 7354	tabular calculation		Rectangular tablet; no copy published.
YBC 7355	tabular calculation		Rectangular tablet; no copy published.
YBC 7356	tabular calculation		Reciprocals; and factors of 48. Box-like format.
YBC 7358	tabular calculation		Square tablet; columnar format.
YBC 10801	squares		No copy published.
YBC 10802	'the technique'		Cf. Sachs (1947).
YBC 11120	geometry		Diagram. Round tablet.
YBC 11125	tabular calculation		Rectangular tablet; no copy published.
YBC 11126	geometry		Diagram. Rectangular tablet.
YBC 11127	tabular calculation		Square tablet; columnar format.

Table 1.2: Published calculations with the problem and answer stated

Museum no.	*Subject*	*Provenance*	*Comments*
2 N-T 30	squares	Nippur	Square tablet. Photograph published by Steele (1951).
2 N-T 116	squares	Nippur	No copy published.
2 N-T 472	squares	Nippur	No copy published.
AOT 304	multiplication	Girsu	Ur III round tablet; obv.: model document on bricks (§4.4).
CBS 11318	squares	Nippur	Square tablet.
NBC 8082	squares		Square tablet.
NCBT 1913	squares		Square tablet. No ruling around text.
UM 29-15-192	squares	Nippur	Square tablet.
YBC 7284	multiplication		Round tablet. Reverse of problem on bricks (see §4.3).

Table 1.3: Problem texts with calculations

Tablet	*Comments*
AO 8862	Rectangular prism from Larsa, with calculations laid out as an integral part of the solutions to six of its eight problems.
BM 80209	'Catalogue' listing the first lines of various problems on squares and circles. At the bottom are scribbled numbers, written in the course of determining appropriate values for the last problems (Friberg 1981: 62).
VAT 8390	Scattered numbers inscribed on the last, blank column of the tablet. It is not clear how they relate to the problems, which concern rectangles.
VAT 8512	Two-column tablet on areas, with a few half-erased numbers on the lower half of the reverse. The signs A.ŠÀ are visible, so the numbers presumably relate to the problems on the tablet.
VAT 8522	Four-column tablet with three columns used. Only the first four or five lines of each column set out the problems, while the remainder is taken up with related diagrams and calculations, interspersed with the solution, neatly laid out.
YBC 7326	First three and a half lines of an unfinished problem on one side and the associated calculation on the reverse. This is laid out between one vertical and one horizontal line forming the shape of a cross. YBC 7273 has a similar layout and set of numbers, partially erased.
YBC 9874	Calculation laid out between four vertical lines on the obverse and a problem on enlarging a canal on the reverse. It is unclear exactly how, or even if, the calculation relates to the solution of the problem.
YBC 10722	Same type of problem as the last three of AO 8862. Although incomplete, the calculation is laid out in a very similar way.

A survey of the rough work in its own right is now needed. Preliminary work on the Ur texts has shown that each type of calculation has a specific format (see §A.5). Whether these same layouts were taught in other scribal schools, and with the same consistency, still remains to be determined. Ideally it should be possible to reconstruct from the preserved calculation the problem the scribe was solving, especially in the case of long series of multiplications and reciprocals like *UET* 6/2 236 and 254.

1.3.4 'Model' documents

It is well known that OB scribal students were taught how to write letters (Kraus 1959–62) and legal contracts (Roth 1979: 14) by copying or adapting 'model' examples.[57] We should then

[57] Niek Veldhuis points out (pers. comm. 1996) that while model contracts — in Sumerian — formed part of the standard OB curriculum and can be found on the same tablets as other 'school' material (cf. Veldhuis 1997), model letters — in Akkadian — are always the only text on a tablet of the standard shape for (real) letters. However, they should not be excluded from consideration as 'school' material solely on the grounds of their use of Akkadian and disposition on tablets any more than mathematical problem texts.

expect to find practice texts for other sorts of documents used by working scribes, such as accounts, field plans, land and canal surveys, and house plans.[58] Few Ur III and OB model mathematical texts have yet been recognized. Part of the problem may be the difficulty of distinguishing between school exercises and the economic and administrative documents themselves (cf. Kraus 1973: 25). Generally speaking, working documents are dated, record the work of real people known from other documents, and may be sealed. School exercises in the main do not have these characteristics,[59] but a working document may not have all of them either. Further, model documents are often on round or misshapen tablets, or on the same tablets as other school material; there may be mistakes in the layout, spelling or arithmetic; the text may be unfinished; there may be complete or partial duplicates.[60] The following list is far from exhaustive: many more exercises of all types remain undetected, both in the published literature and in museums. They are worthy of a full study in their own right (see §10.5).

Table 1.4: Mathematical model documents

Tablet	*Provenance*	*Type*	*Comments*
AOT 304	Ur III Lagash	Survey of brick stack (§4.4)	Round tablet; incorrect calculation on the reverse.
CBS 9044	OB/Ur III Umma	Model account	Reverse of single-column tablet. Obv. and rev. 1–3 contain a list of Sumerian personal names.
HSM 7500	Ur III	Field plans	Roughly shaped tablet. Conspicuously round numbers; no GNs, PNs or date.
PRAK 2 D 30	OB Kish	House plans	Round tablet. Plan on obverse has measurements; plan on reverse does not. (Cf. Heinrich and Seidl 1967: 37.)
PRAK 2 O 107	OB Kish	House plans	Round tablet; plans on obverse and reverse; measurements and the remains of a calculation. (Cf. Heinrich and Seidl 1967: 37.)
UET 5 855	OB Ur	Canal survey	Simplified version of *UET* 5 856. (Cf. Vaiman 1960: 10; Charpin 1986: 481.)
UET 5 856	OB Ur	Canal survey	Cf. Vaiman 1960: 10; Charpin 1986: 481.
UET 5 860	OB Ur	Canal survey	Terminology inconsistent; no GNs, PNs or date.
UET 5 861	OB Ur	Canal survey	Unfinished; undated, no PNs, no total, format errors. Cf. Charpin 1986: 451.
UET 5 862	OB Ur	Field survey	Cf. Charpin 1986: 451.
Walker 47a	Ur III	Survey of agricultural work (§6.4)	No GNs, PNs, or subscript; arithmetic error; odd jottings on reverse. Cf. Civil 1994: 119 n. 21; 123.
YBC 3892[61]	OB/Ur III	Field plan	Round tablet. May equally be considered as a mathematical diagram (§1.3.3), as there is no explanatory text.
YBC 3895	OB/Ur III	Field plan	Round tablet. May equally be considered as a mathematical diagram (§1.3.3), as there is no explanatory text.
YBC 9819	Ur III Umma	Survey of brick walls (§4.5)	Incorrect calculation. Identified as a school text by Dunham (1982: 41).

[58] See already Charpin (1986: 481).

[59] Model legal contracts often conclude with the phrase lú.ki.inim.ma.bi itu.bi mu.bi 'its witnesses, its month, its year' instead of the actual witness list and date (Roth 1979: 101).

[60] Cf. Charpin (1986: 426); Kraus (1959–62: 29–30); Roth (1979: 15–16) and the texts listed here.

[61] Personal communication, David Brown.

1.3.5 Table texts

Table texts are by far the commonest type of mathematical document known.[62] They have been discovered at sites all over the Near East, from Ugarit to Susa, from Nineveh to Ur, and many places in between (*MCT*: 1 n. 3). Neugebauer's huge survey of the types and formats of table texts (*MCT*: 4–36; *MKT*: I 4–96; II 36–7; III 49–51) has never been updated or superseded, but most recently Friberg (1987–90: 542–8) has given a brief survey of the different types of texts: metrological lists and tables,[63] reciprocals, products, squares, square roots and cube roots, quasi-cube roots, exponentials and logarithms, and help tables for market-rate exercises.[64] Tablets may contain just a single table, in one or more copies, or an extract containing two or more tables from the standard series, often in conjunction with other school subjects.[65]

Table texts — or their memorized contents — were a vital tool for even the simplest calculations. For instance, in 2 N-T 30 the problem is to find the area of a square which is $^1/_3$ cubit and $^1/_2$ finger long. First the length of the square is converted to a base unit — 0;01 45 nindan — for the solution of the problem, using a table of length measures. The calculation itself is performed by looking up a table of squares, and then the answer — 0;00 03 03 45 sar_a — is transformed into smaller units with a table of area measures for the final answer.

1.3.6 Coefficient lists

Coefficient lists are lists of numbers which embody a mathematical relationship and are necessary for solving particular types of mathematical problems. Ten of these lists, eight from the OB period (texts A, C–H, M) and two from the first millennium (B, J), have already been published; a further two LB fragments are presented for the first time here (K, L).[66] Each number in the list is given a short name or description. The relationship may be an 'absolute constant', such as that which embodies the relationship between the diameter of a circle and its area, or an 'empirical constant' concerning a relationship fixed by convention, such as the daily work-rate for making bricks (*MCT*: 132). There are also metrological constants, which facilitate conversion between different units of measurement. Coefficient lists, very characteristic in both appearance and content, are clearly recognisable as a distinctive genre of mathematical text.

Old Babylonian coefficient lists must have served a similar purpose for trainee scribes to that of formula-booklets for modern students of mathematics. However, whereas an algebraic formula provides a numerical relationship and the variables involved in that relationship, without necessarily naming it,[67] the coefficient gives the numerical relationship and the object to which it applies, but does not specify which variables that relationship refers to.[68] In other

[62] An exhaustive count and listing is not given, as they are not of central concern to this study. Many hundreds have now been published.

[63] Metrological lists simply give the standard notations for metrological units; tables give their sexagesimal equivalents in a basic unit too (cf. Friberg 1987–90: 542–3).

[64] See also Nemet-Nejat (1993: 17–19; 1995) for a survey of publications on table texts. Add Arnaud (1994: nos. 55, 66, 81); Cavigneaux (1996: nos. 277–82, 286, 288); Sigrist (fc.: nos. 240–3, 250) for OB; Britton (1991–3) and Friberg (1993) for LB. A table from second millennium Hazor will be published by Horowitz.

[65] See §§10.4–5 for more detailed discussion of their role in OB mathematics education.

[66] See §§2.2.3, A.1.

[67] For instance, $A = \frac{1}{2}\, l \times b$ is the standard algebraic expression for the relationship between the area A of a right-angled triangle and its two shorter sides, l and b. Convention dictates the choice of initial letters for variables, upper case for areas and lower case for lengths. The formula alone gives no clue as to the name of the shape.

[68] 30 *ša* SAG.DÙ (Text F 1) is the coefficient for a (right-angled) triangle. The coefficient contains no information on which relationship between which parameters of triangle this represents.

words, one cannot successfully use an algebraic formula or a coefficient without being able to fill the gaps in the information which it provides. This is one of the major obstacles to understanding them: unless a problem text is known which uses a particular coefficient, there is only a name and a number to work with. Even if the name is fully understood, the particular mathematical relationship embodied in that coefficient must still be identified without unequivocal knowledge of the variables involved.

While Mesopotamian scholarship was clearly more than a 'science of lists' (*pace* von Soden 1994: 145) as the problem texts show, the list was a central feature of scribal culture:

> It was usual for almost every type of information to be summarized and recorded by listing pairs of associated terms, arranged in columns. This technique acquired specialized conventions appropriate to the particular subject matter involved. (Livingstone 1986: 2)

Thus on one hand the coefficient lists firmly place mathematics within the mainstream of Mesopotamian intellectual endeavour; on the other hand, they have been seen by some as practical administrative tools. It has already been accepted by some scholars (e.g. Goetze 1962; von Soden 1977; Walters 1970: pp. xiv, 118–19, 147–8) — albeit with scanty evidence — that the constants found in the lists were used by working bureaucrats in quantity surveying and labour calculations. The dilemma that coefficient lists encapsulate, then, is that of OB mathematics in general: do these texts represent the display of scribal virtuosity — or intellectual curiosity — divorced from practical concerns, or are they simply part of a useful training for future overseers, accountants and surveyors? This question is returned to in §10.6, after the lists themselves have been studied in detail.

2. COEFFICIENTS AND COEFFICIENT LISTS

In the first section of this chapter the eight OB coefficient lists which are the main subject of this study are presented and compared. The way the coefficients are grouped within each list, the manner in which the coefficients are written, and the use of logograms, are studied for each.[1] Four later lists, two of which are published here for the first time, are then briefly discussed and contrasted with their OB antecedents. The coefficients found in problem texts are the subject of the third section. This deals with the distribution of coefficients throughout the corpus of OB mathematics, the terminology used in association with them, and the format and language of the coefficients themselves. The chapter ends with a short explanation of the structure of §§3–8, in which the individual coefficients are discussed.

2.1 THE EIGHT OLD BABYLONIAN COEFFICIENT LISTS

Five of the OB coefficient lists were excavated from controlled excavations, and two have been provenanced by their external characteristics; only the origin of list A is unknown. All but C and M are in a similar columnar layout, with the numerical values on the left and the names of the coefficients on the right. Although the lengths of the coefficient lists range from 5 lines to 70, all eight are organized in broadly the same way. Individual coefficients are arranged in groups according to their subject matter: geometry, quantity surveying, storage and metrology, and metals.[2] Parts of some lists, though, have no discernible organization, or contain coefficients which are not yet readable. Occasionally one line of a list contains more than one coefficient.

There is no standard format for writing coefficients.[3] Adapting Civil's typology of the lexical lists as a model (Civil 1976: 124), each component of the coefficient may be assigned a numerical code, so that the format can be represented by a string of one or more numbers:[4]

1 the numerical value of the coefficient;
1* the numerical value of the coefficient with a metrological unit;
2 the word 'coefficient' (IGI.GUB/*igigubbûm*);
3 the word 'of' (*ša*);
4 the name of the coefficient.

[1] Cf. Black (1991: 11) on Babylonian lists of Sumerian grammatical forms: 'The features susceptible of analysis in a two-column, i.e. two-dimensional, list are: (1) the arrangement of the entries, (2) the selection of forms from either language, and (3) the juxtaposition of particular Sumerian forms and Akkadian forms.'

[2] Studied further in §3, §§4–6, §7, and §8.1 respectively.

[3] The mixed format of text F led Neugebauer and Sachs (*MCT*: 133 n. 303) to the proposition, summarized here by Nemet-Nejat (1988: 287), that 'when the noun directly follows the number, the number represents the actual value of the following noun; when the number is followed by *ša* and the noun, the number represents a fixed metrological proportion between units.' This misconception has only recently been explicitly corrected (Nemet-Nejat 1993: 19–20 n. 6).

[4] Civil's code 0 (1976: 124), which represents a vertical wedge at the beginning of each line, is excluded from this typology. An initial wedge is never attested in the coefficient lists, for the obvious reason that it might be mistaken for part of the numerical value of the coefficients.

So, for instance, the coefficient for an earth wall (§6.1) appears in a different format in each of the lists:

A 8	3 45 IGI.GUB IM.DÙ.A	1-2-4
D 43	3 45 *šà pí-ti-iq-ti*	1-3-4
E 13	3 45 IM.DÙ.A	1-4
F 17	3 45 *ša* IM.DÙ.A	1-3-4
G 13	3 45 IGI.GUB «*pí-iq*» *pí-⌈ti⌉-iq-tim*	1-2-4
H 2	3 45 *pí-ti-iq-tum*	1-4.

The coefficient is not found in text C, but might have been expressed there as:

pí-ti-iq-tum 3 *45 i-gi-gu-bu-ša* 4-1-2.

This example also shows that the use of logograms is not consistent from list to list — or even, often, within an individual list.

The numerical values of the coefficient lists are, in general, sexagesimally regular; that is, their only factors are 2, 3, and 5 (the factors of 60), with no more than three sexagesimal places, and have reciprocals with no more than three sexagesimal places. Exceptions are noted below and in §I.2.

2.1.1 List A

The provenance of list A is unknown (Brinkman, pers. comm. 1995). It contains very little syllabic Akkadian, and so cannot be placed on orthographic grounds. The tablet measures 100 × 55 × 23 mm (Kilmer 1960: 275) and contains 50 lines of text. It lists 54 coefficients, one on each line of the obverse and reverse, plus four on the left edge. Lines 1, 2 and 4 give more than one value for the coefficient (see §§3.1–3, §A.1). There are apparently no introductory remarks or a colophon, although the last entry on the reverse, ⌈X⌉ IGI.GUB ŠU.NÍGIN.NA, may in fact give the total number of coefficients on the tablet (see §8.5.1). Its numerical value is too damaged for restoration.

The format of the coefficients is consistently 1-2-4 throughout the list, with the large majority written logographically. There are apparently five irregular coefficients (ll. 9, 20, 26, 34, 38), all of which appear to have arisen through copying or calculation errors. *Ṣiliptum* 'diagonal, rectangle' (see §3.4.3) is the only word indisputably written in syllabic Akkadian (ll. 3, 42); two further possible instances are in the badly damaged lines 23 and 47b (see §8.4.1). All appear to be mimated and in the nominative case; this latter phenomenon may suggest that the entries should be read, for instance, 'coefficient: a rectangle' rather than 'the coefficient of a rectangle'. On the other hand, they may simply be case errors, of which there are several instances in the other lists.

The coefficients are arranged in broad subject groupings as follows:

Table 2.1: Contents of coefficient list A

Lines	*Subject*	*Format*	*Comments*
1–7	geometry	1-2-4	One line per figure; sometimes more than one value; simplest first.
8–17	no clear order	1-2-4	Some quantity surveying and storage; many unknown.
18–23	metals	1-2-4	In ascending numerical order.
24–36	quantity surveying	1-2-4	Some two- or three-line groupings apparent, e.g. bricks.
37–40	storage	1-2-4	One unclear.
41–48	no clear order	1-2-4	Includes one quantity surveying, one geometry.

2.1.2 Lists C and H

Tell Harmal, ancient Shaduppum, is one of a group of mounds about 6 miles east of Baghdad. The whole of this small site was excavated intermittently between 1945–9 and 1958–9, and thereby shown to be an important administrative centre for the kingdom of Eshnunna until its conquest by Hammurabi in 1762 (Postgate 1992: 49). The town was surrounded by a heavily fortified wall with only one entrance gate, close to a temple dedicated to Nisaba, the scribal goddess, and her consort Haya. Four building levels were identified, dating from the Akkadian to the Kassite periods (Baqir 1946; 1948). Preliminary reports were published for the first few excavations, but both published maps of the site are incomplete, showing only rooms 1–162 (Baqir 1946; 1959). Over three thousand cuneiform tablets were found, including commercial documents, letters, administrative documents, religious texts (incantations, hymns, exorcisms, prayers, etc.), lexical and literary texts (lists of toponyms, botanical and zoological terms, divine names, sign-lists, syllables etc.), the Law Code of Eshnunna (Roth 1995: 57–70), as well as mathematical texts. The destruction level identified by Baqir (1946: 25) suggests that all the Shaduppum mathematical tablets predate Hammurabi's conquest.

The tablet containing coefficient list H was found during the 1947 season in Level II of Room 180. Thirty other tablets, of unrecorded contents, were with it, one of which was dated to Ibalpiel II 8 and one to Ibalpiel II 10 (Baqir 1950*b*). The tablet of text C, on the other hand, was dug illicitly and left on the surface of the tell as worthless because of its poor state of preservation (Goetze 1951: 126). There are a further twelve published OB mathematical tablets from Shaduppum as well as ten which are unprovenanced but almost certainly from the same site, as they share many terminological, grammatical and orthographic characteristics with the known Shaduppum texts, and fall within the same range of museum numbers.[5]

The coefficients of List C, of which 47 are extant, are on the same tablet as a 'catalogue' (see §1.3.2).[6] The tablet, now approximately 280 × 85 mm, is in poor condition, and Goetze (1951: 127) estimates that it originally contained 60–65 lines per side, of which 105 are extant or reconstructible. The obverse is entirely devoted to a listing of the first lines of fifty or so related problems on squares, and this 'catalogue' of first lines continues, interspersed with sections of coefficients, on the reverse. The end of the reverse is missing, but five small columns on the left edge are partially preserved.

The text uniquely lists the name of the coefficients at the beginning of each line, followed by their numerical values and the phrase 'its coefficient'. There is one instance of two coefficients occupying a single line (C 13, §3.2), and in general a columnar layout was not used, except in the first two columns of the edge. The extant coefficients use only six common logograms, namely SAG.DÙ 'triangle' (ll. 6–8); SIG_4 'brick', SAḪAR 'earth', ŠE.IN.NU.DA 'straw' (ll. 22–4); the length unit NINDAN (ll. 25–9); and KÙ.SIG_{17} 'gold' (l. II 3); URUDU 'copper' and KÙ.BABBAR 'silver' may also have been used in lines II 1–2 (see §8.1). Three geometrical (ll. 7–9) and one erroneous metal coefficient (l. 64) have sexagesimally irregular values.

[5] Certainly from Shaduppum: IM 53953, IM 53957, IM 53961, IM 53965, IM 54010, IM 54011, IM 54033, IM 54464, IM 54478, IM 54538, IM 54559, IM 55357. On the characteristics of this group see Høyrup (1996*b*: 6–17). Probably from Shaduppum: IM 52548, IM 52672, IM 52879, IM 53963, IM 54216, IM 54346, IM 54472, IM 54486, IM 55111, IM 55292.

[6] IM 52685, a fragment measuring about 120 × 90 mm, was found with this tablet and includes a partial duplicate of the same 'catalogue'. It may also have contained some coefficients (Goetze 1951: 155). However, this text has not been included in this study because the only evidence for coefficients, on lines 56–7 of the obverse, is neither conclusive nor informative:

[...] 20 *i*-[*gi-gu-ub-bu*?]
[...] 30 *i*-[*gi-gu-ub-bu*?].

The organization on the reverse and edge of the tablet is as follows:

Table 2.2: Contents of coefficient list C

Lines	*Subject*	*Format*	*Comments*
1–3	'catalogue'?		Too broken to read.
4–13	geometrical coefficients	4-1-2	Grouped by geometrical figure, with two or three coefficients each.
14–21	'catalogue'		Problems on calculating the areas of geometrical shapes inscribed in other shapes.
22–29	quantity surveying coefficients	4-1-2; 4-1-4	Three on carrying different materials; five on carrying over particular distances, in decreasing numerical order; see §5.2.3 for a discussion of the word *allum*.
30	'catalogue'		Problem on calculating the areas of geometrical shapes inscribed in other shapes.
31–36	coefficients	4-1-2	Broken. Impossible to determine the subject matter.
37–46	'catalogue'		Broken.
47–55	'catalogue'		One possible coefficient (47); subdivision of figures by parallel lines.
col. I	coefficients	[4-1]-2	Broken. Impossible to determine the subject matter.
col. II	metal coefficients	4-1-2	In increasing numerical order?
col. III	coefficients	[4-1]-2	Broken. Impossible to determine the subject matter.
col. IV	mixed	4-1-2	Two storage coefficients; three work-rates.
col. V	'catalogue'		Broken.

List H comprises just seven coefficients listed at the end of a tablet containing three problems, none of which make use of the coefficients in the list. The first coefficient occupies a whole line, and then the list is divided into two columns, going from right to left, the normal order for the reverse of a tablet. The format of the coefficients varies from line to line, with just one logogram, SIG_4 'brick' in line 3. Four of the six syllabically written lines are mimated, two of which (ll. 1, 6) contain case errors. These inconsistencies suggest that the list was compiled by copying individual entries from other, longer lists.

Table 2.3: Contents of coefficient list H

Lines	*Subject*	*Format*	*Comments*
1	storage	1-2-4	Grain coefficient.
2–3 (col. I)	quantity surveying	1-4 1-3-4	Building work-rates.
4–5 (col. I)	geometry	1-3-4	Simple shapes.
6–7 (col. II)	storage	1-3-4 1-4	Grain coefficients.

2.1.3 List D

The twenty-six mathematical tablets published in *TMS*, including coefficient list D, were discovered in 1933 by de Mecquenem. They were under a large paved area, some 30 × 15 m, in Level I of the so-called Ville Royale of Susa, where in the 1934–5 season van der Meer was to find the school texts he published in 1935 (*TMS*: 1). However, because of de Mequenem's 'lack of interest in mud-brick structures' and a badly disturbed archaeological context (Harper *et al.* 1992: 22, 277) it is difficult to date the tablets accurately. Not far away a pot was found inscribed with the name of the Elamite king Adda-hushu, thus dating it to the early eighteenth century BC (Carter and Stolper 1984: 233), but the consensus is that the mathematical texts are a little later in date, probably from the end of the First Dynasty of Babylon. Rutten (*TMS*: 2) used the sign forms of TI, ÚR, AM and GUB as dating criteria. A table of forms of the sign TI datable to between 1650 and 1350 BC (Stève *et al.* 1980: 90 n. 43) later enabled Tanret (1986: 141 n. 3) to reach a similar conclusion:

Les TI que nous avons trouvés dans les textes mathématiques [de Suse] sont tous semblables au deuxième signe de l'évolution graphique; ce qui doit nous situer dans la première moitié du deuxième millénaire, en tout cas avant la fin de la première dynastie de Babylone en Mésopotamie.

List D is both the longest, at 71 lines and 68 coefficients, and the most clearly organized of the eight OB lists.[7] Indeed, the first line claims that the tablet contains 'coefficients for absolutely everything'.[8] Line 53 is probably an introduction to the following section on metals: '9 coefficients of the furnace'.[9] The organizational principles are particularly noticeable in two sections: geometry, where the coefficients are grouped in the order area-diameter-radius for each shape; and metals, which are listed in ascending numerical order. A double ruling between lines 52 and 53 separates the storage from the metal coefficients.

Text D uses the 1-2-3-4 format throughout the obverse and edge, but only once on the reverse, to begin a new section; the 1-3-4 format, used elsewhere on the reverse, seems to serve as an abbreviation of it. The three-column layout is abandoned only when the description of the coefficient is too long to fit into the final column (see §A.1). The final two lines of the text, which together make up one coefficient, deviate from this format, being more reminiscent of C 25–9. In both cases the first part of the coefficient describes the activity to which the coefficient relates, with an active verbal form; the second part states and names the coefficient.

Table 2.4: Contents of coefficient list D

Lines	*Subject*	*Format*	*Comments*
2–32	geometry	1-2-3-4[10]	Starts with coefficients for circles; moves to semicircles and circular segments; other circle-based shapes; polygons; quadrilaterals.
33–48	quantity surveying	1-2-3-4 (ll. 33–8) 1-3-4	Irrigation; carrying and making bricks; construction work-rates.
49–52	storage	1-3-4	In ascending numerical order.
54–62	metals	1-3-4	In ascending numerical order.
63–70	quantity surveying	1-3-4 4-1-4 (line 70)	Brick densities; bitumen; work-rates for damming.

The 32 logograms in list D comprise 17 geometrical terms with 7 quantity surveying expressions, 1 storage word, 2 metrological units and 5 metals.[11] Only 2 of the 32 geometrical

[7] No measurements are given for this tablet in *TMS*. According to Høyrup (pers. comm. 1995) 'lacking measures and photographs in *TMS* are probably due to the fact that some of the Susa tablets were mislaid before the War'; they have recently been rediscovered.

[8] IGI.GUB *ša mimma kališu*; von Soden (1964: 47): 'Konstanter jeglicher Art'; Muroi (1992*b*: 45): 'the fixed coefficient, that of everything in its entirety'. Høyrup (1990*a*: 46-7 n. 39) keeps Bruins' original reading *ullišu* (*TMS*: 25) and translates 'that of making everything high'. However, the published photo and copy of the tablet both show KA, not UL: there are no final Winkelhakens as one finds, e.g., on the UL in line 34 of the same list.

[9] IGI.GUB *ša kūrim*; *contra* '9, la constante de la fourneau' (*TMS*: 28).

[10] Here '2' includes the words *pirkum* 'radius' and DAL 'diameter' as well as IGI.GUB. See §3.

[11] Namely, geometry: DAL, GÚR, ŠÀ, GAR, GÁN, GIŠ.ILLAR, A.ŠÀ.ŠE, IGI.GU$_4$, BAR.TA, SAG.5, SAG.6, SAG.7, SAG.DÙ, UB,

coefficients are in syllabic Akkadian (ll. 9, 24), compared to 16 in the rest of the list. Mimation does not always occur (e.g. ll. 34–5); there is one case error, in line 40. There are 10 irregular coefficients for geometry (ll. 6, 10–11, 13–14, 17–18, 27–8, 31) and 2 for metals (ll. 57, 62); all appear to be numerically correct.

2.1.4 Lists E and F

On the basis of sign use and 'dialect', Goetze (1945: 147 n. 353) suggested that texts E and F, 'written in different hands, have the appearance of Larsa tablets from the time of Rim-Sin'.[12] In particular, Text F shares the orthographic habit of Pleneschreibung with AO 8862, which is also presumed to come from Larsa (*MKT*: I 108).

List E (approx. 126 × 68 mm (*MCT*: pl. 49)) is complete but has suffered much surface damage and is very difficult to read in places. It contains 56 coefficients, on 52 lines, with no introduction or colophon. The organization of the entries is not solely thematic or numerical: one group of coefficients is listed together under acrographic principles. In other places there is no apparent order at all.

The tablet is divided into two columns in the format 1-4, except lines 45 and 47, written 1-3-4. The geometrical entries are written logographically, with the exception of the word *ṣiliptum* 'diagonal', 'rectangle'. All but nine of the remaining extant coefficients also contain at least one logogram. With one exception the Akkadian is mimated and grammatically correct,[13] but a few of the entries appear to be conflations, copied incorrectly from some original (e.g. ll. 4 (§3.7), 34 (§8.1)). There is one irregular geometry coefficient (l. 10), and possibly another (l. 15) which may be the result of an ancient or modern copying error; the photo is unclear here (*MCT*: pl. 49).

Table 2.5: Contents of coefficient list E

Lines	*Subject*	*Format*	*Comments*
1–12	geometry	1-4	The last three entries are obscure.
13–18	acrographic	1-4	Words beginning with IM; many unclear.
19–37	no apparent order	1-4	Storage, quantity surveying, metals, and others.
39–49	quantity surveying	1-4	Bricks and carrying bricks.
		1-3-4 (ll. 45, 47)	
50–52	no apparent order	1-4	

Text F is also a complete tablet, measuring around 100 × 50 mm (*MCT*: pl. 44). Most of its 66 lines contain a single coefficient, although three lines of the reverse (ll. 56, 57, 63) each have two entries. A few coefficients have more than one value attached to them (ll. 42, 44, 49, 53). There is an introductory line, IGI.GUB.BA *ša né-pe-eš$_{20}$-tum* 'coefficients for calculation (?)',[14] and a final dedication to Nisaba, the patron goddess of scribes.

NIGIN, UŠ, SAG; quantity surveying: SAḪAR, ESÍR, ÉŠ.GÀR, LÚ, SIG$_4$, ÍD, PA$_5$.SIG; storage and metrology: GUR$_7$, SÌLA, SAR; metals: URUDU, ZABAR, AN.NA, KÙ.BABBAR, KÙ.SIG$_{17}$.

[12] Høyrup (1996*b*: 33) has little comment to make about Goetze's assignation of provenance to these lists.

[13] In line 10 one would expect the construct state *ṣilipti* ÍB.SI$_8$ instead of the nominative case.

[14] Neugebauer and Sachs translated this difficult and grammatically incorrect phrase as 'the thing in connection with which the operation is to be performed' (*MCT*: 132 n. 298). Von Soden (*AHw*: 778) takes *nēpeštum* to mean 'inventory', 'list' (Aufstellung), with some non-mathematical occurrences. But one would expect it to have a technical meaning here specific to mathematics, as do the related terms *epēšum* 'to calculate', 'to perform (a calculation)', and *nēpešum* 'method (of calculation)'.

The list is composed of seven sections which are clearly differentiated by the use of different formats for the coefficients they contain. Although each forms a coherent group of coefficients (apart from the last, which appears to consist of odds and ends omitted from or irrelevant to the earlier ones), their order is unusual and, indeed, some topics are dealt with in more than one section. This strongly suggests that list F was compiled from excerpts from several different lists, each of which used its own layout, which was also copied. The material from those sources was not regrouped thematically or edited stylistically.

There is a high proportion of logograms in list F. Only thirteen entries are entirely in syllabic Akkadian, which is always mimated, but not always grammatically correct, with a couple of nominative-for-genitive case errors after *ša* (ll. 61, 65). There is just one irregular geometry coefficient (l. 51); both irregular brick coefficients arise from calculation errors (ll. 12, 15, §5.3).

Table 2.6: Contents of coefficient list F

Lines	*Subject*	*Format*	*Comments*
2–15	quantity surveying	1-4	Bricks and carrying bricks.
16–18	quantity surveying	1-3-4	Work-loads for building.
19–21	geometry	1-3-4	Simple geometrical shapes.
22–28	metals	1-4	In two groups, the first ascending numerically.
29–33	storage	1-3-4	
34–46	quantity surveying	1-3-4 ll. 34–5; 1-4 ll. 36–44; 1-3-4 ll. 45–6	Brick sizes; brick weights and carrying. No clear order.
47–54	geometry	1-3-4	More complex geometrical shapes; no clear order.
55–65	no clear grouping	1-3-4 line 55; 1-4 line 56a; 1-3-4 ll. 56b–59; 1-4 ll. 60–4; 1-3-4 line 65	Some minerals; two lines grouped acrographically (ZI).

2.1.5 List G

Tell ed-Dēr, about 25 km south of Baghdad, was a residential suburb of Sippar during the Old Babylonian period, and has now been identified as Sippar Amnanum (van Lerberghe and Voet 1991: XII n. 2). Iraqi government excavations, directed by Taha Baqir, started there in 1941. Five sites on the tell were excavated; Area 2, a large public building, yielded all but two of the tablets. Rooms 6 and 8 of Level III (Baqir and Mustafa 1945: 46) contained an archive of over 100 letters, contracts and economic texts, the so-called 'Anum-pisha archive' named after the creditor of most of the loan contracts. There was also a letter archive connected to one Ikuppisha, whose relation to Anum-pisha is uncertain, as no less than five different Ikuppishas are named in the administrative archive (Edzard 1970: 16). There were also various other administrative documents and the coefficient list G, although the Iraq Museum catalogue contains no further information on its findspot (Edzard 1970: 26). The level III tablets date to 1894–1831 BC, the early Old Babylonian period, during the reigns of Sumu-abum, Sumu-lael and Sabium (Edzard 1970: 14–15). Level II contained tablets from the period 1740–1626 BC, during the reigns of Samsu-iluna to Ammi-saduqa (Edzard 1970: 15). A group of exercise tablets was also found in Area 2, but have not yet been published (Tanret 1982). With no information available on the findspot or contents of these texts it is impossible to determine whether Text G belongs to this scholarly group or to the administrative archive.

List G comprises sixteen entries, one on each line. It has four ruled columns: for the numerical value, for the word 'coefficient', and for one or more versions of its name. There is no introductory matter. The descriptions of the first two coefficients, on geometry, consist of schematic representations of the figures they represent and the names of the figures in syllabic Akkadian. Five of the remaining lines (ll. 3–5, 9, 10, 12) give the name of the coefficient both logographically and syllabically while the rest only give one name. There are a total of seven different logograms. Line 11 is not mimated, and in line 15 a noun in the construct state is followed by a nominative. Of the ten lines not in the construct state all but two (ll. 2, 9) are in the genitive, which suggests that the IGI.GUBs in the second column should also be understood as constructs, e.g. 6 IGI.GUB *i-ga-ri-im* '6, the coefficient of a brick wall', line 7 (§4.5).

Table 2.7: Contents of coefficient list G

Lines	*Subject*	*Format*	*Comments*
1–2	geometry	1-2-4	Simple shapes, with schematic illustrations.
3–8	quantity surveying	1-2-4	Carrying; building with bricks.
9	storage	1-2-4	
10–11	meaning unclear	1-2-4	Metals; storage?
12	storage	1-2-4	
13–16	quantity surveying	1-2-4	Building; carrying.

2.1.6 List M

Old Babylonian mathematical texts were found in two private houses in Ur, named by Woolley 'No. 7, Quiet Street' and 'No. 1, Broad Street'. List M comes from 'No. 1, Broad Street', which, by Woolley's reckoning, was a school run by a priest called Igmil-Sin:

> nearly two thousand [tablets] were found in the building. Some hundreds of these were of the regular 'school exercise' type, that flat bun-shaped tablet used for fair copies etc.; there were very many religious texts, perhaps used for dictation or for learning by rote, some historical texts, mathematical tablets, multiplication tables etc. as well as a quantity of business records apparently referring to temple affairs. (Woolley 1931: 365–6)

Charpin (1986: 419–86) has studied the house and its contents, using information from Woolley and Mallowan (1976) and the *UET* volumes. Only 382 tablets from 'No. 1, Broad Street' are accounted for in the published catalogue, far fewer than Woolley's first estimate even when one adds to them the 149 tablets published in *UET* 6/1–2 without excavation numbers. Charpin shows convincingly that they do not comprise the archives of one individual or a family, but form a chronologically and thematically disparate group: Ur III texts are mixed in with documents dated to the reigns of Warad-Sin and Rim-Sin, archival records with school tablets. He concludes that they were probably used as fill during the restructuring of the house at an unknown date (Charpin 1986: 481–2).[15]

[15] The mathematical texts from rooms 2, 3, and 4 are: U16866 a–f, unedited, except *UET* 7 188 (4 times multiplication table); U16871 a–b, unedited; U16833 = *UET* 5 862, 'model' document; U17900d = *UET* 5 864, problem text; and U17900w = *UET* 5 861, 'model' document (§1.2.4); and, of the round tablets containing calculations which are published in §A.5, the 31 with the excavation number U17207 (as well as, presumably, the further 9 without excavation numbers). From rooms 6, 7, and 9 the mathematical tablets are: U17212b = *UET* 5 855 and U17215f = *UET* 5 856, both 'model' documents (§1.2.4); U17213b = *UET* 5 859, a problem text; U17214r = *UET* 5 881, coefficient list M. Some of the administrative tablets from the house, comprising part of a reed-working archive, have been studied by van de Mieroop (1992*b*).

The list is six lines long and contains just three fully logographic entries, all on carrying bricks. It is not in tabular format, with no introductory or concluding matter, and does not include the word 'coefficient'. For these reasons, it remained unidentified for over forty years (Friberg 1996*a*). The format is 1-4 throughout. M appears to be a short extract from a longer list, an *ad hoc* note written for a specific purpose and not intended as a permanent reference work.

2.1.7 Summary

The basic facts about the OB coefficient lists are as follows:

Table 2.8: Summary of the Old Babylonian coefficient lists

Text	*Provenance*	*Date*	*Lines*	*Coefficients*	*Subjects*[16]
A	unknown	early II mill. BC	48	54	g, q, s, m, o
C	Shaduppum	pre-1762 BC	132 extant	41 extant	g, q, s, m, o
D	Susa	16th century BC	71	68	g, q, s, m
E	Larsa?	{late 19th–early	52	57	g, q, s, m, o
F	Larsa?	{18th century BC	66	75	g, q, s, m, o
G	Sippar Amnanum	early OB	16	16	g, q, s, o
H	Shaduppum	pre-1762 BC	7	7	g, q, s, o
M	Ur	early II mill. BC	6	3	q

The lists, except C and M, are tabular in format, and are subdivided more or less rigidly into groups of coefficients. Typically the geometrical ones are first, followed by coefficients on quantity surveying, storage, and metals, as exemplified by text D. Every list includes parts whose organization is not clear, either because the coefficients themselves are not understood or damaged, or because the connections between them are not apparent. Within each group one of three organizing principles may be used:

- thematic, according to which the coefficients are listed in a sequence logically determined by content; e.g. the geometry section in text D;
- numerical, according to which the coefficients are listed in numerical order; e.g. all four sections on metals (texts A, C, D, F);
- acrographic, according to which coefficients written with the same first sign are grouped together; e.g. the IM section of text E; the ZI section of text E.[17]

All the lists, except the entirely logographic M, contain a few grammatical mistakes, mainly in the form of nominative-for-genitive case errors after a construct or *ša*. The use of logograms in these lists ranges from, at one extreme, A with only a few Akkadian words, to C, which is predominantly syllabically written.

The coefficients can be divided into four subject groups, listed in Table 2.9, plus those whose meaning or mathematical function (or both) is unclear. Those with sexagesimally irregular values are shown in brackets; those with probably erroneous values are in *italics*.

Figure 2.1 shows the same information in a different form. The size of each pie chart is proportional to the length of the list it represents. The large proportion of obscure entries in

[16] g = geometry; m = metals; o = other/obscure; q = quantity surveying; s = storage and metrology

[17] Three Isin-Larsa-OB lexical lists — Proto-Izi, Proto-Kagal, Nigga — were also ordered exclusively or predominantly on acrographic principles, as were various 'unlabelled' lists (cf. Black 1988: 272).

lists A, C, and E is due as much to their bad state of preservation as to lack of mathematical context for the coefficients.

Table 2.9: Subject matter of the coefficients in Old Babylonian lists

List	*Lines*	*Geometry*		*Quantity surveying*		*Storage and metrology*	*Metals*		*Others*		*Total*	
A	48	10		16	(*3*)	4	6	(*1*)	18	(*1*)	54	(*5*)
C	40	11	(3)	8		2	6	(*1*)	14		41	(3 + *1*)
D	71	31	(10)	24		4	9	(2)	0		68	(12)
E	52	12	(1)	17		5	6		17	(*1*)	57	(1 + *1*)
F	66	17	(1)	31	(2)	7	7		13		75	(1 + 2)
G	16	2		10		2	1		1		16	
H	7	2		2		3	0		0		7	
M	6	0		3		0	0		0		3	
Total	299	85	(15)	111	(*5*)	27	35	(2 + 2)	69	(2)	321	(17 + *9*)

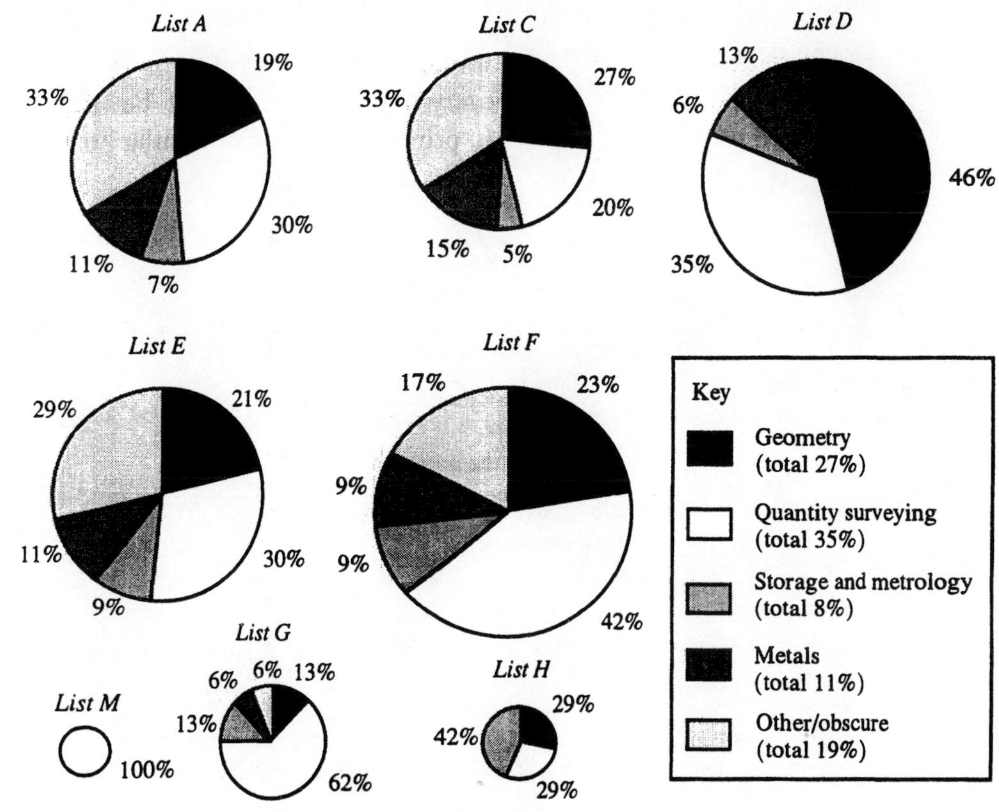

Figure 2.1: Division of the Old Babylonian coefficient lists by subject matter

In most of the lists the coefficients are written using a uniform format, consisting minimally of the numerical value of the coefficient and its name. One or both of the words 'coefficient' and 'of' may appear between them. In the two cases (lists F and H) where the formatting is not consistent, the changes are consonant with a change in subject matter, which suggests that they have been compiled from more than one original. Crucially, though, there is no evidence for a common 'Urtext'; at least, one would be impossible to (re)construct from the extant lists.

2.2 THE FOUR LATER COEFFICIENT LISTS

2.2.1 List B

Text B is on the remains of a three-column tablet, of which parts of four columns are extant. It was excavated at Nippur[18] and apparently dates to the Kassite period, although this has been a matter for dispute.[19] The first column is devoted to a treatise on tuning the strings of a musical instrument. This section alone has generated a huge literature[20] and its contents will not be treated further here.

Column II starts with a list of 'boat' coefficients for different commodities, in fact volume–capacity conversion coefficients, which have close parallels in the OB material (§§7.2, 7.4). Line 16′ contains two entries and 18′–21′ each have three. They are followed by a rather more obscure section on the relation between various standard capacity measures, often with more than one coefficient on each line — separated by double-Winkelhaken markers which Kilmer (1960: 279) did not always recognize (see §A.1). Column III contains coefficients on various practical subjects, some of which are known from OB lists or problem texts (see §§6.5, 8.2). Four of the preserved numerical values are irregular (ll. 15′–17′, 25′) and one coefficient is given two values (l. 19′). The first two columns of the reverse are completely missing; the last consists of around 15 numerical values and a few initial signs of the names of what seem to be geometrical coefficients. The exact meaning of every coefficient may not be clear, but it is apparent that the list is well organized, according to principles which are similar to those found in the OB texts:

Table 2.10: Contents of coefficient list B

Lines	*Subject*	*Format*	*Comments*
II 3′–10′	storage	1-4	Thematic groupings clear; OB precursors (§§4.3, 7.2, 7.4).
II 11′–33′	metrology	1-(2)-4	Exact meaning unclear.
III 1′–8′	linseed/sesame?	1-4	No OB precursors.
III 9′–14′	astronomy	1-4	Clear numerical/thematic groupings (§8.2).
III 15′–20′	pomegranate?	1-4	First four in descending numerical order with constant difference 2 00.
III 21′–28′	irrigation	1-4	Thematic grouping; some OB precursors (§6.5).
III 29′–34′	carrying	1-4	Meaning and ordering unclear.
IV 1′–15′	geometry?	1-4	Broken.

It has been a matter for some comment that this tablet contains a mixture of mathematical and non-mathematical material. Nemet-Nejat clearly has List B in mind when she states that:

the coefficient lists are concerned with both non-mathematical operations and mathematical ones. [...] That is to say, in the tradition of the lexical lists, the coefficient lists may have been padded or expanded in order to achieve the idealized form of scientific organization. (Nemet-Nejat 1993: 19–20)

Høyrup, on the other hand, has implicitly used it as evidence for a break in scribal tradition after the Kassite conquest, after which 'to a certain degree mathematics came to be mixed up with other subjects on the same tablets, having lost its disciplinary autonomy', in contrast to the

[18] No provenance was given on publication (Kilmer 1960), but the Philadelphia University Museum catalogue states: 'Nippur, 2nd Exp., west side of Shatt en-Nil' (unpublished; pers. comm. Dr Linda Bregstein).

[19] 'Kassite' (Kilmer 1960: 273; Duchesne-Guillemin 1984: 6); 'MB' (University Museum catalogue), but see Kilmer (1980–3: 575; 1992: 101): 'NB'.

[20] Works in which this part of the text is discussed are listed in Nemet-Nejat (1993: 279). Most recently, see Kilmer (1992), West (1993–4) and Gurney (1994).

'autonomous concern' of the OB period (Høyrup 1992: 604). However, both these standpoints are a reflection more of the modern availability of mathematical material than of ancient reality. School tablets containing a mixture of 'lessons' have been excavated, although not well published, from Old Babylonian Ur and Nippur (see §§1.3.3, 10.4, A.5). In this light the format of text B is neither unique nor surprising. The musical material is clearly separated from the coefficient list, occupying a separate column. In the same way model contracts, lexical lists, and metrological and mathematical tables appear on Nippur tablets containing the Sumerian Proverb Collections (Gordon 1959). The inclusion of a variety of subjects on one school tablet cannot be seen as a distinctively post-OB phenomenon.

2.2.2 *Text J*

This coefficient list is part of the tablet series I.NAM.GIŠ.ḪUR.AN.KI.A, whose full title is translated 'the ... plans of heaven and underworld, matching pairs of (things in) heaven and underworld, things of the Apsû, as many as were designed' (Livingstone 1986: 35). In this series:

> mythical or religious associations between different items or between names and their explanations are supplemented by pointing out haphazard and artificial connections derived arithmetically or philologically. (Livingstone 1986: 49)[21]

This particular tablet, around a third of which is extant, is dated in the colophon to the 33rd year of Darius I (489 BC), and was copied and collated, almost certainly at Babylon, by Marduk-nadin-apli, a land registrar[22] and butcher (GÍR.LÁ) of Marduk. The other surviving exemplars of tablets belonging to this series were copied by an Assyrian scribe in Ashur at some point(s) during the reigns of Sargon II and Sennacherib and are thus at least 200 years older (Livingstone 1986: 17).

Despite the esoteric and numerological content of the rest of the text, this coefficient list does seem rather prosaic. It contains 15 extant storage coefficients, in 1-4-2 format. These do not appear to concern 'dimensions of granaries and Ekur, the temple of Enlil in Nippur' (*pace* Livingstone 1986: 17). The first 6 lines concern the (floor?) areas of different types of storage facility, including an *ammammu*, some sort of large receptacle (*AHw*: 43); a triangular *araḫḫu* storehouse; two grain-piles; and a floor.[23] Although the exact significance of the coefficients remains undetermined, after collation their numerical relationship is now clear. Table 2.11 shows that he errors must be copying mistakes which occurred in the course of calculating the value of the coefficients: they are cumulative, in that each mistake is carried over into the calculation of the following coefficient.

The remainder of the obverse contains volume–capacity conversion coefficients for different sorts of goods, including grain, water, oil (?), and millet, with numerical values of the kind which are well attested from the OB period (see §7.2); three have irregular numerical values (ll. 2, 10–11). The format of all the entries in the list is 1-2-4. On the reverse a coefficient ⸢2⸣ 40 IGI.GUB.BA is multiplied by 6 in a broken context.[24]

[21] Although those connections were presumably not considered haphazard by the compiler(s).

[22] *šassukku* does not mean 'librarian' (*pace* Livingstone 1986: 23); see *CAD* (Š II: 145).

[23] The *kerṣu* of line 3 must mean something other than its usual 'pinched-off piece of clay' here (cf. *AHw*: I 468). In line 5 the sign after KA remains unreadable after collation. The reading *qaqqari* for GAB-*ri* in line 6 is based on the glosses in BM 96957 + VAT 6598 (§A.4). Stephanie Dalley provided invaluable help in interpreting this list.

[24] See Friberg *et al.* (1990: 503–4) for a discussion of some of the difficult mathematical terminology on the reverse of the tablet.

Table 2.11: Numerical relationships in coefficient list J

Line	*Coefficient*	*Description*	*Comments*
1	4;26 40	*ammammu*	$^{40}/_{9}$
2	0;26 15	*araḫḫu* SAG.3	$^{7}/_{16}$; equal to the OB coefficient for the area of an equilateral triangle (see §3.3)
3	0;33 45	*kerṣu*	$^{9}/_{16}$, or $(^{3}/_{4})^2$ = 1 − 0;26 15 = 2;30 ÷ 4;26 40
4	0;16 42 30	*karû*	Scribal error: 0;33 45 ÷ 2 = 0;16 **52** 30
5	0;04 13 07 30	*ka*-X *karê*	Scribal error: 0;16 42 30 ÷ 4 = 0;04 **10 37** 30
6	0;12 36 22 30	*qaqqari* (?)	Scribal error: 0;04 13 07 30 × 3 = 0;12 **39** 22 30

2.2.3 Lists K and L

Texts K and L are hitherto unpublished fragments of two LB coefficient lists, thought to be from Babylon.[25] Both must have come from multi-columned tablets: K, from the top of a tablet, preserves parts of two columns on each side; L, from a left edge, contains only one, and the ruling for the next column. Only traces of signs remain on the reverse. It seems most unlikely that they are fragments of the same tablet, as the reverse of text K treats the same subject matter of the obverse of text L. The size of the original tablets is unclear.

Although the word IGI.GUB/*igigubbûm* 'coefficient' does not occur in either text it is clear that these are indeed coefficient lists. Both are in list format, with numbers on the left of each column, where preserved, and descriptions of those numbers on the right. The lines in adjacent columns of text K are not aligned to each other, and so cannot be read as a multi-columned table. Further, the subject matter is unprecedentedly similar to that of the OB lists, and even the order and numerical values of individual entries are directly related to them. See §A.1 for handcopies and transliterations of these two fragments.

Table 2.12: Contents of coefficient lists K and L

Text	*Lines*	*Subject*	*Format*	*Comments*
K	I 1–2	broken	[1-...]-4	
	I 3–9	metals	[1-...]-4	Similar order to the OB lists. See §8.1.
	II 1–8	broken	1-[...-4]	
	III 1′–5′	broken	1-[...-4]	
	IV 1′	broken	[1-...]-4	
	IV 2′–5′	bricks	[1-...]-4	Same order as lists L and B. See §4.3.3.
	IV 6′	broken	[1-...]-4	
L	I 0′–13′	bricks	1-4S	See detailed discussion in §4.3.

2.3 COEFFICIENTS IN PROBLEM TEXTS AND ROUGH WORK

2.3.1 General characteristics

Coefficients are not just curious numbers found in lists. They were widely used in problem-solving of all sorts — as the following statistics show. Sixty-two different coefficients are attested 210 times in problems, calculations, and diagrams on 52 tablets. While the majority of

[25] Friberg (1996*a*: 3) dates lists K and L to the OB period, but it is clear from the palaeography and orthography that they are much later than that.

Table 2.13: Coefficients in Old Babylonian problem texts and rough work

Museum number	*Problems using coefficients*	*Number of coefficients*	*Number of attestations*[26]	*Subject matter*
AO 6770	1 of 5	1	1	q
AO 8862	**4** of 8	**1**	**4**	q
AO 10822	**8** of 10 extant	**4?**	**8**	q
Ashmolean 1922.168 (§A.5.2)	1 of 1	1	3	g
AUAM 73.2834	**1 of 1**	**1**	**1**	g
AUAM 73.2841	**1 of 1**	**1**	**1**	g
BM 85194	17 of 35	8	20	g, q, s, o
BM 85196	11 of 18	9	13	g, q, s, o
BM 85200	19 of 19 extant	1	19	s
BM 85210	4 of 12 extant	6	7	g, o
BM 96954 + BM 102366 + SÉ 93 (§A.3)	11 of 16 extant	5	12	g, s
BM 96957 + VAT 6598 (§A.4)	11 of 16 extant	4	13	q
Böhl 1821	1 of 1	2	2	g
Erm 15073	2 of 8	1	2	q
Haddad 104	9 of 10	11	32	g, q, s
IM 53961	1 of 1	1	1	q
IM 54011	2 of 2	1	2	q
IM 54478	1 of 1	2	2	s
IM 54538	1 of 1	1	1	q
M 288	**1 of 1**	**1**	**1**	q
MLC 1354	1 of 1	1	1	g
Str 366	1 of 4	1	1	o
TMS 2	**2 of 2**	**2**	**2**	g
TMS 14	2 of 2	2	3	s
TMS 20	1 of 1	1	1	g
TMS 21	1 of 1	3	3	g
TMS 24	2 of 2 extant	1	2	s
TMS 25	1 of 2 extant	1	1	q
UET 6/2 218 (§A.5.1.4)	1 of 1	1?	1	s
UET 6/2 233 (§A.5.1.4)	1 of 1	2?	2	q, s
UET 6/2 236 (§A.5.1.4)	1 of 1	2?	2	q, s
UET 6/2 247 (§A.5.1.4)	1 of 1	1?	1	s
UET 6/2 250 (§A.5.1.4)	1 of 1	1?	1	s
UET 6/2 254 (§A.5.1.4)	1 of 1	2?	2	q
UET 6/2 269 (§A.5.1.4)	1 of 1	1?	1	s
UET 6/2 290 (§A.5.1.4)	1 of 1	1?	1	s
UET 6/2 293 (§A.5.1.4)	1 of 1	3?	3	q, s
UET 6/2 298 (§A.5.1.4)	1 of 1	2?	2	q, s
UET 6/3 374 (§A.5.1.4)	1 of 1	1?	1	s
UET 6/3 452 (§A.5.1.4)	1 of 1	1?	1	s
VAT 8522	1 of 4	3	3	g, s
VAT 8523	2+**1** of 6	1+**1**	2+**1**	q
YBC 4669	1+**3** of 25 extant	1+**4**	1+**4**	q
YBC 4673	6+**4** of 21 extant	7+**3**	7+**5**	q, m
YBC 4708	**60 of 60**	**1**	**60**	s
YBC 7164	*2* of 18 extant	*1*	*2*	q
YBC 7284	1 of 1	1	1	q
YBC 7289	1 of 1	1	1	g
YBC 7997	1 of 1	3	3	g
YBC 8588	1 of 1	1	1	q
YBC 8600	1 of 1	1	1	s
YBC 9874	1 of 1	1	1	q
YBC 10722	**1 of 1**	**1**	**1**	q

[26] This figure includes repetitions of a coefficient used in two or more problems on the same tablet, but not those cases where a coefficient is mentioned more than once in a single problem.

these coefficients are stated explicitly as part of the method of solution, the use of others can be inferred in cases where only the statement and answer are given. These instances are shown in **bold** in Table 2.14. None of the coefficients yet attested in problems or rough work have irregular numerical values.

As already mentioned in §1.3.1, coefficients are not given explicitly in the statement of the problems, but are intrinsic to the problems themselves. This means that a coefficent can be introduced apparently without explanation in the course of a solution because its relevance has already been signalled implicitly by the choice of a particular type of problem. Thus numerical information which looks like a coefficient given in the statement of a problem is not strictly speaking a coefficient, as it pertains to that problem in particular and not to the type of problem to be solved. All but 3 of the 82 instances known concern work-rates for excavating canals, in VAT 7528, YBC 4657, YBC 4666, YBC 5037 and YBC 7164 (§6.3). The others, from YBC 4673, give the work-load for constructing an earth wall (§6.1).

Al-Rawi and Roaf (1984: 181) noted the frequent use in Haddad 104 (viii)–(x) (§§5.1–2) of the verb *lu-pu-ut* 'write down'[27] to introduce coefficients (which are never mentioned in the statement of a problem) into the solution.

The mathematical use of *lapātum* is not restricted to coefficients. It is more generally found as the instruction to write down a number (*nalpattum*) to be used in the next stage of the calculation.[28] This number may either be part of the numerical data of the problem[29] or an intermediate answer.[30] The Haddad 104 attestations of *lapātum* have exactly the same meaning, only here the number is to be taken from a coefficient list, not from an earlier part of the problem. Indeed *šakānum*/GAR, mathematically synonymous to *lapātum* (Høyrup 1990*a*: 57–8), is also used occasionally to 'put down' coefficients that are about to be used.[31] More generally, though, coefficients are introduced into the solution with no special terminology. Coefficients are multiplied using the verb *našûm* reserved for multiplication of concrete quantities (Høyrup 1990*a*: 46–7).[32]

Three Ur III model documents[33] also use a total of six different coefficients, all of which relate to quantity surveying. Two of the work-rates in W 47a are stated explicitly, but the rest must be inferred from the data in the documents. Both of the brick coefficients in YBC 9819 are used incorrectly (see §4.5).

2.3.2 *Subject matter and format*

All the coefficients occurring in problem texts and rough work are dealt with individually in the relevant sections of §§3–8, but Table 2.14 gives a breakdown of their subject matter. The distribution of these coefficients between the different subjects is rather different from that of the coefficients in the lists (see §2.1.6). There is a much greater incidence of quantity

[27] Reading, with all other mathematical occurrences, *lu-pu-ut* as the simple G-stem a/u imperative *luput*, not *lupput* (D stative) 'used of numbers which "stand written down" ' (e.g. Høyrup 1990*a*: 58 n. 83). This is confirmed by instructions such as 10 *ba-la-lam lu-pu-ut* 'write down 0;10, (the coefficient of) mixing' (Haddad 104 (x): IV 29), and the functionally identical second person singular present/future: *ta-la-pa-at-ma* 'you write down' (AO 8862 (ii): II 5).

[28] The non-mathematical meaning of the verb is closely related: 'eingreifen in, anfassen; schreiben' (*AHw*: II 535).

[29] e.g. AO 6770 (iii): II 5. This is not as rare an occurrence as Høyrup (1990*a*: 58) suggests.

[30] e.g. MLC 1842: I 8.

[31] 1 30 *i-gi-gu-ba-am šu-ku-un-ma* 'put down 1;30, your coefficient', IM 54538: 6–7 (§5); 15 ÉŠ.GÀR GAR.RA 'put down 0;15, the work-rate'. YBC 8588: I 9 (§6).

[32] But see §3.1 for a brief discussion of the verbs used with 0;20, the coefficient for the diameter of a circle.

[33] AOT 304 (§4.4), W 47a (§6.4), and YBC 9819 (§4.5).

surveying and metrological coefficients, in particular, and proportionally fewer on geometry and metals. The smaller incidence of obscure coefficients is a direct consequence of the problems providing contexts for their use.

Table 2.14: Subject matter of the coefficients in OB problem texts and rough work

Subject	*Tablets*	*Problems*[34]	*Coefficients*	*Attestations*	
Geometry (§3)	16	32	14	46	(22%)
Quantity surveying (§§4–6)	28	24	36	86	(41%)
Storage and metrology (§7)	24	58	10	68	(32%)
Metals (§8.1)	1	1	1	1	(1%)
Other/obscure (§8.3)	3	8	4	8	(4%)

While over a third of coefficients used in problems are not found in the lists, the large majority of the coefficients in the lists are not attested in problem texts — which just shows how much there remains to be discovered about Mesopotamian mathematics. The penultimate column of Table 2.15 shows the number of *different* coefficients in use (as opposed to the number of attestions of those coefficients, in the final column). From this it can be seen that we are dealing with a limited number of coefficients, each attested several times, especially in the case of geometry and metrology. However, those coefficients that are found most frequently in the problem texts are also in the lists, with the exception of the three linear metrological conversion coefficients (§7.1).

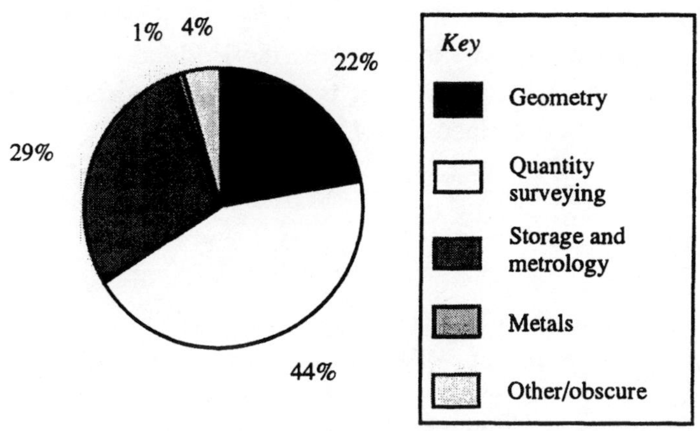

Figure 2.2: Division of the coefficients in OB problem texts and rough work by subject matter

The coefficients occur in several different formats, including some which are not found in the lists themselves, so that the typology set out in §2.1 needs to be expanded:

5 NUMUN a quasi-synonym for IGI.GUB, only in Sippar texts (see §7.1.3)
6 verbal form only.

It is particularly striking that the word 'of' (*ša*) occurs only once in descriptions of coefficients in the problem texts (see table below), although it is very common in the lists. The format of coefficients can vary within a tablet and even within a problem; the most common

[34] The total of this column is larger than the number of problems given above, because some problems use coefficients from more than one category.

scenario is for a coefficient to be stated in 1-2-4 form first and then abbreviated to 1,[35] 1-2,[36] or 1-4[37] format for the rest of the problem or series of problems, as one might expect.

Table 2.16 analyses the coefficients found in problem texts by subject matter and format. Attestations of one coefficient in different problems on the same tablet are counted separately, but multiple attestations of the same coefficient within one problem are not, unless they occur in different formats. Thus the totals are slightly larger than those given in §§2.3.1–2. Data for implicit coefficients are given at the bottom of the table.

Table 2.15: Format of the coefficients in Old Babylonian problem texts and rough work

Format	*Geometry*	*Quantity surveying*	*Metrology and storage*	*Metal or obscure*	*Total*	*Example*	
1	13	14	22	1	50	YBC 7997	(§4.4)
1-2	10	3	7	1	21	IM 53961	(§6.1)
1-2-4	8	11	1	1	21	*TMS* 25	(§6.5)
1-3-4	0	0	0	1	1	BM 85210 (xix)	(§8.3)
1(*)-4	8	29	13	4	54	Haddad 104 (x)	(§5.2)
1-5-4	2	0	20	0	22	BM 85210 (viii)	(§3.2)
2-1	0	2	0	0	2	YBC 7284	(§4.3)
6	7	0	6	0	13	BM 85194 (iv)	(§3.1)
Implicit	4	27	0	1	28	M 288	(§6.2)
Total	52	86	69	9	212		

Frequent attestations of one coefficient in a tablet can distort these figures; for instance the nineteen occurrences of one metrological coefficient in BM 85200 have inflated the 1-5-4 format total. Similarly, it should be borne in mind that the Haddad 104 attestations account for nearly 20% of the total. The eleven attestations of type 6 format represent just two coefficients, only one of which is attested outside Haddad 104.

2.4 INTRODUCTION TO §§3–8

As far as possible the division of the coefficients into groups follows the groupings within the texts themselves. Thus the order of the following chapters, and the sections within them, is similar to the internal sections of the texts listed in §2.1. Geometrical coefficients are dealt with in §3, and quantity surveying in the following three chapters. Although these coefficients make up only about 30% of those known from the lists they are very well attested in problem texts and yet have been studied relatively little. They have been divided into three main topics: bricks and brickworks (§4), making bricks and carrying materials (§5), and earthworks and waterworks (§6). This arrangement is not found in the lists but has been dictated by mathematical considerations. For instance, although *nalbanum* and *nazbalum* brick coefficients are usually listed together (e.g. F 2–15), they are presented separately here, with the *nazbalum* coefficients in the same chapter as those on carrying other materials (§5.3). Although they depend on the mathematics of the *nalbanum* coefficients (§4.1) they are not necessary to the understanding of brickworks; but they are closely related to the mathematics of carrying earth,

[35] e.g. 18 IM.LÁ, BM 85194 (x): III 6, shortened to 18 in problem (xii): III 15 (§8).

[36] e.g. 5 IGI.GUB.BA GÚR, Böhl 1821: diagram, shortened to IGI.GUB in line 15 (§3.1).

[37] e.g. 5 *i-gi-gu-bé-e ki-pa-tim*, Haddad 104 (iii): II 8, shortened to 5 *ki-pa-tim* in (iv): II 43 (§3.1).

which in turn links with making bricks. Chapter 7 deals with storage and metrology while §8 concerns those coefficients for which there is little or no evidence outside the lists themselves.

Each of the following six chapters is arranged in the same way. Each section within a chapter deals with a particular topic. The coefficients from the lists and the problem texts are first listed in transliteration and translation, followed by a mathematical and a philological commentary, in which the mathematical significance and the terminology are discussed. The final section is always a summary of the findings of that chapter.

3. GEOMETRICAL COEFFICIENTS

This chapter concerns the group of coefficients which have been studied most by modern scholars.[1] Much of the basic work on the interpretation of geometrical coefficients was done by Vaiman (1959; 1962; 1963). Relevant sections of his books and articles have been summarized, especially by Friberg (1982; 1987–90) and Kilmer (1990); the Russian originals have not been accessible to me directly.

Sources of geometrical information beside the coefficient lists include other mathematical texts, both problems and metrological tables. Especially important is the famous tablet BM 15285 (§A.2), which contains around thirty extant drawings of figures inscribed within squares, and a few lines of description underneath each. The problem each time is to calculate the areas of the illustrated shapes, although the answer is never given.[2] This document is particularly rich in technical terms, most of which are identifiable from the illustrations.

Because of the success of previous authors in interpreting OB geometry, some parts of this chapter simply survey the established meanings of the coefficients. However, there are still some coefficients and technical terms which have not yet been fully understood, and these are examined more closely.

Seven of the Old Babylonian lists contain geometrical coefficients, with a total of 86 attestations of 49 separate coefficients. Thirteen of them, and one which is not in the lists, are attested 46 times in 32 mathematical problems and diagrams from 16 tablets. Ten of these problems and diagrams are specifically about geometrical figures;[3] the rest concern objects such as wells and reed-bundles which are composed of one or more of these shapes.[4] Thirty-one of the attestations concern circles; the rest are coefficients for semicircles (5), triangles (2), squares (2), polygons (2), and the *apsamikkum* (4). Several of the currently obscure coefficients may also turn out to be geometrical (see §§3.8, 8.3–5).

3.1 CIRCLES

3.1.1 The coefficients

A 2	5 *ù* 6 IGI.GUB KA.KÉŠ	0;05 and 6, the coefficients of a circle.
C 4	⌜*ki*⌝-*pa-ta-am* ⌜5⌝ [*i-gi-gu-bu-ša*]	A circle: 0;05 [is its coefficient].
D 2	5 IGI.GUB *šà* GÚR	0;05, the coefficient of a circle.
E 1	5 KA.[KÉŠ]	0;05, a circle.

[1] See §1.1.1 for the reasons for this.

[2] Problems on drawing figures within figures are catalogued in IM 52916: II 14–22 (§A.1).

[3] AUAM 73.2834: triangle (implicit); AUAM 73.2841: circle (implicit); BM 85194 (xx): circle; MLC 1354: semicircle; YBC 7289: square; *TMS* 2 (i)–(ii): hexagon and heptagon (implicit); *TMS* 20 and 21: concave square.

[4] Ash 1922.168: trapezoidal fields; BM 85194 (iv): circular defence works of a city; (xiv): reed-bundle — truncated cone; (xvi) and (xix): cylindrical wells; BM 85196 (ii): reed-bundle — truncated cone; (vi): cylindrical container; BM 85210 (viii): unnamed solid composed of cube and semicircular channel; Böhl 1821: circular defence works of a city; Haddad 104 (i)–(v): cylindrical grain containers; VAT 8522 (i): cedar log — truncated cone; YBC 7997: cylindrical oven.

F 20	⸢5?⸣ *ša* GÁN GÚR	0;05, of the area of a circle.
G 1	5 IGI.GUB ○ *ki-pa-tim*	0;05, the coefficient of a circle.
H 4	5 *ša ki-pa-ti*	0;05, of a circle.
C 5	*ta-al-li ki-pa-[tim ... i]-⸢gi-gu⸣-[bu-šu]*	The diameter of a circle: ... is its coefficient.
D 3	20 DAL *šà* GÚR	0;20, the diameter of a circle.
D 4	10 *[pi]-ir-ku šà* GÚR	0;10, the radius of a circle.

AUAT 73.2841	(5)	(0;05)
BM 85194 (iv): I 49	5 GÚR	0;05, a circle.
BM 85194 (xiv): III 26, 27	5 GÚR	0;05, a circle.
BM 85196 (ii): I 9	5 IGI.GUB	0;05, the coefficient.
BM 85196 (vi): I 30	5 IGI.GUB.BA	0;05, the coefficient.
BM 96954 (xxii): V 8	5 IGI.GUB	0;05, the coefficient.
BM 96954 (xxiii): V 18	[5 IGI].GUB.BA	[0;05], the coefficient.
Böhl 1821: 15	5 IGI.GUB.BA	0;05, the coefficient.
diagram, 16	5 IGI.GUB.BA GÚR	0;05, the coefficient of a circle.
Haddad 104 (i): I 6	5 *i-gi-gu-bé-e ki-pa-tim*	0;05, the coefficient of a circle.
Haddad 104 (ii): I 15	5 *i-gi-gu-bé-em*	0;05, the coefficient.
Haddad 104 (iii): I 27, II 27:	5 *i-gi-gu-bé-em*	0;05, the coefficient.
II 8	5 *i-gi-gu-bé-e ki-pa-tim*	0;05, the coefficient of a circle.
Haddad 104 (iv): II 43, III 4	5 *ki-pa-tim*	0;05, a circle.
Haddad 104 (v): III 27	5 *ki-pa-tim*	0;05, a circle.
VAT 8522 (i): I 4a	5 IGI.GUB ⸢*ki-pa-tim*⸣	0;05, the coefficient of a circle.
I 3c	5 IGI.GUB	0;05, the coefficient.
YBC 7997: II 2	*a-na* 5 *tu-ub-ba-al-ma*	You multiply by 0;05.

BM 85194 (iv): I 45	IGI.3.GÁL BA.ZI	Take one third.
BM 85194 (iv): I 47	DAL *šu-li-iš*	Triple the diameter.
BM 85194 (xvi): III 44, 50	A.RÁ 3 TAB.BA	Multiply by 3.
BM 85194 (xix): IV 28	IGI 20 BA.ZI	Take 0;20.
BM 85194 (xx): IV 34–5	20 DAL BA.ZI	Take 0;20, the diameter.
BM 96954 (xxii): V 7	*šu-ul-li-iš*	Triple.
Böhl 1821: 8	*a-di* 3 *a-lik*	Go up to 3.
Haddad 104 (i): I 4	*ta-al-lam šu-li-iš-ma*	Triple the diameter.
Haddad 104 (ii): I 14	*šu-li-iš-ma*	Triple (the diameter).
Haddad 104 (iii): I 26, 40, II 7, 20, 26	*šu-ul-li-iš-ma*	Triple (the diameter).
Haddad 104 (iv): II 36, 42	*šu-ul-li-iš-ma*	Triple (the diameter).
Haddad 104 (v): III 14, 20, 26	*šu-ul-li-iš-ma*	Triple (the diameter).
VAT 8522 (i): I 2a	3 *ša-lu-uš-tim*	3, a third [*sic*].

3.1.2 Mathematical commentary

These circle coefficients were first discovered in the 1920s (Neugebauer and Struve 1929) and have been presented in almost every work on Mesopotamian mathematics since. The most frequently attested, in both the lists and the problem texts, is 0;05 or $^1/_{12}$ as, for example, in YBC 7997: I 9–II 3:

1 30 *ki-[ip-pa]-tam*	You square 1;30, the circumference.
[*tu*]-*uš-ta-ka-al*	
a-na 5 *tu-ub-ba-al-ma*	You multiply (the result) by 0;05 and
11 15 *i-li-a-ma*	0;11 15 will come up, and (...).

As the extract illustrates, this coefficient represents the equality:

$$A = 0;05\ c^2,$$

where A is the area of the circle, c its circumference, and 0;05 is equivalent to the modern $1/(4\pi)$.[5] The second coefficient listed depicts the relationship:

$$d = 0;20c, \text{ or } c = 3d,$$

that is, $c = \pi d$, where d is the diameter of the circle. It is also likely that the missing number in C 5 is either 3 or 0;20. Although fully attested only once in the lists, the second coefficient occurs in the problem texts as frequently as the first, and indeed often in the same problems. This is because, to find the area of a circle from its diameter, the method is not to exploit directly the fact that:

$$A = \frac{\pi d^2}{4}, \text{ or } A = 0;45d^2,$$

but to work out the circumference first,[6] and then the area using the two methods shown above (Haddad 104 (iii): II 26–8):

1 40 *mu-uḫ-ḫi i-ṣí-im šu-ul-li-iš-ma*	Triple 1;40, the top of the log, and
5 *ki-pa-at i-ṣí-im i-lí*	5, the circumference of the log, will
5 *šu-ta-ki-il-ma* 25 *i-lí*	come up. Square 5 and 25 will come up.
25 *a-na* 5 *i-gi-gu-bé-em i-ši-ma*	Multiply 25 by 0;05, the coefficient, and
2 05 A.ŠÀ *i-lí*.	2;05, the area, will come up.

The value 6 in A 2 is the reciprocal of 0;10, found in D 4, which explicitly mentions the radius of a circle. This coefficient represents the relationship:

$$r = 0;10c, \text{ or } c = 6r,$$

that is, $c = 2\pi r$, where c is the circumference and r the radius of the circle (*TMS*: 31). No extant OB problems use this coefficient.

The numerical value of F 20 is read here as 5 instead of 6 (*pace MCT*: 134; Kilmer 1960: 284). Both fit the traces on the tablet, but 5 is more likely, as it is so frequently attested. In 1945 the reading and meaning of GÚR were not yet known and so the correct value for the coefficient could not have been recognized.

The constant π embodies several relationships. Smeur (1969–70: 252) states that:

$$\text{const (I)} = \frac{\text{circumference of a circle}}{\text{diameter of the circle}}$$

$$\text{const (II)} = \frac{\text{area of a circle}}{\text{square on radius}}$$

Both relations can be found in ancient mathematics, accepted intuitively, without strict mathematical proof.

[5] The unlabelled diagram on the roughly square tablet AUAM 73.2841 shows the number 45 written inside a circle. Immediately to the left are two numerals 3, on above the other, and in the bottom left corner of the tablet is a 9. If 3 is the circle's circumference, we have $3^2 = 9$, and $0;05 \times 9 = 0;45$, the area of the circle.

[6] But 0;45 is attested as a semicircle coefficient (§3.2), where it has a similar function.

Much deeper awareness lies in the fact that constants (I) and (II) are identical.[7]

How valid is this statement for the OB period? Constant (I) is equivalent to the coefficients 3 or 0;20. However, constant (II) would be equivalent to a circle coefficient 0;45, so far unattested in OB mathematics. Neither is it likely to be found: it is the circumference that is the *defining component* of the OB circle: not only are identical words used to describe the circumference and circle but the circumference is the starting point from which the circle is conceptualized.[8] This is reflected in the names of the coefficients, especially those from list D, which define other components of the circle with respect to circumference. It can be seen in problems too: for instance, whereas BM 85194 (xx) names the diameter coefficient 0;20 explicitly, its reciprocal 3 — which expresses the circumference or circle in terms of the diameter — is never named. In short, constant (II) is outside the conceptual framework of OB mathematics.

If instead we substitute an expression representing the coefficient 0;05:

$$\text{const (II*)} = \frac{\text{area of a circle}}{\text{square on circumference}},$$

then the first sentence of Smeur's analysis does hold true for Mesopotamia. The second sentence poses a more difficult question. On the one hand, the fact that the area of a circle was never calculated directly from the diameter suggests that no attempt was made to assimilate the coefficients 0;05 and 3. On the other hand, using the two separate coefficients has the advantage of giving a useful, and conceptually fundamental, intermediate result: the circumference. Further, all circle and semicircle coefficients share an underlying value of $\pi \approx 3$, which speaks against a complete conceptual separation of the two identities.

Attempts to show that more accurate values were used remain unconvincing.[9] The sole exception is the value $\pi \approx$ 3;07 30 underlying the 'log' coefficient 0;04 48 (Powell 1984: 42–4). It is used as a more accurate alternative to the circle coefficient 0;05 when calculating the cross-sections of some cylindrical grain containers.[10] Powell (1984: 44–5) notes that:

> interpretations which depart from the assumption that the approximation $\pi \approx 3$ is due to ignorance or ease of calculation do not strike at the heart of the matter. The value was maintained after the discovery of 3;07 30, not primarily because 3 was easier to use, but because it simplified the conceptualization of the ratios of areas between circles and squares, and probably because the system had already developed coefficients to deal with this problem [i.e. storage coefficients (§7.2)].

The approximation $\pi \approx$ 3;07 30 is used for storage coefficients only; it is purely for metrological purposes and is not found in other domains of OB mathematics.

3.1.3 Philological commentary

Three words are used for 'circle/circumference' in the lists: the logograms GÚR and KA.KÉŠ (which is not found in the problem text attestations), and the Akkadian *kippatum*. The syllabic readings are found in both texts from Shaduppum, as well as in list G, from Sippar; coefficient

[7] Proved by Archimedes (third century BC) in Proposition I of *Measurement of the Circle* (Heath 1981: II 50).

[8] Compare this to the modern view of the circle, in which the radius is the defining component: we are taught to calculate using formulae based on the radius.

[9] On the approximation $\pi \approx$ 3;07 30 supposedly implicit in D 30 (*TMS*: 33; Neugebauer 1952: 47, 52) see §3.3; see also Smeur (1969–70: 264), whose attempt to show, using 'direct measurement' and BM 85194 (xiv), that a more accurate value for constant (II*) existed, is unconvincing.

[10] This coefficient appears in five lists, but never in geometrical sections, so that it does not appear that it was considered to relate to geometrical figures. See §7.2.

C 4 gives accusative *kippatam* instead of the expected nominative. A small picture of a circle is actually inscribed in second column of list G (Kilmer 1985: 2). The other lists use logograms.

DAL/*tallum* and *pirkum* are translated 'diameter' and 'radius' here, but their more general mathematical meanings are 'long transversal' and 'short transversal' respectively;[11] they are perpendicular to one another (see §3.7). The diameter coefficient is nameless in all but one of the problem texts, and even within BM 85194 it is expressed in five different ways. The use of the verb *šullušum* 'to triple'[12] is not restricted to this coefficient: in Haddad 104 (ix)–(x) the combined daily work-rate for making bricks is tripled for reasons that remain obscure,[13] and in BM 96954 + ... (xiii) and (xiv), tripling is one stage in the solution to difficult problems on grain-piles.[14]

GÁN has a very specific mathematical meaning, restricted to the area of a geometrical figure. Kilmer (1960: 227) translates this term as 'plane' to distinguish it from the more general A.ŠÀ, but the more natural English 'area' is retained here. Its Akkadian equivalent is unknown.

3.2 SEMICIRCLES

3.2.1 The coefficients

D 7	15 IGI.GUB *šà ús-ka$_4$-ri*	0;15, the coefficient of a semicircle.
E 2	15 GÁN [U$_4$.SAKAR?]	0;15, of the area [of a semicircle?].
E 3	10 GÁN ⸢U$_4$⸣.[SAKAR KI.2?]	0;10, of the area [of a semicircle, 2nd?].
F 54	45 *ša* GÁN U$_4$.SAKAR KI.3	0;45, of the area of a semicircle, 3rd.
D 8	40 DAL *šà ús-ka$_4$-ri*	0;40, the diameter of a semicircle.
D 9	20 *pi-ir-ku šà ús-ka$_4$-ri*	0;020, the radius of a semicircle.
C 13b	*ta-al-li ús-ka-rum* [... *i-gi-gu-bu-šu*]	The diameter of a semicircle: [... is its coefficient].
C 13a	*ús-ka-rum* 2 30 *i-gi-gu-bu-šu*	A semicircle: 0;02 30 is its coefficient.
A 4	10 15 36 IGI.GUB GÁN U$_4$.SAKAR	0;10, 0;15, 36, coefficients of the area of a semicircle.
E 8	15 36 10 GÁN ⸢U$_4$.SAKAR KI?⸣.IN.ŠUB	0;15, 36 and 0;10, the area of a semicircle ... .
F 21	⸢1⸣5 *ù* 10 *ša* GÁN U$_4$.SAKAR	0;15 and 0;10, of the area of a semicircle.[15]
F 52	4 20 *ša* GÁN U$_4$.SAKAR KI.2	4 and 0;20, of the area of a semicircle, 2nd.
F 53	4 36 10 *ša* GÁN GÚR KI.2	4, 36 and 0;10, of the area of a circle, 2nd.[16]

BM 85210 (viii): III 3, 6, 10	IGI.6.GÁL	One sixth.
BM 85210 (viii): III 5, 8–9	1 30 BAL GÁN U$_4$.SAKAR	1;30, the ratio of the area of a semicircle.

[11] DAL/*tallum* has a very wide semantic range including 'diaphragm' and 'lintel' (Jeyes 1989: 77–9; *AHw*: 1311).

[12] Non-mathematical meaning: 'to do for a third time' (*CAD*: Š I 236).

[13] e.g. Haddad 104: IV 26 *šu-ul-li-iš*. See §5.1.

[14] BM 96954 + BM 102366 + SÉ 93: V 12'; 24' *šu-ul-li-iš*. See §A.3.

[15] Reading 15 instead of 45 (*MCT*: 134; Kilmer 1960: 285). The traces on the photo show one large Winkelhaken; the reading of the number 45 in F 54 is clear, and it is unlikely that two entries in the same coefficient list would share a numerical value.

[16] This circle coefficient has been included here because of the characteristics it shares with the others in the group.

BM 85210 (viii): III 18	15 IGI.⌜GUB.BA⌝	0;15, the coefficient.
BM 96954 (xix): IV 30	[15 IGI].GUB.BA GÁN U$_4$.SAKAR	[0;15], the coefficient of the area of a semicircle.
MLC 1354: 6	15 IGI.GUB.BA	0;15, the coefficient.

3.2.2 Mathematical commentary

There are many more coefficients attested for the semicircle than for the circle, although only three of these are actually found in use, all in difficult or broken contexts.[17] If A is the area of the semicircle, s its semicircumference, d its diameter, r its (perpendicular) radius, and $\pi \approx 3$ then we have:

$A = 0;15s \times d$ (Thureau-Dangin 1932: 139)
$A = 0;10s^2$, or $s^2 = 6A$ (Thureau-Dangin 1932: 139)
$A = 0;45r \times d$ (Lewy 1947: 318)

$d = 0;40s$, or $s = 1;30d$ (Thureau-Dangin 1932: 139)

$r = 0;20s$ (*TMS*: 31).

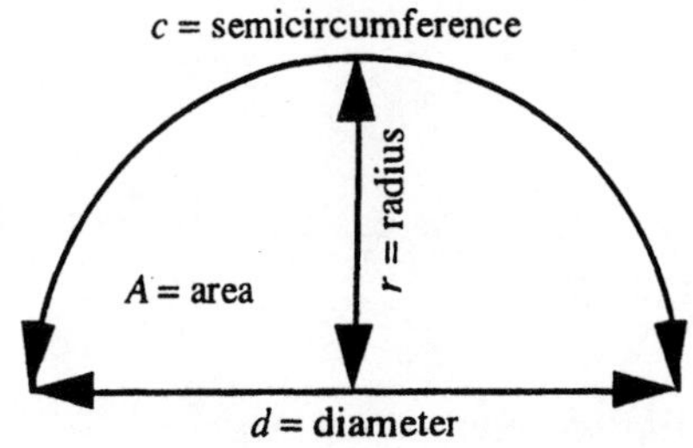

Figure 3.1: The components of a semicircle

It has been supposed that this multiplicity of coefficients for the semicircle, many of which have no counterparts for the circle, implies that the two figures were conceptualized completely independently (Smeur 1969–70: 264). However, the numerical value of C 13a speaks against this. It almost certainly gives the ratio of the area of a semicircle to the circumference of the whole circle (Friberg 1987–90: 555), showing that the relationship between the two was known, at least in the region of Eshnunna:

$$A = \frac{0;05\,c^2}{2} = 0;02\,30.$$

Friberg's tentative hypothesis that the numerical values of A 4 should be read separately (Friberg 1987–90: 555) is certainly correct. The second listing of coefficients here shows all those commonly transcribed with more than one 'sexagesimal place'. These multi-place numbers are in fact composed of the numerical values attested in other coefficients for semicircles:

0;10 (A 4, E 8, F 21, F 53)
0;15, and its reciprocal 4 (A 4, E 8, F 21, F 53)
0;20 (F 52).

The only otherwise unattested value is 36 (A 4, E 8, F 53). A separate reading of 0;30 and 6 is unlikely for two reasons. Firstly, in all three occurrences the numerical signs are written closely together in the same order; there are no attestations of the 6 preceding the 0;30, or of them being separated by another numerical value. Secondly, in each instance 36 is listed with 0;10.

[17] Although the top of BM 85210 (viii) is missing, so that the name of the shape concerned has been lost, it is possible to identify it from the procedures and terminology in the extant calculation. The text concerns a cube out of which a semicircular channel has been cut, whose diameter is half the length of the cube (Thureau-Dangin 1932: 136–9). Vogel called this figure 'das Tor in der Mauer' (1959: 80 Abb. 35).

MLC 1354 concerns a figure — either $^2/_3$ of a circle or $^2/_3$ of a semicircle — from the bottom of which a slice 5 nindan wide has been removed. The new area is given; the problem is to work out the length and width of the (old or new?) figure. The text was erased below the eleventh line, and parts of the first three lines are missing.

See §A.3 for BM 96954 + ... (xix).

This is the reciprocal of 6 and therefore an equivalent numerical value for the same coefficient. It is highly improbable that two equivalent coefficients would be included in entries on three separate lists.

There remains the problem of interpretation. The obvious choice for the absolute value would be 36 ≈ $(2\ )^2$, but there is no clear candidate for the identity that this would represent. One possibility *might* be the area of a third of a circle defined with respect to the square of the circumference of the whole circle, rather like coefficient C 13a:

$$A = {}^1/_3 \times 0;05c^2 = 0;01\ 40c^2, \text{ or}$$
$$c^2 = 36\,A.$$

The third-circle is the basis for three other geometrical shapes (see §3.5).

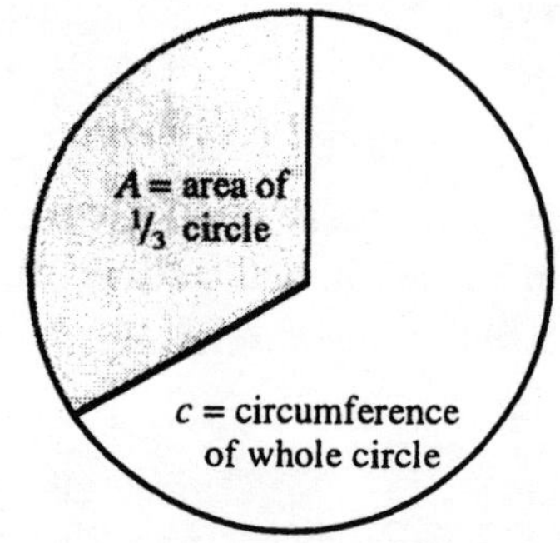

Figure 3.2: Possible interpretation of the coefficient 36

3.2.3 Philological commentary

The general OB mathematical meaning for GÁN U_4.SAKAR/*uskarum* is 'segment of a circle',[18] but unless the coefficient 36 proves otherwise, with coefficients it is restricted to 'semicircle'. In lists C and D the word is written syllabically (with a nominative-for-genitive case error in C 13b); in all other coefficients the logogram, with GÁN 'area', is used. See §7.1.3 for a brief discussion of the meaning of NUMUN in BM 85210 (viii).

The meaning of the last three signs in E 8 is unclear. Both hand-copy and photo (*MCT*: pls. 23, 49) show a clear final RU, which is preceded either by IN, GÌR, ANŠE or some similar sign, or by ŠE+NI or ŠE+ER. Before that the traces suggest U_4, DI, KI or something comparable. For want of better, Neugebauer and Sachs' reading ⸢KI⸣.IN.ŠUB (*MCT*: 136) is retained here without, however, an attempt at translation.[19]

3.3 TRIANGLES

3.3.1 The coefficients

A 1	30 *ù* 1 IGI.GUB SAG.DÙ	0;30 and 1, the coefficients of a triangle.
H 5	30 *ša sa?-ta?-⸢ki?⸣*	0;30, of a triangle?.
F 19	30 *ša* SAG.DÙ	0;30, of a triangle.
C 7	SAG.DÙ-*kum ša sa-am-na-[tu na]-ás-ḫa* 26 15 *i-[gi-gu-bu-šu]*	A triangle from which an eighth part is taken: 0;26 15 [is its coefficient].
J 2	26 15 GÁN UR.RU.UḪ SAG.3 KI.MIN IGI.GUB-*ú*	0;26 15, the area of a triangular storehouse, ditto [i.e. storage], the coefficient.
C 8	*ta-al-li* SAG.DÙ-*ki* 52 30 *i-[gi-gu-bu-šu]*	The long transversal of a triangle: 0;52 30 [is its coefficient].

[18] e.g. MLC 1354, where it may mean ²/₃ circle; more generally '(Mond-)Sichel' (*AHw*: 1438).

[19] If the phrase is logographic, then ŠUB has the basic meanings 'to fall (upon), to throw' (Thomsen 1984: 318); a reading in syllabic Akkadian is doubtful as there are no roots Cnr, and no appropriate words ending in **-šerru* (Hecker 1990: 24, 237).

D 29 52 30 IGI.GUB *šà* SAG.DÙ	0;52 30, the coefficient of a triangle.
G 2 7 30 IGI.GUB ▷ SAG-*ga-kum*	0;07 30, the coefficient of a triangle.
C 6 SAG.DÙ-*kum i-*[...] *i-gi-gu-*[*bu-šu*]	A triangle ...: ... is its coefficient.
D 30 57 36 IGI.GUB *šà* UB	0;57 36, the coefficient of a corner.

Ash 1922.168	30	0;30 (three occurrences).
AUAM 73.2834	(30)	(0;30).

3.3.2 Mathematical commentary

The first three coefficients simply express the fact that the area of a triangle is equal to half the width multiplied by its (perpendicular) height:[20]

$$A = 0;30 \text{ width} \times \text{height}.$$

The first coefficient expresses this most fully: the area is equal to the product of 0;30 × one side and 1;00 × the other. This is the only instance in which a coefficient with several numerical values should be read as one coefficient.[21]

The second set of coefficients concerns equilateral triangles, those whose *sides* are all of equal length. Thinking of them as composed of two right-angled triangles of hypotenuse 1 and short side 0;30, then (using an OB approximation later attributed to Heron of Alexandria)[22] the height of the triangle is:

$$\begin{aligned} h &= 1 - \frac{(0;30)^2}{2} \\ &= 1 - 0;07\ 30 \\ &= 0;52\ 30 \end{aligned}$$

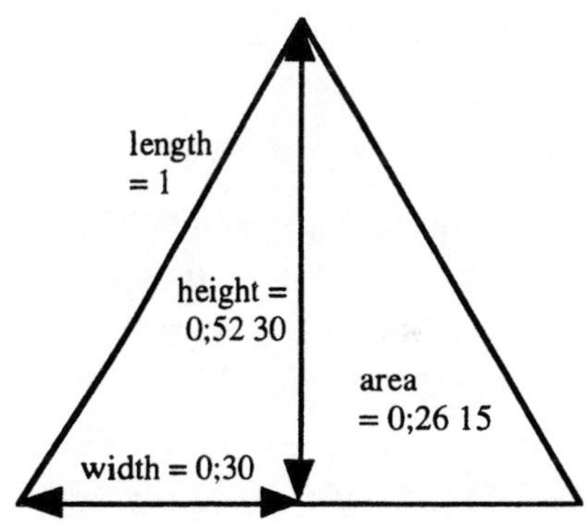

Figure 3.3: An equilateral triangle

the value of the second two coefficients in this group. The area of this triangle is then (Friberg 1987–90: 554–5):

$$A = 0;52\ 30 \times 0;30 = 0;26\ 15,$$

the value of both C 7 and the LB coefficient J 2.

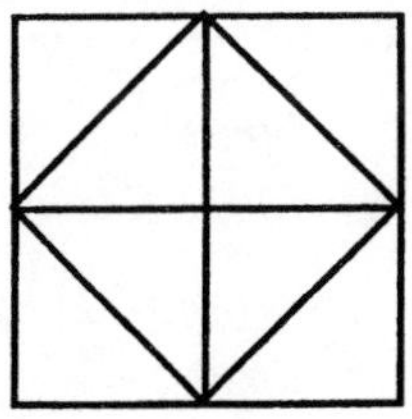

Figure 3.4: BM 15285 (x)

The value of G 2 may represent the area of one of the eight right subtriangles defined by the diagonals and diameters in a square of side 1 (Friberg 1987–90: 555), as shown in BM 15285 (x), illustrated in Fig. 3.4. On the other hand, it may equally stand for the 'eighth part' (i.e., 0;07 30) named in C 7 which is subtracted from the length of an equilateral triangle to give its height.

[20] Cf. AUAM 73.2834, a lenticular tablet which shows a picture of a triangle, above and below which are written the numbers 25 and 16 40! respectively. 3 28 20 is recorded in the centre of the triangle. If we suppose that 25 and 16;40 are the lengths of mutually perpendicular sides of the triangle (although this is not the impression given by the picture), then its area is indeed 25 × 16;40 × 0;30 = 3 28;20.

The triangle coefficient 0;30 is recorded three times on Ash 1922.168 (§A.5.2), in a diagram of trapezoidal areas.

[21] Compare lines 2 and 4 in the same list (§§3.1, 3.2).

[22] As used, for instance, in BM 96957 + VAT 6598 (xviii)–(xix), §A.4. See, most recently, Fowler and Robson (fc.).

Four coefficients found in BM 85194 and BM 85196, which are provisionally classified as miscellaneous here (§8.4), might well turn out to be coefficients for the area of a triangle:[23]

BM 85194 (ix): III 2	IGI.2.GÁL BA.ZI	'take a half'	(area of an *a-aš-ba-tum*)
BM 85196 (iii): I 15	$^1/_2$ *ḫe-pé*	'break in half'	(volume of **tamliātum*, 'dam'?)
BM 85196 (iv): I 22	$^1/_2$ *ḫe-pé*	'break in half'	(volume of **tamliātum*)
BM 85196 (vii): I 34	[$^1/_2$] *ḫe-pé*	'break in [half]'	(volume of a *sikkatum*, 'peg')

The first of these objects is said to be 20 (nindan) square (*im-ta-ḫar*), but the expected square area is halved to give the result, which suggests that the figure is not square after all but an isosceles right triangle. The text of the problem is unhelpfully short, at 4 lines, and the meaning — and reading — of *ašbatum* is uncertain (cf. Nemet-Nejat 1993: 74, 159). The remaining three problems all concern three-dimensional constructions. The 'dams' appear to be long and wall-like with triangular and trapezoidal cross-sections respectively. The base of the 'peg' appears to be the same shape as the *ašbatum*, with vertical sides, and may contain grain.

There have been two rival theories for the meaning of D 30, one based on the numerical value of the coefficient and the other on the reading SÁR for the single cuneiform sign used to write its name. Bruins and Rutten proposed that the number was to be understood as 0;57 36, or $^{24}/_{25}$:

Ce facteur transformant l'anneau GAM en cycle SAR, rend évident que la valeur plus précise de $\pi = 3^1/_8$ était connue des Babyloniens. En effet,

$$\frac{1}{4\pi} = \frac{1}{12} \times \frac{24}{25} = 3\frac{1}{8}$$

Ce facteur [...] que O. Neugebauer (*MCT*: 59) n'a pas voulu accepter est donc incontestible. (*TMS*: 33)

No justification is given in this passage for interpreting SÁR as 'cycle (cercle plus parfait)' (*TMS*: 26),[24] although the translation may have been derived from the Akkadian reading *kiššatum*, 'Gesamtheit, Welt' (*AHw*: I 492, *k*. I). There is no comparable mathematical word.

A second interpretation is based on the fact that the sign SÁR resembles a diamond shape (Vaiman 1963: 84; Kilmer 1990: 86 & n. 19). BM 15285 (vii), illustrated right, (viii)–(xii), (xiv), and (xvii) all contain illustrations of squares inscribed obliquely within squares and it is this construction that the coefficient is thought to represent. The significance of the numerical value is not clear, though, as the area of the smaller square is 0;30 of the larger, and the length of the smaller side $1/\sqrt{2}$ or 0;42 30.

Figure 3.5: BM 15285 (vii)

Muroi (1992*b*: 48) has more recently suggested the reading UB 'corner' for the final sign of the entry. The mathematical meaning of the word would then be 'either a right triangle in general or the right triangle whose sides are 1, $^7/_{25}$ and $^{24}/_{25}$ (= 0;57 36) in particular.' The coefficient would give the length (UŠ) of the triangle. However, the rest of the list D geometrical coefficients which begin with the word IGI.GUB all give the areas (GÁN) of the figures they describe, with respect to an outer side with length 1. One would thus expect this coefficient to do so too, so that Muroi's interpretation is not quite correct.

[23] A complete listing of all certain occurrences of halving for the areas of triangles and trapezoids has not been attempted here. See also BM 96954 (xviii), in which the procedure for calculating the volume of a triangular grain-pile is corrupt and the coefficient missing.

[24] It might appear that the whole motivation for this interpretation was to prove another scholar wrong, something for which Bruins was renowned (Høyrup 1996*a*: 15).

A plausible — though by no means certain — alternative explanation retains Muroi's idea of the right triangle. A 3-4-5 right triangle with hypotenuse 1 has area

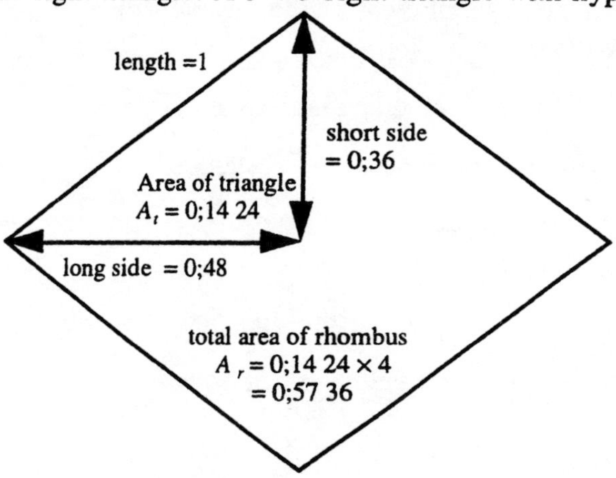

Figure 3.6: Interpretation of the coefficient D 30

$$A_t = \frac{0;36 \times 0;48}{2} = 0;14\ 24.$$

Four of these triangles placed back to back produce a rhombus of outer length 1 and area

$$A_r = 4 \times 0;14\ 24 = 0;57\ 36,$$

the exact value of the coefficient D 30. It is unclear whether Muroi's reading UB should be retained or whether, following Vaiman (1963: 84), the final sign should be seen as a pictorial representation of the figure it describes.

F 65 has the same numerical value but unclear meaning (see §8.5.4). It is in a section of the list with a rather obscure, partly acrographic, organization which includes at least one geometrical coefficient, F 63a, and it may be possible that F 65 is related to D 30.

3.3.3 Philological commentary

The second column of G 2 contains what may be either a picture of a triangle, like the drawing of a circle just above it in the list, or a unique occurrence of the sign KAK used as a logogram for *santakkum* (Friberg *et al.* 1990: 503), parallel to the use of LAGAB for *mitḫartum* 'square' (Høyrup 1990*a*: 51).[25]

The varying degrees of phoneticization of the spelling of this word apparent throughout OB mathematics can be seen in these coefficients. *Santakkum* is never preceded by the logogram GÁN, and never refers to a component of a triangle, in the way that *kippatum, mitḫartum* and *ṣiliptum* can each refer to both the defining component and the whole of a figure.

The LB coefficient J 2, translated by Livingstone as 'three headed Urruh' (1986: 21),[26] refers to a triangular *araḫḫum* storehouse: just as SAG.5, SAG.6 and SAG.7 (D 26–8, §3.6) are the names of regular, i.e. equal-sided, polygons, so too is SAG.3 a logogram for a regular, i.e. equilateral, triangle as its numerical value suggests.

The word *santakkum* is used for both types of triangle attested in the coefficient lists,[27] and recognized in OB geometry generally: the (nearly) right-angled triangle (half a rectangle), and the symmetric triangle, which is two right-angled triangles back to back (Friberg 1987–90: 554). The equilateral triangle is a special case of the symmetric triangle.

3.4 SQUARES AND RECTANGLES

3.4.1 The coefficients

A 42 1 21 IGI.GUB *ṣí-li-ip-tum* — 1;21, the coefficient of a diagonal.

D 31 1 25 IGI.GUB *šà* BAR.TA *šà* NIGIN — 1;25, the coefficient of the diagonal of a

[25] Cf. Klein (1986) on the etymology of the word *santakkum*.

[26] He presumably took SAG to stand for *rēšu(m)* 'head' instead of *pūtu(m)* 'short side, width'.

[27] *Pace* Kilmer (1990: 90 n. 14), who 'expect[s] different names for isosceles, equilateral and right triangles'.

		square.
D 32	1 15 IGI.GUB *šà* BAR.TA UŠ *ù* SAG	1;15, the coefficient of the diagonal of a rectangle.
E 10	1 ⌜24 51 10⌝ ⌜*ṣi-li*⌝-*ip-tum* ‹*ša*› ÍB.SI$_8$	1;24 51 10, the diagonal ‹of› a square.
A3	30 IGI.GUB *ṣí-li-ip-tum*	0;30, the coefficient of a rectangle.
E 11	1 *ṣi-li-ip-tum*	1, a rectangle.
F 63a	1 *ṣí-li-ip-tum*	1, a rectangle.

YBC 7289	1 24 51 10	(Along the diagonal of a square.)
BM 85210 (viii): III 6	1 NUMUN ÍB.SI$_8$	1, the ratio of a square.

3.4.2 Mathematical and philological commentary

The coefficients for squares and rectangles are all very straightforward, complicated only by the fact that *ṣiliptum*, a *pirist* formation from *ṣalāpum* 'to strike through', means both 'diagonal'[28] and 'rectangle'.[29] In common with the circumference of a circle and the side of a square,[30] the diagonal is seen as the *defining component* of a rectangle, as though only one right-angled quadrilateral (and not infinitely many) could be constructed around a diagonal of given length.

D 31 and E 10 are both approximations for √2, the length of the diagonal of a square of length 1 (*MCT*: 43). The occurrence of this number on the famous diagram YBC 7289 is almost certainly a coefficient too; it is multiplied by the length of the square to give the diameter.[31] The method used to calculate these square roots, according to Neugebauer and Sachs, 'consists in the alternating approximation of [the diagonal] by arithmetic and harmonic means of previously found approximations' (*MCT*: 43). Such an interpretation, however, is far from certain: there is no direct evidence for iterative methods such as this in OB mathematics.[32] It may be that A 42, whose value is 1;21, is an arithmetically convenient but inaccurate version of the same coefficient: it does not derive from any known OB method of approximation for squares or rectangles with length 1 but, rather, gives the closest two-place sexagesimal approximation to √2 (Fowler and Robson fc.: n. 14).

D 32 gives the diagonal for the 'favorite rational rectangle' (Friberg 1987–90: 553) with length 1 and width 0;45 (i.e. a quarter-sized version of the 'Pythagorean' 3-4-5 triangle). This is the rectangle found in several 'algebraic' problem texts[33] composed of length = 1 and width = 0;45, although there are many other texts which use a rectangle of length 0;30 and width 0;20.[34] The algebraic origin of the coefficient is reflected in the use of the term UŠ *ù* SAG, literally 'length and width', instead of *ṣiliptum*, which is more commonly found in geometrical contexts, such as the descriptions in BM 15285. Muroi (1992*b*: 47) interpreted BAR.TA as a variant of Sumerian **bar-da** 'cross-piece, cross bar' (*PSD*: B 117).

[28] e.g. in BM 96957 + VAT 6598 (xix).

[29] e.g. in BM 15285 (xxxiv) (§A.2).

[30] Indeed, *mitḫartum* means not only 'square' and 'side of square' but also (geometrical) 'square root' (Høyrup 1990*b*; Fowler 1992).

[31] Likewise 0;15, a coefficient for the semicircle (§3.2), is unlabelled on the diagram in MLC 1354, but is explicitly described as a coefficient in the accompanying problem.

[32] See most recently Fowler and Robson (fc.). The first stage of the so-called 'Heron's approximation' is used in BM 96957 + VAT 6598 (xviii) (§A.4).

[33] e.g. Db$_2$-146, from Tell Dhibai near Shaduppum, with diagram.

[34] e.g. A 24194, A 24195. In these cases, though, no diagonal is involved.

The values of E 11, F 63a and BM 85210 (viii), although unexplained until now (Friberg 1987–90: 554), have an extremely simple function: they record that the area of a square or rectangle is equal to 1 × length × width.[35] The value of A 3 may well denote the area of a rectangle whose length is twice the width.[36]

3.5 FIGURES IN $^1/_3$ AND $^1/_4$ CIRCLE

3.5.1 The coefficients

D 10	6 3⌜3⌝ 45 IGI.GUB *šà* GÁN *pa-na-ak-ki*	0;06 33 45, coefficient of the area of a bow.
D 11	52 30 DAL *šà* GÁN *pa-na-ak-ki*	0;52 30, the diameter of the area of a bow.
D 12	15 ⌜*pi-ir-ku*⌝ *šà* GÁN *pa-na-ak-ki*	0;15, the radius of the area of a bow.
D 13	13 ⌜07⌝ 30 IGI.GUB *šà* GÁN IILLAR	0;17 07 30, the coefficient of the area of a bow.
D 14	52 30 DAL *šà* GÁN ILLAR	0;52 30, the diameter of the area of a bow.
D 15	30 ⌜*pi*⌝*-ir-ku šà* GÁN ILLAR	0;30, the radius of the area of a bow.
C 12	*i-in al-pí-im* 16 52 30 *i-*⌜*gi*⌝*-[gu-bu-šu]*	An ox-eye: 0;16 52 30 is [its coefficient].
D 19	1⌜6⌝ 52 30 IGI.GUB *šà* IGI.GU$_4$	0;16 52 30, the coefficient of an ox-eye.
D 20	52 30 DAL *šà* IGI.GU$_4$	0;52 30, the diameter of an ox-eye.
D 21	30 *pi-ir-ku šà* IGI.GU$_4$	0;30, the radius of an ox-eye.
A 7	⌜45⌝ GÁN MÁ.GUR$_8$	0;45, the area of a barge.
D 16	1⌜3⌝ 20 IGI.GUB *šà* A.ŠÀ ŠE	0;13 20, the coefficient of a grain-field.
D 17	5⌜6⌝ 40 DAL *šà* A.ŠÀ ŠE	0;56 40, the diameter of a grain-field.
D 18	2⌜3⌝ 20 *[pi]-ir-ku šà* A.⌜ŠÀ⌝ ŠE	0;28 20, the radius of a grain-field.
E 6	⌜13 20⌝ GÁN ⌜MÁ⌝.GUR$_8$	0;13 20, the area of a barge.
F 47	[13]⌜20⌝ *ša* GÁN MÁ.GUR$_8$	[0;13] 20, of the area of a barge.
A5	12 30 IGI.GUB GÁN GIŠ.PAN	0;12 30, the coefficient of the area of a bow.
E 5	12 30 GÁN PAN	0;12 30, the area of a bow.
F 49	⌜5⌝2 30 *ù* 12 30 *ša* GÁN PAN	0;52 30 and 0;12 30, of the area of a bow.
F 51	3 30 *ša* GÁN ⌜PAN⌝ KI.2	3;30, of the area of a bow, 2nd.

3.5.2 Mathematical commentary

The first three figures have all been fully identified (Fig. 3.7). They form a series of shapes inscribed in a third-segment of a circle, of arc $a = 1$ and thus of area 0;15.[37] As with the circle and semicircle, the arc seems to be the starting point for area calculations. In each instance the IGI.GUB gives the area A of the figure, the DAL coefficient the (same) chord d, and the *pirkum* the length of the perpendicular p, through the chord to the origin of the circle (Vaiman 1963).

The derivations of these values can be found in Friberg *et al.* (1990: 489–90) (where the terms 'ox-eye' and 'grain-field' have been inadvertently swapped, except in fig. 2a).[38] None of them are the subject of known OB problems, although the geometrical problem 'to construct

[35] Compare 0;30, the coefficient of the area of a triangle (§3.3).

[36] e.g. the rectangle in Str 368.

[37] Cf. the discussion of the coefficient 36 in §3.2.1.

[38] See also Muroi (1992*b*: 45–6), who arrives at the same interpretation independently of other recent publications, but reading GÁN GIŠ.MÁ for GÁN ILLAR (GIŠ.RU).

a "brick-mould" inside an ox-eye' may be described in the 'catalogue' entry IM 52916: II 30′ on the same tablet as C 12 (see §A.1).

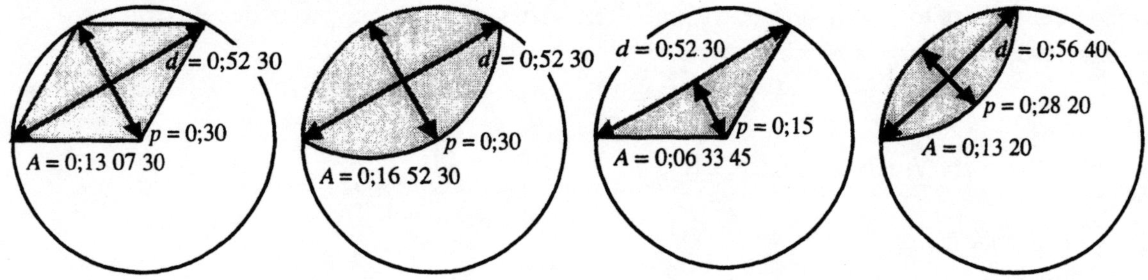

GÁN ILLAR *IGI.GU₄* *GÁN* panakki *'grain-field' or 'barge'*

Figure 3.7: The shapes inscribed in $^1/_3$ and $^1/_4$ circles

The 'grain-field' is constructed on the same principles as the 'bow' shapes, but within $^1/_4$ instead of $^1/_3$ of a circle. If the outer arc is 1, then the coefficients give the area of the figure and its two transversals (Vaiman 1963; Friberg *et al.* 1990: 489–90). Friberg (1987–90: 558) identifies it with the 'barge', which is illustrated on BM 15285 (xl) (§3.7), and is the name given to the numerically identical coefficients E 6 and F 47.

Only coefficient A 7 has been hitherto unidentified. It may refer to the fact that the 'grain-field' or 'barge' occupies approximately 0;45 of the area of a quarter circle. This value is much closer to the correct fraction (where r represents the radius of the circle):

$$\left(\frac{\pi r^2}{4} - \frac{r^2}{2}\right) \Big/ \frac{\pi r^2}{4} = 2 - \frac{4}{\pi} \approx 0;43\ 36\ \ldots$$

than that obtained using the OB method of approximating π by 3 and taking the quarter-arc a as the defining component and starting point:

$$\frac{0;13\ 20a^2}{\frac{1}{4} \times 0;05 \times (4a)^2} = \frac{0;13\ 20}{0;20} = 0;40.$$

A 7 would appear to represent a different mathematical tradition to the other coefficients for the 'barge' then: it is at once more accurate and simpler to calculate than those of lists D, E, and F. It is impossible to determine which coefficients the author of BM 15285 was working with.

The final group of coefficients describes a 'composite bow' shape of three arcs and a straight line. This interpretation was first proposed by von Soden (1964: 47) on the basis of BM 15285 (xxxiv)–(xxxv) (see §A.2). His figure seems to show two outer arcs of 135° and a central one of 90°, each with the same radius. The coefficients have not yet been explained.

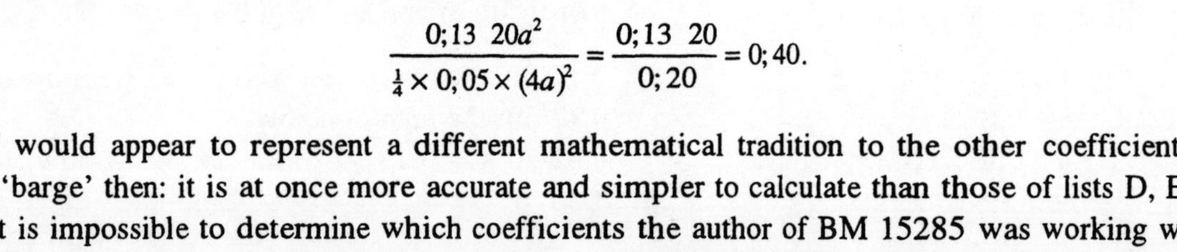

(xxiv) *(xxxiv)* *(xxxv)* *(xxxviii)*

Figure 3.8: Diagrams in BM 15285 related to the 'bow'

All the BM 15285 illustrations are based on a square, further subdivided into a grid of 16 squares, as (xxiv), shown in Fig. 3.8, far left, makes clear. The figures for problems (xxxiv) and (xxxv), Fig 3.8 centre left and right, both show that the GIŠ.PAN fits into a space a quarter of the size of the square.

Problem (xxxviii), which starts the next column of the tablet, shows semicircles in an arrangement very suggestive of the GIŠ.PAN (Fig. 3.7, far right). Whereas von Soden's interpretation assumes that the two 135° arcs are formed from circles which do not touch, the shapes in (xxxiv) meet the sides of the rectangle in which they are enclosed. The simplest construction would be based on the pairs of semicircles shown on either side of (xxxviii). Two construction lines forming an equilateral triangle from the origins of the semicircles would then provide the origin for a facing arc to complete the figure (cf. Fig. 3.8). Indeed, the traces of construction lines and compass marks on BM 15285 (xxxiv)–(xxxv) seem to indicate that was exactly how the figures were drawn (see copy in §A.2).

If the length of the rectangle is 1, and the width 0;15 then the GIŠ.PAN is composed of two semicircles and a central shape. The area of the rectangle is 0;15 and the area of the two semicircles is:[39]

Figure 3.9: Possible construction of the GIŠ.PAN

$$A_s = 2 \times 0;45\, r \times d$$
$$= 0;11\ 15.$$

The coefficient in A 5, E 5 and F 49, 0;12 30, must then represent the slightly larger area of the GIŠ.PAN. Dividing the figure into two outer third-circle segments and three central equilateral triangles of length 0;15 (approximating the central arc by a straight line) gives:

$$A_b = 2r^2 + 3(0;26\ 15r^2)$$
$$= (2 + 1;18\ 45)r^2$$
$$= 3;18\ 45 \times 0;03\ 45$$
$$= 0;12\ 25\ 18\ 45 \approx 0;12\ 30,$$

the nearest regular two-place approximation. This construction is essentially the same as von Soden's except that the arcs are different lengths. A more accurate method of calculation would have been to subdivide the figure into two semicircles and a 'convex equilateral triangle' (D 25; §3.7), which would give an area of:

$$A_b = 3r^2 + 0;15r^2$$
$$= 3;15 \times 0;03\ 45$$
$$= 0;12\ 11\ 15,$$

but there is no evidence for the coefficient of the 'convex equilateral triangle' being known outside Susa. The meanings of F 49 and 51 remain unclear; they may both refer to some unknown which is both $3^1/_2$ times greater than the radius, 0;15, and 0;52 30 of the length, 1.

3.5.3 Philological commentary

Three of these figures are described as different types of bows, because of their obvious schematic resemblance to them. GIŠ.PAN/*qaštum* is the so-called composite bow, which was present in Mesopotamia from the third millennium (Miller *et al.* 1986: 182). ILLAR/*tilpānum* seems to have originally meant a throwstick, but by the Neo-Assyrian period had become a literary synonym for *qaštu(m)* (*AHw*: 1359). The geometrical figures these two words depict

[39] See §3.2 for this semicircle coefficient, and §3.3 for the triangle coefficient used below.

suggest that they may even have been synonymous as early as the OB period. Miller *et al.* (1986: 182) state that the composite bow

> ... can be recognized as composite by the recurved ends standing above the bowstring when the bow is braced and by the full *semicircular* curve the centre of the bow takes when fully drawn [italics added].

However, Miller *et al.* (1986: 186 pl. 4) shows a photograph of the angular composite bow at full draw describing an arc of nearer 120° than 180°, like the figure called ILLAR in OB maths. It would seem, then, that the two mathematical terms depict the same type of bow in different ways. The GIŠ.PAN drawn in BM 15285 shows the composite bow at rest, while the synonymous ILLAR refers to it at full draw.[40] The **panakkum*, an otherwise unattested Akkadian word (*AHw*: II 818), is surely a Sumerian loanword PAN 'bow' with the common genitive ending *-akkum* (cf. Kilmer 1990: 85); it too refers to a figure derived from the fully drawn composite bow.

The name of coefficient A 7 was previously read GÁN SA.LI, which Kilmer (1960: 287 n. 2) equated with the *pagûm* lyre. But this would have been the only known OB occurrence of the word (*AHw*: II 809), and it was not at all clear which geometrical figure it might refer to.[41]

Kilmer (1980–3: 574) has shown that the *īn alpim* 'ox-eye' is a schematic representation of exactly that object, 'based on highly stylized art forms and artifacts related to bovine faces and bovine lyres' (Kilmer 1990: 86). It seems that the 'boat' and the 'grain-field' are also named by stylized visual analogy, although one would expect *GÁN ŠE, '(geometrical) area of a grain',[42] as the representation is of a grain, not a field. A.ŠÀ is normally reserved for areas in non-geometrical contexts. The Akkadian for MÁ.GUR$_8$ 'barge' is presumably *makurrum*, as it is outside OB mathematics (*AHw*: 591).

3.6 REGULAR POLYGONS

3.6.1 The coefficients

D 26 1 40 IGI.GUB *šà* SAG.5	1;40, the coefficient of a pentagon.
D 27 2 37 30 IGI.GUB *šà* SAG.6	2;37 30, the coefficient of a hexagon.
D 28 3 41 IGI.GUB *šà* SAG.7	3;41, the coefficient of a heptagon.

3.6.2 Mathematical commentary

The polygon coefficients occur only in the Susa list, and are closely related to a fragment of tablet also found at Susa, *TMS* 2.[43] The obverse shows a hexagon composed of six equilateral triangles, and the reverse a heptagon of seven symmetric triangles. The coefficients are based on the assumption that the base of each figure has length 1 (Fig. 3.9).[44] Then the hexagon will be composed of 6 equilateral triangles of length 1. The area of each of these is 0;26 15 (using coefficient C 7 (§3.3)) and so the area of the whole hexagon is:

$$A = 0;26\ 15 \times 6 = 2;37\ 30.$$

[40] An earlier interpretation was that it represented two facing boomerangs of 120° (von Soden 1964: 47), and the figure was referred to as a 'double throw-stick' (Kilmer 1990: 85).

[41] The new reading GÁN MÁ.GUR$_8$ is based on collation by John Brinkman in May 1997.

[42] Cf. GÁN *panakki* and GÁN ILLAR in the same list.

[43] Bruins (1986–90: 125) most recently states that in OB Susa 'the sides of the regular n-gon with n = 2 [*sic*], 3, 4, 5, 6, 7, 10, 12, 24 were computed'. List D gives evidence for n = 3–7; n = 24 presumably refers to Bruins' interpretation of coefficient D 30 (see §3.3); he leaves n = 2, 10, and 12 unaccounted for.

[44] It is not necessary to assume that the polygons are inscribed in circles (*pace TMS*: 32; Friberg 1987–90: 556–7); indeed the drawings on *TMS* 2 show no traces of them.

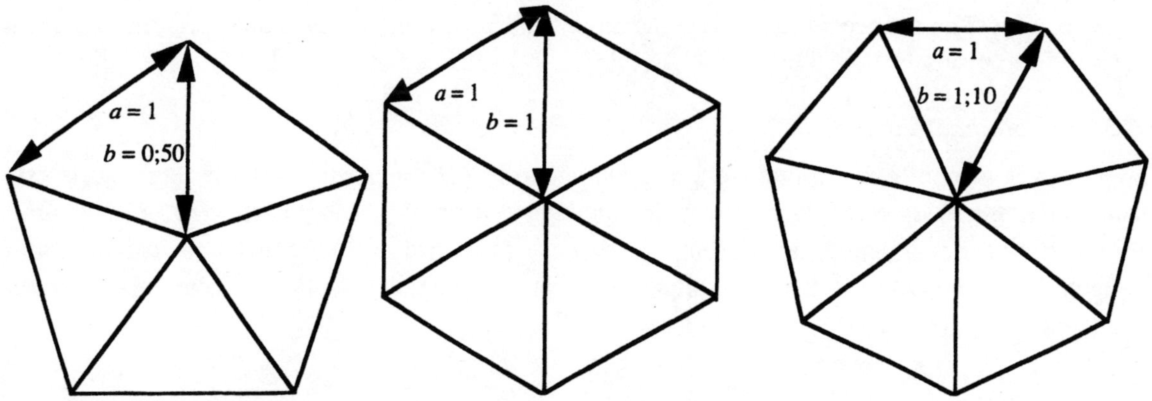

Figure 3.10: The pentagon, hexagon, and heptagon

This is confirmed by the hexagon drawn on *TMS* 2, in which the radius and segment length are both marked 0;30. The numeral 6 33 45 within the segment must then be its area:

$$A_6 = (0;30)^2 \times 0;26\ 15 = 0;06\ 33\ 45.$$

The total of the hexagon is not given, but must then be:

$$T_6 = 6 \times 0;06\ 33\ 45 = 0;39\ 22\ 30.$$

The heptagon illustrated on the other side of *TMS* 2 shows one internal side marked 35 UŠ '0;35, the length' and although there are no visible markings on the external sides it is possible that they were assumed to be 0;30, like the hexagon. If so, the radius of the heptagon would have been considered to be $^7/_6$ = 1;10 of the base and, by analogy, the radius of the pentagon $^5/_6$ = 0;50 of its base. Working on that assumption, the height of the segments can be calculated, using the standard square-root approximation and 'Pythagoras' theorem', to be, respectively:[45]

$$\begin{aligned} h_7 &= \sqrt{(1;10)^2 - (0;30)^2} \\ &= \sqrt{1;06\,40} \\ &\approx 1 + \frac{0;06\,40}{2 \times 1} = 1;03\,20 \end{aligned} \qquad \begin{aligned} h_5 &= \sqrt{(0;50)^2 - (0;30)^2} \\ &= 0;40. \end{aligned}$$

The areas of the segments are thus:

$$\begin{aligned} A_7 &= 1;03\,20 \times 0;30 \times 1 \\ &= 0;31\,40 \end{aligned} \qquad \begin{aligned} A_5 &= 0;40 \times 0;30 \times 1 \\ &= 0;20 \end{aligned}$$

and the total areas:

$$\begin{aligned} T_7 &= 0;31\,40 \times 7 \\ &= 3;41\,40 \approx 3;41 \end{aligned} \qquad \begin{aligned} T_5 &= 0;20 \times 5 \\ &= 1;40 \end{aligned}$$

Inside one of the segments of the heptagon on *TMS* 2 is written:

⌜SAG⌝.7 *a-na* 4 *te-[ṣi]-ip-ma*	A heptagon. You multiply (it) by 4 and
ší-in-šé-ra-ti	you subtract a twelfth and (you will see)
ta-na-as-sà-ḫa-ma A.ŠÀ	the area. (cf. *CAD*: Š III 54)

Although the restoration of the first few signs is tentative, it is clear that these concise instructions give a further approximation of the heptagon coefficient, equivalent to

[45] Discussed in more detail by Fowler and Robson (fc.).

$T_7 = 3;40 \times b^2$. The procedure loses little accuracy and is arithmetically much simpler, requiring easy multiplications by 4 and 0;05 (i.e. division by 12) instead of by 3;41.

3.6.3 Philological commentary

The polygons are not attested outside Susa, although the LB logogram SAG.3 for an equilateral triangle (J 2, §3.3), is certainly of direct terminological descent. Brick pillars were built around hexagonal and octagonal designs from at least the late third millennium onwards in south Mesopotamia (see §9.3), so that it seems likely that the geometry of these figures was known there too.

3.7 THE CONCAVE SQUARE AND RELATED FIGURES

3.7.1 The coefficients

A 6	26 40 IGI.GUB GEŠTÚ ZÀ.MÍ	0;26 40, the coefficient of a concave square.
C 9	*a-ap-sà-mi-kum* 26 15 *i-[gi-gu-bu-šu]*	A concave square: 0;26 15 is [its coefficient].
D 22	26 40 IGI.GUB *šà a-pu-sà-am-mi-ki*	0;26 40, the coefficient of a concave square.
F 50	26 40 *ša* GEŠTÚ ZÀ.MÍ	0;26 40, of a concave square.
C 11	*ṣi-li-ip-ti a-ap-sà-mi-ki* 1 20 *i-gi-⸢gu⸣-[bu-šu]*	The diagonal of a concave square: 1;20 is its coefficient.
D 23	1 20 BAR.TA *šà a-pu-sà-mi-ki*	1;20, the diagonal of a concave square.
C 10	*ta-al-‹li› a-ap-sà-mi-ki* 48 *i-gi-⸢gu⸣-[bu-šu]*	The diameter of a concave square: 0;48 is its coefficient.
D 24	3⸢3⸣ 20 *pi-ir-ku šà a-pu-sà-mi-ki*	0;33 20, the radius of a concave square.
D 25	15 IGI.GUB *šà a-pu-sà-mi-ik-ki šà* 3	0;15, the coefficient of a concave triangle.
E 4	⸢2⸣6 40 GÁN ⸢ZÀ.MÍ⸣	0;26 40, the area of a harp/lyre.
F 48	⸢5⸣3 20 *ša* GÁN ZÀ.MÍ	0;53 20, of the area of a harp/lyre.
TMS 20: 2, 24	26 40 IGI.[GUB ...]	0;26 40, the coefficient [*of a concave square*].
TMS 21: 9	26 40 IGI.GUB *a-pu-sà-mi-ki*	0;26 40, the coefficient of a concave square.
TMS 21: 6	1 *a-pu-sà-mi-ka*	1, (the length of) a concave square.
TMS 21: 6	1 20 DAL *šà a-pu-sà-mi-ka*	1;20, the diameter of a concave square.

3.7.2 Philological commentary

Opinions on the possible mathematical meanings of the *apsamikkum* coefficients are so bound up with the debate over the identification of the logogram ZÀ.MÍ as 'harp' or 'lyre' that some preliminary philological discussion is called for. At one extreme is Kilmer (1980–3: 574), who explains each of four logograms as a different object, but does not give a mathematical interpretation for all of them:

GÁN.ZÀ.MÍ	= *eqel sammîm*	'plane of the lyre'
ÁB.ZÀ.MÍ	= *apsamikkum*	'cow('s nose) of the lyre'
GEŠTÚ.ZÀ.MÍ	= *ḫasīs sammîm*	'ear of the lyre'
GÁN.GEŠTÚ.ZÀ.MÍ		'plane of the ear of the lyre'.

The phrases *eqel sammîm* and *ḫasīs sammîm* do not occur in OB mathematics, and can thus be ignored for our purposes. The logogram GÁN GEŠTÚ ZÀ.MÍ is synonymous with GEŠTÚ ZÀ.MÍ, as GÁN in means 'geometrical area' and is not an essential part of the sign group (cf. (GÁN) GÚR, §3.1). So we have already simplified matters considerably.

Kilmer (1990: 88) takes the ZÀ.MÍ to be a lyre because she understands ÁB.ZÀ.MÍ — and hence *apsamikkum* — to mean 'cow of the lyre', with reference to the shape of the nose of the bovid often depicted on third millennium sound-boxes. This appealing theory is unlikely for two reasons. First, the logogram ÁB.ZÀ.MÍ is found only in first millennium extispicy texts, where it is probably a late pseudo-logogram for *apsamikkum*, based on a false etymology. Second, all attested lyres show *bulls'* heads, with horns and beards (Collon 1980–3: 578).[46] The other geometrical terms which Kilmer correctly associates with the lyre (1990: 86) both contain the sign GU_4 'bull', not ÁB 'cow'.[47]

Goetze (1951: 139) suggested that the true etymology of *apsamikkum* is AB.ZÀ.MÍ, 'window of the lyre', which should be equated with GEŠTÚ ZÀ.MÍ, 'ear of the lyre', i.e. a sound-hole.[48] Lawergren and Gurney (1987), arguing that ancient Near Eastern lyres do not have sound-holes, have concluded that the ZÀ.MÍ/*sammûm* must be a harp. Leaving aside the identity of the musical instrument for the moment, let us examine the possibility that, *as OB mathematical terminology*, GEŠTÚ ZÀ.MÍ and *apsamikkum* are equivalent.

There are a total of six coefficients listed in §3.7.1 with the numerical value 26 40. The three from Susa (D 22, *TMS* 20, *TMS* 21) are for the syllabically written *ap(u)samikkum*.[49] There are two unprovenanced GEŠTÚ ZÀ.MÍ's (A 6, F 50), and a GÁN ZÀ.MÍ (E 4). This last, however, may be explained as an erroneous conflation of the numerical value and name of two separate coefficients (cf. E 34: §8.1). We might imagine that the originals from which it was copied were something like F 48 and 50, and that the numerical value 26 40 should have been paired with GEŠTÚ ZÀ.MÍ while the name GÁN ZÀ.MÍ belonged to some other number. We also have C 9 from Shaduppum to deal with, in which an *apsamikkum*-coefficient takes the value 26 15. Goetze (1951: 138) has already suggested that this value was accidentally copied from the triangle coefficient on the previous line (§3.3), and that 26 40 should be understood here instead.

Provisionally, then, we have Goetze's hypothesis that GEŠTÚ ZÀ.MÍ = *apsamikkum* to test against the mathematical evidence. There remains the GÁN ZÀ.MÍ of lists E and F, which we would expect to be identical to the GÁN GIŠ.ZÀ.MÍ of BM 15285.[50] We resume the philological analysis after a discussion of the coefficients themselves.

3.7.3 Mathematical commentary

When the first ZÀ.MÍ coefficients were published in texts E and F Neugebauer and Sachs understandably limited themselves to commenting on the possible derivation of their names (*MCT*: 135). Since then, as further evidence has come to light from coefficient lists, the Susa texts *TMS* 20 and 21, and BM 15285, there have been three major schools of interpretation,

[46] The beards may indicate that they are artistic survivals of representations of the (by then extinct) bison. Bison bulls are bearded; wild bulls (aurochs), like domestic bulls, are not.

[47] Namely, IGI.GU_4 'ox-eye' (§3.5) and SAG.KI.GU_4 'ox's forehead', a trapezoid (e.g. YBC 4608 (i), and BM 15285 (xxxv), where it refers to a quadrilateral, one of whose sides is an arc).

[48] Krispijn's (1990: 7) interpretation of the *ÁB.ZÀ.MÍ/*apsamikkum* exclusively as a three-sided sound-hole, i.e. concave equilateral triangle, is to be discounted: see §3.7.3.

[49] Although the name of the coefficient in *TMS* 20 is missing its identity is implied by *a-pu-sà-mi-ik-ki* in line 15.

[50] Cf. §3.5, in which GÁN GIŠ.PAN and GÁN PAN both describe a bow figure, also written with the GIŠ sign in BM 15285 (xxxiv)–(xxxv).

two of which can be discounted completely.

Goetze and Neugebauer (Goetze 1951: 138–9) used the text C values (with 26 15 amended to 26 40) to argue for a trapezoidal shape. This idea was taken up by Kilmer (1960: 285–6), *AHw* (I 61a), and adapted most recently by Spycket (1983). According to her, the GEŠTÚ ZÀ.MÍ and the *apsamikkum* are the names of two separate figures, 'bien que leur constante fixe (IGI.GUB) soit la même' (Spycket 1983: 46). While she sees the former as the concave square (see below), she supposes the latter to be a trapezium representing the sound-box of the lyre. However, not only does this explanation depend on a misinterpretation of *TMS* 21 (Lawergren and Gurney 1987: 46) but, further, it does not in any way explain why 0;26 40 should be the value of the coefficient.[51]

Lawergren and Gurney (1987: 44) agree with Spycket both that 'the fact that the *ḫasīs sammû* [*sic*, = GEŠTÚ ZÀ.MÍ] and the *apsamikkum* have one coefficient in common cannot be said to prove that they are the same figure' and that the GEŠTÚ ZÀ.MÍ is a concave square. At this point their explanations diverge. They too consider the *apsamikkum* to be a trapezium (Fig. 3.10), but suppose it to represent the frame of a harp, over which the strings are stretched (Lawergren and Gurney 1987; cf. Goetze 1951: 139 n. 15). There are several flaws in their interpretation. First, the proportions of the shape are implausible, both as a representation of a harp and as an OB geometrical figure *per se*.[52] Second, the mathematical meaning of *pirkum* has been well established as the shorter transversal through a figure at right angles to the longer *tallum*. It runs parallel to the short side, SAG, of the figure (*TMB*: 223),[53] whereas in this construction it is supposed to be parallel to the long side.

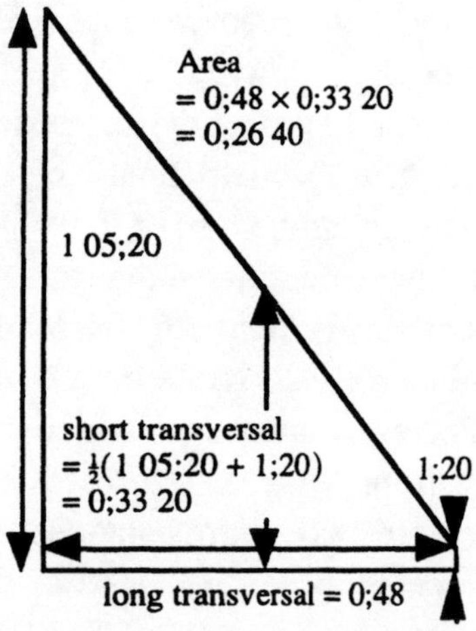

Figure 3.11: Erroneous interpretation of the apsamikkum

The conclusive argument against this theory, though, comes from the Susa problem text *TMS* 20. Although badly damaged, enough of it is reconstuctible to show that it sets up a relationship between the parameters of the *apsamikkum*, equivalent to the equation:

$$A + u + d = 1;16\ 40,$$

where A is the area, u the length (UŠ) and d the diameter (DAL). The coefficients of *TMS* 21 (listed above) show that $A = 0;26\ 40u^2$ and $d = 1;20u$. The solution of this equation is $u = 0;30$. The trapezoidal reconstruction, however, does not meet these conditions, whichever measurement one takes as the length (*pace* Lawergren and Gurney 1987: 47).

Figure 3.12: BM 15285 (xxvii) and (xl)

Vaiman (1963: 81), on the other hand, suggested that the *apsamikkum* and (GÁN) GEŠTÚ ZÀ.MÍ were both names for the 'concave square'. This idea, based on *TMS* 20 and BM 15285 (xxvii) and (xl), shown in Figure 3.11, was taken up by Kilmer (1990), Friberg (1987–90: 557) and *CAD* (A II 192–3). The interpretation goes as follows:

[51] *TMS* 21 in fact concerns a concave square, of (arc) length 0;30, inscribed in an unnamed figure of length 0;50 and width 0;45. The text is too broken for a certain restoration.

[52] In that, for instance, one of the vertical lengths is almost 50 times longer than the other.

[53] But note in *TMB* that all writings DAL should be read *tallum*.

If each of the four circular arcs bounding this figure has the length 1, then the area is $^4/_9$ [0;26 40], and the lengths of the longest and shortest transversal lines are $^4/_3$ [1;20] and $^5/_9$ [0;33 20]. (Friberg 1987–90: 557)

In other words, as the curved lines form a sort of inside-out circle (Fig. 3.12), with circumference $c = 4$ and quarter-arc $a = 1$, then the diameter, d, or longest transversal line, is

$$d = c/\pi = {}^4/_3.$$

The area A of the shape would be equal to twice the area of the square minus the area of the circle, or

$$A = d^2 - \frac{\pi d^2}{4} = {}^{16}/_9 - {}^4/_3 = {}^4/_9$$

and the shortest transversal s is equal to

$$s = \sqrt{2}d - d = {}^4/_3(\sqrt{2}-1) \approx {}^4/_3 \times {}^5/_{12} = {}^5/_9.$$

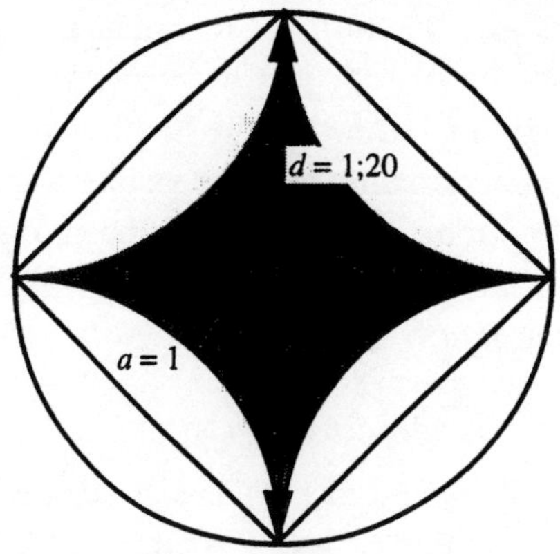

Figure 3.13: The components of an apsamikkum

This argument can be taken further. It follows that the area of the *apsamikkum* is exactly half that of the square in which it is inscribed. Denoting the length of the square by s, then by 'Pythagoras' Theorem':

$$d^2 = s^2 + s^2$$

$$s^2 = \frac{d^2}{2} = \frac{(1;20)^2}{2} = 0;53\ 20.$$

Similarly the area A of the circle surrounding that square (as in BM 15285 (xl), Fig. 3.12 above) is three times the size of the *apsamikkum*:

$$A = \frac{\pi d^2}{4} = 1;20.$$

Thus 0;53 20, the 'area of the harp/lyre' of F 48,[54] *might* refer either to the square around the *apsamikkum*, and including the area of that *apsamikkum*, or to the area of external circle minus the *apsamikkum* itself. In either case, E 4 is almost certainly a conflation of the two separate coefficients given in list F (§3.7.2).

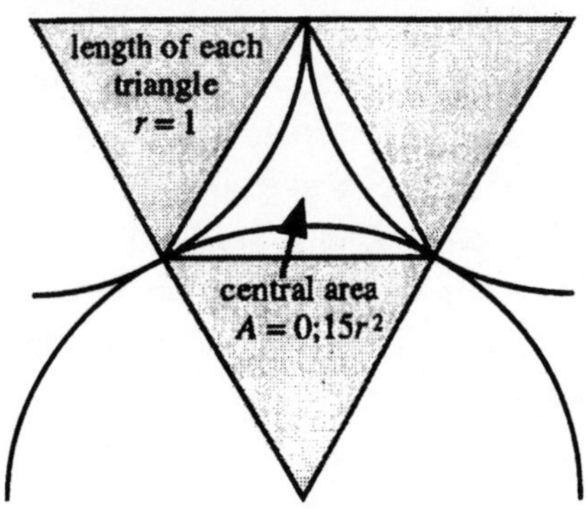

Figure 3.14: The three-sided apsamikkum

The same concept can also be extended to the three-sided *apsamikkum* (Kilmer 1990: 86), but in this case the basic unit can be interpreted either as the curved side of the figure or as the side of the equilateral triangle, because of the approximation $d = 0;20c$ (§3.1). The area of the *apsamikkum* is equal to the area of the large equilateral triangle minus a semicircle composed of the three shaded segments (Fig. 3.13):

$$\begin{aligned} A &= 0;26\ 15 \times (2r)^2 - 0;45 \times 2r \times r \\ &= (1;45 - 1;30)r^2 \\ &= 0;15r^2. \end{aligned}$$

[54] Reverting to Neugebauer and Sach's original reading of the number (*MCT*: 135), against more the recent 0;33 20 (e.g. Kilmer: 1960: 286; Nemet-Nejat 1993: 214). The remains of only two Winkelhakens, one above the other, are visible on the copy and photo (*MCT*: pl. 18, 44); one would expect rather different traces if the first digit were 30.

However, the matter is not completely solved. BM 15285 (xxx) incontrovertibly shows a shape called the GÁN GIŠ.ZÀ.MÍ which is apparently unrelated to the concave square (Fig. 3.14). The diagram on the tablet is badly damaged, but it is clear that the figure is based on a rectangle occupying the middle half of the square. Traces of a horizontal construction line are just visible a quarter of the way down, and at the top and bottom a crude arc has been inscribed between the two verticals;[55] the apex of the upper one is half way between the top of the square and the construction line.

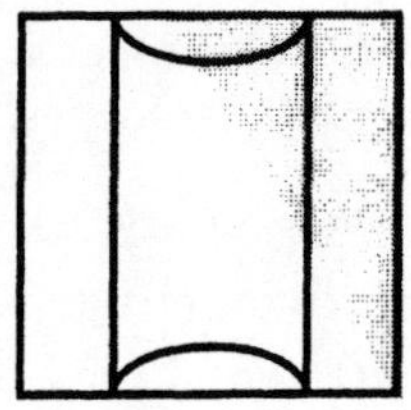

Figure 3.15: BM 15285 (xxx)

3.7.4 Concluding remarks

Thus the concave square described as a GÁN GEŠTÚ ZÀ.MÍ can be firmly identified with the *apsamikkum* of the Susa texts *TMS* 20, 21 and list D, as well as the GEŠTÚ ZÀ.MÍ of the coefficients A 6 and F 50. The figure which surrounds it — be it a square or circle — may be the GÁN ZÀ.MÍ of F 48. Further, if the value 0;26 15 of C 9 was indeed copied in error from the triangle coefficient C 7 (§3.3), it could also be argued that the *tallum* attested in the following line is a theoretical construction, an imaginary transversal perpendicular to the *pirkum*, which gives the area of the *apsamikkum* when multiplied by the latter:

$$0;48 \times 0;33\ 20 = 0;26\ 40.$$

In short, we have no fewer than six coefficients, all expressed with respect to the defining component of the concave square, the quarter-circle arc (see Table 3.1).

Table 3.1: Coefficients of the apsamikkum

Parameter	*Coefficient*	*Attestations*
Length	1	*TMS* 21
Area	0;26 40 (× arc length^2)	A 6, C 9, D 22, E 4, F 50; *TMS* 20, *TMS* 21
Diagonal	1;20	C 11, D 23; *TMS* 21
Diameter	0;48	C 10
Transversal	0;33 20	D 24
Area of external circle	0;53 20 (× arc length^2)	F 48
Area of concave triangle	0;15 (× arc length^2)	D 25

On the other hand, the identification GÁN ZÀ.MÍ = GÁN GIŠ.ZÀ.MÍ does not appear to hold. The one sure coefficient for the former seems to refer to the circle enclosing the concave square, while the sole illustration of the latter shows a rectangular figure with two concave sides.[56]

So what do these conclusions tell us about the ZÀ.MÍ as harp or lyre? Probably very little. As the rest of this chapter has shown, many OB geometrical figures are named for certain visual attributes of real objects. The author's instinct is that the mathematical *apsamikkum* has as much to do with the musical ZÀ.MÍ as do the moons, bows and barges of preceding sections with their respective real-life ancient counterparts.

[55] Almost certainly a badly drawn curve and not two straight lines (*contra* Kilmer 1990: 90 n. 16); see copy in §A.2.

[56] The author has not been able to reconcile F 48 with the BM 15285 illustration — whether reading the number as 0;53 20 or as 0;23 30. Reading 0;33 20, one could say that the coefficient described the area between the external square and the figure, but this is rather far-fetched.

3.8 OTHER GEOMETRICAL SHAPES

3.8.1 The coefficients

D 5 16 [IGI].GUB *šà* GÚR *šà* 2 ŠE *i-na* ŠÀ GÚR GAR
D 6 16 26 46 40 IGI.GUB *šà* GÚR *šà* 3 ŠE *i-na* ŠÀ GÚR GAR

E 7 5 ⌜24⌝ GÁN X X X
E 9 ⌜45?⌝ GÁN IN

3.8.2 Mathematical and philological commentary

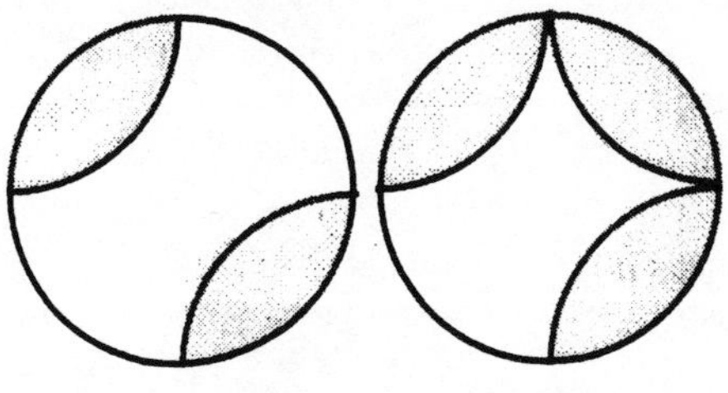

Figure 3.16: Possible interpretation of D 5–6

The first two coefficients are for 'circles in which 2 (or 3) grains have been put'. No satisfactory interpretation has been proposed for their numerical values, but if the geometrical meaning of the word ŠE 'grain' were the same as A.ŠÀ ŠE in D 16–18, i.e. two facing 90° arcs (see §3.5), then the figures described might look something like those shown in Figure 3.16 to the left. It is not obvious, though, how they would lead to the numerical values given.[57]

There is little that can be said about the final two coefficients. Their places in the lists, and the initial sign GÁN, indicate that they must be geometrical, but the reading of none is certain and there are no attestations outside the lists. E 7 has been read GÁN.UD.A.SAR (*MCT*: 136; Kilmer 1960: 285), but this does not quite fit the traces on the tablet, and it is not clear what the phrase would mean.[58] A satisfactory reading for E 9 has never been proposed. E 12 (numerical value 0;45) may also be geometrical, reading ⌜GÁN. X A.TI⌝, but the signs are too damaged for a sure reading (see §8.5).

Various other coefficients have in the past been called geometrical, either because they have a geometrical component[59] or for reasons which are not made clear.[60] As they do not appear in the geometrical sections of the lists they were not classified as geometrical in the Old Babylonian period and so are not dealt with in this chapter.

3.9 SUMMARY

The geometrical coefficients can be divided roughly into two groups: those for the simpler figures — circles, semicircles and triangles — which can be found in all seven lists; and those for more complex shapes, which depend on knowledge of the simpler ones, and are attested only in the more extensive texts, listed after the simple coefficients. This division is explicitated in text F, where four basic geometrical coefficients occur in lines 19–21, and the more complex ones in lines 47–51, immediately followed by further coefficients for the circle and semicircle. Indeed, it is only the simple geometrical coefficients which are attested in problem texts on

57 Friberg *et al.* (1990: 491) discuss similar shapes within circles in LB maths.

58 Compare GÁN.U$_4$.KAK.SAKAR in BM 15285 (xxxi), which is a semicircle and right-angled triangle forming the shape of an ice-cream cone. Its area is 0;52 30 times the square on the circle's diameter, or the triangle's height.

59 e.g. E 35, F 58, the coefficient of a log (Kilmer 1960: 285) (see §3.1.2; §7.2).

60 e.g. A 45, F 64, the coefficient of a cut reed (§8.4); A 41, the coefficient of a mortared wall (§4.3) (Kilmer 1960: 286–7); *rāṭum* and *ruqqum* coefficients (§8.1) (Kilmer 1990: 85). Friberg (1987–90: 553) reads H 3 as ÍB.‹SI$_8$› 'square', but this is a daily rate for building brick walls attested implicitly in YBC 4673 (vi) (§5.6).

non-geometrical subjects.

Table 3.2: Summary of the geometrical coefficients[61]

Figure	*Defining component* a	*Area* A	*Other components*	*Coefficients*
Circle	circumference	$A = 0;05a^2$	diameter d	$d = 0;20c$
			radius r	$r = 0;10c$
Semicircle	semicircumference	$A = 0;10a^2$	diameter d	$d = 0;40a$
				$A = 0;15a \times d$
			radius r	$r = 0;20a$
				$A = 0;45r \times d$
Right triangle	width	$A = 0;30a \times l$	length, l	
Equilateral triangle	width	$A = 0;26\ 15a^2$	height h	$h = 0;52\ 30a$
Rhombus	length	$A = 0;57\ 36a^2$		
Square	length	$A = a^2$	diagonal d	$d = 1;21a$
				$d = 1;25a$
				$d = 1;24\ 51\ 10a$
Rectangle	diagonal?		length, l; width, w	$A = l \times w$
panakkum	third-circle arc	$A = 0;06\ 33\ 45a^2$	long transversal, l	$l = 0;52\ 30a$
			short transversal, s	$s = 0;15a$
ILLAR	third-circle arc	$A = 0;13\ 07\ 30a^2$	long transversal, l	$l = 0;52\ 30a$
			short transversal, s	$s = 0;30a$
Ox-eye	third-circle arc	$A = 0;16\ 52\ 30a^2$	long transversal, l	$l = 0;52\ 30a$
			short transversal, s	$s = 0;30a$
Barge or grain-field	quarter-circle arc	$A = 0;13\ 20a^2$	long transversal, l	$l = 0;56\ 40a$
			short transversal, s	$s = 0;28\ 20a$
			area of quarter-circle, Q	$A = 0;45Q$
Pentagon	outer width	$A = 1;40a^2$		
Hexagon	outer width	$A = 2;37\ 30a^2$		
Heptagon	outer width	$A = 3;41a^2$		

All the geometrical coefficients — unless they explicitly state otherwise[62] — give the area GÁN of the figure they describe with respect to its defining component, often all or part of the perimeter. The diagonal of the rectangle is one exception to this. This defining component, which may be an arc or a straight line,[63] either remains nameless[64] or shares the same term as the shape it defines, such as the circle, square and rectangle. Two of the simpler rectilinear figures are logographically written stylised depictions: while the well-known LAGAB for a square is not found in the coefficients,[65] we find KAK for a triangle and UB for a rhombus.

[61] It should be stressed that the letters used in this table (and elsewhere in this work) to represent parameters are for ease of reference only; there was no equivalent in the ancient texts. See Table 3.1 for the *apsamikkum*.

[62] The only exception is the D 29 (§3.3), the coefficient for the height of an equilateral triangle with respect to its length: it is described as IGI.GUB, and not DAL as expected in this list.

[63] That there was no conceptual distinction between rectilinear and curved figures is shown in BM 15285 (xl), which denotes the spaces in between the circles and the corner of the square as 'triangles' SAG.DÙ.

[64] The Susa polgyons SAG.*n*, in which the SAG is the defining component, take their names from the fact that they are composed of triangles whose shorter side is always denoted SAG.

[65] But cf. NIGIN = LAGAB+LAGAB in list D.

4. BRICKS AND BRICKWORKS

This chapter first explores the metrology of bricks: their standardized sizes and methods of counting them. It then shows how this system was used to derive coefficients and apply them to problems concerning brick structures.

Brick metrology was first correctly interpreted in two studies in *MCT* (91–9, 137–8), based on the texts and coefficients published there and a revision of some texts from *MKT*. Lewy (1949) took up the discussion, with partial success. Powell's analysis (1982*a*: esp. 119) presented a brick typology, a refinement of Neugebauer and Sachs' work in *MCT*, using coefficient lists A, E, F and G, as well as problem texts concerning bricks. Although later works on OB mathematics do not seem to have adopted this typology in its entirety, Powell's study and terminology are taken as the starting point for the following two chapters. The sole post-*MCT* studies of the mathematics of brick structures are Dunham (1982) and Robson (1996*a*), which present and discuss the Ur III school text YBC 9819 and the brick problems in BM 96957 + VAT 6598 respectively.[1]

Useful sources of information on Mesopotamian brick metrology include: OB problem texts and a few Ur III school texts; datable stamped and baked bricks, in particular from the British Museum collection; administrative texts (especially from the Ur III period); ancient house plans; and modern data on excavated OB houses. There is good evidence that this branch of OB mathematics was used in everyday scribal practice (see §9.4).

There are 21 different brick coefficients in 34 attestations from texts A, D, E, F, and G, plus a further 12 attested in the LB list L. None are extant in lists C or M; text H includes one related coefficient (H 3) which is dealt with in §5.6. Ten of these coefficients (14 attestations, 9 of which are from text F) concern brick metrology; the rest relate to brick structures. Fifteen problems and calculations from five tablets explicitly use a total of seven brick coefficients:[2] four of these are found in the coefficient lists. A further four tablets contain problems on bricks but do not give the methods used to solve them; the answers show that coefficients must have been necessary;[3] a roughly written numerical table VAT 7002 may also be related to brick metrology. Two school texts from the Ur III period, which are in the form of model administrative documents, contain numerical information which must have been calculated using brick coefficients.[4] The coefficients for making and carrying bricks are discussed in §5.1 and §5.3.

[1] And see now Friberg (1996*a*), which covers essentially the same ground as the following three chapters but differs in many points of detail, classification, and typology from the following account. It came to my attention too late to be taken into systematic consideration here.

[2] Haddad 104 (viii) — laying plaster; AO 6770 (iv) — laying bitumen; YBC 7284 — weight of a brick; YBC 7997 — a brick-kiln; BM 96957 + VAT 6598 (vi), (vii) — planning a house; (i)–(iii) — building a brick wall; (ix′)–(xiv′): building a wall of baked bricks.

[3] AO 10822 (iv)–(v) — brick walls, (vi)–(xi) — brick-piles; YBC 4669 (xii) — brick-pile?; YBC 4708 (i)–(lx) — brick-piles. See §5.6 for YBC 4673 (v) — building a brick-pile and carrying earth.

[4] AOT 304 — brick wall; YBC 9819 — brick walls. See §1.3.4 for more on model documents.

4.0 BRICK SIZES ATTESTED EXPLICITLY IN OB MATHS

The measurements of seven different sizes of bricks are given explicitly in the OB mathematical problem texts listed below. The table shows the measurements and volume of the bricks, with approximate decimal equivalents; the problem texts in which they are found; and the name of the bricks given in the texts. The first column lists the number assigned to each brick by Powell (1982*a*: 119), which will be used here as a convenient shorthand when discussing the bricks.

Table 4.1: Names and sizes of bricks attested in OB mathematical problems

Type	*Measurements*	*Volume*	*Attestations*	*Name given*
2	15 × 10 × 5 fingers	0;00 00 41 40 sar$_v$	BM 85194 (xxx)	SIG$_4$
	(25 × 17 × 8 cm)	(3470 cm^3)	YBC 4607 (i)	SIG$_4$
			YBC 4607 (vi)	SIG$_4$
			YBC 4673 (i)	SIG$_4$
			YBC 7284	SIG$_4$
3	20 × 10 × 5 fingers	0;00 00 55 33 20 sar$_v$	YBC 4607 (iii)	SIG$_4$.ÁB
	(33 × 17 × 8 cm)	(4630 cm^3)	YBC 4607 (viii)	SIG$_4$.ÁB
4	18 × 12 × 5 fingers	0;00 01 sar$_v$	YBC 4607 (ii)	SIG$_4$
	(30 × 20 × 8 cm)	(5000 cm^3)	YBC 4607 (vii)	SIG$_4$
8	20 × 20 × 5 fingers	0;00 01 51 06 40 sar$_v$	YBC 4607 (iv)	SIG$_4$.AL.ÙR.RA
	(33 × 33 × 8 cm)	(9260 cm^3)	YBC 4607 (ix)	SIG$_4$.AL.ÙR.RA
9	20 × 20 × 6 fingers	0;00 02 13 20 sar$_v$	Haddad 104 (ix)	SIG$_4$
	(33 × 33 × 10 cm)	(11,100 cm^3)	Haddad 104 (x)	SIG$_4$
11	30 × 30 × 5 fingers	0;00 04 10 sar$_v$	YBC 4607 (v)	SIG$_4$.AL.ÙR.RA
	(50 × 50 × 8 cm)	(20,800 cm^3)	YBC 4607 (x)	SIG$_4$.AL.ÙR.RA

In addition, BM 85194 (xvi) and (xix) concern wells made of trapezoidal bricks 20 fingers long, 10 to 15 fingers wide, and of unknown thickness (see §9.3).

The bricks are all the same thickness, except the thicker Type 9, which has the same surface area as the Type 8 'baked bricks'. Type 3 bricks are half the size of Type 8s, and are thus called 'half-bricks', literally 'cows'. All the others are simply called 'bricks' except for Type 11, which are also 'baked bricks'. Both baked brick types have a square upper surface. All measurements are in multiples of five fingers, with the exception of Type 4, whose dimensions seem to have been chosen to give an easily calculable volume. The brick sizes attested in OB maths are compared in §9.4 and §A.6 with those found in excavations; see also §4.3.3.

4.1 COUNTING BRICKS USING VOLUME MEASURE

4.1.1 The coefficients

F 13	9 *na-al-ba-an* $^1/_3$ KÙŠ SIG$_4$	9, the brickage of a $^1/_3$-cubit brick.
F 14	8 20 *na-al-ba-an* IM.DUGUD	8;20, the brickage of clay lumps.
F 3	7 12 *na-al-ba-an-ša*	7;12, its brickage [i.e. of bricks].
L −1′	[7 12 SIG$_4$ ...]	[7;12, a brick ...]
A 25	5 24 IGI.GUB PEŠ SIG$_4$.AL.ÙR.RA	5;24, the coefficient of ... of a baked brick.
F 4	5 24 *na-al-ba-an* SIG$_4$.ÁB	5;24, the brickage of a half-brick.
L 0′	[5 24 SIG$_4$.ÁB ...]	[5;24, a half-brick ...]

D 45	5 *šà na-al-ba-ni*	5, of brickage.
L 2′	5 SIG$_4$.$^2/_3$-*ti* [...]	5, a $^2/_3$-brick [...]
F 11	4 48 *na-al-ba-an* $^1/_2$ KÙŠ SIG$_4$	4;48, the brickage of a $^1/_2$-cubit brick.
F 34	3 20 *ša* SIG$_4$.AL.ÙR.RA	3;20, of a baked brick.
A 26	2! 42! IGI.GUB SIG$_4$.AL.ÙR.RA	2;42, the coefficient of a baked brick.
E 44	⸢2⸣ 42 ⸢SIG$_4$ *a-gu*⸣*-ru-um*	2;42, a baked brick.
F 6	2 42 *na-al-ba-an* SIG$_4$.AL.ÙR.RA	2;42, the brickage of a baked brick.
L 1′	[2] ⸢42 SIG$_4$⸣.[AL.ÙR.RA ...]	[2];42, a [baked] brick [...]
F 35	1 52 30 *ša* SIG$_4$.AL.ÙR.RA KI.2	1;52 30, of a baked brick, 2nd.
E 39	1 12 [SIG$_4$ 1].⸢KÙŠ⸣	1;12, a [1]-cubit [brick].
F 9	1 12 *na-al-ba-an* 1 KÙŠ SIG$_4$	1;12, the brickage of a 1-cubit brick.

4.1.2 Mathematical commentary

As Neugebauer and Sachs showed (*MCT*: 94), bricks are counted in groups of 720, with a system of units analogous to OB volume and area measure. Thus:

1 gín	=	$^1/_{60}$ sar	=	12 bricks	(12)
		1 sar	=	12 00 bricks	(720)
1 iku	=	100 sar	=	20 00 00 bricks	(72,000)
1 ešè	=	600 sar	=	2 00 00 00 bricks	(432,000)
1 bùr	=	1800 sar	=	6 00 00 00 bricks	(1,296,000).

In other words, the brick-sar is not a measurement but a number, comparable to modern English commodity-specific number words such as the ream (480 sheets of paper). It does not depend on the size or quality of the bricks. Throughout this work, the brick-sar is written sar$_b$, to distinguish it from the area-sar (sar$_a$) and the volume-sar (sar$_v$).

Table 4.2 shows the volume of 1 sar$_b$ of each of the brick sizes given in §4.0, and the reciprocal of that figure: the number of sar$_b$ in a sar$_v$ for each type of brick. This quantity is called the *nalbanum* or 'brickage' of each brick type (*MCT*: 138; Powell 1982*a*: 120). The final column lists the entries in the coefficient lists where each number is found.

Table 4.2: Brickage coefficients found in the lists

Type	*Volume of 1 brick*	*Volume of 1 sar$_b$*	*Brickage*	*Coefficients*
2	0;00 00 41 40 sar$_v$	0;8 20 sar$_v$	7;12	F 3, L –1′
3	0;00 00 55 33 20 sar$_v$	0;11 06 40 sar$_v$	5;24	A 25, F 4, L 0′
4	0;00 01 sar$_v$	0;12 sar$_v$	5;00	D 45, L 2′
8	0;00 01 51 06 40 sar$_v$	0;22 13 20 sar$_v$	2;42	A 26, E 44, F 6, L 1′
9	0;00 02 13 20 sar$_v$	0;26 40 sar$_v$	2;15	—
11	0;00 04 10 sar$_v$	0;50 sar$_v$	1;12	E 39, F 9

Using this definition of brickage, one can then work back from the remaining *nalbanum* coefficients — all from list F — to determine the volumes of the bricks they represent, and their likely measurements, assuming that the thickness is either 5 or 6 fingers. They are listed in Table 4.3, with their position in Powell's typology (1982*a*: 119) shown in the final column. According to Powell (1982*a*: 117; 1987–90: 491) the whole system is based on the notion of a brick 1 cubit square and 6 fingers thick (Type 12), whose volume is thus 0;00 05 sar$_v$. There are exactly 1 sar_b, or 720, of these bricks in a sar_v, yielding a brickage coefficient of 1. The brick itself is not attested in any extant OB mathematics.

Table 4.3: Further brickage coefficients

Brickage	*Volume of 1 sar$_b$*	*Volume of 1 brick*	*Possible measurements*	*Type*
9;00	0;06 40 sar_v	0;00 00 33 20 sar_v = 600 $fingers^3$	10 × 10 × 6 fingers	1
8;20	0;07 12 sar_v	0;00 00 36 sar_v = 648 $fingers^3$	12 × 9 × 6 fingers	1a
4;48	0;12 30 sar_v	0;00 01 02 30 sar_v = 1125 $fingers^3$	15 × 15 × 5 fingers	5
3;20	0;18 sar_v	0;00 01 30 sar_v = 1620 $fingers^3$	18 × 18 × 5 fingers	7
1;52 30	0;32 sar_v	0;00 02 40 sar_v = 2880 $fingers^3$	24 × 24 × 5 fingers	10

The brickage coefficient is in essence a metrological conversion coefficient. Its role is to transform a volume into a number of bricks by the following operation:

volume (in sar_v) × Type *n* brickage coefficient = number of Type *n* bricks (in sar_b).

The coefficient is inversely proportional to the size of the brick: the larger the brick, the smaller the brickage. It is primarily used in 'brick-pile' problems in extant OB mathematics (§4.4), and as a basis for the mathematics of carrying bricks (§5.3). There are no problems known which concern brickage calculations alone. An illustration of brickage, and summary tables of the various brick sizes and their coefficients, are included in §4.7 and §5.7.

4.1.3 Philological commentary

Nalbanum is a *mapras* nominal formation from the verb *labānum* 'to make bricks', and thus may be expected to mean 'the place where, or the means by which bricks are made'. Reiner (1970) and Heimpel (1987) have both drawn attention to its distinction from *nalbattum* 'brick mould'. Surveying the use of their Sumerian equivalents in Gudea Cylinder A, Heimpel (1987: 207) concluded that *nalbattum*/^{giš}**ù-šub** means 'the whole brick mold understood as an actual implement', whereas *nalbanum*/**pisan$_x$** ^{giš}**ù-šub-ba** refers to '*that which determines the form of the brick*, that is in particular the inner sidings of the brick mold' [italics added]. The *nalbanum* coefficient, conversely, *is determined by* the form of the brick, or more particularly its size: each type of brick has a unique *nalbanum* coefficient by which its volume is represented mathematically.

Powell (1982*a*: 121) coined the translation 'brickage' for *nalbanum*, which, like the analogous modern term 'tonnage', conveys the idea of measuring volume or capacity in terms of the number of bricks (or tons) it contains. Only the coefficients in texts D and F actually mention the term 'brickage'. No mathematical logogram is known.

Some of the brick sizes are explicitly named in list F: bricks $^1/_3$, $^1/_2$ and 1 cubit square. A 'half-brick' SIG$_4$.ÁB is also named (F 4). All the baked bricks SIG$_4$.AL.ÙR.RA/*agurrum* are square, except the one in A 25, which is half a Type 8 baked brick. The meaning of the sign PEŠ is unclear here (cf. Kilmer 1960: 289 n. 2). It may be an otherwise unattested logogram for *nalbanum*, or have a meaning like 'half'; there is no non-mathematical evidence for either.[5]

The comment 'its brickage' in F 3 refers back to the previous line of the text, 'carriage of bricks', just as many of the following 'carriage' coefficients refer back to preceding 'brickage' (§5.3). Although the hand-copy of the text and Neugebauer and Sachs' transliteration (*MCT*:

[5] Whatever the reading of that one sign, the line itself almost certainly refers to a Type 3 brick, as does the corresponding carriage coefficient given in line 30 of the same list (§5.3). But for another possible interpretation, see §4.2.

pl. 18; 132) both show the final sign as SAḪAR 'earth', they acknowledge that this does not make sense (*MCT*: 137 n. 312). A close examination of the photograph (*MCT*: pl. 44) shows this sign to be a rather sloppily written *ša*, with both Winkelhakens before the final vertical visible. The lower of the two is above the lower horizontal, which presumably led to the initial erroneous reading.

The IM.DUGUD coefficient F 14 should perhaps be fitted into the existing brick typology, although excluded by Powell (1982*a*: 119). It occurs at the end of the list of brick coefficients in text F and shares with them the *nalbanum* terminology. The most probable theoretical measurements would be 12 × 9 × 6 fingers (20 × 15 × 10 cm) or 9 × 9 × 8 fingers (15 × 15 × 14 cm). It is designated Type 1a here, being larger only than the $^1/_3$-cubit square brick.

The only mathematical evidence for brick Types 7 and 10 is the two brick coefficients F 34 and 35, and Powell's typology is based solely on these lines. They occur in the middle section of Text F, completely separated from the other brick coefficients in the first part of the text. The phrase KI.2 '2nd' or 'ditto' occurs at other places in this text (e.g. F 51–4), where it refers to a second coefficient about an object already listed. It may be, then, that coefficients F 34–5 refer to some otherwise unknown properties of the Type 8 baked brick. On the other hand, Text F does attest three further otherwise unknown brick types (F 11, 13, 14), and the brick sizes associated with coefficients F 34 and 35 are plausible.

The numeral in A 26 is 3 44, as confirmed by recent collations. (The photo shows the final 4 clearly, but the first sign of the line is not visible (Kilmer 1960: pl. LXXXIV).) The numeral 3;44 is not sexagesimally regular and thus could not refer to a brick with integer measurements. It is assumed here that the number is a scribal error for 2;42, the brickage for Type 8 bricks: the corresponding carriage coefficient is found a little further down the list, in A 31 (§5.3), and the two other brickage coefficients in the text are also paired with their corresponding carriages.

The extensive restoration of the LB coefficients L –1′ – L 2 is assured by the following twelve lines of the list, which are discussed in §4.3. See also §2.2.3 and §A.1.

4.2 COUNTING BRICKS USING AREA MEASURE

4.2.1 The coefficients

F 38 14 24 *ta-ad-di-tum ša* SIG_4 14 24, a layer of bricks.

4.2.2 Mathematical and philological commentary

Neugebauer and Sachs (*MCT*: 94) determined that the aim of YBC 4607 (vi)–(x) is to calculate the *taddītum* of five sizes of brick (Types 2, 3, 4, 8 and 11) introduced in §4.0. As the text shows, this turns out to be the number of bricks that occupy 1 sar_a: the reciprocal of the surface area of the largest face of a brick. See also Fig. 4.2 in §4.7.

Table 4.4: Layers of bricks in YBC 4607

Type	*Problem*	*Surface area*	*A layer of bricks*
2	YBC 4607 (vi)	150 fingers^2 = 0;00 04 10 sar_a	14 24 bricks (864)
3	YBC 4607 (vii)	216 fingers^2 = 0;00 06 sar_a	10 48 bricks (648)
4	YBC 4607 (viii)	200 fingers^2 = 0;00 05 33 20 sar_a	10 00 bricks (600)
8	YBC 4607 (ix)	400 fingers^2 = 0;00 11 06 40 sar_a	5 24 bricks (324)
11	YBC 4607 (x)	900 fingers^2 =0;00 25 sar_a	2 24 bricks (144)

F 38 gives the *taddītum* for Type 2 bricks, which matches the answer in problem (vi) of the text. Because the 'layer' coefficient is dependent only on the surface area of the brick, it follows that different brick sizes may have the same coefficient (e.g. Types 3 and 6, 8 and 9, 11 and 12), where they differ only in their thickness. The 'layer' coefficient differs from the 'brickage' coefficient in that the bricks are counted individually, not in sar$_b$.

Coefficients A 25 (§4.1) and E 38 (§8.1) might refer to the 'layer' of brick types 8 and 11 respectively, as they fit numerically and the exact reading of their names is unclear. 'Layer' coefficients for all brick types are included in the summary table in §4.7.

The word *taddītum* has been regarded hitherto as a *nomen actionis* formation from the D-stem of *nadûm* III, 'to throw down, lay' (e.g. *MCT*: 94), although the D-stem of *nadûm* is not attested (*AHw*: 705–8). A philologically more convincing derivation would be from *naṭûm* IV = *nadûm* V, D ~ G 'to hit' (*AHw*: 768; *CAD*: N II 132–3). One could imagine the bricks being knocked or flattened into a pavement. *Taddītum* has a purely metrological meaning and is not a synonym of the more general *tibkum*, 'a course of bricks' (e.g. BM 85194 (xvi): III 45).

4.3 THE DENSITY OF BRICKS

4.3.1 The coefficients

G 8	5 IGI.GUB *mi-nu-ut li-bi-tim*	0;00 00 05, the coefficient of brick-counting.
L 11′	[8] 20 KI.LÁ SIG$_4$ *ṣa-rip-ti*	[8];20, the weight of a baked brick.
L 12′	[11 06 40 KI].LÁ SIG$_4$.ÁB *ṣa-rip-ti*	[11;06 40], the weight of a baked half-brick.
L 13′	[22 13 20 KI.LÁ SIG$_4$].AL.ÙR.RA KI.MIN	[22;13 20, the weight of a] baked [brick], ditto.
L 14′	[12 KI.LÁ SIG$_4$.$^2/_3$]-*ti* KI.MIN	[12, the weight of a] $^2/_3$-[brick], ditto.
L 3′	12 30 KI.ḶÁ SIG$_4$ [...]	12;30, the weight of a [... brick].
L 4′	16 40 KI.LÁ SIG$_4$.⌈ÁB⌉ [...]	16;40, the weight of a [... half-brick].
L 5′	33 20 KI.LÁ SIG$_4$.AL.⌈ÙR⌉.[RA ...]	33;20, the weight of a [... baked brick].
L 6′	18 KI.LÁ SIG$_4$.$^2/_3$-*ti* [...]	18, the weight of a [...] $^2/_3$-brick.
L 7′	10 KI.LÁ SIG$_4$.ḪA$_5$.⌈RÁ⌉	10, the weight of a dried brick.
L 8′	13 20 KI.LÁ SIG$_4$.ÁB.ḪA$_5$.RÁ	13;20, the weight of a dried half-brick.
L 9′	26 40 KI.LÁ SIG$_4$.AL.ÙR.RA ḪA$_5$.RÁ	26;40, the weight of a dried baked brick.
L 10′	14 24 KI.LÁ SIG$_4$.$^2/_3$-*ti* ḪA$_5$.RÁ	14;24, the weight of a dried $^2/_3$-brick.
D 64	14 24 *šà ki-ir-ba-ni*	14 24 00, of earth clods.
D 65	12 *šà ḫa-aṣ-bi*	12 00 00, of potsherds.
F 36	‹1›4 24 *ma-aš-qá-al-tum ša* SIG$_4$	‹1›4 24 00, the weight of a brick.
F 37	«1» 12 *ša-ḫa-al-lu-ú-um*	«1» 12 00 00, ...
YBC 7284: II 3	IGI.GUB.BA.BI 12	Its coefficient is 12 00 00.

4.3.2 Mathematical commentary

YBC 7284 is a small lenticular tablet which uses an unnamed density coefficient to calculate the weight of a Type 2 brick from its volume (*MCT*: 97–8):

$$0;00\ 00\ 41\ 40 \times 12\ 00\ 00 = 8;20 \text{ mina},$$

around 4 kg, or $9^1/_2$ lb. A density of 12 00 00 mina/sar$_v$ is approximately 1200 kg/m^3, or 1.2 times the density of water. The coefficient G 8 may be the reciprocal of this number.

This weight is identical to that given in the LB list L for an unnamed brick which has been oven-baked (line 11′). Multiplying the weights given for the bricks in L 12′–14′ by 0;00 00 05 (G 8) gives the volumes of these bricks:

$$11;06\ 40 \times 0;00\ 00\ 05 = 0;00\ 00\ 55\ 33\ 20\ \text{sar}_V$$
$$22;13\ 20 \times 0;00\ 00\ 05 = 0;00\ 01\ 51\ 06\ 40\ \text{sar}_V$$
$$12;00\ 00 \times 0;00\ 00\ 05 = 0;00\ 01\ \text{sar}_V,$$

which are identical to those of the OB types 3, 8, and 4 respectively (§4.0). Therefore the unnamed brick in L 11′ must also be identical to the OB type 2, and the OB coefficients YBC 7284 and G 8 must refer to the density of baked bricks.

Coefficients L 3′–10′ list the weights of the same four brick types when they have been sun-dried (7′–10′) and in a broken context (3′–6′). Dividing these weights by the volume of one brick produces the density coefficient for each:

Table 4.5: The brick coefficients of list L

Line	*Brick*	*Weight*	*Volume*	*Coefficient*	
L 3′	Type 2	12;30 minas	0;00 00 41 40 sar_v	18 00 00	
L 4′	Type 3	16;40 minas	0;00 00 55 33 20 sar_v	18 00 00	1;30 × 12 00 00
L 6′	Type 4	18;00 minas	0;00 01 sar_v	18 00 00	(baked bricks)
L 5′	Type 8	33;20 minas	0;00 01 51 06 40 sar_v	18 00 00	
L 7′	Type 2	10;00 minas	0;00 00 41 40 sar_v	14 24 00	
L 8′	Type 3	13;20 minas	0;00 00 55 33 20 sar_v	14 24 00	1;12 × 12 00 00
L 10′	Type 4	14;24 minas	0;00 01 sar_v	14 24 00	(baked bricks)
L 9′	Type 8	26;40 minas	0;00 01 51 06 40 sar_v	14 24 00	

In short, these coefficients imply that bricks which have undergone some process, missing in text L, are $1^1/_2$ times as heavy as baked bricks, and that sun-dried bricks are rather lighter, but still $1^1/_5$ times the weight of baked bricks. If we presume that the weight change is due to water loss through the drying process, it may be that the first set of coefficients (L 3′–6′) refers to bricks when freshly made, and thus still very moist, DUR_5/*raṭbum*. In other words, bricks lose a fifth of their weight through sun-drying, and a full third by baking.

D 64 and 65 almost certainly refer to the density of sun-dried and baked clay products respectively, as perhaps do F 36 and 37. The name of F 36 certainly means 'the weight of a brick', although that of F 37 is otherwise unknown and of obscure derivation. It may be that when these two entries were copied from an original text — as is almost certainly the case; see §2.1.4 — the first Winkelhaken of 14 24 was mistaken for the head of a 1 in the following line.[6]

One would expect, though, to find all three density coefficients listed together, and not just two, as is the case in lists D and F. A coefficient with the expected numerical value, and which may concern the weight of freshly made bricks, is indeed listed near both pairs, and also occurs in a few obscure mathematical problems:

A 17 18 IGI.GUB IM.LÁ	18, the coefficient of a ....
D 63 18 *šà im-le-em*	18, of a ....
E 14 18 IM.LÁ	18, a ....
F 40 18 *i-im-lu-ú-um*	18, a ....

[6] Winkelhakens are occasionally written almost as horizontals; cf. the numeral 1 52 30 in F 35; also YBC 7164 (xviii): II 20, where 12 is written identically to the $^1/_3$ sign (§6.5; cf. *MCT*: 82).

BM 85194 (x): III 6	18 IM.LÁ	18, a ...
BM 85194 (xi): III 11	18 IM‹LÁ›.BI	18, its clay.
BM 85194 (xii): III 15	18	18.
BM 85194 (xiii): III 20	[1]8 IM.LÁ	18, a ...

BM 85210 (xviii) and (xix) also concern IM.LÁ, but do not use this coefficient (see §8.3). It is tempting to conclude that the phrase 18 IM.LÁ means '18 00 00, weighed clay', from the logograms IM/*ṭīdum* 'clay, earth', and LÁ/*šaqālum* 'to weigh' — especially considering its proximity to the two other brick weight coefficients in lists D and F, and the fact that it is once written KI.LÁ, which may be read *šuqultum* 'weight', in BM 85194 (xi): IV 10. However, it is difficult to reconcile this interpretation with the evidence of the problem texts themselves, which are discussed further in §8.3. It *may* be, then, that IM.LÁ is not a logogram for *imlûm*, and that we are dealing with two separate coefficients, albeit with the same numerical value: the latter is a brick density coefficient, which occurs only in proximity to others in the series, while the former describes some as yet undetermined structure.

The weight of 1 (baked) brick is numerically identical to 60 times the reciprocal of its brickage. This is because:

$$\frac{1}{\text{brickage}} = \text{volume of 1 sar}_b$$
$$= 12\ 00 \times \text{volume of 1 brick}$$

and thus:

$$\frac{60}{\text{brickage}} = 12\ 00\ 00 \times \text{volume of 1 brick}$$
$$= \text{weight of 1 brick}$$

4.3.3 Philological commentary

The reading *mi-nu-ut* for the construct state of *minûtum* 'calculation' (*AHw*: 656) in G 8 is Kilmer's (1985: 15).

The meaning of *šaḫallûm* (F 37) is unclear. It has been suggested that it derives from *šaḫālum* 'to sift' and refers to a milled or ground product (*CAD*: Š I 77), although **parassûm* is not a known Akkadian word shape.[7] Kilmer (1960: 301) describes it as 'operational(?)' with no further comment. Neither interpretation is particularly convincing, considering the position of the coefficient in the middle of a section on bricks and brick-making. It is more likely to be a loanword from Sumerian, but what the original might be it is impossible to say.

Comparison of the brick names in the first millennium list L with those given in YBC 4607 (§4.0) shows how little this particular terminology changed between the OB and LB periods. Only the Type 4 brick has changed its name, from simply SIG_4/*libittum* 'brick' to $\text{SIG}_4.{}^2/_3$-*ti*/*libitti šinnipēti* (*AHw*: 1242) '${}^2/_3$-brick', indicating that one side of the brick is two-thirds of the length of the other. These four brick names are attested in identical order — although in a different sequence from YBC 4607 — in the two other known LB coefficient lists, B and K:[8]

B II 3′	[6 GIŠ.MÁ.LÁ] SIG_4	[6, a cargo-boat] of bricks.
B II 4′	[4] ⌜30⌝ GIŠ.MÁ.LÁ SIG_4.ÁB	[4] 30, a cargo-boat of half-bricks.
B II 5′	[2] 15 GIŠ.MÁ.LÁ SIG_4.AL.ÙR.RA	[2] 15, a cargo-boat of baked bricks.
B II 6′	4 10! GIŠ.MÁ.LÁ $\text{SIG}_4.{}^2/_3$-*ti*	4 10, a cargo-boat of ${}^2/_3$-bricks.

[7] A *parass* formation (*GAG*: 62) is excluded because of the repeated vowel indicating a long final syllable.
[8] See also §7.2 for the list B coefficients.

K IV 2′	[...] SIG$_4$	[...] brick.
K IV 3′	[...] ⸢SIG$_4$⸣.ÁB	[...] half-brick.
K IV 4′	[... SIG$_4$].⸢AL⸣.ÙR.RA	[...] baked [brick].
K IV 5′	[... SIG$_4$].$^2/_3$-*ti*	[...] $^2/_3$-[brick].

There seems to have been a strong continuity of tradition for the mathematics of bricks which is not (yet) in evidence for other mathematical genres. There have been two LB problems published on the laying of bricks (cf. §4.2), W 23291-x (8a) and (8b), which attest to bricks of Type 8, called *a-gur-ru* (Friberg *et al.* 1990: 528–31).

4.4 STACKING BRICKS

4.4.1 The coefficients

A 24	7 12 IGI.GUB SIG$_4$.ANŠE	7;12, the coefficient of a brick-pile.
E 40	7 12 ⸢SIG$_4$.ANŠE⸣	7;12, a brick-pile.
F 16	7 12 *ša* SIG$_4$.ANŠE	7;12, of a brick-pile.
G 6	7 12 IGI.GUB.BI *a*[!]*-ma-ri-im*	7;12, the coefficient of a brick-pile.
D 48	5 *šà a-ma-ri-im*	5, of a brick-pile.
YBC 7997: II 7	7 12	7;12.

4.4.2 Mathematical commentary

The brick-pile coefficient is numerically and conceptually identical to the Type 2 brickage (§4.1). In other words, it converts the volume of a pile of Type 2 bricks into the number of bricks in that pile, counted in sar$_b$:

$$\text{volume of brick-pile } (\text{sar}_v) \times \text{coefficient} = \text{number of bricks in pile } (\text{sar}_b).$$

The problems using this coefficient are often straightforward. In YBC 7997 the volume of a cylindrical kiln is calculated. This is multiplied by 7;12 to give the bricks that will fit in it to be baked.[9] The text YBC 4708 (i)–(lx) lists the statements and numerical answers to sixty problems which involve calculating the brickage of various brick-piles in the shape of rectangular blocks and truncated pyramids (*MCT*: 94). As the full solutions are not given, neither is the coefficient, but in each case the answer shows that the Type 2 brickage coefficient 7;12 was intended. However, the problems on this tablet are primarily exercises in 'algebraic' manipulation (see §10.5).

D 48, from Susa, gives a different value for the brick-pile coefficient, which is identical to the only brickage coefficient in the same text (D 45; §4.1). This suggests that Type 4 bricks were used as the basis for calculation in Susa, though are no extant mathematical texts on bricks from that city.

AO 10822 (vi)–(xi) concern piles of bricks and half-bricks. The volume of the piles is to be found first, and then the length or height that must be removed (*harāṣum*) from the piles in order to take away 10 sar$_b$ of bricks. The answers are not given, so it is impossible to determine

[9] But cf. Joannès (1993–7: 102), who states that the *utūnum* denotes a metal-smelting furnace and never refers to a brick-kiln.

which coefficients were meant to be used. YBC 4673 (vii)–(viii) may also involve brick-piles, but are too damaged to read satisfactorily.

One problem whose coefficient can be determined from the statement and solution alone is YBC 4669 (xii).[10] The text reads:

III $3^{1}/_{2}$ NINDAN 3 ⸢KÙŠ UŠ⸣	The length is $3^{1}/_{2}$ nindan, 3 cubits.
$^{1}/_{3}$ KÙŠ SIG$_4$ IN.⸢TA?⸣	The bricks are $^{1}/_{3}$ cubit *square*.[11]
4 KÙŠ ‹SUKUD› ANŠE.GUB?	The ‹height› of the *pile* is 4 cubits.
2 KÙŠ DAGAL ANŠE.GUB?	The width of the *pile* is 2 cubits.
SAḪAR SIG$_4$.ANŠE.GUB? ‹EN›.NAM	What is the volume of the brick-pile
Ù SIG$_4$ EN.NAM	and how many bricks are there?
2 15 SAHAR ANŠE.GUB?	The volume of the *pile* is 2;15 sar.
20 SAR 3 *šu-ši* SIG$_4$	There are 20 sar and 3 sixties of bricks.

The subject of the problem is apparently a brick-pile in the form of a rectangular block. Its volume is 3;45 × 4 × 0;10 = 2;30 sar$_V$, although the text gives it as 2;15 sar$_V$. This may be because the scribe looked up the calculations in a multiplication table, reading off 15 (the product of 3;45 and 4) by 9, instead of 10.[12] We are told that the bricks in the pile are $^{1}/_{3}$ cubit square (line 2). According to the coefficient F 13 (§4.1) these have a brickage of 9, and indeed the number of bricks in the pile tallies with this figure:

2;15 × 9 = 20;15 sar$_b$, or 20 sar$_b$ 3 00 bricks.[13]

AOT 304 is a lenticular school text which may date back as far as the Ur III period (Friberg 1987–90: 451). On the obverse the calculation is laid out in the form of a model document (§1.3.4), imitating the layout of administrative texts of the time, except that the height and width are given in reverse order to the usual practice:

6 nindan $4^{1}/_{3}$ kùš gíd	Length: 6 nindan, $4^{1}/_{3}$ cubits.
$^{1}/_{2}$ nindan sukud	Height: $^{1}/_{2}$ nindan.
2 kùš 5 šu-si dagal	Width: 2 cubits, 5 fingers.
⸢saḫar⸣-bi $3^{1}/_{2}$ sar $2^{1}/_{2}$ gín	Its volume: $3^{1}/_{2}$ sar, $2^{1}/_{2}$ shekels.
sig$_4$-bi $25^{1}/_{2}$ sar	Its bricks: $25^{1}/_{2}$ sar.

On the reverse the three measurements are written in sexagesimal place value notation,[14] although the scribe has mis-transcribed the length and height as 6;⸢31⸣ 50 and 3 instead of 6;21 40 nindan and 6 cubits respectively. The width is correctly given as 0;10 50 nindan. These produce a volume of 3;32 14 35 sar$_v$, rounded up on the obverse to 3;32 30. This volume is multiplied by 7;12, the Type 2 brickage coefficient, to give 25;30 sar$_b$, or 18,360, bricks. This last calculation is not recorded on the tablet. The name of the object is not given, but the coefficient implicit in the text suggests that it concerns a pile of Type 2 bricks. The correct volume and bricks are 6;53 28 20 sar$_v$ and 49;37 sar$_b$, or 35, 724 bricks.

Two Ur III administrative texts — one of which may in fact be a school model document — are known which employ brick-pile coefficients; they are discussed in §9.4. For a coefficient for the daily rate of building a brick-pile see §5.6.

10 For a very different reading and analysis of this problem, see now Friberg (1996*a*: 23–4).

11 Same phrase as coefficient A 30 (§5.3)?

12 For further examples of this sort of error, see §5.2.

13 Or 2;30 × 9 = 22;30 sar$_b$, using the correct volume for the brick-pile.

14 See Powell (1976*a*: 420–1) for further evidence of place notation used in Ur III documents.

4.4.3 Philological commentary

It may be that the logogram SIG$_4$.ANŠE, literally 'brick donkey', was used for *amarum* 'brick-pile' because of similarity in sound between *amarum* and *imērum* 'donkey', whose logogram is also ANŠE; the origin of the word *amarum* is itself unclear. It must be connected with the verb (occurring only in Old Assyrian) *emērum* 'to stack (bricks)', and it is tempting to connect it with the Assyrian word *nāmaru* 'tower', which may be a *mapras* derivative of the same verb.

4.5 BRICK WALLS AND HOUSES

4.5.1 The coefficients

A 41 6 IGI.GUB IZ.ZI	6, the coefficient of a brick wall.
G 7 6 IGI.GUB *i-ga-ri-im*	6, the coefficient of a brick wall.
A 10 20 IGI.GUB É.DÙ.A	0;20, the coefficient of a built house.

BM 96957+VAT 6598

(i): I 3–4	6 [IGI].GUB É.GAR$_8$	6, the coefficient of a brick wall.
(ii): I 7	6 IGI.GUB É.GAR$_8$	6, the coefficient of a brick wall.
(v): [I 3']	[6 IGI.GUB É.GAR$_8$]	[6, the coefficient of a brick wall.]
(vi): II 6	IGI.3.GÁL *le-qé*	Take a third (of the area).
(vii): II 11	[20 IGI.GUB É.DÙ].⸢A⸣	[0;20, the coefficient of a built house.]
(xii): III 2	2 15 IGI.GUB É.GAR$_8$	2;15, the coefficient of a brick wall.
(xiii): III 7	2 15 IGI.GUB É.GAR$_8$	2;15, the coefficient of a brick wall.
(xiv): III 14–15	2 15 IGI.GUB SIG$_4$.AL.UR$_5$.RA	2;15, the coefficient of baked bricks.
(xv): III 21	2 15 IGI.GUB SIG$_4$.AL.UR$_5$.RA	2;15, the coefficient of baked bricks.
(xvi): III 31	2 15 IGI.GUB É.GAR$_8$ SIG$_4$.AL.UR$_5$.RA	2;15, the coefficient of a wall of baked bricks.
(xvii): IV 8	[2] 15 IGI.GUB SIG$_4$.AL.UR$_5$.‹RA›	[2];15, the coefficient of baked bricks.

4.5.2 Mathematical commentary

The existence of separate coefficients and problems for brick walls suggests that they were perceived as something other than simple brick-piles in Old Babylonian mathematics. Two statements of problems on a brick wall or IZ.ZI in AO 10822 (iv)–(v) immediately precede several on brick-piles (§4.4).

(iv) IZ.ZI 5 NINDAN UŠ 5 ⸢50⸣ *ku-bu-ur-re* ⸢IZ.ZI 12 SUKUD⸣ SIG$_4$ EN.NAM

A wall. The length is 5 nindan, the thickness of the wall is 0;05 50 (nindan), the height is 12 (cubits). What are the bricks?

(v) IZ.ZI 5 NINDAN UŠ *a-na* $^1/_2$ NINDAN SUKUD 12$^1/_2$ SAR [SIG$_4$]
ku-bu-ur-re-e IZ.ZI EN.NAM

A wall. The length is 5 nindan by $^1/_2$ nindan, the height. The [bricks] are 12$^1/_2$ sar. What is the thickness of the wall?

Although neither the size of the bricks, the value of the coefficient, nor the answers, are given in the problems, the value 6 given in A 41 and G 7 gives plausible answers using the method:

$$\text{length} \times \text{width} \times \text{height} \times \text{coefficient} = \text{number of bricks.}$$

For problem (iii) this gives 35 sar_b bricks; and for problem (iv) a thickness of 0;04 10 nindan or $^5/_6$ cubit.

This same method and coefficient are used explicitly in BM 96957 + VAT 6598 (i)–(v) (§A.4), where the various parameters of a brick wall are calculated in turn (Robson 1996*a*). On the same tablet is a parallel set of problems (ix′)–(xiv′) on a wall of baked bricks. The method and parameters are the same as before, but the coefficient is now 2;15.

The most commonly attested brick sizes in OB mathematics are Types 2 and 8 (§4.0), with brickage coefficients of 7;12 and 2;42 respectively (§4.1). The two brick wall coefficients, 6 and 2;15, are both five-sixths of these values. Two inferences are possible: either the bricks in the wall are proportionally bigger than Types 2 and 8, or they occupy only five-sixths of the wall. The problems Haddad 104 (ix)–(x) mention Type 9 bricks, which are $^2/_3$ cubit square, like the Type 8 bricks, but 6, not 5, fingers thick (§4.0). It is possible that thicker bricks such as these are meant here, although there is no evidence for a thicker equivalent to the Type 2 brick. However, that same tablet also includes two problems about spreading mortar on a wall to a depth of 1 finger (Haddad 104 (vii)–(viii); §4.6). As both Type 2 and Type 8 bricks are 5 fingers thick, the coefficients must imply that only five-sixths of the volume of a wall is brick, and that the rest is mortar. In short, for every course of bricks 5 fingers thick there is a layer of mortar 1 finger thick, as illustrated in §4.7. This estimate ignores the small amounts of mortar placed vertically between the bricks on each course (Robson 1996*a*).[15]

Support for this theory comes from an Ur III school text in the form of a model document. YBC 9819 purports to be about bricks for the walls of the temples of Shara and Ninurra in Umma (Dunham 1982). The measurements of four walls (**iz-zi**) are given, as well as the number of bricks, counted in sar_b, contained in each:

Table 4.6: Summary of the contents of YBC 9819

Line	*Volume*	*Number of bricks*	*Coefficient implied*	*Brick name*
I 1–2	37 02;00 sar_v	2 00 00 sar_b	3;14 24	**ukurrum**
II 8–9	9 33;50 sar_v	31 00 sar_b	3;14 24	**ukurrum**
III 1–3	10 24;51 40 sar_v	33 42 sar_b	3;14 24	**ukurrum**
III 4–6	2 15;40 sar_v	20 00 sar_b	8;50 24	**zarin**

Three walls are made of **ukurrum** bricks, written **ù-ku-ur-ru-um**. The brickage coefficient of the **ukurrum** bricks in this text is exactly $^6/_5$ of 2;42, the brickage of Type 8 bricks called *agurrum* in the coefficient E 44. It seems that **ukurrum** and *agurrum* bricks are in fact the same,[16] but that the student writing this text made a mistake, calculating that $^6/_5$, not $^5/_6$, of the walls were made of bricks!

The evidence from the fourth wall, made of **zarin** bricks (**za-rí-in**), is a little more complicated. The number of bricks in this wall is more than $^6/_5$ of what would be expected from Type 2 bricks. However, the coefficient used to calculate this quantity is 8;50 24, exactly $^6/_5$ of the number written sexagesimally as 7;22, which differs by only one Winkelhaken from the correct *nalbanum* value 7;12 when written in cuneiform. So the unfortunate student made two mistakes here: not only did he divide, instead of multiplying, the brickage by $^5/_6$, but he

[15] The table on the small, re-used tablet VAT 7002 may be related to this subject. The numbers in the left-hand column run from 1 to 1;30 in increments of 0;05, while on the right they run from 1;12 to 1;48 in increments of 0;06. In other words, the table contains seven pairs of numbers in the ratio 5:6. I thank Jeremy Black for collations.

[16] It has previously been disputed whether **ukurrum** is a loan into Sumerian from Akkadian *agurrum*, 'baked brick' (e.g. *MCT*: 96 n. 248a), or from *waqrum*, 'top quality' (e.g. Goetze 1948*a*: 34). This new evidence now precludes the latter interpretation.

even copied the coefficient down wrongly in the first place.[17] **Zarin** bricks can thus be identified with Type 2 unbaked bricks, which seem to be cheap low-grade housing bricks. This accords well with the established meaning of *zarinnum*, an Akkadian term for 'poor quality', which refers to commodities such as bronze and wool (*AHw*: 1515).[18]

BM 96957 + VAT 6598 (vi)–(vii) concern the number of bricks required to build a house of a given area, assuming that brickwork takes up one third of that area. This is the meaning of coefficient A 10. The method of solution is:

$$\tfrac{1}{3} \times \text{area of house} \times \text{height} = \text{volume of walls in nindan}^3.$$

At this period volume is usually measured in sar_v, i.e. 1 nindan^2 area × 1 cubit height, but in this case the height of the house is given explicitly in nindan. The house, of area 5 sar_a and height $2^1/_2$ nindan, has walls of volume 4;10 nindan^3. In the problem (vii) this is converted directly to $2^1/_2$ iku of bricks, without explicit calculation. The steps behind this conversion seem to be:

$$4;10\ \text{nindan}^3 = 50\ \text{sar}_\text{v}$$
$$50\ \text{sar}_\text{v} = 250\ \text{sar}_\text{b}$$
$$250\ \text{sar}_\text{b} = 2^1/_2\ \text{iku}_\text{b}.$$

This explains the unusual step of working out the volume in nindan^3. The expected method would be to calculate the height in cubits by multiplying the measurement in nindan by 12, and then to multiply the volume by 5, the coefficient implicit here, to give the number of bricks. Instead the scribe has used the fact that 12 × 5 = 60. He has simply multiplied the volume in nindan^3 by 60 (i.e. moved the 'sexagesimal place'), so that 4;10 nindan^3 becomes 4 10 (= 250) sar_b, or $2^1/_2$ iku_b.

The relationship 50 sar_v = 250 sar_b implies the use of Type 4 bricks, which have a brickage of 5 (§4.1), whereas all the other brick problems in this text concern Types 2 and 8. It may be valid in this context to note that $5 = (^5/_6)^2 \times 7;12$. In other words, the reasoning behind the use of 5 as the coefficient may have been that if one calculates the area of a house by multiplying the lengths of its walls, then the proportion of bricks to mortar in that house must also be calculated by multiplying the fractions of each wall made up of bricks (in this case $^5/_6$). This is not logically correct, but would be consistent with the use of Type 2 bricks throughout the rest of the text except where Type 8 (SIG_4.AL.UR_5.RA) are explicitly mentioned.

4.5.3 Philological commentary

Elizabeth Stone (1981: 20) has conclusively shown that in house sale contracts from OB Nippur the meaning of É.DÙ.A/*bītum epšum*, literally 'built house', is 'roofed floor-space', excluding the area of the walls and central courtyard. However, in brick texts it is precisely the area of the walls that is under consideration, so É.DÙ.A must have a broader sense there. When a built house is sold the most important consideration is the amount of living space it provides; the ground-area of the walls is largely irrelevant, not to mention very difficult to measure. With an empty building plot, on the other hand, the surveyor is concerned with the site as a whole. The proportion of walls to roofed floor-space would be meaningless in this context: É.DÙ.A seems to signify the whole area of the house in OB mathematical texts. See further §9.4.

[17] The numerical errors in AOT 304 are also caused by miscopied numerals (see §4.4).

[18] AO 7667 (Scheil 1915; see also *MCT*: 94–5) is an administrative list of constructions made of **ukurrum** and **zarin** bricks, dated to Amar-Suen year 4 (2043 BC). It provides further evidence for the identification of **ukurrum** and *agurrum* bricks. The brickage of the **ukurrum** bricks is 2;42, the same as the Type 2 bricks. Unfortunately only the total number of **zarin** bricks is extant and this does not provide any information on their size.

4.6 LAYING MORTAR, PLASTER AND BITUMEN

4.6.1 The coefficients

D 67	12 *šà ku-up-ri*	0;12, of refined bitumen.
A 13	15 IGI.GUB ESÍR.ḪA₅.RÁ	0;15, the coefficient of dried bitumen.
D 66	15 *šà* ESÍR	0;15, of bitumen.
E 31	15 ⌈ESÍR.ḪI⌉.A	0;15, mixed? bitumen.
F 18	15 *ša* ESÍR.ḪA₅.RÁ.A	0;15, of dried bitumen.
A 14	16 IGI.GUB ESÍR	0;16, of bitumen.
AO 6770 (iv): II 13	IGI.GUB-*bu-um* 15	The coefficient: 0;15.
Haddad 104 (viii): IV 8	1 SAR *ṭì-dam ša ka-la u₄-mi-im* 1 LÚ *i-se-ru*	1 sar_a, the earth that 1 man plasters every day.

4.6.2 Mathematical commentary

AO 6770 (iv) is a short text about finding the volume of bitumen needed to cover an area of 1 40 sar_a. Only the statement of the problem, the coefficient needed to solve it and the answer — 12 gur — are given. If, as the text implies, 12 gur of bitumen is used to cover an area of 100 sar_a, then for 1 sar_a the bitumen needed is 12 ÷ 100 gur, or 36 sila. For an area of 1 square cubit the quantity is $^1/_{144}$ of this amount, or 0;15 sila (Thureau-Dangin 1936: 80). This is roughly equivalent to 1 litre of bitumen per square metre. The depth to which the bitumen is spread is:

$$\text{depth in cubits} = \frac{\text{volume of bitumen}}{\text{area of cover}}$$
$$= \frac{36 \text{ sila}}{1 \text{ sar}_a} \approx \frac{0;00\ 07\ 12 \text{ sar}_v}{1 \text{ sar}_a}$$
$$= 0;00\ 07\ 12 \text{ cubits,}$$

or 0;03 36 fingers, approximately 1 mm. It is not apparent why this coefficient is given in terms of a square cubit instead of a sar_a, the standard unit of measure in the Ur III and Old Babylonian periods. The bitumen is quantified in capacity measure, as is usual also in non-mathematical contexts.

Three of the four lists which include bitumen coefficients agree with the value given in AO 6770 (iv). Text D, though, lists this same number as the coefficient of unrefined bitumen — ESÍR/*iṭṭûm*, while refined bitumen — ESÍR.HA₅.RÁ.A/*kuprum* — is spread more thinly, with only 0;12 sila covering a square cubit (to a depth, therefore, of less than 0;03 fingers, less than a millimetre (D 66–7)). Text A too suggests that unrefined bitumen spreads more thickly than the refined sort, with a coverage of 0;16 sila per square cubit, a depth of almost 0;04 fingers. These numerical differences are all so tiny as to be negligible if the coefficients were ever used for small-scale practical purposes. AO 6770 (iv) makes no distinction between the two types, referring to both interchangeably in the same line (rev. 12).

Haddad 104 (vii) and (viii) are both on a subject which is reflected only indirectly in the coefficient lists: spreading plaster on a wall,[19] at a daily rate of 1 sar_a to a thickness of 1 finger,

[19] Although it is not stated explicitly, this may be deduced from the fact that the 'spreading' is vertical: it has a height *mūlûm*, length UŠ and thickness *mūbûm* (Haddad 104: IV 6–7).

approximately 17 mm (ll. IV 6–7). Unlike the problem on bitumen, the thickness is stated explicitly, so that the volume of the plaster may be calculated directly from its measurements. It is then transformed with a conversion coefficient (§7.1) to an equivalent capacity: 2 gur or 600 sila per sar_a.[20] The same thickness of mortar was assumed in the mathematics of brick walls (see §4.5).

4.7 SUMMARY

Although several of the brick coefficients occur in more than one list, each attestation is written uniquely: no coefficient has an exact word-for-word duplicate in another list. Unlike the geometry coefficients (§3), there seems to be no core set of basic brick coefficients: the only one the lists all share is the brick-pile, and even list G contains two rarer coefficients. There is no evidence for a common ordering of the lists.

The table below shows the various sizes of bricks attested in OB mathematics and the coefficients associated with each. Entries in bold type are directly attested coefficients or measurements mentioned in this chapter; the rest are calculated according to the methods below, which summarize the results discussed in the preceding sections.

Table 4.7: Summary of the brickage, layer, and brick wall coefficients

Type	*Size in fingers*	*Volume in* sar_v	*Brickage*	*Layer*	*Brick walls*	*Friberg 1996a*[21]
1	**10 × 10 × 6**	0;00 00 33 20	9;00	21 36	7;30	S2v
1a	12 × 9 × 6	0;00 00 36	8;20	20 00	6;56 40	
2	**15 × 10 × 5**	**0;00 00 41 40**	7;12	14 24	6;00	R1
3	**20 × 10 × 5**	**0;00 00 55 33 20**	5;24	10 48	4;30	H1
4	**18 × 12 × 5**	**0;00 01**	5;00	10 00	4;10	R2
5	15 × 15 × 5	0;00 01 02 30	4;48	9 36	4;00	S3
[6]	20 × 10 × 6	0;00 01 06 40	4;30	10 48	3;45	H1v
7	18 × 18 × 5	0;00 01 30	3;20	6 40	2;46 40	S5
8	**20 × 20 × 5**	**0;00 01 51 06 40**	2;42	5 24	2;15	S6
[8a]	30 × 15 × 5	0;00 02 05	2;24	4 48	2;00	
9	**20 × 20 × 6**	**0;00 02 13 20**	2;15	5 24	1;52 30	S6v
10	24 × 24 × 5	0;00 02 40	1;52 30	3 45	1;33 45	S7
11	**30 × 30 × 5**	**0;00 04 10**	1;12	2 24	1;00	S8
[12]	30 × 30 × 6	0;00 05	1;00	2 24	0;50	U, or S8v

$$\text{Volume of 1 brick (in } sar_v) = \frac{\text{length} \times \text{width} \times \text{thickness (in fingers)}}{18\ 00\ 00\ 00}$$

$$\text{Brickage} = \frac{1}{12\ 00 \times \text{volume of 1 brick}} = \frac{1\ 00}{\text{weight of 1 brick}}$$

$$\text{Brick wall coefficient} = \tfrac{5}{6} \times \text{brickage}$$

$$\text{Floor area of walls} = \tfrac{1}{3} \times \text{floor area of house}$$

[20] Or 4;10 sila per square cubit, for comparison with AO 6770 (iv).

[21] That is, the sigla used by Friberg (1996*a*).

$$1 \text{ layer of bricks} = \frac{1}{\text{surface area of 1 brick (in sar}_a\text{)}}$$

$$= \frac{36\ 00\ 00}{\text{length} \times \text{width (in fingers)}}$$

The most frequently attested bricks are Types 2 and 8, which, along with Types 3 and 4, continued to be used in LB mathematics. There is no direct evidence for Types 6 and 12 (listed in square brackets here), but Powell (1982*a*: 119) inferred their existence because Type 6 is half the size of a Type 9, and Type 12 is the theoretical brick upon which the whole system is based (Powell 1987–90: 491). The existence of Type 8a depends on a doubtful reading of E 38 and the fact that it is half a Type 11 brick. Similarly, the only implications that Types 1a existed are the list F coefficients; unlike Types 6 and 12 it does not share the same proportions of the remaining bricks (below), and for this reason its existence must remain unconfirmed for the time being.

Table 4.8: Summary of the proportions and sizes of OB bricks

Length	$^1/_3$ *cubit*	$^1/_2$ *cubit*	$^3/_5$ *cubit*	$^2/_3$ *cubit*	$^4/_5$ *cubit*	*1 cubit*
Square brick	Type 1	Type 5	Type 7	Type 8 Type 9	Type 10	Type 11 Type 12
Two-thirds-brick	—	Type 2	Type 4	—	—	—
Half-brick	—	—	—	Type 3 Type 6	—	Type 8a

The proportions of the bricks are all based on the square with lengths which are simple fractions of the cubit; the width may be equal to two-thirds or half of the length. The bricks are all 5 fingers — $^5/_6$ cubit — thick, except Types 6, 9 and 12, which are 6 fingers — $^1/_5$ cubit — thick. See §9.4 and §A.6 for a comparison with excavated baked bricks.

Table 4.9: Weight of bricks after processing

Type	*Volume (sar$_v$)*	*Fresh weight (mina)*	*Dried weight (mina)*	*Baked weight (mina)*
1	0;00 00 33 20	10;00	8;00	6;40
1a	0;00 00 36	10;48	8;38 24	7;12
2	**0;00 00 41 40**	**12;30**	**10;00**	**8;20**
3	**0;00 00 55 33 20**	**16;40**	**13;20**	**11;06 40**
4	**0;00 01**	**18;00**	**14;24**	**12;00**
5	0;00 01 02 30	18;45	15;00	12;30
[6]	0;00 01 06 40	20;00	16;00	13;20
7	0;00 01 30	27;00	21;36	18;00
8	**0;00 01 51 06 40**	**33;20**	**26;40**	**22;13 20**
[8a]	0;00 02 05	37;30	30;00	25;00
9	**0;00 02 13 20**	40;00	32;00	26;40
10	0;00 02 40	48;00	38;24	32;00
11	**0;00 04 10**	1 15;00	1 00;00	50;00
[12]	0;00 05	1 30;00	1 12;00	1 00;00

The density of freshly made bricks changes as they dry out, and are then baked:

Weight of 1 fresh brick (in mina) = 12 00 00 × volume of 1 brick
Weight of 1 dried brick (in mina) = 14 24 00 × volume of 1 brick
Weight of 1 baked brick (in mina) = 18 00 00 × volume of 1 brick.

There is a clear distinction between mortared walls and piles of bricks. The problem of the *utūnum* kiln in YBC 7997 does not concern the number of bricks needed to build it, but the quantity it can contain for baking them (*pace MCT*: 99). Perhaps most importantly there is clear evidence from AOT 304 and YBC 9819 that the metrology and mathematics of bricks was well developed by the Ur III period, when the place value system was already established.

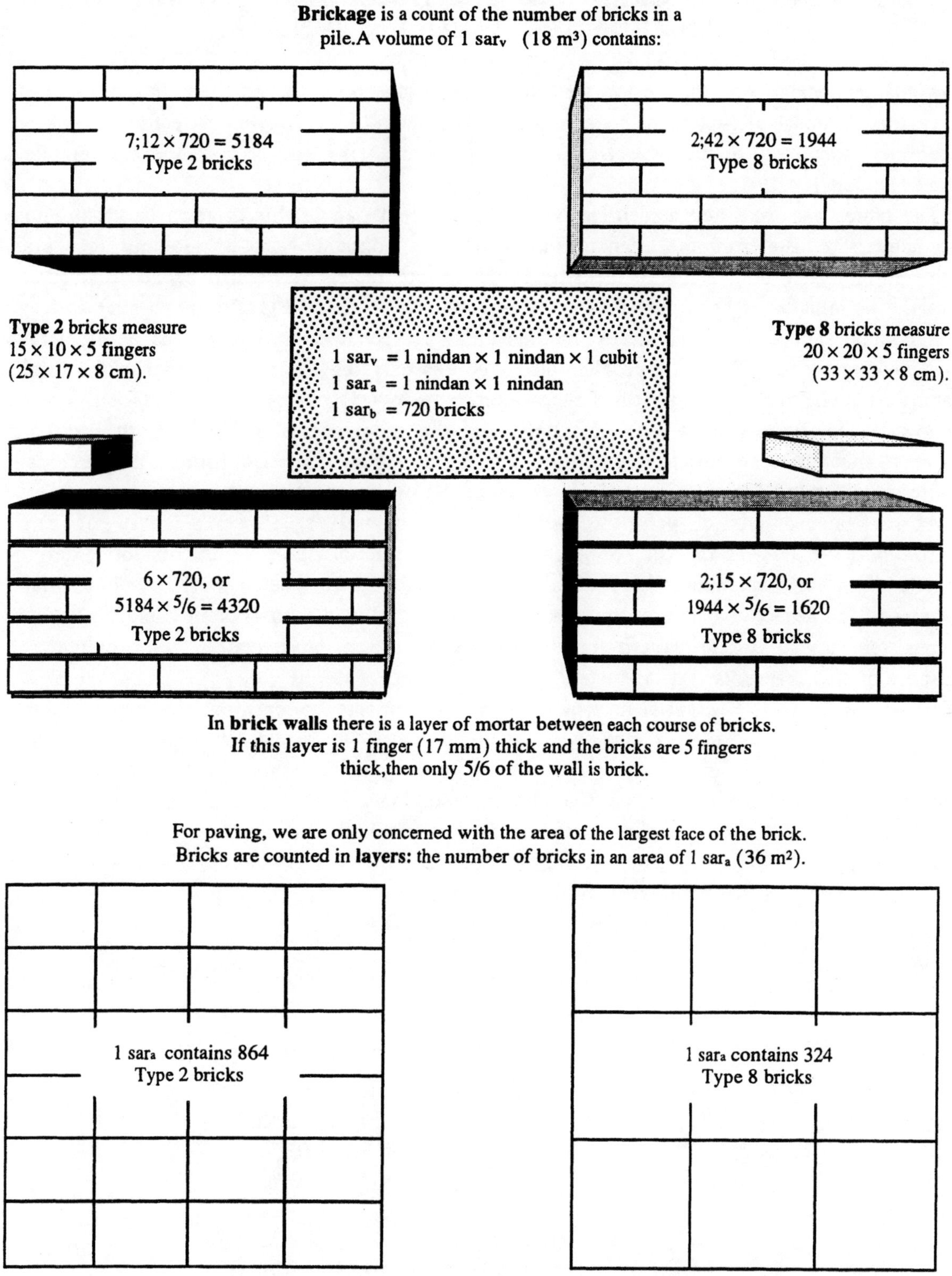

Figure 4.1: Type 2 and Type 8 brickage, brick wall, and layer coefficients

5. MAKING BRICKS, CARRYING BUILDING MATERIALS, AND COMBINING COEFFICIENTS

This chapter tackles some of the less well understood aspects of the mathematics of quantity surveying. Although at first sight the three topics may seem to have little connection, they share a common source: all of the subjects arise from problems (ix) and (x) of the mathematical tablet Haddad 104.

The tablet was excavated at Tell Haddad, ancient Me-Turan in the kingdom of Eshnunna, and appears to date from the reign of Ibal-pi-el II, a contemporary of Hammurabi (mid-eighteenth century BC). Other mathematical and school tablets were also found at this site, although none have yet been published (Al-Rawi and Roaf 1984: 176). Problems (ix) and (x) both concern an otherwise unattested mathematical subject: the work-rates for a three-stage brick-making process. The latter problem also involves carrying the raw material to the brick-making site. The mathematical methods used illuminate several more problem texts and a large number of coefficients from Susa, Eshnunna and the more central regions. Both problems utilize a specific technique for combining coefficients which is also found in previously uninterpreted texts from south central Mesopotamia on other quantity surveying matters.

Apart from editions of pertinent problem texts themselves, Powell's study of brick metrology (1982*a*) was the first work done on this subject. Much of this chapter derives directly from the mathematics of Haddad 104, published two years after Powell's article. Because of the deductive nature of this chapter and the fact that the coefficient lists do not show a consistent organizational pattern for the subject of bricks and quantity surveying, the structure of this chapter is text-orientated, rather than coefficient-led.

Only one of the OB coefficient lists contains coefficients concerning the brick-making process, but all except text H include some on carrying materials: in total 29 different coefficients, attested 49 times. Haddad 104 uses two carrying coefficients, and three for brick-making, in a total of eight occurrences. Only two carrying coefficients are stated explicitly in two other problem texts,[1] but there are eight further implicit attestations of another.[2] Two further coefficients in the lists (three attestations) concern quantity surveying operations which are used implicitly in two 'combined coefficient' problem texts.[3]

5.1 MAKING BRICKS

5.1.1 The coefficients

D 38 20 ⸢*šà*⸣ [...]	0;20, of [...].
D 39 20 *šà* ⸢SAḪAR⸣ [*ḫa-ba*]-*ti*	0;20, of [digging] earth.
D 40 10 *šà* SAḪAR *ba-la-lu*	0;10, of mixing earth.
D 41 25 *šà* SIG_4 *šà-ha-a-ṭì*	0;25, of *moulding* bricks.

[1] IM 54538 — carrying bricks; BM 85196 (xiii) — carrying sheaves of corn.

[2] AO 8862 (v)–(viii); YBC 4673 (ii)–(iv); YBC 10722 — all carrying Type 2 bricks.

[3] YBC 4669 (B x) — demolishing wall, carrying rubble; YBC 4673 (v) — carrying earth, making brick-pile.

Haddad 104 (ix): IV 16–17	20 *al-li ḫa-ba-tim*	0;20, the daily rate of digging.
Haddad 104 (x): IV 29	20 *ḫa-ba-tam*	0;20, digging.
Haddad 104 (ix): IV 17	20 *al-li la-ba-nim*	0;20, the daily rate of moulding.
Haddad 104 (x): IV 29	20 *la-ba-nam*	0;20, moulding.
Haddad 104 (ix): IV 17	10 *al-li ba-la-lim*	0;10, the daily rate of mixing.
Haddad 104 (x): IV 29, 33	10 *ba-la-lam/-lim*	0;10, mixing.

5.1.2 Mathematical commentary

Haddad 104 (ix) is the simplest of the two extant OB problems on making bricks. Some of the technical terms are discussed in §5.1.3 and §5.2.3, and problem (x) is presented in §5.2.2.

IV *na-al-ba-nu-um ne-eq-re-bu-um al-li la-ba-nim mi-nu-um*
ù iš-ka-ar 1 LÚ *mi-nu-um i-na e-pe-ši-ka* 20 *al-li*
ḫa-ba-tim 20 *al-li la-ba-nim* 10 *al-li ba-la-lim lu-pu-ut*
pa-ni 20 *pu-ṭur-ma* 3 *i-lí pa-ni* 20 *pu-ṭur-ma* 3 *i-lí*
pa-ni 10 *pu-ṭur-ma* 6 *i-lí ku-mu-ur-šu-nu-‹ti›-ma* 12 *i-lí*
pa-ni 12 *pu-ṭur-ma* 5 *i-lí* 5 *al-li ne-eq-re-bi-im*
tu-úr 3 20 *mi-it-ḫa-ar-ti* SIG₄ *šu-ta-ki-il-ma*
11 06 40 A.ŠÀ SIG₄ *i-lí tu-úr* 1 *mu-bé-e* SIG₄ *šu-pi-il-ma*
12 *i-lí* 12 *a-na* 11 06 40 *i-ši-ma* 2 13 20 SAḪAR SIG₄ *i-lí*
pa-ni 2 13 20 *pu-ṭur-ma* 27 *i-lí* 27 *a-na* 5 *al-lim*
i-ši-ma 2 15 *šu-ši* SIG₄ *iš-ka-ar* 1 LÚ *i-lí* 2 15
šu-ul-li-iš 6 45 *šu-ši iš-ka-ar al-lim i-lí*

Brick-making; combined coefficient.[4] What is the daily rate of making bricks and what is the output of one man? In your working, put down 0;20, the daily rate of digging, 0;20, the daily rate of brick-making, 0;10, the daily rate of mixing. Take the reciprocal of 0;20 and 3 will come up. Take the reciprocal of 0;20 and 3 will come up. Take the reciprocal of 0;10 and 6 will come up. Add them and 12 will come up. Take the reciprocal of 12 and 0;05 will come up. 0;05 is the daily rate of the combined coefficient. Return. Square 0;03 20, the square side of a brick, and 0;00 11 06 40, the area of the brick, will come up. Return. Convert 0;01, the thickness of the brick, and 0;12 will come up. Multiply 0;12 by 0;00 11 06 40 and 0;00 02 13 20, the volume of the brick, will come up. Take the reciprocal of 0;00 02 13 20 and 27 00 will come up. Multiply 27 00 by 0;05,the daily rate, and 2;15 sixties of bricks, the output of one man, will come up. Triple 2;15. 6;45 sixties, the output of the daily rate, will come up.

Lines 16–17 give the following daily work-rates:

Table 5.1: Work-rates for making bricks attested in Haddad 104 (ix)

Name	*Meaning*	*Earth processed per day*
ḫabātum	digging	$^1/_3$ sar$_v$ (approx. 6 m^3)
balālum	mixing	$^1/_3$ sar$_v$ (approx. 3 m^3)
labānum	moulding bricks	$^1/_6$ sar$_v$ (approx. 6 m^3)

So to process completely 1 sar$_v$ of earth it would take 3 + 3 + 6 = 12 days (line 19). Therefore $^1/_{12}$ sar$_v$ of earth (1.5 m^3) can be processed completely in one day (line 20). This combination of the three coefficients — a third of the harmonic mean — is called the *alli neqrebim* (line 20) or 'daily rate of the combined coefficient' (see §5.1.3):

[4] See §5.1.3 for an explanation of this translation.

$$\text{daily rate of the combined coefficient} = \frac{1}{\frac{1}{\text{digging}} + \frac{1}{\text{mixing}} + \frac{1}{\text{moulding}}}$$

To convert this daily volume into the number of bricks made in one day, the volume of one brick must be calculated. We are told that the square side of the brick is 0;03 20 nindan, or 20 fingers, and that the thickness is 0;01 nindan or 6 fingers:[5] a Type 9 brick according to Powell's typology (1982*a*: 119; see §4.1), and otherwise unattested in Old Babylonian mathematics.[6] Once the volume of one brick is found (ll. 21–3),[7] the number of bricks made is calculated by dividing the volume of earth processed in one day by the volume of a brick (ll. 24–5):

$$\frac{0;05}{0;00\ 02\ 13\ 20} = 0;05 \times 27\ 00 = 2\ 15, \text{ or } 135.^{8}$$

This is the work-load of one man for one day, but as there are three different tasks there seem to be three men in a brick-making team, with a total output of 3 × 2 15 = 6 45 (405) bricks a day (line 26).[9] This result is called the 'output of the daily rate', *iškar allim* (see §5.2.3).

5.1.3 Philological commentary

Ḫabātum, labānum and *balālum* must all be stages in the brick-making process: the latter terms are already well attested as meaning 'making bricks' (line 15), and more specifically, 'filling a brick mould' (line 17) (*AHw*: 522, *labānum* II; *CAD*: L 8, *labānum* A) and 'mixing' (*AHw*: 92; *CAD*: B 39) respectively. *Ḫabātum*, however, is rather more problematic. In Haddad 104 it is the first task mentioned, but does not necessarily denote the first stage of the processing: *labānum* (moulding the bricks) is listed before *balālum* (mixing the clay), but clearly must have been done after the mixing. In list D, however, *ḫabātum* is the first of the three pertinent coefficients. In short, the role of *ḫabātum* within brick-making cannot be deduced from its relative position within the mathematical texts.

In a detailed study of the verb *ḫabātum*, Kraus (1975: 40) concluded that:

> Meine Darlegungen zusammenfassend schlage ich vor, die von unseren Wörter-büchern angenommen Verben *ḫabātu* A–E bzw. *ḫabātu* I–IV und *ḫapātu* zu ersetzen durch die zwei Verben
> (I) *ḫabātum* (*a*/*u*), (I) 'rauben'; (II) 'leihen'; (III) 'sich begeben';
> (II) *ḫabātum* (*i*/*i*), etwa 'gewichtig/übermächtig sein/werden'.

Civil (1994: 95 & n. 123) adds a further meaning to this list, namely 'to thresh'. But none of these translations would be appropriate in the context under discussion.

In their edition of Haddad 104 Al-Rawi and Roaf (1984: 211) tentatively suggested that *ḫabātum* meant 'digging' here, presumably on the evidence both of the context and of *AHw* (304b) *ḫabātu* IV 'abraümen', in which a possible Aramaic cognate *'bṭ* 'aufgraben?' is

[5] *Contra* Nemet-Nejat (1993: 31), who understands this quanitity to mean 1 cubit thickness.

[6] Powell's suggestion (1982*a*: 120) that the obverse of VAT 6598 (i.e. BM 96957 + VAT 6598 (xii)–(xviii); §A.4) refers to Type 9 bricks has now been disproved with the discovery of the adjoining fragment BM 96957 (see §4.5).

[7] To obtain the volume the thickness must be 'made deep', or converted from nindan into cubits. See §7.1.3.

[8] The gloss 'sixties' is put after the number to indicate the absolute value of the number, and does not mean 135 × 60. It is not a common practice in Old Babylonian maths, and can be rather confusing, but it is used consistently throughout this text.

[9] For further mathematical attestations of the verb *šullušum* 'to triple' see the circle coefficients in BM 85194 (iv) and Haddad 104 (i)–(v) (§3.1) and the grain-pile problems BM 96954 (xiii) and (xiv) (§A.3).

mentioned. However, von Soden cites only two examples of this verb, one of which Kraus (1975: 33) shows to be *a'-bu-ut* 'I destroyed', and the second of which has first radical *q*, not *ḫ*. Yet an Akkadian meaning 'to dig', based on this cognate — bearing in mind that its sole mathematical attestations are from the peripheral regions of Susa and north Eshnunna — would have numerical support from several OB mathematical problems on digging out canals. YBC 4666 (i)–(xiii), YBC 7164 (ii)–(vii) and BM 85196 (xvi) all list a daily rate of 0;20 sar_v, when digging to a depth of 1 cubit.[10] It is only when the digging gets deeper that the daily rate decreases.

The word *neqrebum,* which Al-Rawi and Roaf translate 'start of building work' (1984: 211), appears to be otherwise unattested. The non-mathematical meaning of *qerēbum*, 'to come or be close' (*AHw*: 915; *CAD*: Q 236), and the mathematical context in which the word is used, together provide a better sense for *neqrebum,* a *mapras* nominal form of the verb, as 'combined (coefficient)'. For the meaning of *allum* see §5.2.3.

The reading and restoration of D 39–41 are new. The last three signs of line 40 had been interpreted as *ka-ru* 'quai' (*TMS*: 26, 29), but a close examination of the photograph (*TMS*: pl. IV) shows the final sign to be much more like the LUs of lines 55 and 60 than the RUs of lines 13–15: the middle horizontal does not in fact extend in front of the first vertical, and in both lines 55 and 60 the lower horizontal extends beyond the final vertical. The spacing of the two preceding signs is rather longer than the KAs of lines 68 and 70; indeed the first central horizontal of the LA may just be made out next to the vertical of the BA. One would of course expect SAḪAR *ba-la-li* instead of *ba-la-lu.*[11] Although this is the only such grammatical mistake on the tablet, this sort of nominative-for-genitive case error is frequent throughout the coefficient lists and OB mathematics in general.[12] The traces of the sign SAḪAR are also visible on line 39 in the photo.

The last signs of D 41 were previously read *šà ḫa-a-ti* (*TMS*: 26; trans. 'de la brique concassée' p 29; Nemet-Nejat 1993: 190). However, the comment as a whole is of the same format as D 40 and 42, 'of mixing earth' and 'of carrying earth'. The coefficient must refer to a daily rate of 0;25 sar_v for a stage in the brick manufacturing process, as do all those from D 36 to D 42. There are two possible interpretations, both of which are problematic. 'To glaze bricks' (*CAD*: Š I 84) seems the most appropriate of the *established* meanings of *šaḫātum* and *šaḫāṭum* in this context, although it is not otherwise documented in OB. If we allow it, on the grounds that list D is very late OB (§2.1.3), D 38 can be restored as 20 ⸢*šà*⸣ [SIG₄ *la-ba-ni*] '0;20, of [moulding bricks]'. The other possibility, following Postgate (1974: 53), is to understand *šaḫāṭum* as 'moulding bricks' — the Assyrian synonym for *labānum*, which is not attested in that dialect. But its numerical value would be different from the moulding coefficient in Haddad 104, and would result in an irregular combined coefficient (the reciprocal of 11;24). The three brick-making coefficients would be in logical sequence, but it is not clear how D 38 ought to be restored in that case. So we are left with an unsatisfactory choice between the technologically implausible glazing and the numerically awkward moulding and must wait on further evidence to make a firm decision.

[10] See §6.3, and coefficient E 21.

[11] The reading SAḪAR *ka-ru* would be grammatically just as incorrect; *iš-ka-ru* is excluded because of the spacing of the signs and the preferred writing ÉŠ.GÀR in this text (cf. line 46). Note the alternation between mimated and non-mimated endings throughout the list.

[12] e.g. the coefficients G 15 and H 6 (see in general §2.1); BM 15285 *passim* (§A.2).

5.2 CARRYING EARTH

5.2.1 The coefficients

D 37	2 13 20 [IGI.GUB *šà tu-up-ši-ki-im*]	0;00 02 13 20, [the coefficient of a basket].
F 45	2 13 ⸢20 *ù* 1⸣ 15 *ša* GIŠ «*tup*» DUSU	0;00 02 13 20, and 1 15, of a basket.
G 3	2 13 20 IGI.GUB GIŠ.DUSU *šu-up-ši-ki-im*	0;00 02 13 20, the coefficient of a basket.
A 28	45 IGI.GUB A.RÁ	45 00, the coefficient of going.
D 36	45 IGI.GUB *šà* ⸢*na-az-ba-li*⸣-*im*	45 00, the coefficient of carriage.
E 46	45 *ta-la-ak-tum*	45 00, going.
F 39	45 *mu-ut-ta-li-ik-tum*	45 00, going.
G 16	45 IGI.GUB *mu-ta-li-ik-tim*	45 00, the coefficient of going.
C 23	*na-az-ba-al* SAḪAR 1 ⸢40⸣ *i-gi-gu-bu*	Carriage of earth: 1;40 is the coefficient.
F 8	1 40 *na-az-ba-al* SAḪAR	1;40, carriage of earth.
G 4	1 40 IGI.GUB SAḪAR *na-az-ba-al* SAḪAR	1;40 the coefficient of carriage of earth.
C 25	*a-na* 40 NINDAN *a-za-bi-il* 2 13 20 *al-lu-um*	I carry for 40 nindan: 0;02 13 20 is the daily rate.
C 26	*a-na* 20 NINDAN *a-za-bi-il* 4 *al-lu-um*	I carry for 20 nindan: 0;04 is the daily rate.
C 27	*a-na* 15 NINDAN [*a-za-bi-il* 5] *al-lu-um*	[I carry] for 15 nindan: [0;05] is the daily rate.
C 28	[*a-na*] ⸢10⸣ NINDAN *a*-⸢*za*⸣-[*bi-il*] 6 ⸢40⸣ *al-lu-um*	I carry [for] 10 nindan: 0;06 40 is the daily rate.
C 29	[*a-na* 5] NINDAN *a-za*-[*bi-il*] 10 *al-lu-um*	I carry [for 5] nindan: 0;10 is the daily rate.
D 42	20 *šà* SAḪAR *za-ba-li*	0;20, of carrying earth.
D 46	10 *šà* SAḪAR *za-ba-li* ÉŠ.GÀR LÚ.1	0;10, of carrying earth; 1 man's work-load.
D 47	5 *šà* SAḪAR	0;05, of earth.
E 32	6 40 *na-az-ba-al* SAḪAR	6 40, carrying earth.

Haddad 104 (x): IV 31	2 13 20 *tu-up-ši-ki*	0;00 02 13 20, a basket.
BM 85196 (xiii): IV 5	[45]	[45 00.]
Haddad 104 (x): IV 30	45 *mu-ta-lik-tim*	45 00, the going.

5.2.2 Mathematical commentary

Haddad 104 (x) adds an additional stage to the previous problem: the earth must be carried to the brick-making site before it can be processed.

IV 27 *al-li la-ba-nim* [*iš*]-*tu* 5 *ṣú-up-pa-am* ‹*a*›-*za-bi-lam-ma*
a-la-bi-in iš-ka-ar 1 LÚ *mi-nu-um i-na e-pé-ši-ka*
20 *ḫa-ba-tam* 20 *la-ba-nam* 10 *ba-la-lam lu-pu-ut tu-úr-ma*
[*pa*]-*ni* 5 *mé-se-tim pu-ṭur-ma* 12 *i-lí* 12 *a*-[*na* 4]5 *mu-ta-lik-tim*
i-ši-ma 9 *i-lí* 9 *a-na* 2 13 20 *tu-up-ši-ki i-ši-ma*
20 SAḪAR *iš-tu* 5 *ṣu-up-pa-am ta-az-bi-lam* 20 SAḪAR
a-na i-di 10 *ba-la-lim lu-pu-ut tu-úr pa-ni* 20 *pu-ṭur-ma*
3 *i-lí pa-ni* 20 *pu-ṭur-ma* 3 *i-lí pa-ni* 10 *pu-ṭur-ma*
«*pu-ṭur-ma*» 6 *i-lí pa-ni* 20 SAḪAR *pu-ṭur-ma* 3 *i-lí*
ku-mu-ur-ma 15 *i-lí pa-ni* 15 *pu-ṭur-ma* 4 *i-lí*
4 *a-na* 1 u_4-*mi-im i-ši-ma* 4 *al-lum i-lí*
tu-úr 3 20 *mi-it-ḫa-ar-ti* SIG_4 *šu-ta-ki-il-ma*
11 06 40 A.ŠÀ SIG_4 *i-lí* 11 06 40 *a-na* 12 *mu-bé-e* SIG_4

i-ši-ma 2 13 20 SAḪAR SIG$_4$ *i-lí pa-ni* 2 13 20 SAḪAR SIG$_4$
pu-ṭur-ma 27 *i-lí a-na* 4 *al-lim i-ši-ma*
1 48 SIG$_4$ *iš-ka-ar* 1 LÚ *i-lí* 1 48 *šu-li-iš-ma*
5 24 *iš-ka-ar al-lim i-lí ki-a-am*
né-pe-šu-um

Daily rate of brick-making. Over 5 (nindan), a *ṣuppam*, I carry here and I make bricks. What is the output of one man? In your working, put down 0;20, the digging, 0;20, the moulding, 0;10 the mixing. Return and take the reciprocal of 5, the distance, and 0;12 will come up. Multiply 0;12 by 45 00, the going, and 9 00 will come up. Multiply 9 00 by 0;00 02 13 20, the basket, and 0;20 is the volume you carried here over 5 (nindan), a *ṣuppam*. Put down 0;20, the volume, by the side of 0;10, the mixing. Return. Take the reciprocal of 0;20 and 3 will come up. Take the reciprocal of 0;20 and 3 will come up. Take the reciprocal of 0;10 and 6 will come up. Take the reciprocal of 0;20, the volume, and 3 will come up. Add them and 15 will come up. Take the reciprocal of 15 and 0;04 will come up. Multiply 0;04 by 1 day and 0;04, the daily rate, will come up. Return. Square 0;03 20, the square side of a brick, and 0;00 11 06 40, the area of the brick, will come up. Multiply 0;00 11 06 40 by 0;12 (cubits), the thickness of a brick, and 0;00 02 13 20, the volume of a brick, will come up. Take the reciprocal of 0;00 02 13 20, the volume of a brick, and 27 00 will come up. Multiply by 0;04, the daily rate, and 1 48 bricks, the output of one man, will come up. Triple 1 48 and 5 24, the output of the daily rate comes up. The working.

This is essentially the same problem as the earlier one, with the added complication that the earth must be carried 5 nindan before it is processed. To calculate the daily rate of carrying, two coefficients are used: *muttalliktum*, the 'going', and *tupšikkum*, the 'basket'. Von Soden (1977: 107; *AHw*: 137) has argued that the 'basket' represents a daily carrying rate of 0;02 13 20, or $_1/_{27}$, volume-sar (i.e. approx. $^2/_3$ m^3, a volume equal to 72 Type 8 bricks), carried over 10 nindan. This volume is also exactly 60 Type 9 bricks, as this text shows. However, the calculations in lines IV 30–2 of Haddad 104 show that 'basket' cannot be a daily rate. These calculations can be summarized as:

$$\frac{\text{going} \times \text{basket}}{\text{distance}} = \text{daily rate of carrying.}$$

The units of measurement of the daily rate, 20 SAḪAR (line 32), are sar$_v$/day. These units are the same for the three other *allum* coefficients too. Therefore, if the units of measurement for the *tupšikkum* were also sar$_v$/day, then the *muttalliktum* would have to be measured in nindan only and it is difficult to imagine what this distance would represent.

If, however, the 'going' were a daily rate (nindan/day), then the *tupšikkum* coefficient would be the size of the basket in which the earth was carried:

$$\frac{\text{going (nindan/day)} \times \text{basket (sar}_v\text{)}}{\text{distance (nindan)}} = \text{daily rate (sar}_v\text{/day)}$$

Thus we have a daily distance of 45 00 nindan, around 16 km, over which earth could be carried in a day, i.e., a 32 km or 20 mile round trip. This daily distance is divided by the actual distance of each trip (ll. 30–1) to give the number of trips made: 9 00, or 540. The number of trips is then multiplied by the 'basket' to give the total volume of earth to be carried in one day (ll. 31–2). A volume of 0;02 13 20 sar for the basket would be around $^2/_3$ m^3 of earth, arguably too much for one man to carry alone. A power of sixty smaller is equivalent to $^1/_{90}$ m^3, or a cube of about 22.5 cm each side, a more reasonable size (and coincidentally exactly the volume of a Type 9 brick, those mentioned in this text). This smaller volume gives:

$$\frac{45\ 00 \times 0;00\ 02\ 13\ 20}{5} = 0;20.$$

That is, one man can carry $^{1}/_{3}$ sar_{v} of earth over 30 m in a working day.

The product of the 'going' and the 'basket' is 1;40, or $1^{2}/_{3}$. This coefficient occurs in three lists and is shown in §5.2.1 just below the 'going' and 'basket' coefficients. It is similar to the *nazbalum* coefficients for bricks (see §5.4):

$$\frac{\text{brickage coefficient}}{\text{distance of each trip}} = \text{number of bricks carried in a day;}$$

$$\frac{\text{earthcarrying coefficient}}{\text{distance of each trip}} = \text{volume of earth carried in a day.}$$

In their discussion of Haddad 104 (x), Al-Rawi and Roaf (1984: 210) noted that a section of the coefficient list C also contained daily rates for carrying. But the numbers given there did not correspond to the result and method of the present problem, which would give the following figures:

Table 5.2: Carrying coefficients in List C

Length of trip (nindan)	*Daily rate of carrying (sar_v/day)*	*List C value*
100	0;01 ($^{1}/_{60}$)	
60	0;01 40 ($^{1}/_{36}$)	
45	0;02 13 20 ($^{1}/_{27}$)	
40	0;02 30 ($^{1}/_{24}$)	0;02 13 20
30	0;03 20 ($^{1}/_{18}$)	
25	0;04 ($^{1}/_{15}$)	
20	0;05 ($^{1}/_{12}$)	0;04
15	0;06 40 ($^{1}/_{9}$)	[0;05]
10	0;10 ($^{1}/_{6}$)	0;06 40
5	0;20 ($^{1}/_{3}$)	0;10
1	1;40 ($1^{2}/_{3}$)	

In all cases list C gives an *allum* value which corresponds to a distance 5 nindan longer than that given in the list. It seems that this section of the text was miscopied from some original table, in the process of which the second column of figures was consistently transferred down one line.[13]

The exact significance of the earth coefficients D 42, 46 and 47 is not clear, but they may well refer to the daily rates of carrying over 5 nindan, 10 nindan and 20 respectively.[14]

BM 85196 (xiii) is a particularly interesting problem about carrying sheaves of grain.

[several lines missing from previous column] ...
IV *i-na* 5 NINDAN UŠ BA.[ZI ...]
1 LÚ *ku-ru-li-šu* [*i-na* U_4 EN.NAM]
i-za-bi-lam 30 UŠ *ù* [5 UL.GAR 30 05 *ta-mar*]
$^{1}/_{2}$ 30 05 *ḫe-pé* 15 02 30 [*ta-mar* 15 02 30 *a-na*]
6 NUMUN $^{1}/_{2}$ ⌜DANNA⌝ *i-ši* 1 30 15 *ta-*[*mar* IGI 45 DU_8.A]
1 20 *ta-mar a-na* 1 30 15 *i-ši* 2 00 20 ⌜*ta*⌝-[*mar*]
i-na ITI.4.KAM U_4.⌜$^{1}/_{3}$.KAM⌝ *iz-bi-*[*lam*]
ki-a-am ne-pé-[*šum*]

... *Subtract* from 5 nindan length [...]. [In how many days] does 1 man carry his sheaves here? [Add] 30 00,

[13] Cf. the ten times multiplication table on YBC 7303, in which 'from 6 to 9, the results are one line too low' (*MCT*: 28), and the eighteen times table on IM 55111, where the results are one line too high (Bruins 1954).

[14] YBC 4673 (v), which also involves carrying earth, is discussed in §5.6.

the length, and [5. You will see 30 05.] Break 30 05 in half. [You will see] 15 02;30. Multiply [15 02;30 by] 6 00, the ratio of $^1/_2$ danna. You will see 1 30 15 00. [Take the reciprocal of 45 00.] You will see 0;01 20. Multiply by 1 30 15. [You will see] 2 00;20. He carried here in 4 months and $^1/_3$ day. This is the method.

Friberg (1987–90: 578) ascertained that this solution involved an arithmetic progression, but did not explain it fully. Three hundred and sixty sheaves are arranged in a line, each 5 nindan apart. The furthest away is thus 30 00 nindan or 1 danna (11 km) away. The carrier has to transport them, one at a time, back to the same place. The problem is to work out how long one man will take to do it.

In lines 3–5 the sum of the distances is calculated, using the correct formula for an arithmetic progression (Friberg 1987–90: 578):

$$5\sum_{r=1}^{360} r = 5\left(\frac{360}{2}\times 361\right)$$
$$= \frac{6\ 00 \times 30\ 05}{2} = 1\ 30\ 15\ 00,$$

i.e. 324,900 nindan or just under 2000 km. To find the total time taken, this total distance is divided by the daily distance walked, or *muttalliktum*, 45 00 nindan (line 6):

$$\frac{1\ 30\ 15\ 00}{45\ 00} = 2\ 00;20,$$

i.e. the sheaves are moved in 120 days, or 4 months, and $^1/_3$ of a day.

This problem confirms that the 'daily distance' is indeed only the length over which the goods are carried, and that the worker actually covers twice this distance, namely around 32 km, or 20 miles, as he makes the round trips.

5.2.3 Philological commentary

In non-mathematical contexts, *muttalliktum* means 'road' or 'way' (*AHw*: 689, *muttalliktum* (2)), the *muptarrist*, feminine participle of the Gtn-stem of *alākum* 'to go' (*GAG*: 66, §56g). The translation 'going' is derived from the *taprāst* form *tallaktum* in E 46. This word is almost certainly synonymous with *muttalliktum*, as it is in non-mathematical contexts (*AHw*: 1310–11): it occurs in a section of the list entirely concerned with bricks and carrying operations, and shares the same numerical value and verbal root. Likewise A.RÁ (A 28) is a common logogram for another derivative of *alākum, alaktum,* which has a similar meaning (*AHw*: 31). It is the iterative element of the Gtn-stem which invokes the repeated nature of the individual journeys made over a fixed daily distance. Kilmer's interpretation of A 28 (1960: 290 n. 4), however, depends on the fact that the volumes of the excavations in YBC 4663 are all 0;45 sar_V (see §6.3). Obviously this is not true for all excavations and thus the volume in question cannot be a coefficient.

Goetze (1951: 142) correctly deduced that:

> the numbers in front of the *allum* [in list C] refer to the rate of loading or digging necessary to employ the same number of workers. The smaller the distance grows over which the material in question is to be carried, the faster the loading must become unless the workers go idle and waste their time.

In mathematical contexts, *allum* is found solely in list C and Haddad 104 (ix)–(x), both tablets from the Eshnunna region. Table 5.3 shows the variant terminology found in parallel passages of the two brick-making problems. It is clear from this table that *allum* may either be used in conjunction with another word referring to a coefficient, or omitted, or used on its own. In each case the phrase refers to a volume of earth to be processed in a day. This is made explicit by problem (x): IV 37 in which the *allum* is multiplied by the time over which the work is carried out. In this case it is only one day, but could equally well be several days or a fraction of a day. The result gives the output of earth for that period of work.

Table 5.3: Brick-making terminology in Haddad 104

Line	*Haddad 104 (ix)*	*Line*	*Haddad 104 (x)*
IV 15	*nalbanum*	IV 27	*alli labānim*
IV 15	*alli labānim*		—
IV 16–17	*alli ḫabātim*	IV 29	*ḫabātam*
IV 17	*alli labānim*	IV 29	*labānam*
IV 17	*alli balālim*	IV 29	*balālam*
IV 20	*alli neqrebim*	IV 37	*allum*
IV 24	*allim*	IV 41	*allim*
IV 26	*iškar allim*	IV 43	*iškar allim*

The volume of earth to be processed in a day is elsewhere called the *iškarum* or ÉŠ.GÀR,[15] but here it refers to the output in terms of the number of bricks produced (IV 25, 42). Problem (ix) asks for both the *iškarum* and *allum* to be found (IV 15–16), confirming that they are indeed different in this text.[16] To differentiate them both from the standard mathematical usage of *iškarum* '(daily) work-load', in texts from Eshnunna *allum* is translated here by 'daily rate' and *iškarum* by 'output'. Goetze (1951: 142) proposed that *allum* was a loanword from the Sumerian **al**, 'hoe' (cf. *AHw*: 37). Although the derivation looks plausible, there is no firm evidence for it.

In C 26 Goetze (1951: 141) reads 5, but his copy shows only one lower vertical on the numeral. Reading it here as 4 eliminates the problem of supplying the missing number on the line below (cf. Goetze 1951: 141 n. 20). See §5.2.2.

D 36 may possibly be the carriage coefficient for Type 2 bricks (see §5.3). However, no other brick carriage coefficients are attested in text D, and Type 4 is the sole brick size mentioned (§4.1). As it immediately precedes a coefficient of value 2 13 20 — probably a 'basket' coefficient — and a series on making bricks (D 37–40) it is most likely that the carriage of earth is meant.

It appears that in F 45, the scribe began to write out *tupšikkim* syllabically but, running out of space, chose to write the logogram DUSU instead, without erasing the sign DUB. For the meaning of 1 15 in this coefficient, see §5.4.

5.3 CARRYING BRICKS

5.3.1 The coefficients

A 29 4 30 IGI.GUB SIG_4 ÍL.ÍL — 4 30 00, the coefficient of carrying bricks.

[15] e.g. YBC 4673 (x): III 12 — building an earth wall (§6.1); Erm 15073 (iv): IV 7′ — digging a canal; see §§6.2–3 for many further examples.

[16] Compare IM 53961 and IM 54011, also from Eshnunna, with YBC 4673 (ix), which all concern building an earth wall (§6.1). The volume which can be built in a day is called the ÉŠ.GÀR in YBC 4673 (e.g. III 12), but *igigubbûm* in IM 53961: 9 and IM 54011: 11. The resulting length of wall built — which depends on its width and height — is referred to as UŠ, 'length', in YBC 4673: III 14, but *iškarum* in the other two texts (ll. 11, 21 respectively); cf. §6.1.

E 41	4 30 *na*-⌜*áz*⌝-[*ba*]-⌜*lu*⌝-*um*	4 30 00, *carriage*.
F 2	4 30 *na-az-ba-al* SIG$_4$	4 30 00, carriage of bricks.
G 5	4 30 IGI.GUB SIG$_4$ *na-az-ba-al* SIG$_4$	4 30 00, the coefficient of bricks; carriage of bricks.
A 30	3 22 30 IGI.GUB SIG$_4$ IN.⌜X⌝	3 22 30, the coefficient of ... bricks.
F 5	3 22 30 *na-az-ba-al-ša*	3 22 30, its carriage.
E 47	3 *ša* $^1/_2$ KÙŠ ‹SIG$_4$›	3 00 00, of $^1/_2$-cubit ‹bricks›.
G 15	3 IGI.GUB *na*-[*az*]-*ba-al ar-ḫu-um*	3 00 00, the coefficient of carriage of half-bricks.
A 31	1 41 15 IGI.GUB SIG$_4$ ÍB.⌜SI$_8$⌝	1 41 15, the coefficient of square bricks.
E 45	1 41 15 *ša* [SIG$_4$] ÍB.‹SI$_8$›	1 41 15, of square [bricks].
F 7	1 41 15 *na-az-ba-al-ša*	1 41 15, its carriage.
G 14	[1] 30 IGI.GUB *na-az-ba-al a-gu-ri-im*	[1] 30 00, carriage of baked bricks.
C 22	*na-az-ba-al* SIG$_4$ 45 *i-gi-gu-bu*	Carriage of bricks: 45 00 is the coefficient.
F 10	45 *na-az-ba-al-ša*	45 00, its carriage.
F 12	4 23 36 *na-az-ba-al-ša*	4 23 36, its carriage.
F 15	7 38 *na-az-ba-al-ša*	7 38 00, its carriage.
IM 54538:6	1 30 [*i*]-*gi-gu-ba*$^!$-*am*	1 30, the coefficient.
7	1 30 *i-gi-gu-bi-ka*	1 30, your coefficient.

5.3.2 Mathematical commentary

Several OB mathematical problems concern carrying a load of 540 bricks over 30 nindan in a day, although coefficients are never explicitly stated in them. AO 8862 (v)–(vii) and YBC 10722 use carrying bricks as a pretext for the investigation of 'remarkable numbers' (cf. Høyrup 1993*b*), while YBC 4673 (ii)–(iv) involve more straightforward calculations on payment and labour. The aim of YBC 4673 (i) is to calculate the volume of a Type 2 brick (§4.0), so it is reasonable to assume that the following three problems also concern the same brick sizes.

If the daily load of Type 2 bricks over 30 nindan is 540 bricks, then the load over 1 nindan will be 30 times greater: 16,200, or 4 30 00, bricks. This is the meaning of the *nazbalum* or 'carriage' coefficient in A 29, E 41, F 2, and G 5 (Powell 1982*a*: 121). It may equally be thought of as the number of bricks a single worker can carry over 30 nindan in a month of 30 days;[17] we may call it the 'monthly carriage', to distinguish it from the daily carriage mentioned in the brick problems. Unlike the *nalbanum* 'brickage' coefficient, the bricks are counted individually, not in sar$_v$.

The total volume of 4 30 00 Type 2 bricks carried is:

$$4\ 30\ 00 \times 0;00\ 00\ 41\ 40 = 3;07\ 30\ \text{sar}_v$$

for the monthly carriage: the 'monthly volume'. If this is true for all sizes of brick, then the 'carriage' for the other brick types can be calculated, by dividing these volumes by the volume

[17] There is no practical difference between these numbers: carrying over 30 nindan instead of 1 means that one can carry $^1/_{30}$ of the original amount, but over 30 days one can carry 30 times as much again. It is important to distinguish between them only for the purpose of determining which was the theoretical basis for the carrying calculations.

of a single brick. They are listed in Table 5.4 below. The table shows that 'carriage' is proportional to the 'brickage' and unique to each type of brick:

$$\begin{aligned}\text{carriage} &= 37\ 30 \times \text{brickage}\\ &= \text{monthly volume} \times 12\ 00 \times \text{brickage},\end{aligned}$$

where 12 00 is the number of bricks in a sar_b.

The number of bricks carried in a day over a certain distance (the 'daily load') may be calculated from the monthly carriage, as the latter is identical to the daily carriage over 1 nindan (see above):

$$\text{daily load} = \frac{\text{monthly carriage}}{\text{length of trip in nindan}}.$$

Because it is easier to calculate with, and more universally applicable, the monthly carriage is listed in the coefficient lists, whereas the daily carriage is not (but see §5.4).

Table 5.4: Monthly carriage of the different bricks

Type	*Volume* (sar_v)	*Brickage*	*Monthly carriage*	*Coefficients*
1	0;00 00 33 20	9;00	5 37 30 (20,250)	
1a	0;00 00 36	8;20	5 12 30 (18,750)	F 15[18]
2	0;00 00 41 40	7;12	4 30 00 (16,200)	A 29, E 41, F 2, G 5
3	0;00 00 55 33 20	5;24	3 22 30 (12,150)	A 30, F 5
4	0;00 01	5;00	3 07 30 (11,250)	
5	0;00 01 02 30	4;48	3 00 00 (10,800)	E 47, F 12, G 15
6	0;00 01 06 40	4;30	2 48 45 (10,125)	
7	0;00 01 30	3;20	2 05 00 (7500)	
8	0;00 01 51 06 40	2;42	1 41 15 (6075)	A 31, E 45, F 7
8a	0;00 02 05	2;24	1 30 00 (5400)	G 14
9	0;00 02 13 20	2;15	1 24 22;30(5062.25)	
10	0;00 02 40	1;52 30	1 10 18;45(4218.75)	
11	0;00 04 10	1;12	45 00 (2700)	C 22, F 10
12	0;00 05	1;00	37 30 (2250)	

IM 54538 is the only brick-carrying text which poses problems in interpretation. Indeed, it has not been discussed at all in the literature since Baqir's (1951) and von Soden's (1952*a*: 51–2) initial attempts at decipherment.[19] At first sight it belongs to the same genre of texts as those above: a number of bricks are to be carried 10 nindan; how many men will it take to complete the task in a day? The answer given is 6. There are two irresolvable ambiguities though: the coefficient 1 30 is not named (ll. 7, 8); and the bricks are described as 54 *e-še-re-et mu-ša-ri* '54 (00), ten sar' (ll. 2, 10–11). It is unclear exactly what this phrase means, but it is clearly not the case that both parts refer to brick-sar, as 10 sar_b = 2 00 00 bricks. If the numeral were a number of brick-sar and the syllabic phrase a volume measure then

$$\begin{aligned}&54\ sar_b = 10\ sar_v\\ \Rightarrow\ &1\ sar_v = 5;24\ sar_b\\ \Rightarrow\ &\text{Type 3 half-bricks.}\end{aligned}$$

although these have a *nazbalum* of 3 22 30, not 1 30 00, as the text seems to suggest.[20]

[18] See §5.3.3 for an explanation of the numerical errors in F 15 and F 12.

[19] Nemet-Nejat's statement that the text concerns transporting 540 bricks over 30 nindan, and is solved with a quadratic equation (1993: 31), is clearly to be discounted.

[20] But for a new interpretation see now Friberg (1996*a*: 31–2).

5.3.3 Numerical errors and carriage coefficients

The sequence of brick coefficients in Text F includes (ll. 11–12, 14–15):

4 48	*na-al-ba-an* $^{1}/_{2}$ KÙŠ SIG$_4$	4;48, the brickage of a $^{1}/_{2}$-cubit brick.
4 23 36	*na-az-ba-al-ša*	4 23 36, its carriage.
8 20	*na-al-ba-an* IM.DUGUD	8;20, the brickage of a lump of clay.
7 38	*na-az-ba-al-ša*	7 38 00, its carriage.

They are unusual in that the *nazbalum* coefficient of each pair is not $^{5}/_{8} \times 60^2$ (or 37 30) of the *nalbanum*. Working from the premiss that the *nalbanum* values are correct and that the *nazbalum* values are calculated from them,[21] it is attempted here to work out how the numbers given in the text were arrived at. In both cases the errors are shown in **bold** type. For line 12:

Correct calculation	*Calculation performed*
4;48	4;48
× 37 30	× 37 30
24 00	**24 00 00**
2 00 00	2 00 00
5 36 00	5 36 00
28 00 00	28 00 00
24 00 00	24 00 00
2 00 00 00	2 00 00 00
3 00 00;00	3 23 36;00

The incorrect answer has arisen simply from putting the first intermediate answer in the wrong sexagesimal place — not an unusual occurrence in Old Babylonian maths.[22] From here it is easy to imagine a copying or calculation error which transformed the initial 3 into a 4, or perhaps the 2 of the final intermediate answer to a 3, to give the number as it appears in the text: 4 23 36. The 4 is clear on the photo of the tablet. For line 15:

Correct calculation	*Calculation performed*
8;20	8;20
× 37 30	× 37 30
10 00	**10 00 00**
20 00	**20 00 00**
2 00 00	**2 00 00 00**
4 00 00	4 00 00
56 00 00	56 00 00
10 00 00	10 00 00
4 00 00 00	4 00 00 00
5 12 30;00	7 38 00;00

Again the mistake has occurred over the sexagesimal place of the intermediate answers. This time, however, the scribe has changed sexagesimal place half-way through the calculation, after

[21] Because the *nalbanum* coefficients are always given before the *nazbalum* where they occur in pairs and there is a coefficient for converting *nalbanum* to *nazbalum* (see §5.4) this is a reasonable assumption to make.

The *nazbalum* for a half-cubit brick (Type 5) is given elsewhere as 3 00 00 (E 47; G 15), which is consistent with a *nalbanum* of 4;48. There are no other coefficients for IM.DUGUD.

[22] See for instance AO 8862 (vii) on carrying bricks, in which the number of workers, bricks and days are to be added. 30 + 20 + 1 30 00 ≠ 2 20, the answer given in the text. Instead the scribe has added 30 + 20 + 1 30.

performing 0;20 × 30 and 0;20 × 7 00, rather than getting just one place wrong, as in the first example.

5.3.4 Philological comments

The word *nazbalum* is a *mapras* nominal formation from *zabālum*, 'to transport, carry'. Powell (1982*a*) coined the translation 'carriage', to reflect its parallels with the related 'brickage' coefficients (§4.1). Indeed, all the text F comments of the type 'its carriage' refer to the brickage coefficients in the immediately preceding lines, namely as if the figure for 'carriage' were (mathematically) derived from that for 'brickage'. The word *nazbalum* is used more frequently than *nalbanum,* though: it is not restricted to text F but is found in four of the coefficient lists: C, E, F, and G.[23]

SIG$_4$ ÍL.ÍL in A 29 should be understood as *libittim zabālim* 'carrying bricks', rather than *nalban libittim* 'carriage of bricks'.[24] Some of the comments to the coefficients are rather damaged. The last sign of A 33 is unclear on the photo and could not be collated; the sign preceding it may not be IN.[25] The restorations to A 31, E 45, E 47 and G 14 are new.

Although the word *arḫum* is generally understood to mean 'half-brick' in the sense of half a square brick (*AHw*: 67), the value of G 15 suggests that, in this context at least, it means a (square) half-*cubit* brick, Type 5. This is the only syllabically written occurrence of the word in OB mathematics. Bricks referred to by the logogram SIG$_4$.ÁB, on the other hand, are indisputably half-bricks (F 4; YBC 4607 (ii), §§4.0–1).

5.4 THE THEORETICAL LINK

5.4.1 The coefficients

E 49	1 15 *na-az-ba-lu-um*	1 15, carriage.
F 45	2 13 20 *ù* 1 15 *ša* GIŠ «*tup*» DUSU	0;00 02 13 20, and 1 15, of a basket.
F 41	6 *ma-aš-⸢šu-ú-um⸣* [*ša*] ⸢SIG$_4$⸣	6, a basket [of] bricks.
M 1	6 SIG$_4$ SI.SÁ GÚ.UN LÚ.1.E	1 man's load is 6 normal bricks.
M 3	2 15 SIG$_4$ $^2/_3$ KÙŠ.TA ÍB.SI$_8$ GÚ.UN LÚ.1.E	1 man's load is 2;15 bricks $^2/_3$-cubit square.
M 2	1 SIG$_4$ KÙŠ ÍB.SI$_8$ GÚ.UN LÚ.1.E	1 man's load is 1 brick a cubit square.
F 42	1 40 *ma-aš-šu-ú-um ša* SAḪAR	1;40, a basket of earth.

5.4.2 Mathematical commentary

The basic distance for carrying bricks is not 1 or 60, but 30 nindan, as all the extant OB texts on in this type of problem show (§5.3.2). E 49 is not the monthly carriage of any known brick type. However, it is one thirtieth of 37 30, the factor by which the brickage is converted to the monthly carriage. This suggests that 1 15 is the factor by which the brickage is multiplied to give the daily carriage, the number of bricks carried over 30 nindan in one day:

$$\text{daily carriage} = 1\ 15 \times \text{brickage}$$
$$= \frac{3;07\ 30 \times 12\ 00}{30} \times \text{brickage.}$$

[23] Cf. D 37, where it has a different meaning (§5.2.3).

[24] Cf. D 68–9; F 2.

[25] But cf. YBC 4669 (xii): III 2 (§4.4.2), in which $^1/_3$ cubit square bricks are described as IN.⸢TA?⸣.

A connection may also be made between carrying bricks and the Haddad 104 method for carrying earth. Since the *nazbalum* represents the total number of bricks carried over 1 nindan in a day (see §5.3.2), then:

nazbalum × volume of 1 brick = total volume of bricks.

For all types this gives 3;07 30 sar_v, nearly twice the amount of earth that can be carried in a *tupšikkum* over the same distance in a day: 1;40 sar_v (see §5.2). We may assume that this is because the size of the basket limits the amount of earth that can be carried at a time, and that the total distance covered is the same for both earth and bricks, namely one 45 00 nindan trip or its equivalent (§5.2).[26] This is the *muttalliktum* or 'going' coefficient. Taking Type 2 bricks as an example, if a man can carry 4 30 00 bricks over 1 nindan per day, he can therefore carry

$$\frac{4\ 30\ 00}{45\ 00} = 6 \text{ bricks}$$

over 45 00 nindan per day. As, by definition of this daily distance, he can only make one journey of this length daily, 6 is thus also the number of Type 2 bricks he can carry on any one journey of whatever length. This is the meaning of coefficients F 41 and M 1. Likewise, M 3 and M 2 respectively give the number of Types 8 and 11 bricks (with *nazbalum* 1 41 15 and 45 00) that constitute a single load:

$$\frac{1\ 41\ 15}{45\ 00} = 2;15 \text{ bricks} \qquad \frac{45\ 00}{45\ 00} = 1 \text{ brick.}$$

A load of six Type 2 bricks, or its equivalent, is equal to a volume of 0;00 04 10 sar_v (approximately 0.02 m^3), compared with a volume of 0;00 02 13 20 sar_v (just over 0.01 m^3), the volume of earth in a *tupšikkum* basket.[27] The weight of a single load of baked bricks is 6 × 8;20 = 50 minas;[28] the total monthly load over 30 nindan is thus 4 30 00 × 8;20 = 37 30 00 minas and the daily load 9 00 × 8;20 = 1 15 00 minas. Thus the coefficient E 49 — and almost certainly the second numerical value of F 45 — could equally represent a conversion factor between daily carriage and brickage, or the daily load of baked bricks (cf. Powell 1982*a*: 119).[29]

If we assume that 50 minas is the standard weight of a load, then for dried and fresh Type 2 bricks (cf. §4.3.2), with weights 10 and 12;30 minas respectively, the number in each load is:

$$50 \div 10 = 5 \text{ dried bricks} \quad \text{and} \quad 50 \div 12;30 = 4 \text{ fresh bricks.}$$[30]

F 42 must have a similar meaning to F 43, but its exact significance is still uncertain. It is unlikely to refer to the daily load of carrying earth (cf. §5.2), as it would then be identical to F 8, from the same list.

5.4.3 Philological commentary

Maššûm 'basket' is a *mapras* (instrumental) noun from the verb *našûm* 'to carry', 'to raise': it is the means by which goods are carried.

[26] For instance, three 15 00 nindan trips, sixty 45 nindan trips, etc.

[27] It is possible that this volume of bricks is represented by the coefficient H 3; but see §5.6 below for another, equally plausible interpretation.

[28] Or about 25 kilos, just under $3\frac{1}{2}$ stone. See §4.3 for brick density coefficients.

[29] See §5.2 for the first numerical value of F 45.

[30] See the table in §5.7 for other (theoretical) numbers of bricks per journey, but note that some other brick types (5, 8a) also have integer values for baked bricks.

5.5 CARRYING OTHER MATERIALS

5.5.1 The coefficients

C 24	*na-az-ba-al* ŠE.IN.NU.DA 3 20 *i-gi-gu-bu*	Carriage of straw: 3;20 is the coefficient.
D 69	5 *šà šu-ri za-ba-li*	5, of carrying reeds.
F 44	1 20 *ù* ⌜15⌝ *na-az-ba-al* A	1;20 and 15, the carriage of water.

5.5.2 Mathematical commentary

These daily rates are the products of the 'going', 45 00 nindan/day, and the load for each commodity, giving the daily rate of carrying analogous to that of earth (§5.2). This is shown in Table 5.5:

Table 5.5: Carrying rates for building materials

Commodity	*Daily rate (sar$_v$)*	÷ *muttalliktum* =	*Volume of one load (sar$_v$)*	*Comparison with the volume of a* tupšikkum-*basket*
straw	3;20	÷ 45 00 =	0;00 04 26 40	twice a *tupšikkum*-basket.
reeds	5;00	÷ 45 00 =	0;00 06 40	three times a *tupšikkum*-basket
water	1;20	÷ 45 00 =	0;00 01 46 40	$^4/_5$ of a *tupšikkum*-basket

The water coefficient may be $^4/_5$ the volume of a *tupšikkum*-basket because the basket cannot be carried when completely full. Compare the irrigation coefficients 0;48, which mean that only $^4/_5$ of a canal is available for irrigation (D 33, E 26: §6.5). The daily rate of carrying reeds is exactly 3,600 times greater than 0;00 05 sar$_V$, the volume of one reed-bundle (cf. BM 85196 (ii), (v), §6.4). It is unclear what the number 15 (if correctly read) in F 44 refers to.

BM 85196 (xiii) is the only extant problem on carrying goods other than earth or bricks (see §5.2.2). It does not use a daily rate coefficient, because it is assumed that the sheaves to be moved are carried one at a time.

5.6 OTHER COMBINED COEFFICIENT PROBLEMS

5.6.1 The coefficients

D 44	15 *šà pí-ti-iq-ti*	0;15, of an earth wall.
E 25	15 ⌜*ni-qí*⌝*-ir-tum*	0;15, demolition.
H 3	4 10 *ša* SIG$_4$	0;04 10, of bricks.
A 34	11 30 IGI.GUB SAḪAR KI.LÁ	11 30, the coefficient of excavating earth.
A 35	1 20 IGI.GUB SAḪAR ÍL.ÍL SIG$_4$ DÙ	1 20, the coefficient of carrying earth, making bricks.
A 36	11 06 40 IGI.GUB SAḪAR ÍL.ÍL IM.DÙ.A	11 06 40, the coefficient of carrying earth, earth walls.

5.6.2 Mathematical and philological commentary

The coefficients discussed in this section are not specifically 'combined coefficients'. They occur in, or are related to, two 'combined coefficient' problems YBC 4669 (B x) and YBC 4673 (v). They appear in densely written compilations, and consist of a brief statement of the problem and a numerical answer. As coefficients are by definition not stated in the setting of mathematical problems (§1.3.1), the methods used in their solution have remained elusive (cf. Lewy 1949: 151; 156–7). However, working back from the answers given in the texts, it can be shown that both demand the use of two Haddad 104 techniques for their solution: the mathematics of carrying earth (§5.2), and the 'combined coefficient' method (§5.1).

YBC 4669 (B x) reads:[31]

VI 7 LÚ.1.E 20 NINDAN UŠ
ḪÉ.GUL.GUL
ḪÉ.ÍL.ÍL
⌈i⌉-*na* IGI.TE.EN U$_4$
ḪÉ.GUL.GUL
i-na IGI.TE.[EN U$_4$]
ḪÉ .ÍL.[ÍL]
i-na IGI.5.GÁL U$_4$ 12
4 GÍN GUL.GUL
i-na 48 ÍB.TAG$_4$ U$_4$
⌈4⌉ GÍN ÍL.ÍL

Let a man carry (earth) for a length of 20 nindan and let him demolish (a wall). For what proportion of the day shall he demolish (the wall), and for what proportion [of the day] shall he carry (the earth)? In a fifth of the day, or 0;12, he demolished 4 shekels (of earth); in 0;48, the remainder of the day, he carried 4 shekels (of earth).

According to the Haddad 104 method, a man can carry 1;40 ÷ 20 = 0;05 sar_V earth over 20 nindan in a day. So, in 0;48 (= $^4/_5$) of day a man can carry 0;05 × $^4/_5$ = 0;04 sar_V or 4 shekels, which is the answer given here. Working back from the answer for the first part, if a man demolishes 4 shekels or 0;04 sar_V earth in 0;12 (= $^1/_5$) day (knocking down a wall perhaps?), then in a whole day he demolishes 5 × 0;04 = 0;20 (= $^1/_3$) sar_V.

The problem itself must have been solved by the Haddad 104 'combined coefficient' method. The man processes the same amount of earth in the two operations: he must destroy only as much earth as he can carry away the same day. In other words, the problem is to find the fractions of the day spent on each task to process the same amount of earth.[32]

So, with the daily rates of carrying over 20 nindan, and of destroying, this gives:

$^1/_3$ sar_V destroyed per day	⇒	3 days to destroy 1 sar_V earth
$^1/_{12}$ sar_V carried per day	⇒	12 days to carry 1 sar_V earth
	⇒	15 days to destroy and carry 1 sar_V earth
	⇒	$^1/_{15}$ = 0;04 sar_V earth processed in one day.
$^1/_3$ sar_V destroyed in 1 day	⇒	$^1/_{15}$ sar_V destroyed in $^3/_{15} = ^1/_5$ day
$^1/_{12}$ sar_V carried in 1 day	⇒	$^1/_{15}$ sar_V carried in $^{12}/_{15} = ^4/_5$ day,

which are the answers given in the text.

While there are no exact matches for this demolition coefficient in the lists, D 44 and E 25 may both refer to a similar operation, but with a daily rate of 0;15 sar_V ($^1/_4$ sar_V) a day instead of 0;20 sar_V. D 44 is immediately preceded by the coefficient for *building* an earth wall, which is given the same description (D 43, §6.1). Both are described simply as 'earth wall' with no further indication of the exact significance of the coefficients.

YBC 4673 (xii) also seems to be about destroying an earth wall. It too follows on from three problems on building them (§6.1):

IV 8 IM.DÙ.A
5 UŠ UŠ.BI
2 KÙŠ DAGAL
$^1/_2$ NINDAN SUKUD.BI

An earth wall. Its length is 5 UŠ, the width is 2 cubits, its height is $^1/_2$ nindan. In 1 cubit (height) it decreases $^1/_3$ cubit in width and a man demolishes and $1^1/_2$ cubits height remains. How long is the length

[31] I thank Paul-Alain Beaulieu for collating this tablet and YBC 4673.

[32] For the mathematical meaning of IGI.TE.EN see §5.6.3.

i-na 1 KÙŠ $^1/_3$ KÙŠ KÚ Ì.KÚ-*ma*
GURUŠ GUL.GUL-*ma*
$1^1/_2$ KÙŠ SUKUD DIRI.⌈GA?⌉
UŠ EN.NAM ḪÉ.GÍD 21⌈$^1/_3$⌉ [NINDAN?]

(demolished in 1 day)? *0;21 20* [*nindan*].

The wall tapers, such that its cross-section is triangular (the base is 2 cubits, the height 6 cubits and the wall gets $^1/_3$ cubit narrower in every 1 cubit height); the wall is partially destroyed. The signs at the end of lines 14 and 15 are damaged, but it appears that the wall is left standing to a height of 1;30 cubits, and a length of 21;20 nindan is cleared daily. The daily rate of destroying is thus:

$$\text{daily rate} = 21;20 \times \frac{4;30 \times 1;30}{2} \times 12$$
$$= 6 \text{ sar}_V.$$

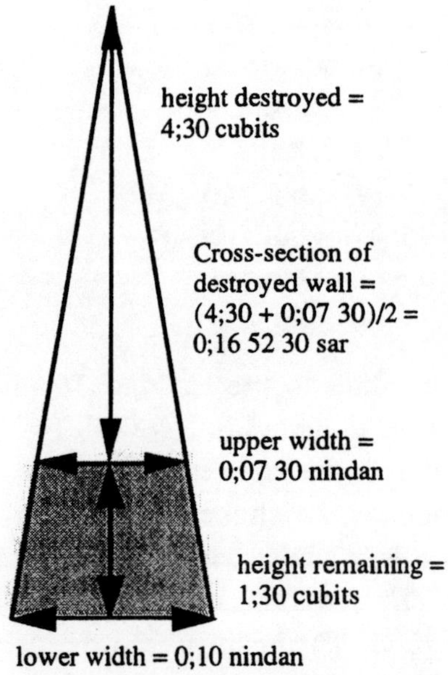

Figure 5.1: The wall destroyed in YBC 4673 (xii)

But this is much larger than either of the attested demolition coefficients, or indeed any of the OB work-rates. In fact, it is 24 times greater than 0;15, the rate in the lists, and this may be the key to its interpretation. The problem would have been solved thus:

$$\text{daily length} = \frac{\text{daily rate}}{\text{height in cubits} \times \text{average width in nindan}}.$$

For the particular numerical values in YBC 4673 (xii) this method would yield:

$$\text{daily length} = \frac{0;15}{(4;30 \times 1;30)/(2 \times 12)}$$
$$= \frac{0;15 \times 24}{4;30 \times 1;30}.$$

It may just be that this product was inadvertently multiplied by 24 twice, producing the answer 21;20 nindan instead of the expected 1;46 40 nindan. However, because the problem does not give the method of solution it is unlikely that a firm explanation can ever be given.

YBC 4673 (v) combines coefficients in a problem about carrying earth:

II LÚ.1.E
a-na 30 NINDAN UŠ
SAḪAR ÍL.ÍL-*ma*
ù SIG_4.ANŠE DU_8.DU_8
i-na IGI.TE.EN U_4
SAḪAR ÍL.ÍL
i-na IGI.TE.EN U_4
SIG_4.ANŠE DU_8.DU_8
ù SIG_4 EN.NAM 2 ⌈40⌉

One man carried earth over a distance of 30 nindan, and he built a brick-pile. For what proportion of the day did he carry earth; for what proportion of the day did he build the brick-pile; and how many bricks were there? 2 40.

Again using the Haddad 104 method, the daily rate of carrying over 30 nindan is 1;40 ÷ 30 = 0;03 20 sar_V. This time we are not given the fractions of the day needed to complete each task, so working back from the answer is slightly more complicated. First, the number of bricks in sar_b is:

$$2\ 40 \text{ bricks} = 160$$
$$160 \div 720 = 0;13\ 20 \text{ sar}_b.$$

Assuming they are Type 2 bricks, with *nalbanum* 7;12, this is equivalent to a volume of:

$$0;13\ 20 \div 7;12 = 0;01\ 51\ 06\ 40\ \text{sar}_V.$$

Using the 'combined coefficient' method,

0;03 20 sar_V earth carried in a day	⇒ 1 sar_V carried in 18 days
0;01 51 06 40 sar_V processed in a day	⇒ 1 sar_V processed in 32;24 days
⇒ time taken to build brick-pile	= 32;24 – 18 = 14;24 days
⇒ daily rate of piling bricks	= 1 ÷ 14;24 = 0;04 10 sar_V,

which is around 1.25 m^3. We know from other texts and from other coefficients that brick-piles are usually made from Type 2 bricks — in mathematical problems, at least (see §4.4). The daily rate for piling bricks, given here in terms of volume, translates to

$$0;04\ 10\ \text{sar}_V \times 7;12 = 0;30\ \text{sar}_b = 360\ \text{bricks}.$$

In other words, a worker can pile up $^1/_2$ sar_b of Type 2 bricks in a day. Such round-numbered coefficients are typical of OB mathematics, and help to confirm this interpretation of the problem.

The coefficient H 3 may refer to this same daily rate, or perhaps the volume of bricks that one man can carry at once (see §5.4). Gundlach and von Soden's reading (1963), followed by Friberg (1987–90: 553), as ÍB.‹SI$_8$› 'a square', has almost certainly to be discounted as no explanation has been put forward for its numerical value (see §3.4). The last sign is right on the edge of the tablet and cannot be seen clearly in the photo.

It has been suggested that the coefficients A 34–6 refer to combined coefficients (Friberg 1987–90: 545). If so, they do not fit into the mathematics outlined in this chapter. For instance, A 36 is supposedly the combined coefficient for carrying earth and building an earth wall. If the number given is a daily rate, 0;11 06 40 sar_V, then its reciprocal, 5;24, gives the number of days for 1 sar_V earth to be put through *both* processes. However, the daily rate of building an earth wall is 0;03 45 sar_V (§6.1); i.e. it takes 16 days to build 1 sar_V of earth wall. This is significantly longer than the 5;24 days that A 36 implies. A 34 and 35 — which has an erased 10 sign written under the initial 1[33] — are similarly contradictory. Their true meaning is still unclear.

5.6.3 Philological commentary

The reading *niqirtum* for E 25 is given in *AHw*: 792, but with no translation. It is a *pirist* nominal formation from *naqārum* 'to destroy' and as such may be expected to function as a *nomen actionis* (*GAG*: §55c), hence the translation 'demolition'.

The word IGI.TE.EN/*igitennûm* is a term exclusive to the OB period.[34] In 'combined coefficient' problems[35] it refers to the proportion of the day in which the same quantity of one task is completed as in the proportions of the day occupied by other tasks. The following expression is also found in a set of 'algebraic' problems YBC 4668 and YBC 4712:

IGI.TE.EN SAG UŠ.ŠÈ = UŠ/SAG	'the proportion of the width to the length'
IGI.TE.EN UŠ SAG.ŠÈ = SAG/UŠ	'the proportion of the length to the width',

although in modern mathematics one would use the phrase with the opposite meanings (cf.

[33] Collation by John Brinkman, 1995.

[34] But cf. the Standard Babylonian word I.BÍ.TE.EN/*itannu* 'interstice (of a net)' which 'is to be considered an Emesal form containing IGI "eye"' (*CAD*: I/J 293).

[35] Namely YBC 4669 (B x) and YBC 4673 (v), discussed here, and YBC 7164 (vii), §6.3.

Friberg 1987–90: 558, 577). IGI.TE.(EN) means 'proportion' in non-mathematical — primarily legal — contexts too (cf. *CAD*: I/J 43). The shape of the word suggests that it derives from an (unknown) Sumerian original.

5.7 SUMMARY

Table 5.6 compares the carrying coefficients for each type of commodity discussed in this chapter (the other daily rates are summarized in §6.7). Coefficients directly attested are in bold type.

Table 5.6: Work-rates for carrying building materials

	Daily distance	*Volume of load*	*Daily rate over 1 nindan*	*Volume compared to basket of earth*
Earth	**45 00 nindan**	**0;00 02 13 20 sar$_v$**	**1;40 sar$_v$**	1
Bricks	**45 00 nindan**	0;00 04 10 sar$_v$	3;07 30 sar$_v$	1;52 30 ($1^7/_8$)
Straw	**45 00 nindan**	0;00 04 26 40 sar$_v$	**3;20 sar$_v$**	2
Reeds	**45 00 nindan**	0;00 06 40 sar$_v$	**5;00 sar$_v$**	3
Water	**45 00 nindan**	0;00 01 46 40 sar$_v$	**1;20 sar$_v$**	0;48 ($^4/_5$)

The daily rate of carrying over a certain distance is calculated by:

$$\frac{\text{daily distance} \times \text{volume of load}}{\text{distance of trip}} = \text{daily rate in sar}_v$$

The brick-carrying coefficients for each type of brick are shown in Table 5.7 below:

Table 5.7: Summary of the brick-carrying coefficients

					Bricks per load		
Type	*Brickage*	*Monthly carriage*	*Daily carriage*	*Weight of load*	*Baked*	*Dried*	*Fresh*
1	9;00	5 37 30	11 15	50 minas	7;30	6;15	5
1a	*8;20*	*5 12 30*	*10 25*	*50 minas*	*6;56 40*	*5;47 13 20*	*4;37 46 40*
2	7;12	**4 30 00**	9 00	50 minas	**6**	5	4
3	5;24	**3 22 30**	6 45	50 minas	4;30	3;45	3
4	5;00	3 07 30	6 15	50 minas	4;10	3;28 20	2;46 40
5	4;48	**3 00 00**	6 00	50 minas	4	3;20	2;40
[6]	4;30	2 48 45	5 37;30	50 minas	3;45	3;07 30	2;30
7	3;20	2 05 00	4 10	50 minas	2;46 40	2;18 53 20	1;51 06 40
8	2;42	**1 41 15**	3 22;30	50 minas	**2;15**	1;52 30	1;30
8a	2;24	**1 30 00**	3 00	50 minas	2	1;40	1;20
9	2;15	1 24 22;30	2 48;45	50 minas	1;52 30	1;33 45	1;15
10	1;52 30	1 10 18;45	2 20;37 30	50 minas	1;33 45	1;18 07 30	1;02 30
11	1;12	**45 00**	1 30	50 minas	**1**	0;50	0;40
[12]	1;00	**37 30**	1 15	50 minas	0;50	0;41 40	0;33 20

Daily work-rates may be combined to calculate the daily rate for performing *n* tasks at once, using the *n*th fraction of the harmonic mean:

$$\text{combined coefficient} = \frac{1}{\dfrac{1}{\text{daily rate 1}} + \dfrac{1}{\text{daily rate 2}} + \ldots + \dfrac{1}{\text{daily rate } n}}.$$

6. EARTHWORKS AND WATERWORKS; OTHER WORK-LOADS

The remainder of the quantity surveying coefficients concern, for the most part, earthworks of various types: earthen walls, irrigation ditches and dykes, and other earthworks. There are also work-rates for cutting reeds and making cloth. Several problems on digging canals contain coefficient-like data in the formulation of the questions. These have been listed and form part of the study in §6.2.

The problems and coefficients concerning irrigation have been the subject of a short study by Powell (1988); the canal problems have recently been examined by Muroi (1992*a*). Much of the Ur III terminology of levees, ditches and dykes has recently been closely examined by Civil (1994: 109–40). The other coefficients in this chapter have not been subject to particular attention outside the general surveys (Kilmer 1960; Friberg 1987–90; Nemet-Nejat 1993).

There are 13 different coefficients for earthworks and waterworks in the OB lists, attested a total of 18 times, in all lists except C. In addition, 4 of the irrigation coefficients in the LB text B are directly based on OB antecedents (§6.5). Twenty-five problems on 11 tablets, and five calculations, use 15 separate coefficients, all but five of which are attested in the lists.[1] A further 7 tablets include numerical data in the formulation of 74 problems which are identical to 3 earthwork coefficients in the lists,[2] and 2 further problems on one of those tablets imply the use of a coefficient attested in the lists.[3]

One further coefficient and a problem concern two work-loads apparently unrelated to the wider mathematical scheme of quantity surveying.[4] Four further labouring problems are mentioned in a 'catalogue',[5] and an earthwork coefficient mentioned in the lists is implicit in a calculation and diagram from Mari.[6] One Ur III model document uses four earthwork and agriculture coefficients, only one of which is identical to those of OB mathematics.[7]

6.1 EARTH WALLS

6.1.1 The coefficients

A 8 3 45 IGI.GUB IM.DÙ.A	0;03 45, the coefficient of an earth wall.
D 43 3 45 *šà pí-ti-iq-ti*	0;03 45, of an earth wall.

[1] BM 85194 (i)–(v) — various earthworks; BM 85196 (i) — damming a canal; (v) — reed-bundles; (xv) — enlarging a canal; (xvi) — digging a canal; Erm 15073 (iv)–(v) — digging a canal; IM 53961 — earth wall; IM 54011 (i)–(ii) — earth wall; VAT 8523 (i), (v)–(vi) — damming a canal; TMS 25 (i) — irrigating a field; YBC 4669 (B i) — unnamed subject; YBC 4673 (xvii)–(xxi); YBC 8588 — excavation; YBC 9874 — enlarging a canal. *UET* 6/2 tablets 233, 236, 254, 293, and 298 (§A.5.1.3) — unspecified labour problems.

[2] VAT 7528 (i)–(iv) — small canals; YBC 4657 (with answers on YBC 4663 and YBC 4662) (i)–(viii), (xxii)–(xxxi) — excavations; YBC 4666 (i)–(xxiii) — small canals; YBC 4673 (x)–(xii) — earth wall; YBC 5037 (i)–(xvi) — excavations; YBC 7164 (i)–(vii), (ix)–(xii), (xiv), (xvi) — small canals.

[3] YBC 7164 (xvii)–(xviii) — irrigation.

[4] BM 85194 (xxiii) — weaving cloth.

[5] Text C: 75–8 — all agricultural?

[6] M 288 — building a city wall.

[7] W 47a — building an earth and reed levee, and hoeing.

E 13 3 45 IM.DÙ.A	0;03 45, an earth wall.
F 17 3 45 *ša* IM.DÙ.A	0;03 45, of an earth wall.
G 13 3 45 IGI.GUB «*pí-iq*» *pí-*⌈*ti*⌉*-iq-tim*	0;03 45, the coefficient of an earth wall.
H 2 3 45 *pí-ti-iq-tum*	0;03 45, an earth wall.

IM 53961: 9	3 45 *i-gi-gu-bi-ka*	0;03 45, your coefficient.
IM 54011 (i): I 11	3 45 *i-gi-gu-[ub-bi]*	0;03 45, the coefficient.
IM 54011 (ii): II 9	3 45	0;03 45.

6.1.2 Coefficient-like data in statements of problems

YBC 4673 (ix): III 12	3 45 ÉŠ.GÀR	0;03 45, the work-load.
YBC 4673 (x): III 19	3 45 ÉŠ.GÀR	0;03 45, the work-load.
YBC 4673 (xi): IV 5	3 45 SAḪAR ÉŠ.GÀR	0;03 45, the volume of the work-load.

6.1.3 Mathematical commentary

Four OB problems concern the daily rate of building a wall.[8] YBC 4673 (ix), simply the statement of a problem and its answer, reads:

III IM.DÙ.[A] 1 KÙŠ [DAGAL] [1] KÙŠ SUKUD [3 4]5 ÉŠ.GÀR LÚ.1.E UŠ EN.NAM AL.DAB$_5$ $^1/_2$ NINDAN 3 KÙŠ UŠ	An earth wall. The [width] is 1 cubit, the height is [1] cubit, the work-load is [0;03 4]5. What length does one man take? The length (taken) is $^1/_2$ nindan 3 cubits.

Although the method of solution is not given, it is clear that the answer must be obtained by dividing the daily volume (the work-load) by the height and width of the wall. This gives the length one worker is expected to build in a day:

$$\text{length} = \frac{0;03\ 45}{0;05 \times 1} = 0;45 \text{ nindan.}$$

The method is confirmed by IM 53961, which fully solves an almost identical problem:

šum-ma ki-a-am i-ša-al-ka um-ma šu-ú-ma *pí-ti-iq-*⌈*tum*⌉ *ši-ta am-ma-tim* *ru-up-šu-um am-ma-at me-li-um* *iš-ka-ar iš-te-en a-wi-li-im* *mi-nu-um at-ta i-na e-pé-ši-ka* [10] *a-na* 1 *me-li-ka i-ši-i-ma* 10 *i-li* 10 *ša i-li-a-ku-um* *i-gi* 10 *pu-ṭú-ur-ma* 6 *i-li* [6] *ša i-li-‹a›-kum a-na* 3 45 *i-gi-gu-bi-ka* [*i*]*-ši-i-ma* 22 30 *i-li* 22 30 *ša i-li-kum* *iš-ka-ar iš-te-en a-wi-li-im* «*i-li*»	If (someone) asks you, saying as follows: 'An earth wall. The width is two cubits, the height is one cubit. What is the output of one man?' You, in your working, multiply [0;10] by 1, your height, and 0;10 will come up. 0;10 that came up for you: take the reciprocal of 0;10, and 6 will come up. Multiply [6] that came up for you by 0;03 45, your coefficient, and 0;22 30 will come up. 0;22 30 that came up for you is the output of one man «will come up».

This wall is twice as thick as the one in YBC 4673 (ix), and so the length built is halved. It is not a particularly difficult text and was fully interpreted by its first editor (Baqir 1951: no. 4). It is included here as an aid to understanding IM 54011 (i)–(ii), to which it is closely related:

[8] Although the statements of YBC 4673 (x)–(xi) mention the daily rate of building an earth wall, it is not actually needed in the calculation of the problems.

[šum-ma ki-a-am i]-ša-al-ka um-ma šu-ú-ma
pí-ti-iq-tum a-ša-al ši-du-um ši-ta am-ma-tim
ru-up-šu-um KÙŠ *ḫe-pí a-na e-le-nu-um [ku-bu]-ur*
ni-ka-ás mu-lu-um e-pé-ru-ka! *ù [ṣa-bu-ka]*
ú-ma-ka-lu-tum mi-nu-um at-ta i-na [e-pé-ši-ka]
ši-ta am-ma-tim ru-up-ša-am ù KÙŠ *ḫe-pí*
ku-mu-ur ḫe-pé-e-ma 6 15 *ḫe-pu-šu [ta-mar]*
[6] 15 *a-na ni-ka-ás mu-li-im i-ši-[ma]*
[1]8 45 *ta*!*-mar*! 18! 45 *a-na a-ša-[al ši-di-im]*
i-ši-ma 3 [07 30] *ta-mar* 3 07 30 *[e-pé-ru-ka]*
na-ás-ḫi-ir-ma 3 45 *i-gi-gu-[ub pí-ti-iq-tim]*
[i]-gi 3 45 *pu-ṭú-ur-ma a-na* [3 07 30]
e-pé-ri!*-ka i-ši-ma* 50 *ta-[mar* 50 *ṣa-bu-ka]*

[If] (someone) asks you, [saying as follows]: 'An earth wall. The length is a rope; the width is two cubits. It is half a cubit wide at the top. The height is half a reed. What are your volume and [your workers] for a whole day?' You, in [your working], add two cubits, the width, and half a cubit. Halve (the result) and [you will see] 0;06 15, its half. Multiply [0;06] 15 by half a reed, the height, [and] you will see 0;18 45. Multiply 0;18 45 by a rope, [the length], and you will see 3;[07 30. Your volume is] 3;07 30. Return, and 0;03 45 is the coefficient [of an earth wall]. Take the reciprocal of 0;03 45 and multiply (it) by [3;07 30], your volume, and you will see 50. [Your workers are 50 strong.]

šum-ma ki-a-am i-ša-al-ka um-ma šu-ú
[iš]-ka-ar a-wi-lim iš-te-en [ki] ma-[ṣi]
at-ta i-na e-pé-ši-ka ši-ta [am-ma-tim]
[ru-up-ša]-am ù KÙŠ *ḫe-pí [ku-mu-ur]*
[ḫe]-pé-ma 6 15 *ḫe-pu-šu ‹ta-mar›* [6 15] *a-na*
[3 *mu]-⌈li⌉-im i-ši-i-ma* [18] 45 *ta-mar*
[i-gi 18] 45 *pu-ṭú-ur-ma a-na* 3 45 *[i-ši-i-ma]*
[12 *iš-ka-ar]-ka-ma*

If (someone) asks you, saying as follows: '[How] much is the output of one man?' You, in your working: [add] two [cubits, the width], and half a cubit; break (them) and ‹you will see› 0;06 15, its half. Multiply [0;06 15] by [3], the height, and you will see [0;18] 45. Take [the reciprocal of 0;18] 45 and [multiply] by 0;03 45 [and 0;12] is your [output].

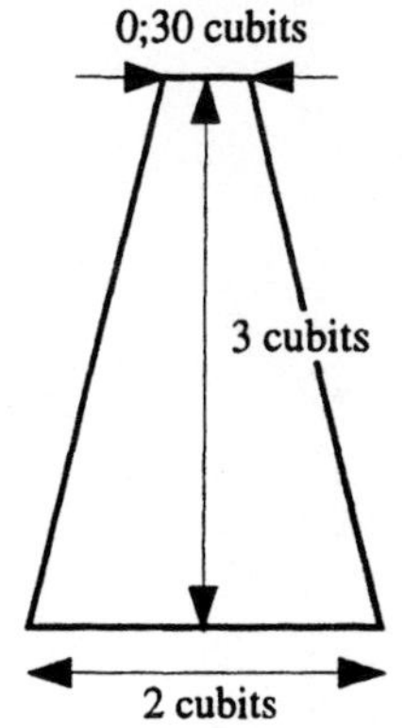

Figure 6.1: The wall in IM 54011

This text has proved very difficult to interpret, partly because of the bad state of preservation of the tablet and partly because of the phonetic writing and unusual units of measurement. This led von Soden (1952*a*: 54–5) to suggest the wall was much wider at the top than at the bottom, and to posit the existence of a length unit called a 'hand' equal to $^{5}/_{12}$ of a cubit (reading *qa*!*-ta*!*-tim* for *am-ma-tim*, lines 2 and 6). Bruins' interpretation (1953: 248) relied on translating *rupšum* as the 'length of the segment measured from the end of the projection of the upper width to one end of the projection of the lower width' instead of simply 'width'. This produced a more realistic shape for the wall, but depended on retaining von Soden's often implausible reading of the signs.

A satisfactory explanation of the text (Fig. 6.1) can be given by keeping the established meanings. The only strange term in the text is the phrase KÙŠ *ḫe-pí*, literally 'broken cubit'. In OB mathematical texts *ḫepûm* means 'to break in half' (see lines 7 and 18 of this text, for instance), so that KÙŠ *ḫe-pí* means 'half a cubit', more usually written $^1/_2$ KÙŠ. This gives a cross-section as shown in the diagram on the left, in which the width decreases $^1/_2$ cubit for every cubit height.[9] The solution is then straightforward:

$$\text{volume} = \frac{\text{width} + \text{top}}{2} \times \text{height} \times \text{length}$$

$$\text{men} = \frac{\text{volume}}{\text{daily rate}}$$

$$\text{daily length} = \frac{\text{daily rate}}{\frac{1}{2}(\text{width} + \text{top}) \times \text{height}}$$

This coefficient is one of the few for which there is unequivocal evidence in Ur III period administrative documents: see §9.6.1.

6.1.4 Philological commentary

Whereas YBC 4673 is written entirely logographically, the Shaduppum tablets IM 54011 and IM 53961 are exclusively in syllabic Akkadian, with the exception of the sign KÙŠ. Syllabic writings of numerals and metrological units are usual among mathematical texts from the Eshnunna region, but rare in those from Babylonia proper.[10] Many of the technical terms normally written logographically are either given in their Akkadian forms (e.g. *rupšum* for DAGAL, line 3 in both problems) or 'Akkadianized' by syllabic writing and a case ending where appropriate (e.g. *i-gi* 10 for *IGI 10 'the reciprocal of 0;10', in line 8).

Iškarum also has a slightly different meaning in the two texts. In the YBC text, and commonly in Old Babylonian mathematics, it refers to the volume of work, measured in sar_v, that a worker is expected to do in one day. In the Eshnunna texts *iškarum* means the length of wall completed in a day, and *igigubbûm*, 'coefficient', has taken over the original meaning 'daily volume' of *iškarum* (see §5.2.3). In non-mathematical contexts, it can refer to a day's march, an allotment of materials, delivery quotas for grain, clothing, bricks, etc., all of which concern work to be done; it can also mean a distribution of rations — payment for the allotted work done — and a type of field. It is only in the first millennium that *iškaru* comes to mean a series of literary texts (*AHw*: 396).

The phrase KÙŠ *ḫe-pí* must be in the absolute state, as is common with units of measure (*GAG*: 79, §62f); otherwise one would expect *KÙŠ *ḫe-pí-tum*/*-tam*.

6.2 OTHER EARTH CONSTRUCTIONS

6.2.1 The coefficients

F 43 10 GÍN ÉŠ.GÀR LÚ.1 — 10 shekels, the work-load of 1 man.

BM 85194 (i): I 9	10 ÉŠ.GÀR	0;10, the work-load.
BM 85194 (ii): I 20	10 ÉŠ.GÀR	0;10, the work-load.
BM 85194 (iii): I 31	10 ÉŠ.GÀR	0;10, the work-load.
BM 85194 (iv): II 1	10 ÉŠ.GÀR	0;10, the work-load.

[9] The same slope (for the sides of a canal) can be found in YBC 4669 (xxi)–(xxii): rev. 25, 27.

[10] Writings of 1, 2 and 60 are commonest. AO 8862, a four-sided prism from Larsa, contains syllabic writings of 3, 4, 5, 7, 11, 13, and 14 (*TMB*: 229–30). This is more likely to be a geographical rather than a chronological phenomenon.

BM 85194 (v): II 22	10 ‹ÉŠ›.GÀR	0;10, the work-load.
YBC 4669 (B i): IV 3	10 ÉŠ.GÀR	0;10, the work-load.
UET 6/2 233	10	0;10.
UET 6/2 236	10	0;10.
UET 6/2 254	10	0;10.
UET 6/2 293	10	0;10.
UET 6/2 298	10	0;10.

6.2.2 Mathematical and philological commentary

BM 85194 (i)–(v) are all concerned with earth works of various sorts: *arammum* (a ramp), *kigurrum* (foundations?), URU^{ki} (city walls), BÀD (a wall with trapezoidal cross-section). In each case, once the volume has been calculated it is divided by the daily work-load 0;10 sar$_v$ to give the number of work-days needed to build it (F 43). In YBC 4669 (B i), which uses the same work-rate, the object being built must have been named in the previous column, now missing. It has a volume of 100 sar$_v$, and takes 600 men to construct.

The interpretation of the five *UET* 6/2 calculations as labouring problems is highly conjectural; see §A.5.1.4.

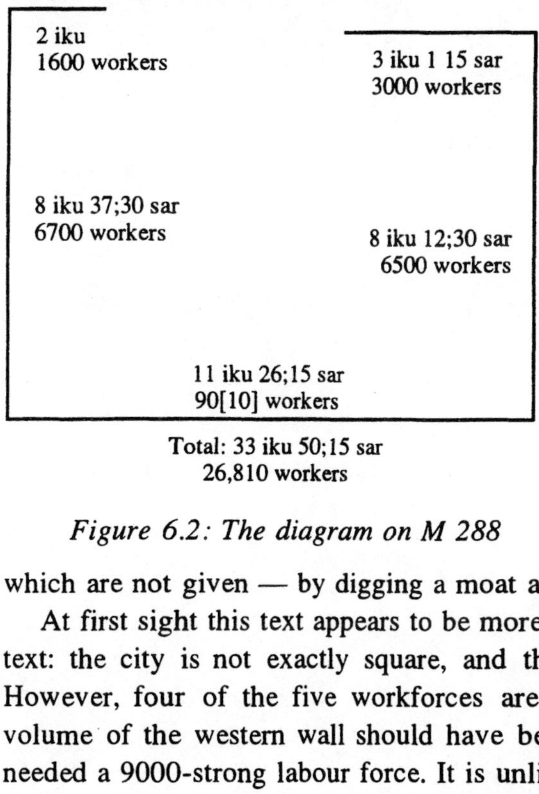

Figure 6.2: The diagram on M 288

M 288 is a lenticular tablet from OB Mari which constitutes the first direct evidence that mathematics in that city went beyond the learning of tables. It shows a drawing of an approximately rectangular city below a square *kirḫum* 'acropolis' (cf. Dossin 1972). Next to each wall the volume of earth (*e-pé-ru*) and workforce needed (*ṣa-bu-šu*) are stated.[11] The totals are given below the 'west' wall.

In each case the daily work-rate — found by dividing the volume by the workers — is 0;07 30 sar$_v$, otherwise attested only for digging deep canals (cf. D 70; §6.3). Charpin (1993: 195–6) suggests that the task was to build a lower city wall around the 'acropolis' — the dimensions of which are not given — by digging a moat and piling it up into banks.

At first sight this text appears to be more typical of a draft economic document than a school text: the city is not exactly square, and the volumes of the walls are not round numbers. However, four of the five workforces are integer multiples of 100; and it may be that the volume of the western wall should have been calculated as 11 iku 25 sar.[12] This would have needed a 9000-strong labour force. It is unlikely that a professional surveyor's plan — that is,. one based on a real city — would have resulted in such neat figures. One would also expect some sort of identifying remarks on the tablet, be it the overseer's name, the toponym or the date, etc.

[11] While the volumes are all in standard Ur-OB measure, the numbers of workers are all given decimally, e.g. 6 *li-mi* 7 ME (= 6700), the workforce for the southern wall. To the number of workers on the southern wall is appended the phase *i-nu-ma i-ip-pé-šu* 'when they construct'.

[12] Such a calculation error could arise through writing down an intermediate result $1^1/_4$ twice, and interpreting it first as $1^1/_4$ iku and then (superfluously) as $1^1/_4$ sar, adding both to 10 iku.

6.3 CANALS AND EXCAVATIONS

6.3.1 The coefficients

D 70 ⌜1⌝ SAR *a-na qa-ti sé-ki-ri iš-šà-ka-[an]* ⌜7⌝ 30 SAḪAR *i-za-ab-bi-il*		1 sar_v is put into the hands of a canal-worker. He transports 0;07 30 (sar_v) earth.
E 20 10 ⌜ÉŠ⌝.GÀR ‹KI?›.LÁ		0;10, the coefficient of an excavation.
E 21 20 ⌜PA₅⌝ *ši-⌜lu⌝-tum*		0;20, the 'throw-out' level of a canal.[13]
BM 85196 (xvi): V 3	6 40 (*ḫa-da-lum aḫ-di-il*)	0;06 40 (... ...)
BM 85196 (xv): IV 33	10 GÍN ÉŠ.GÀR	0;10 shekels, the work-load.
BM 85196 (xvi): V 2	10 ([...])	0;10 ....
Erm 15073 (iv): IV 7'	10 ÉŠ.GÀR	0;10, the work-load.
Erm 15073 (v): IV 18'	10 ÉŠ.GÀR	0;10, the work-load.
YBC 8588: I 9, 12, 19, II 5	15 ÉŠ.GÀR	0;15, the work-load.
BM 85196 (xvi): V 1	20 (*li-lum*)	0;20 (the ...).
YBC 9874: I 6	20 *ši-i-lu-ú-tim*	0;20, the 'throw-out' level.

6.3.2 Coefficient-like information in the statements of problems

VAT 7528 — subsidiary canals		
(i): I 3; (ii): I 9; (iii): I 14; (iv): I 20	$^1/_3$ SAR SAḪAR ÉŠ.GÀR	$^1/_3$ sar, the volume of the work-load.
YBC 4666 — subsidiary canals		
(i): I 2; (ii): I 5; (iii): I 7; (iv): I 9; (v): I 11; (vi): I 13; (vii): I 15; (xi): I 23; (xii): II 1; (xiii): II 4	$^1/_3$ GÍN ÉŠ.GÀR	$^1/_3$ (shekels), the work-load.[14]
(viii): I 17; (ix): I 19; (x): I 21	$^1/_3$ GÍN SAḪAR ÉŠ.GÀR	$^1/_3$ (shekels), the volume of the work-load.
(xiv): II 8; (xv): II 11; (xvi): II 13; (xvii): II 15; (xviii): II 17; (xix): II 19; (xx): II 21; (xxi): II 23; (xxii): II 24; (xxiii): II 26	10 GÍN ÉŠ.GÀR	10 shekels, the work-load.
YBC 7164 — subsidiary canals		
(i): I 1	10 GÍN ÉŠ.GÀR	10 shekels, the work-load.
(ii): I 4; (iii): I 6 (iv): I 8; (v): I 10; (vi): I 12; (vii): I 15	1 KÙŠ *šu-‹up-lim›* $^1/_3$ MA.NA SAḪAR *ši-lu-tum*	For 1 cubit depth $^1/_3$ mina (sic) is the 'throw-out' work.
(ii): I 5; (iii): I 7; (iv): I 9;	2 KÙŠ *šu-‹up-lim›* 10 GÍN DUSU	For 2 cubits depth 10 shekels is

[13] This term is explained in §6.3.4.

[14] i.e. $^1/_3$ mina, or 20 shekels. For an explanation of this type of gloss, see Friberg (1987–90: 536–7).

(v): I 10; (vi): I 13; (vii): I 16		the 'basket' work.[15]
(vi): I 13	2 KÙŠ *šu-‹up-lim›* $7^{!1}/_{2}$ GÍN DUSU	For 2 cubits depth 7 1/2 shekels is the 'basket' work.
(ix): I 22; (x): I 24; (xi): I [27]; (xii): II ‹1›; (xiv): II 8; (xvi): II 15	$^{1}/_{3}$ GÍN SAHAR ÉŠ.GÀR	$^{1}/_{3}$ (shekels), the volume of the work-load.

YBC 4657 (with answers on YBC 4663 and YBC 4662) — excavations

(i): I 1; (ii): I 4; (iii): I 6; (iv): I 8; (vi): I 12; (vii): [I 14]; (viii): I 16; (xxii): II 7; (xxiii): II 9; (xxiv): II 11; (xxv): II 13; (xxvi): II 15; (xxvii): II 16; (xxix): II 20; (xxx): II 22; (xxxi): II 24	10 GÍN ÉŠ.GÀR	10 shekels, the work-load.

YBC 5037 — excavations

(i): I 1; (ii): I ‹3›; (iii): I 4	$7^{1}/_{2}$ GÍN ÉŠ.GÀR	$7^{1}/_{2}$ shekels, the work-load.
(iv): I 6; (v): I 7; (vi): I 8; (vii): I 10; (viii): I 12; (ix): I 14; (x): I 16; (xi): I 17; (xii): I 19; (xiii): I 21; (xiv): I 22; (xv): I 23; (xvi): I 24	10 GÍN ÉŠ.GÀR	10 shekels, the work-load.

6.3.3 Mathematical commentary

There is no distinction between the work-loads used in problems on canals (PA$_5$, PA$_5$.SIG) and those in excavation problems (KI.LÁ/*kalakkum*). They function in the same way as other work-loads, naming a total volume of work to be done in a day for a particular task. However, as a group these problems do not seem to use a consistent set of coefficients for the work-loads, as Table 6.1 shows.

Two more complex sets of problems, involving different work-loads for different depths of work, present a similarly disparate picture at first sight:

YBC 7164 (ii)–(vii):	top level	0;20 sar$_V$ a day	to 1 cubit depth
	second level	0;10 sar$_V$ a day	1–3 cubits depth
	third level[16]	0;07 30 sar$_V$ a day	3–4.5 cubits depth
BM 85196 (xvi):	top level	0;20 sar$_V$ a day	to 1 cubit depth
	second level	0;10 sar$_V$ a day	1–2 cubits depth
	third level	0;06 40 sar$_V$ a day	2–3 cubits depth.

The only clear fact that immediately emerges is that for the first cubit or so of digging the work-load is 0;20 sar$_V$ a day, and is called the 'throw-out' work *šilūtum*, from *šalûm* 'to throw out'. The lower levels are named after the carrying basket DUSU in which the excavated earth is brought to the surface (cf. *MCT*: 88). The key to understanding these problems lies with the

[15] This term is explained in §6.3.4.
[16] Problem (vi) only.

Table 6.1: Excavation coefficients listed in order of magnitude

Tablet	*Subject*	*Depth*	*Work-load*
YBC 9874	silted canal — *atap taraḫḫim*	1 cubit	0;20 sar$_v$ a day
YBC 4666 (i)–(ix), (xii)–(xiii)[17]	subsidiary canal — PA$_5$.SIG	1 cubit	0;20 sar$_v$ a day
YBC 7164 (ix)–(xvi)	subsidiary canal — PA$_5$.SIG	1 cubit	0;20 sar$_v$ a day
YBC 4666 (x)–(xi)	subsidiary canal — PA$_5$.SIG	1.5 cubits	0;20 sar$_v$ a day
VAT 7528 (i)–(iv)	subsidiary canal — PA$_5$.SIG	1.5 cubits	0;20 sar$_v$ a day
YBC 8588	excavation — KI.LÁ/*kalakkum*	3 cubits	0;15 sar$_v$ a day
YBC 4666 (xiv)–(xxii)	subsidiary canal — PA$_5$.SIG	2 cubits	0;10 sar$_v$ a day
BM 85196 (xv)	silted canal — *atap taraḫḫim*	3 cubits	0;10 sar$_v$ a day
YBC 4666 (xxiii)	subsidiary canal — PA$_5$.SIG	3 cubits	0;10 sar$_v$ a day
YBC 7164 (i)	subsidiary canal — PA$_5$.SIG	3 cubits	0;10 sar$_v$ a day
Erm 15073 (iv)–(v)	river — ÍD	4 cubits	0;10 sar$_v$ a day
YBC 4657	excavation — KI.LÁ	6 cubits	0;10 sar$_v$ a day
YBC 5037 (iv)–(xvi)	excavation — KI.LÁ	6 cubits	0;10 sar$_v$ a day
YBC 5037 (i)–(iii)	excavation — KI.LÁ	2.5 cubits	0;07 30 sar$_v$ a day

method of 'combined coefficients' (§5.6). YBC 7164 (vi), which uses the same IGI.TE.EN terminology as other combined coefficient problems (§5.6.3), makes this clear:

I 15 PA$_5$.SIG 5 UŠ UŠ 3 KÙŠ DAGAL 3 KÙŠ BÙR.BI 1 KÙŠ *šu-‹up-lum›* $^1/_3$ MA.NA *ši-lu-tum*
2 KÙŠ *šu-‹up-lum›* 10 GÍN DUSU LÚ.1.E IGI.TE.EN U$_4$ *ši-lu-ta-am iḫ-re*
IGI.TE.EN U$_4$ DUSU *iḫ-re ù* SAḪAR EN.NAM IGI.5.GÁL U$_4$ *ši-lu-ta-am*
4 GÍN SAḪAR *is-sú-uḫ* $^2/_3$ U$_4$ *ù* IGI.5.GÁL $^2/_3$ U$_4$ DUSU *iḫ-re* 8 GÍN SAḪAR

A subsidiary canal. The length is 5 UŠ, the width is 3 cubits, its depth is 3 cubits. 1 cubit's depth is $^1/_3$ mina,[18] the 'throw-out' work. 2 cubits' depth is 10 shekels, the 'basket' work. One man dug the 'throw-out' work for what proportion of a day? He dug the 'basket' work for what proportion of the day? And what is the volume? In a fifth of a day he excavated the 'throw-out' work: 4 shekels (volume). In $^2/_3$ of a day and a fifth of $^2/_3$ of a day[19] he dug the 'basket' work: 8 shekels (volume).

This solution is slightly more complicated than the 'combined coefficient' problems as the second layer of the excavation is twice as deep as the first. In other words, for every sar$_V$ dug from the 'throw-out' layer, 2 sar$_V$ of the 'basket' layer needs to be removed. Thus:

1 sar$_V$ of 'throw-out' work	at 0;20 sar$_V$ a day	takes 3 days to complete;
2 sar$_V$ of 'basket' work	at 0;10 sar$_V$ a day	takes 12 days to complete;
the whole 3 sar$_V$		takes 15 days to complete.
Thus 1 sar$_V$		takes 5 days to complete

⇒ a combined work-load of 0;12 sar$_V$ a day.

Thus a day's 'throw-out' work is $^1/_3 \times$ 0;12 = 0;04 sar$_V$, in $^3/_{15} = {}^1/_5$ of the day,
and the 'basket' work is $^2/_3 \times$ 0;12 = 0;08 sar$_V$, and takes $^{12}/_{15} = {}^4/_5$ of the day.

Although this particular combined coefficient does not figure in other problems or the coefficient lists, those of the three-layer excavation problems do. Powell (1988: 163–4; cf. *MCT*: 85–6; 88–90) briefly discussed this type of problem. He touched on the matter of combined coefficients without mentioning them explicitly, and included a summary table of YBC 7164 (vi) and BM 85196 (xvi), with some of the following data:

[17] These last two problems concern canals which are *meant* to be dug to a depth of 1 cubit but instead are made $^2/_3$ cubit and 1$^1/_3$ cubit deep respectively. The work-load is not used in solving the problems. VAT 7528 (vii)–(viii) are similar (Muroi 1992*a*: 175–6).

[18] Error for sar$_V$, because shekels are used for sexagesimal fractions of both the mina and sar$_V$.

[19] i.e. $^4/_5$ of a day. There is no simple way of expressing this fraction in Akkadian without writing 0;48. The number has been written in the form of an ascending continued fraction here: see Goetze (1946: 212).

Table 6.2: Combined excavation coefficients

Problem	*Level*	*Work-load*	*Depth*	*Time taken*	*Output*	*Baskets*
YBC 7164 (vi)	top level	0;20 sar_v	to 1 cubit	$^1/_9$ day	0;02 13 20 sar_v	60
	second level	0;10 sar_v	1–3 cubits	$^4/_9$ day	0;04 26 40 sar_v	120
	third level	0;07 30 sar_v	3–4.5 cubits	$^4/_9$ day	0;03 20 sar_v	90
	total		to 4.5 cubits	1 day	0;10 sar_v	270
BM 85196 (xvi)	top level	0;20 sar_v	to 1 cubit	$^1/_6$ day	0;03 20 sar_v	90
	second level	0;10 sar_v	1–2 cubits	$^1/_3$ day	0;03 20 sar_v	90
	third level	0;06 40 sar_v	2–3 cubits	$^1/_2$ day	0;03 20 sar_v	90
	total		to 3 cubits	1 day	0;10 sar_v	270

In other words, although the numerical data of the two problems seem very different, they both produce a combined coefficient of 0;10 sar_V a day — the total output. This is the coefficient found in all but two problems[20] on excavations 2 or more cubits deep which use only one coefficient. Neither can it be coincidental that the output of each of the three layers in these two problems is an exact multiple of the volume of the 'basket' coefficient, the volume of a load of earth (§5.2). The problems must have been designed to be compatible with those on carrying earth — a necessary adjunct to digging a canal.

6.3.4 Philological commentary

The word KI.LÁ/*kalakkum* is discussed in *MCT*: 65. More recently, Heimpel (1987: 207) has suggested that its origin is in the Sumerian word **ka-al**, literally meaning 'mouth of the hoe', which is attested on Ur III field plans and refers to loam pits dug for making bricks (see §5.1). *Kalakkum* would be a typical formation for a Sumerian loanword in Akkadian, and KI.LÁ possibly a back-loan into Sumerian or a pseudo-logogram.

The reading of E 20, and the addition of KI, is supported by its pairing with E 21. The word used to describe the upper level of work, *šilûtum* can reasonably be understood as a *pirist* derivative of the verb *šalûm* 'to throw out' (cf. *MCT:* 88), and thus refers to the act of throwing out the earth which has been dug from the canal. The lower levels are called DUSU 'carrying basket' because the hole is so deep the earth has to be lifted out.

The first signs of D 70 have hirtherto been read ⌜$KIRI_6$⌝, but it is unclear how an orchard could have volume, and why $^1/_8$ of it should be taken. Putting together von Soden's readings, of *sēkirum* 'canal- and dam-builder' (*AHw*: 1036) from *sekērum* 'to block, to dam' (cf. VAT 8523, §6.4), and *i-za-ab-bi-il*, 'he shall carry (earth)' (1964: 47) instead of **iṣabbit*!, leads to a more plausible interpretation. It is *not* a carrying coefficient, despite the presence of the verb *zabālum* 'to transport': the numerical value would imply an improbable carrying distance of 13;20 nindan (cf. §5.2); further, one would expect it to be with the other earth-carrying coefficients in lines 36–48 of the list. Instead, it comes straight after coefficients for cutting and carrying reeds (cf. §6.4) and specifically mentions that it pertains to a canal-worker. Accordingly, it appears that D 70 states that the daily work-load for earth moving (i.e. digging and/or piling it up) is 0;07 30 sar, the same as in M 288 (§6.2) and YBC 5037 (i)–(iii) (§6.3).

The terminology used for the three levels of excavation in BM 85196 (xvi) is obscure. Von Soden (*AHw*: 306 *ḫadālum* II) suggested the translation 'etwa "zurückgehen"?' for the bottom layer, although it is unclear what this would mean in relation to canal excavation.

[20] YBC 8588: 0;15 sar_V a day; YBC 5037 (i)–(iii): 0;07 30 sar_V a day: the same work-rate as D 70.

6.4 REEDS AND DAMS

6.4.1 The coefficients

D 68	10 *šà šu-ri ka-sà-mi*	0;10, of cutting reeds.
E 48	5 20 Ú.SAG₁₁	0;05 20, reeds.[21]
F 59	40 *ša* DÙ GIŠ.*šu-rum*	0;40, of building reeds (for dams).

BM 85196 (i): I 6	$^1/_3$	$^1/_3$.
BM 85196 (v): I 26	5 SAḪAR.ḪI.A 1 GI.SA.ḪI.A	0; 05, the volume of 1 00 reed-bundles.
VAT 8523 (i): I 12	20 *ša-lu-uš-tam*	0;20, a third.
VAT 8523 (v): III 14	3 KÙŠ UŠ SIG$_4$.ANŠE 3 KÙŠ SAG *ù* 3 KÙŠ SUKUD *ša* ÉŠ.GÀR 1! LÚ	3 cubits length of a brick-pile, 3 cubits width, and 3 cubits height, of 1 man's work-load.
(vi): III 23, 24	3 45 ÉŠ.GÀR 1 LÚ	0;03 45, 1 man's work-load.
YBC 4673 (xvii): VI 1	$^2/_3$ Ú.SAG$_{11}$	$^2/_3$, the reeds.
(xvii): VI 4	$^1/_3$ SAR SAḪAR	$^1/_3$ sar, the earth.
(xviii): VI 6	5 GÍN Ú.SAG$_{11}$ <KU$_5$>	5 shekels, cutting reeds (?).
(xix): VI 10–11	6 GÍN Ú.SAG$_{11}$ ÍL.ÍL	6 shekels, carrying reeds.
(xx): VI 14	$^2/_3$ Ú.SAG$_{11}$	$^2/_3$, reeds.
(xxi): VI 17	10 ÉŠ.GÀR	0;10, the work-load.

6.4.2 Mathematical commentary

There are four problems on the volume of a reed-bundle GI.SA,[22] which all give the same data: the bundles are 6 cubits (3 m) high, with an upper and lower circumference of 0;02 and 0;04 nindan (20–40 cm) respectively, and volume 0;00 05 sar$_V$. The absolute values of these figures are fixed by twò pieces of information. First, BM 85196 (ii) describes the upper circumference as 2 ŠU.SI GÚR AN.NA, i.e. 0;02 nindan, in the range of fingers (as opposed to 2 nindan, which would be the obvious interpretation without the gloss) (cf. Friberg 1987–90: 536–7). Second, BM 85196 (v) states that a boat of 4 sar$_V$ contains 48 00 of these bundles, where the initial 4 is written as four ŠÁR signs, i.e. 4 × 3600. Thus 1 sar$_V$ contains 12 00 bundles, each of whose volume is 1/12 00 = 0;00 05 sar$_V$.

At first sight the dimensions of these bundles may seem rather odd, with a diameter of a few centimetres and a height of 3 metres, but it cannot be coincidental that the metrological unit GI 'a reed' is defined as 6 cubits, exactly the height of these bundles (Powell 1987–90: 463).[23] Further, these measurements seem to have been widely assumed for a bundle of reeds in OB maths, as they also underlie the mathematics of carrying and storing reeds (§5.5, §7.4).

Detailed studies of BM 85196 (i) and VAT 8523 (i)–(vi) have been made by Powell (1988: 166–8). He has shown that both sets of problems concern dams which are composed of two different layers in the proportion 2:1.

The top third of the *appum* (literally 'nose') of BM 85196 (i) is 'earthworks' SAḪAR.ḪI.A, while the lower part, presumably underwater, is called 'reed-bundle work' GI.SA.KIN. The shape

[21] For the reading Ú.SAG$_{11}$ for Ú KIN see §6.4.3.

[22] BM 85194 (xiv) and (xv); BM 85196 (ii) and (v).

[23] Jens Høyrup (pers. comm. 1995) notes that this is also the length of the 'original reed' in the 'broken reed' problems of VAT 7532 and Str 368.

of the construction must be a triangular block, although the method for calculating its volume is corrupt (Powell 1988: 167).

Note that the lower part, at 6 nindan high, is exactly the height of the reed-bundles in the problems just discussed — one of which is the next problem on the same tablet.

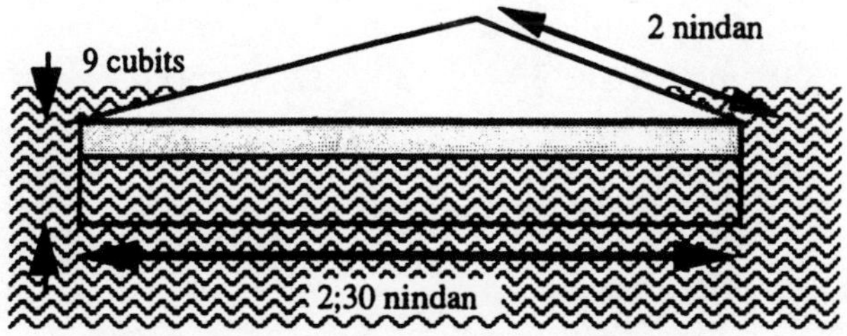

Figure 6.3: The dam in BM 85196 (i)

YBC 4673 (xvii)–(xxi), not dealt with by Powell, concern the same sort of problem:

VI $^{2}/_{3}$ Ú.SAG$_{11}$ *a-na* $7^{1}/_{2}$ SAR SAḪAR ÍL-*ma* 5 SAR Ú.SAG$_{11}$ $^{1}/_{3}$ SAR SAḪAR *a-na* $7^{1}/_{2}$ SAR ÍL-*ma* $2^{1}/_{2}$ SAR SAḪAR	Multiply $^{2}/_{3}$, the reeds, by $7^{1}/_{2}$ sar, the volume, and (you will see) 5 sar$_V$, the reeds. Multiply $^{1}/_{3}$ sar$_V$, the earth, by $7^{1}/_{2}$ sar$_V$, and (you will see) $2^{1}/_{2}$ sar$_V$, the earth.
IGI 5 GÍN Ú.SAG$_{11}$ ‹KU$_5$› DU$_8$ *a-na* 5 SAḪAR Ú.SAG$_{11}$ ÍL-*ma* 1 *šu-ši* ERÍN.ḪI.A IGI.DU$_8$	Take the reciprocal of 5 shekels, *cutting*? reeds. Multiply (it) by 5, the volume of the reeds, and you will see 60, the workers.
IGI 6 GÍN Ú.SAG$_{11}$ ÍL.ÍL DU$_8$ *a-na* 5 SAR SAḪAR ÍL-*ma* 50 ERÍN.ḪI.A IGI.DU$_8$	Take the reciprocal of 6 shekels, carrying reeds. Multiply (it) by 5 sar$_V$, the volume, and you will see 50, the workers.
IGI $^{2}/_{3}$ Ú.SAG$_{11}$ DU$_8$ *a-na* 5 SAR SAḪAR ÍL-*ma* 7 30 ERÍN.ḪI.A IGI.DU$_8$	Take the reciprocal of $^{2}/_{3}$, the reeds. Multiply (it) by 5 sar$_V$, the volume, and you will see 7;30, the workers.
IGI 10 ÉŠ.GÀR DU$_8$ *a-na* $2^{1}/_{2}$ SAR SAḪAR ÍL-*ma* 15 ERÍN.ḪI.A IGI.DU$_8$	Take the reciprocal of 0;10, the work-load. Multiply (it) by $2^{1}/_{2}$ sar$_V$, the volume, and you will see 15, the workers.

The name and shape of the object are not mentioned. All we are told is that it has a volume of 7;30 sar$_V$, and is composed of $^{1}/_{3}$ SAḪAR 'earth' and $^{2}/_{3}$ Ú.SAG$_{11}$ 'reeds'. This suggests very strongly that we are dealing with a dam, as in BM 85196 (i). The first problem of the sequence establishes the volume of each part, multiplying the coefficients $^{2}/_{3}$ and $^{1}/_{3}$ by the total volume. The second part calculates the workers needed for some operation connected with the reeds. As it precedes the problem on carrying, it must refer to something like cutting them down; compare 0;10, the coefficient for cutting reeds (D 68). Although the general meaning of D 68 is clear, it is not apparent whether this daily rate refers to the volume of reeds cut, or the area of ground covered, the criterion used in administrative texts (see §9.8.1).

The third part states that the daily rate for carrying reeds is 0;06 sar$_V$. No distance is given, but, if we assume with coefficent D 69 (§5.6) that the volume of each load is 0;00 06 40 sar$_V$, then the distance d of each trip must be:

$$\mathrm{d} = \frac{45\ 00 \times 0;00\ 06\ 40}{0;06} = 50 \text{ nindan}$$

with 45 00 ÷ 50 = 54 trips made by each worker in a day.

Finally, the text works out the labour force for building the dam. The plant-work is assembled relatively quickly, at a rate of 0;40 sar$_V$ a day. This is the meaning of the coefficient

F 59. The earthwork takes rather longer, at a rate of 10 sar$_{\text{v}}$. This is the usual rate for earthwork in OB maths: see §6.2. In all, $132^1/_2$ working days are needed to complete the dam.[24]

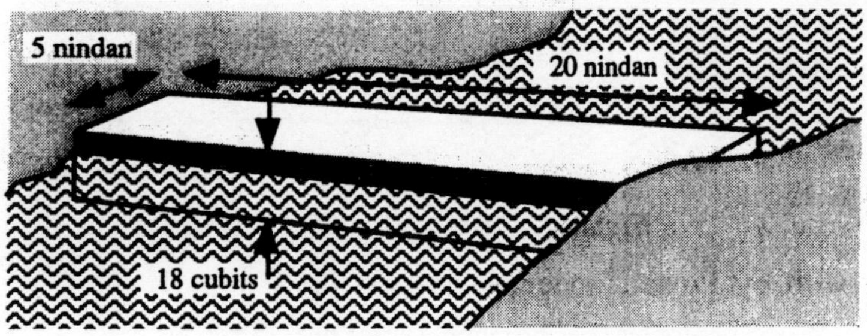

Figure 6.4: The dam in BM 85196 (i)

VAT 8523 (i) involves finding the volume of a rectangular dam (*kāsirum*) which stretches the whole width of a river. Problems (ii)–(iv) go on to find the depth, length and width of the dam respectively. The upper third is called 'earth' SAḪAR, while the lower two-thirds is referred to as KIN 'work'. These terms are almost certainly abbreviations of those used in BM 85196 (i) and mean the same thing: 'earthworks' in the top layer and '(reed-bundle) work' in the bottom.[25]

VAT 8523 (v)–(vi), the final problems of the tablet, ask how long the dam will take to build, and how much reed-work is needed for it. Once again the height of the reed-work is an integer multiple of the height of a reed-bundle. Problem (v) reads:

II 29 *šum-ma* 20 DAGAL ÍD
5 SAG *ka-sí-ri-im*
18 *iš-pi-il* 20 KIN
III 1 ÍD *ši-i i-na* UŠ *ki ma-ṣi*
is-sé-ek-ke-er
20 DAGAL ÍD GAR.RA
5 SAG *ka-sí-ri-im* GAR.RA
18 *šu-up-lam* GAR.RA 20 KIN GAR.RA
3 KÙŠ UŠ SIG$_4$.ANŠE 3 KÙŠ SAG
ù 3 KÙŠ SUKUD *ša* ÉŠ.GÀR 1! LÚ GAR.RA
aš-šum ḫe-pi-tum
ša-lu-uš-ti 3 SUKUD EN.NAM
1 SUKUD *sà-an-qum*
⌜15⌝ UŠ SIG$_4$.ANŠE *it-ti* 15 SAG SIG$_4$.ANŠE
šu-ta-ki-il-ma 3 45
3 45 *a-na* 1 SUKUD *sà-an-qi*
ÍL 3 45 ÉŠ.GÀR 1 LÚ
ša ÍD *sé-ke-ri-im*
IGI 3 45 ÉŠ.GÀR 1 LÚ
pu-ṭur-ma 16
16 *a-na* 20 KIN ÍL 5 20
i-na 5 20 UŠ *is-sé-ke-er*

If the width of the river was 20, the width of the dam was 5, it was 18 deep, the reed-work was 20 00, in how many (daily) lengths is that river dammed? Put down 20, the width of the river. Put down 5, the width of the dam. Put down 18, the depth. Put down 20 00, the reed-work. Put down 3 cubits, the length of a brick-pile, 3 cubits, the width, and 3 cubits, the height, of 1 man's work-load. Because it is broken, what is a third of 3, the height? The corrected height is 1. Multiply 0;15, the length of the brick-pile, by 0;15, the width of the brick-pile, and (you will see) 0;03 45. Multiply 0;03 45 by 1, the corrected height. 1 man's work-load for damming a river is 0;03 45. Take the reciprocal of 0;03 45, and (you will see) 16. Multiply 16 by 20 00, the reed-work. (You will see) 5 20 00. It is dammed in 5 20 00 (daily) lengths.

Problem (vi) reverses the last part of the calculation to find the KIN again. Some of the terminology is difficult; past editors have had to twist the meaning of the text to produce a consistent interpretation. In particular, the phrase *i-na* UŠ *issekker* in the first and last lines of the

[24] The dimensions of the construction are not given, but it could plausibly be a rectangular block 5 nindan × 2 nindan × 9 cubits high, i.e. four times the size of the dam in BM 85196 (i) and 40 times smaller that that in YBC 8523 below. The height of the reeds would once again be 6 cubits, as expected.

[25] Cf. Powell (1988: 266–7), who concludes that the top layer is Ú.SAḪAR 'vegetation and earth' while the bottom is brickwork, because of the phrase SIG$_4$.ANŠE 'brick-pile' in problem (v). But, as argued below, this is part of a later, spurious addition to the text; further, YBC 4673 (v) and coefficient H 3 (§5.6) show the daily rate of building a brick-pile to be 0;04 10 sar$_{\text{v}}$, or 0;30 sar$_{\text{b}}$ of Type 2 bricks.

problem[26] refers neither directly to the men needed (*TMB*: 126–7; Nemet-Nejat 1993: 38), nor to the length built (*MKT*: I 377). We have seen how the primary concern of other work-rate problems (e.g. in §6.1) is to find the length one worker can complete in a day. The *number* of these (daily) lengths is the quantity to be found here, equivalent to the days needed to finish.

The details of the first half of the solution are difficult to understand. Lines 6–7 state that the daily rate of damming is 0;15 × 0;15 × 3 = 0;11 15 sar$_V$. However, only a third of this work-rate is used (lines 11–14), because of an unclear justification in line 8. This may refer to a 'broken' day in that other tasks have to be carried out as well. The final lines follow the usual method of solution for labour calculations:

$$\text{time taken} = \text{volume} \div \text{daily rate.}$$

This work-load does not appear to relate to those found in the VAT 4673 problems, but it is the reciprocal of the coefficient E 48. This may be the key to a plausible interpretation of the problem. 0;05 20 sar$_V$ is attested twice in Ur III administrative documents as the daily rate of piling up dykes (**ég si-ga**, §9.6.3). If we assume, for the moment, that this is the meaning of E 48, and apply it to the dimensions of the reed part of the dam in this problem, then the daily length constructed would be:

$$\frac{0;05\ 20}{12\times 5} = 0;00\ 05\ 20 \text{ nindan,}$$

or just over half a finger. One may imagine a situation in which VAT 8523 (v)–(vi) are expanded versions of simple statements of problems and their answers, such as many of those on YBC 4669 and YBC 4673. The original task of (v), to find the daily length, was misunderstood, and was furnished with a spurious or corrupt solution in order to justify the answer.

W 47a concerns the work involved in constructing an irrigation dyke. It was first published as an Ur III administrative document of unknown provenance (Jones and Snyder 1961: no. 38), but Civil (1994: 119 & n. 21) has recently suggested that it may be a school exercise. It contains no date, totals, toponyms or personal names, and appears to be unfinished. Further, some of the phrasing is rather odd, and a mistake has been made in one of the calculations.

1 05 nindan uš 1 nindan	1 05 nindan length,
$^{2}/_{3}$ <sar> saḫar-ta	$^{2}/_{3}$ sar volume per nindan,
ég-šè ì-íb-si	piling up levees:
saḫar-bi 43$^{1}/_{3}$ sar	its volume is 43$^{1}/_{3}$ sar;
á-guruš-bi 4 20 u$_4$-1-šè	its workers are 4 20 days.
1 05 nindan uš ég-e ú-sag$_{11}$-e	1 05 nindan length, reeds on a levee;
guruš-e u$_4$-1-šè 1 nindan uš in-ak	a worker does 1 nindan length a day:
á-guruš-bi 1 05 u$_4$-1-[šè]	its workers are 1 05 days.
5 iku guruš-e u$_4$-1-šè	5 iku; a worker for 1 day
al-bi 2$^{1}/_{2}$ sar-ta in-ak	does 2$^{1}/_{2}$ sar of hoeing:
á-guruš-bi 3 20 u$_4$-1-šè	its workers are 3 20 days
mu-1-a-kam	in the first year.
5 iku guruš-e u$_4$-1-šè	5 iku; a worker for 1 day
al-bi 3 sar-ta in-ak	does 3 sar of hoeing;
á-guruš-bi 2 53 u$_4$-1-šè	its workers are 2 53 days
mu-2-kam	in the second year.
giš-apin u$_4$-1-šè	Ploughing for one day ...

[26] Also in problem (vi): III 20 *i-na* 5 20 UŠ *is-sé-ke-er* and III: 20, 25 5 20 UŠ *ša* ÍD *is-sé-ke-er*.

No coefficients are stated for the first two tasks, but these can be reconstructed from the figures in the text. The first job is to pile up earth for the dyke at a rate of:

$$43;20 \div 43\ 20 = 0;10 \text{ sar}_V \text{ a day;}$$

this rate is identical to that used most commonly in OB mathematics both for digging earth (§6.3) and for heaping it up (§6.2). Reeds are then placed on the same length of dyke at a rate of 1 nindan a day. Civil (1994: 123) has calculated that if the reeds were $^1/_3$ of the volume of the earth bank, then the work would be done at a daily rate of:

$$0;40 \div 3 = 0;13\ 20 \text{ sar}_V.$$

However, assuming still that we are dealing with a school text, it might be more reasonable to assume that the volume of reeds is a quarter of that of the earth, or a fifth of the levee as a whole, whose cross-section would then be 0;50 sar_a. In that case, the work would be carried out at a rate of:

$$0;40 \div 4 = 0;10 \text{ sar}_V \text{ a day,}$$

a quarter of that given in YBC 4673 (xx) but identical to the rates given in Ur III administrative documents for reinforcing dykes with reeds (§9.6.3).[27] The proportions of earth to reeds are different to those found in the OB dam problems because we are dealing with two different structures, but the measurements in the text are all round numbers, commensurate with the figures typically found in school texts.

The final sections are unique amongst known model documents in that they deal with hoeing. The 5 iku_a (= 8 20 sar_a) area is hoed twice, at $2^1/_2$ and 3 sar_a a day. The last number of workers is incorrect, as

$$8\ 20 \div 3 = 2\ 46;40 \neq 2\ 53.$$

There is no obvious explanation for this error.

6.4.3 Philological commentary

Civil (1994: 70, 122) established the reading Ú.SAG$_{11}$ for Ú KIN on the basis of a variant Ú.SÁG (SÁG = PA+GAN) in line 8 of the *Farmer's Instructions* (cf. §9.6). He identifies GI.Ú.SAG$_{11}$ with Akkadian *šūrum* 'reeds cut [...] and piled to reinforce the levees [...] and barrages'.

The word order of F 59 (if correctly read) is unusual: one would expect **ša šūrī epēšim*. Nevertheless, the interpretation given here is more satisfactory than Neugebauer and Sachs' untranslated *ša*-KAK-UD-*šu-rum* (*MCT*: 135), which was not improved on either by Kilmer (1960: 301) or by Nemet-Nejat (1993: 221).

The constructions in BM 85196 (i) and VAT 8523 (v) are described as having a depth, BÙR/*šuplum* respectively, where one would expect them to have a height SUKUD/*mēlûm* (cf. Waetzoldt 1990: 16–17). This may be because the majority of the dam was under the surface of the water.

6.5 IRRIGATION

6.5.1 The coefficients

D 33 48 IGI.GUB *šà* PA$_5$.SIG 0;48, the coefficient of a subsidiary canal.

[27] One could also argue for the reeds being half of the volume of the earth, or a third of the total. The cross-section would be 1 sar_a, and the work-rate 0;20 sar_V.

E 26	48 *me-e* A.ŠÀ	0;48, the waters of a field.
D 34	1 12 ⌈IGI.GUB⌉ *šà šu-pu-ul-ti*	1;12, the coefficient of a ....
D 35	30 ⌈IGI⌉.[GUB] *šà pu-ut* ÍD$^{!?}$*he-ri-im*	0;30, the coefficient of the width of a dug river.
B III 24′	48 A.MEŠ *pat-ti*	0;48, the waters of a canal.
B III 21′	1 36 A.MEŠ *mi-lu-ti*	1;36, the waters of flooding.
B III 22′	1 12 A.MEŠ GIŠ.*zi-rí-qum*	1;12, the waters of a sprinkler.
B III 23′	1 48 A.MEŠ *pat-ti*	1;48, the waters of a canal.
TMS 25 (i): I 7	48 IGI.GUB PA$_5$	0;48, the coefficient of a canal.

6.5.2 Mathematical commentary

TMS 25 (i) is the only extant problem on irrigation which explicitly uses a coefficient; the statement of the problem, and the first few lines of its solution, are missing:

[...] *a-na* X X [...]
[...] UŠ$^?$ 5 TA.ÀM *šà* 6 12 [ŠÁR ...]
IGI [5] *pu-ṭú-úr* ⌈12 *ta*⌉*-mar* 12 *a-na* 6 12 ŠÁR *i-ší-ma*
1 12 *ta-mar* 1 ŠÁR 12 *šu-ši mu-ú a-na* 40 ERÍN.ḪI.A ⌈GAR⌉
IGI 40 ERÍN *pu-ṭú-úr* 1 30 *ta-mar* 1 30 *a-na* 1 12 *i-ší-ma*
1 48 *ta-mar* ‹1 48› *mu-ú šà* ERÍN.1.⌈KAM⌉ [...] GAR 30 DAGAL
[1] 48 *mu-ú šu-up-lu mi-nu* IGI 48 IGI.GUB PA$_5$ *pu-ṭú-*[*úr*]
[1 15 *ta*]*-mar* 1 15 *a-na* 1 48 *me-e*$^!$ ⌈*šà* ERÍN.1.KAM *i*⌉-[*ší-ma*]
[2 15 *ta*]*-mar* IGI 30 DAGAL *pu-ṭú-úr* 2 *ta-mar*
[2 15 *a*]*-na* 2 *i-ší-ma* 4 30 *ta-mar* 4 30 *šu-up-lu*

[... ... ...] 5 each, of 6 12 00 [00 00 ... ...] Take the reciprocal of [5 00 00]. You will see 0;00 00 12. Multiply 0;00 00 12 by 6 12 00 00 00 and you will see 1 12 00. Put down 1 12 00, the water for 40 00 workers. Take the reciprocal of 40 00, the workers. You will see 0;01 30. Multiply 0;01 30 by 1 12 00 and you will see 1;48. Put down ‹1;48›, the water for 1 worker [...]. (If) the width is 0;30 and the water is [1];48, what is the depth? Take the reciprocal of 0;48, the coefficient of a canal. You will see [1;15. Multiply] 1;15 by 1;48, the water for 1 man, and you will see [2;15]. Take the reciprocal of 0;30, the width. You will see 2. Multiply [2;15] by 2 and you will see 4;30. The depth is 4;30.

Powell (1988: 165–6) deduced that the problem concerns a canal containing 1 12 00 sar$_V$ of water,[28] which is to be shared out amongst 40 00, or 2400, workers. The water available to each man is 1;48 sar$_V$. The available water is divided by the coefficient 0;48. In other words, only 0;48, or $^4/_5$, of the total volume of the canal allocated to each man, 2;15 sar$_V$, is available for irrigation. Either the canal is only four-fifths full, or it may not be drained completely to the bottom. The depth of the canal is calculated by dividing this volume by the width.[29]

The same irrigation coefficient, 0;48, is implicit in YBC 7164 (xvii)–(xviii):

II 17 PA$_5$.SIG 1 KÙŠ DAGAL 1 KÙŠ BÙR.BI *me-e pa-ta-a-tim*
EN.NAM *a-ra-nim* 4 ‹*a*›*-ra-nim*

A subsidiary canal. The width is 1 cubit, its depth is 1 cubit. What is the water of the canals? ... 0;04 (sar$_V$) ....

[28] Although the first few lines are unclear, they seem to refer to the conversion of the water from capacity measure (sila) to volume measure (sar$_V$), using the metrological conversion coefficient 5 00 00 (§7.1), although it is unclear how this can be related to the extant traces.

[29] Division by the length is omitted, because it is 1 nindan. Trivial operations such as this are normally included in the solution for paradigmatic reasons; the omission suggests that a length of 1 nindan per worker is an instrinsic feature of this type of problem.

II 19 PA$_5$.SIG 2 KÙŠ DAGAL $1^1/_2$ KÙŠ BÙR.BI *me-e pa-ta-a-tim*
EN.NAM *a-ra-‹nim›* 12 *a-ra-‹nim›*

A subsidiary canal. The width is 2 cubits, its depth is $1^1/_2$ cubits. What is the water of the canals? ... 0;12 (sar$_V$) ....

The volumes of a 1 nindan stretch of the canals are 0;05 sar$_V$ and 0;15 sar$_V$ respectively. The answer in each case is 0;48 of the volume. This confirms both that the irrigation coefficient represents 0;48 of the total volume of the canal, and that the parameters of the problem imply that the length of the canal is 1 nindan.

The coefficients of text B are clearly related to this type of problem. B III 24′ is identical to D 33 and E 26, while B III 23′ seems to refer to the available water per nindan length as shown by *TMS* 25: 6. The two preceding entries in the list must refer to analogous coefficients for different types of irrigation. If in both cases the water available is a constant four-fifths of the volume of the canal, then these two coefficients reflect the usable volume of smaller irrigation channels. For B III 21′ and 22′ a 1 nindan length of canal would have volume 1;36 ÷ 0;48 = 2 sar$_V$, and 1;12 ÷ 0;48 = 1;30 sar$_V$ respectively.

It would appear that the hitherto uninterpreted coefficients D 34 and 35 also concern irrigation, if only on the grounds that their immediate predecessor in such a well-organized list clearly does, and that they do not appear to refer to brick-making as their successors D 36–8 do (§5.1). While the first, like B III 22′, may give the available water in a canal of cross-section 1;30 sar$_a$, the second may be the width of such a canal: 0;30 nindan, as in *TMS* 25 (i). This would lead to depths of 4, 3 and 4;30 cubits for the watercourses of B III 22′–24′ respectively; on the other hand, YBC 7164 (xvii)–(xviii) give much smaller widths, and in general one would expect the width of a canal to be a variable rather than a constant in OB mathematics. The reading and interpretation of these two lines remain uncertain while there is no published photograph[30] and the tablet is unavailable for collation.

E 12 (§8.5.4) is unlikely to be an irrigation coefficient with value 0;48 (*pace* Kilmer 1985: 27), because it would then be duplicating E 26 from the same text.

6.5.3 Philological commentary

Text B uses a number of different words to describe the irrigation coefficients. *Mīlûtum* 'flooding' seems to be an abstract noun derived from the widely attested *mīlum* 'flood'. *Zirīqum*, in non-mathematical contexts a 'wooden or clay pipe', is from the verb *zarāqum* 'to sprinkle, to scatter', from which the more apposite translation 'sprinkler' is taken. This coefficient takes the least water from the canal and provides the lightest irrigation.

The word *a-ra-nim* in YBC 7164 (xvii)–(xviii) is particularly obscure. Both dictionaries read *awirānum* 'meadows' (*AHw*: 1546), but the reading *-num* for NIM is not common in OB. The position of the word after EN.NAM would normally lead one to expect a verb whose object is 'the water of the canals', or a noun in a question like '(As for the) water of the canals, what is the *a-ra-nim*?'.

The interpretations of D 34–5 given here are very tentative, being dependent solely on the published copy. *Šupultum* (or *šuppultum* 'very deep'?), derived from *šapālum* D 'to make deep', while admittedly otherwise unattested, seems a more promising reading for an irrigation coefficient than *šubultum*, 'an ear of corn' or *šūbultum* 'a gift' (*AHw*: 1258; cf. *TMS*: 29). It is not yet clear exactly how to translate this word. The verbal adjective *ḫerûm* is a derivative of *ḫerûm* 'to dig', also found in the problem YBC 7164 (vi), §6.3.

[30] *TMS*: pl. IV is of the obverse and reverse only; these lines are on the lower edge of the tablet.

6.6 OTHER WORK-LOADS

6.6.1 The coefficients

A 27 6 IGI.GUB ‹ÉŠ›.GÀR LÚ.1.KAM — 6, the coefficient of 1 man's work-load.

BM 85194 (xxiii): IV 47 — *i-na* U$_4$.1.KAM 20 *im-ḫa-aṣ* — In 1 day he wove 0;20.

6.6.2 Mathematical and philological commentary

It is unclear which task A 27 refers to; it may be the reciprocal of 0;10 sar$_V$, the well-attested earthwork coefficient (§§6.2–3), or the work-rate 0;06 sar$_V$ for some other activity.

In BM 85194 (xxiii) it is shown that a length of *ṣubātum* cloth 48 (nindan?) long, woven at a rate of 0;20 (nindan?) a day, takes 4 months, 24 days to finish (Waschow 1936–7: 246).

The 'catalogue' part of text C refers to some agricultural work-rates which are otherwise unattested in OB mathematics:

75	*[iš]-ka-ar e-⸢ṣe⸣-[di] ⸢e-pé-ša⸣-am*	To calculate the harvest work-rate.
	[iš]-ka-ar za-ri-i ša še-im e-pé-⸢ša-am⸣	To calculate the winnowing work-rate.
	[iš-ka-ar ...] e-pé-ša-⸢am⸣	To calculate [the work-rate of ...].
	iš-ka-ar [... e-pé-ša-am]	[To calculate] the work-rate of [...].

Nemet-Nejat (1993: 87) suggests that BM 85196 (xii), YBC 4673 (xvi) and YBC 9856 (i) all concern work-rates, for making rope, refining wool, and for combining unnamed coefficients repectively. The texts are not clear enough to confirm this; BM 85196 (xii) is probably related to the problem which follows it on the tablet, which is about carrying sheaves of corn (§5.2). Erm 15073 (viii) seems to involve labour calculations but contains no explanation of their purpose. It is not at all clear that the number 3 45 refers to the daily rate for building an earth wall (§6.1) (cf. Nemet-Nejat 1993: 51).

6.7 SUMMARY

Dams are constructed in two sections, with the lower two-thirds consisting of reed-bundles below water, and the upper third made of earth (§6.4). It is assumed that 0;48 of the water in a canal is used for irrigation, whatever its size (§6.5).

Table 6.3: Summary of daily work-rates in OB mathematics

Task	*Daily rate*	*Reciprocal*	*Section*
Building a brick-pile	0;04 10 sar$_v$	14;24 days	5.6
Building a dam?	0;05 20 sar$_v$	11;15 days	6.4
Building an earth wall	0;03 45 sar$_v$	16 days	6.1
Building earthworks	0;10 sar$_v$	6 days	6.2
Cutting reeds	0;05 sar$_v$; 0;10 sar$_v$	12 days; 6 days	6.4
Digging deeper than 1 cubit[31]	0;10 sar$_v$	6 days	6.3
Digging to a depth of 1 cubit	0;20 sar$_v$	3 days	5.1; 6.3
Knocking down walls	0;15 sar$_v$; 0;20 sar$_v$	4 days; 3 days	5.6
Mixing brick mixtures	0;10 sar$_v$	6 days	5.1
Moulding bricks	0;20 sar$_v$; 0;25 sar$_v$?	3 days; 2;24 days?	5.1
Stacking reed-bundles	0;40 sar$_v$	1;30 days	6.4
Weaving cloth	0;20 nindan?	3 days	6.6

[31] For exceptions see the table in §6.3.3.

The procedure for simple labour calculations is the same whatever the task is:

$$\frac{\text{total volume in sar}_v}{\text{daily rate in sar}_v} = \text{days needed to complete task.}$$

When two are more tasks are to be done in a day the combined coefficient method is used (see §5.7). The daily rates, or work-loads, attested in OB mathematical problems and coefficients are listed in Table 6.3 above. The reciprocal of the daily rate is the time taken for one labourer to complete 1 sar_v of work. The daily rates for carrying are listed in §5.7.

7. STORAGE AND METROLOGY

For the most part measurements were converted from one unit to another by referring to metrological lists and tables (see §1.3.4). Under certain circumstances, though, these conversions were explicitly mentioned as a step in the solution of a problem, carried out by using coefficients. This chapter discusses those coefficients and explores the conditions under which they were, or were not, used. The large topic of brick metrology has already been dealt with in §4, but is touched on again in §7.2.

The relationships between metrological units have long been the subject of study, first brought together accurately by Neugebauer and Sachs (*MCT*: 4–10) and most recently summarized by Powell (1987–90). A more detailed study of certain aspects of Ur III and OB period metrology, including conversion between volume and capacity measures, can be found in Powell (1984: 41–6). Earlier works on the same subject include Lewy (1947: 310–12), Muroi (1986) and Vaiman (1976).

All the OB lists except M include at least one metrological coefficient, with 11 different coefficients attested a total of 26 times. Six of these are found in a total of 8 calculations[1] and 19 problems from 7 tablets,[2] while one is implied in a catalogue entry.[3] Three further metrological coefficients not attested in the lists are found in 33 problems from 9 different tablets.[4] In addition, there are 17 metrological entries in the LB texts B and J which have OB antecedents. The many occurrences of implicit metrological coefficients have not been included in this chapter.[5]

The terminology of storage and metrology coefficients is not consistent from list to list. For instance, there are four different values attested for 'boat' coefficients, all of which have other names too (see §7.2, 7.3). Similarly, storage coefficients for particular commodities may take differing values — once even within the same list.[6] The coefficients are grouped by value in §7.2–4, as this is what determines their function, but they are also listed by name in the summary section, §7.5.

[1] *UET* 6/2 218, 247, 250, 269, and 290; *UET* 6/3 374 and 452 — grain containers? *UET* 6/2 293 — earthworks? (§A.5.1.4).

[2] BM 85194 (xxx) — bricks in a boat; BM 85196 (xiv) — basket; BM 96954 + 102366 (vi), (viii)–(xi), (xiii)–(xix) — grain-piles; Haddad 104 (i)–(v) — grain containers, (vii)–(viii) — plastering (§4.6); *TMS* 14 (ii) — a grain-pile; VAT 8522 (i) — a grain container; YBC 8600 — a grain container.

[3] Text C 74′ — a grain-pile. See §A.1.

[4] BM 85194 (xvii) — a city wall; BM 85200 (i)–(xix) — holes in the ground; BM 96957 + VAT 6598 (xvii) — a brick wall; Haddad 104 (i)–(v) — grain containers, (vii) — plastering (§4.6); IM 54478 — a hole in the ground; *TMS* 14 (i)–(ii) — a grain-pile; *TMS* 24 (i)–(ii) — irrigation (§6.4); YBC 7997 — bricks in a kiln (§4.1).

[5] Nor have the possible month–day conversion coefficients (30 and 0;02) in *UET* 6/2 233, 236, 254, 293, and 298 (§A.5.1.4), as their interpretation is highly conjectural and there is no corroborative evidence for them in the rest of the OB mathematical corpus.

[6] e.g. E 22 and E 23 — see §7.4.

7.1 CONVERSION BETWEEN LINEAR MEASURES

7.1.1 The coefficients

D 52 10 *šà* 1 2 *ù* 1		0;10, of 1, 2 and 1.
BM 85194 (xvii): IV 8	12 NUMUN SUKUD	12, the ratio of the height.
BM 85200: *passim* (19 times)	12 NUMUN BÙR	12, the ratio of the depth.
BM 96957 (xvii): IV 14	5 1 KÙŠ NUMUN[!]	0;05, 1 cubit, the ratio.
Haddad 104 (i): I 3	*šu-pí-il-ma*	Deepen, and ....
Haddad 104 (ii): I 12	*šu-pí-il-ma*	Deepen, and ....
Haddad 104 (iii): I 23; II 6, 25	*šu-pí-il-ma*	Deepen, and ....
Haddad 104 (iv): II 41; III 3	*šu-pí-il-ma*	Deepen, and ....
Haddad 104 (v): III 25	*šu-pí-il-ma*	Deepen, and ....
Haddad 104: (vii): IV 1	*šu-pí-il-ma*	Deepen, and ....
IM 54478: 7, 11	12	12.
TMS 14 (i): I 13	12 *šu-up-li*	12, the depth.
TMS 14 (ii): II 2′	[12]	12.
TMS 24 (i): 24	12 *šu-up-li*	12, the depth.
TMS 24 (ii): 39	12 *šu-up-li*	12, the depth.
YBC 7997: 16	12 *šu-up-li-im*	12, the depth.
BM 96957 (xvii): IV 17	[6]	6 00.
IM 54478: 10	1	1.
UET 6/2 293	5 12	0;05, 12.

7.1.2 Mathematical commentary

There is no 'sexagesimal point' in the system of place value notation used in OB mathematics (Friberg 1987–90: 533–4). It is common practice to give the measurements of the subject of the problem in a mixture of units, to establish their absolute value. These mixed measurements are then converted into base units for calculating with:

length:	nindan	capacity:	sila
area:	$\text{sar}_a = 1\ \text{nindan}^2$	weight:	mina
volume:	$\text{sar}_v = 1\ \text{nindan}^2 \times 1$ cubit		

This is because each metrological system has an inconvenient set of relationships between units, in much the same way as the British imperial system has, which makes arithmetical operations unnecessarily complicated. Once converted, these measurements can be treated sexagesimally as pure numbers with no unit attached, and reconverted into appropriate units at the end of the calculation. The conversions are invariably made without comment, presumably with the aid of metrological lists and tables.[7]

However, whenever volumes are calculated, any height or depth stated in nindan needs to be converted into cubits, because 1 sar_V is the product of 1 sar_a area and 1 cubit height. Heights and depths measured in nindan are almost never used,[8] just as in the British imperial system

[7] These conversions are explicit in the little metrological problems such as NBC 8082. The problem and answer are given in mixed measure on the bottom right of the tablet; the working — in sexagesimal notation — is shown on the top left (*MCT*: 10; Neugebauer and Sachs 1984). See also §1.3.5.

[8] But see BM 96957 + VAT 6598 (vi)–(vii), in which the height of the house is kept in nindan to simplify the

vertical measurements are more usually given in feet than in yards.[9] In both cases the vertical unit is the smaller of the two. The transformation of nindan into cubits through multiplication by 12 is often stated explicitly, as shown by the examples above.[10] At other times heights are given in cubits at the beginning of the problem and then, of course, do not need to be converted.[11] Horizontal measurements stated in cubits, however, such as the thickness of a wall, or the width of a canal, are always given in nindan as soon as they are needed in the calculation; the conversion is never stated explicitly.[12]

A particularly clear example of this coefficient is given in IM 54478, about finding the side of a cube. Its volume is described in the statement of the problem as *mu-ša-ar ù zu-uz*$_4$ *mu-ša-ri*, 'a sar$_v$ and half a sar$_v$' (line 3). The solution reads:

7	*i-gi* 12 *pu-ṭú-ur-ma*	Take the reciprocal of 12, and [you will see] 0;05.
	⌜5⌝ [*ta-mar* 5 *a-na* 1] ⌜30⌝ *e-pé-ri-ka*	Multiply [0;05 by 1];30, your volume, and you will
	i-ši-ma 7 30 *ta-mar* 7 30	see 0;07 30. What is the cube root of 0;07 30? The
10	*mi-nam* ÍB.SI$_8$ 30 ÍB.SI$_8$ 30 *a-na* 1	cube root is 0;30. Multiply 0;30 by 1, and you will
	i-ši-ma 30 *ta-mar* 30 *a-na* 12	see 0;30. Multiply 0;30 by 12, and you will see 6.
	i-ši-ma 6 *ta-mar* 30 *mi-it-ḫa-ar-ta-ka*	Your square side is 0;30 (nindan). Your depth is 6
	6 *šu-pu-ul-ka*	(cubits).

First the volume in sar$_v$ is converted into nindan^3 by division through the height coefficient (ll. 7–9). This enables the cube root of the volume — in nindan — to be found (ll. 9–10). This is then multiplied by 1 (because there is 1 nindan in a nindan!) to give the length of the cube in nindan — a unique occurrence in published OB mathematics — and by 12 to give the depth, in cubits.

Towards the end of BM 96957 + VAT 6598 (xvii) two metrological conversions are performed in the course of finding the slope of a brick wall. The height of the wall is transformed from nindan to cubits, as one would expect, but, unusually, the coefficient is given as 0;05. Its reciprocal, 12 — the more usual form — has to be taken before the calculation can proceed. A few lines later the slope in 1 cubit height is found to be 0;00 12 30 nindan. It is multiplied by 6 00 to express this figure more appropriately as $1^1/_4$ fingers.

Although the height conversion coefficient is never stated in the lists, it is implicit in D 52. This states that a rectangular block measuring 1 nindan by 2 cubits by 1 cubit has a volume of 0;10 sar$_v$ (*TMS*: 29).

The reciprocal pair 0;05 12 in *UET* 6/2 293 should probably be understood as a linear conversion coefficient if the understanding of the whole calculation as a computation of earthworking labour rates is correct (see §A.5.1.4).

7.1.3 Philological commentary

The verb *šuppulum* (*šapālum* D) has a wide use in the non-mathematical meanings 'to make low' 'to dig down', 'to deepen', etc. 'To deepen' is used here as a simple one-word translation for the mathematical sense of 'convert into a vertical measurement'. The associated *purs-*

calculation (§4.5) (Robson 1996*a*).

[9] We would never describe someone as two yards tall, for instance.

[10] But exceptions are not hard to find: e.g VAT 8523 (i), where the height of a dam is given as $1^1/_2$ nindan in the statement of the problem (line I 4) and as 18 (cubits) in the course of the solution (e.g. line I 8). Cf. §6.5.

[11] e.g. BM 96954 (§A.3), where the grain-piles are all 48 (cubits) tall. Metrological units are not mentioned at all in problems (vi)–(xvi) but can be inferred by analogy with (xvii)–(xix).

[12] e.g. YBC 8600 — the thickness of a log (§7.2); cf. BM 96957 + VAT 6598 (i), in which the thickness of the wall, in cubits, and its height, in nindan, are both converted without coefficients during the calculation. Similarly, in VAT 8389 and VAT 8391, the area 1 bur is converted into sar$_a$ without an explicit multiplication: e.g. 30 *pu-ra-am* GAR.RA 'put down 30 00, the bur' VAT 8389: I 6.

formation noun *šuplum*, logogram BÙR, is the usual OB mathematical term for depth. Thus 12 *šuplum* can either mean '12 (cubits), the depth', or '12, the depth (coefficient)', depending on the context (cf. *MCT*: 98–9, YBC 7997). The word 'coefficient' has not yet been attested in association with this coefficient, but the fact that it has is own particular noun and verb suggests that it was considered as such. The verbal form of the circumference coefficient, *šullušum* 'to triple', is also a D-stem formation.[13]

The logogram NUMUN 'ratio' is only found in the Sippar mathematical texts.[14] It is occasionally used to denote geometrical coefficients,[15] as well as the metrological conversions listed here. It may also mean 'ratio' in other mathematical contexts: in VAT 6597[16] it is not simply a synonym of IGI.GUB. The Akkadian equivalent might be *zērum* 'seed', which also has the logographic writing NUMUN, or there may have been a separate technical term.

7.2 CONVERSION FROM VOLUME TO CAPACITY MEASURE

7.2.1 The coefficients

A 37	4 48 IGI.GUB É.KIŠIB IN.NU.⌜DA⌝	4 48 00, the coefficient of grain storage.
E 35	4 48 *ku-⌜bu-ur i-ṣi-im⌝*	4 48 00, the thickness of a log.
F 58	⌜4⌝ 48 *ša ku-bu-ur-re-e* GIŠ	4 48 00, of the thickness of a log.
F 61	4 48 É.KIŠIB ŠE	4 48 00, grain storage.
G 12	4 [4]8 IGI.GUB GIŠ.MÁ *e-le-pí-im*	4 48 00, the coefficient of a boat: a boat.
E 19	5 ⌜GIŠ⌝.MÁ	5 00 00, a boat.
D 49	6 *šà na-aš-pa-ak šà* SÌLA X-*di-im*	6 00 00, of storage, of ...ing silas.
F 29	6 *ša* GIŠ.MÁ.LÁ	6 00 00, of a cargo-boat.
G 9	6 IGI.GUB É.ÙR!.RA *na-aš-pa-ku-um*	6 00 00, the coefficient of storage.
H 6	6 *ša na-aš-pa-kum*	6 00 00, of storage.
J 8	6 É.Ì.DUB *me-e* IGI.GUB-*ú*	6 00 00, the storage of water: the coefficient.
J 9	6 É.Ì.DUB ⌜Ì?⌝ IGI.GUB-*ú*	6 00 00, the storage of oil?: the coefficient.
J 10	6 12 É.Ì.DUB ŠE *duḫ-nu* IGI.GUB-[*ú*]	6 12 00, the storage of millet: the coefficient.
J 11	6 12 É.Ì.DUB ⌜ŠE⌝ [...] X ⌜IGI.GUB⌝-[*ú*]	6 12 00, the storage of ... grain: the coefficient.
J 12	6 24 É.Ì.[DUB ... IGI.GUB-*ú*]	6 24 00, the storage [of ... : the coefficient].
J 13	6 24 ⌜É⌝.[Ì.DUB... IGI.GUB-*ú*]	6 24 00, the storage [of ... : the coefficient].
A 11	6 40 IGI.GUB É.KIŠIB	6 40 00, the coefficient of storage.
D 50	6 40 *šà na-aš-pa-ki-im*	6 40 00, of storage.
F 30	6 40 *ša* É.KIŠIB ŠE	6 40 00, of grain storage.
H 1	6 40 *i-gi-gu-ub-bi-im qù-up-pí-im*	6 40 00, the coefficient of a box.
J 7	6 40 É.Ì.DUB *še-im* IGI.GUB-*ú*	6 40 00, the storage of grain: the coefficient.
J 14	6 45 [É.Ì.DUB ... IGI.GUB-*ú*]	6 45 00, [the storage of ... : the coefficient].
YBC 8600: I 4	4 48 IGI.GUB.BA	4 48 00, the coefficient.

[13] Haddad 104 (i)–(v): §3.1.

[14] i.e. BM 85194, BM 85196, BM 95200+VAT 6599, BM 85210, BM 96954, BM 96957 + VAT 6598, VAT 6597.

[15] e.g. in BM 85210 (viii): see §§3.2, 3.4.

[16] e.g. in VAT 6597 NUMUN refers to the share each 'brother' receives in the division of silver.

BM 85194 (xxx): VI 9	5 GÁN	5 00 00, ...
BM 85196 (xiv): IV 17	5	5 00 00.
Haddad 104 (vii): IV 4	5 *mi-id-di-im*	5 00 00, ...
Haddad 104 (viii): IV 13	5 *mi-id-di-im*	5 00 00, ...
Haddad 104 (i): I 8	6 *na-aš-pa-ki-im*	6 00 00, storage.
(ii): I 17	6 *na-aš-pa-ki-im*	6 00 00, storage.
(iii): I 29, 38; II 10, 18, 29	6 *na-aš-pa-ki-im*	6 00 00, storage.
(iv): II 34; III 1, 5	6 *na-aš-pa-ki-im*	6 00 00, storage.
(v): III 12, 18, 29	6 *na-aš-pa-ki-im*	6 00 00, storage.
VAT 8522 (i): 6a	6 40 IGI.GUB *na-aš-pa-ki-im*	6 40 00, coefficient of storage.
2b	6 40 IGI.GUB	6 40 00, the coefficient.
UET 6/2 218	6 40	6 40 00.
UET 6/2 247	6 40	6 40 00.
UET 6/2 269	6 40	6 40 00.
UET 6/2 290	6 40	6 40 00.
UET 6/3 374	⸢6 40⸣	6 40 00.
UET 6/3 452	⸢6⸣ 40	6 40 00.
UET 6/2 250	7 12	7 12 00.

7.2.2 Mathematical commentary

The function of the first coefficient listed here has been explained comprehensively by Powell (1984: 42–6; 1987–90: 491). The problem text YBC 8600 calculates the capacity of a cylindrical container (a 'log') by a procedure analogous to that used for finding the area of a circle. The first half of the text reads:

1 KÙŠ *ki-ip-pa-at i-ṣí-im*
i-na ki ma-ṣi ik-bi-ir
at-ta 5 *it-ti* 5 *šu-ta-ki-il-ma*
25 *a-na* 4 48 IGI.GUB.BA
ta-ta-na-aš-ši-ma 2
2 SÌLA *ku-bu-ur* GIŠ

The circumference of a log was 1 cubit. How thick was it? You: square 0;05 by 0;05, and multiply 0;00 25 by 0;04 48, the coefficient. (You will see) 0;00 02. The thickness of the log was 2 sila.

Lines 6–12 give instructions for performing the inverse operation.

In lines 3–5 the area A of the log's cross-section is obtained from the circumference c:[17]

$$A = \frac{c^2}{4\pi} = 0;04\ 48 \times c^2$$

The coefficient implies an approximation $\pi \approx 3^1/_8$, much more accurate than the value used for geometrical purposes (see §3.1). At first sight the next line makes the seemingly nonsensical statement that an area measure is equivalent to a capacity:

$$0;00\ 02\ \text{sar}_a = 2\ \text{sila}.$$

Powell (1984: 43) has shown, however, that every 6-finger length (i.e. 0;01 nindan, or 0;12 cubits) of the 'log' has a capacity of 2 sila, i.e. that

$$0;00\ 02 \times 0;12 = 0;00\ 00\ 24\ \text{sar}_V = 2\ \text{sila}.$$

[17] Converted from cubits to nindan: see §7.1.

In short, 1 sar_v = 5 00 00 sila, the standard volume–capacity equivalence (Powell 1987–90: 493). This relationship is also expressed in E 19 and the problems Haddad 104 (vii)–(viii),[18] BM 85194 (xxx) and BM 85196 (xiv).

BM 85194 (xxx) concerns calculating the capacity measure of 1 sar_b Type 2 bricks in a 'boat'. The volume of 1 brick is multiplied by 5 00 00 to give its capacity in sila.[19] The total capacity of 1 sar_b is then 12 00 times greater (*MCT*: 96).[20] This means that the capacity of 1 sar_b of bricks in sila is conveniently 1 00 00 00 00 times greater than the volume of one of those bricks in sar_v: the two stages in the procedure just given amount to no more than moving the 'sexagesimal point' four places.

On the other hand, the 'boat' coefficients from the LB list B are rather different:

B II 3′[6 GIŠ.MÁ.LÁ] SIG₄	[6, a cargo-boat] of bricks.
B II 4′[4] ⌜30⌝ GIŠ.MÁ.LÁ SIG₄.ÁB	[4];30, a cargo-boat of half-bricks.
B II 5′[2] 15 GIŠ.MÁ.LÁ SIG₄.AL.ÙR.RA	[2];15, a cargo-boat of baked bricks.
B II 6′4 10! GIŠ.MÁ.LÁ SIG₄.²/₃-*ti*	4;10, a cargo boat of ²/₃-bricks.

These values are identical to the OB 'brick wall' coefficients (§4.5). They seem to give the number of bricks in sar_b per 1 sar_v for each brick type, allowing for a sixth of the volume to be occupied with packing material — reed mats for instance — to prevent damage during transport.[21]

The two remaining coefficients listed here are both attested in problems about the grain capacity of cylindrical containers.[22] In Haddad 104 (i) the volume of the cylinder is calculated as 0;18 45 sar_v. The text continues:

I 8 18 45 *a-na* 6 *na-aš-pa-ki-im i-ši-ma* 1 52 30 *i-li*
1 (PI) 5 BÁN 2¹/₂ SÌLA *še-a-am i-ṣú-um i-ṣa-ba-at ki-a-am ne-pé-šum*

Multiply 0;18 45 by 6 00 00, the storage, and 1 52;30 will come up. The 'log' holds 1 bariga, 5 ban, 2¹/₂ sila of grain. This is the method.

The same procedure occurs at the end of problems (ii)–(v) as well. According to the standard equivalence, 1 sar_v = 5 00 00 sila, but these texts — and several of the coefficents in the lists — use the ratio 1 sar_v = 6 00 00 sila. VAT 8522 (i) uses yet another coefficient — 6 40 00 sila — in essentially the same circumstances. Powell (1984: 45) analysed the problem and interpreted this value as 1;20 × 5 00 00. The area A_c of the cross-section of a cylinder of diameter d is

$$A_c = \pi d^2/4 = {}^3/_4 d^2,$$

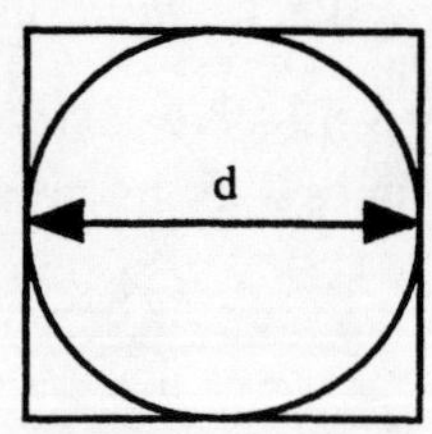

Figure 7.1: Circle and circumscribed square

when $\pi \approx 3$, as is the case in VAT 8522 (i), and Haddad 104 (i)–(v).[23] He concluded that the factor 1;20 served to convert A_c into the area of its circumscribed square A_s:

$$A_s = 0;45\ d^2 \times 1;20 = d^2,$$

so that the coefficient as a whole gave the capacity not of the cylinder but of a square-sectioned block with the same diameter. If that were its function, it is unclear why the method was not simply to square d and

[18] By which a volume of plaster is converted into a capacity measure: see §4.6.

[19] Also implicitly in YBC 4607 (i)–(v), in which the capacity measure of five different sizes of bricks is asked for and given; no method of solution is shown (see also §4.0).

[20] This simple interpretation remains more convincing than Powell's (1987–90: 491).

[21] See §4.3 for the identity of these brick types.

[22] See §A.5.1.4 for a discussion of the possible storage coefficients in the seven *UET* calculations.

[23] See §3.1.

multiply by the standard conversion coefficient 5 00 00.

Powell's analysis was made before the publication of Haddad 104, and so understandably fails to explain the use of the coefficient 6 00 00, which itself could be considered as 1;12 × 5 00 00. This would give a theoretical cross-section of:

$$A_t = 1;12 \times 0;45d^2 = 0;54d^2$$

whose significance is unclear. Further, the storage coefficients in the LB list J, all in the range of 6 00 00 to 6 45 00,[24] also suggest that a different interpretation is needed (see Table 7.1).

Table 7.1: Commodity storage coefficients in List J

Coefficient	*Cross-section*	*Commodity*	*Entries*
6 00 00 = 1;12 00 × 5 00 00	$^{9}/_{10}\ d^2$	water; oil?	J 8; J 9
6 12 00 = 1;14 24 × 5 00 00	$^{93}/_{100}\ d^2$	millet; another grain (broken)	J 10; J 11
6 24 00 = 1;16 48 × 5 00 00	$^{24}/_{25}\ d^2$	both commodities missing	J 12; J 13
6 40 00 = 1;20 00 × 5 00 00	d^2	barley	J 7
6 45 00 = 1;21 00 × 5 00 00	$^{81}/_{80}\ d^2$	commodity missing	J 14

When thought of as variants on the standard coefficient 5 00 00 these values do not produce theoretical cross-sections which conform to any obvious geometrical patterns, as the table shows. This late list hints, rather, at the possibility that the coefficients are commodity-specific: i.e. that liquids such as water and oil have a smaller capacity equivalence to fine grains such as millet, and that there are coefficient for cereals of three further grades of coarseness too.[25] This evidence alone would not be enough to discredit Powell's thesis of a theoretically square grain container embodied in the grain capacity coefficient 6 40 00; it is over a thousand years younger than the OB texts, and from a badly copied source whose purpose was not primarily mathematical (see §2.2.2). Yet, taken together with Haddad 104 (i)–(v), it suggests that the OB linkage between capacity and volume measure is rather more complex than it might seem, as Powell (1984: 45) himself acknowledged.[26] Friberg (1987–90: 568) attributes the existence of differing capacity coefficients to the use of different sizes of sila. Although such a phenomenon is undoubtedly documented in economic records and, indeed, would be unavoidable without some system of rigorously imposed and generally accepted standards (Powell 1987–90: 493), the possible reasons Friberg (1987–90: 568) gives for its reflection in OB maths are unconvincing.[27]

7.2.3 *Philological commentary*

Like many other mathematical terms, *našpakum* 'storage' is a *mapras* noun, from *šapākum* 'to heap up'. In non-mathematical contexts it means 'storeroom' or 'granary'. The logogram for

[24] Published two years after Powell's article.

[25] In this respect OB capacity measure may resemble the (now virtually obsolete) imperial fluid ounce more than the litre. Culinary measuring jugs were until relatively recently marked with different fluid ounce calibrations for flour, water, rice, raisins, etc., in proportion to the size of the individual particles and therefore also to the amount of space between them. See also §7.3 below.

[26] Powell's survey of Mesopotamian metrology (1987–90) does not discuss storage coefficients.

[27] 'Different regional or ethnic origins' can be discounted because of the widespread use of both 6 40 00 and 6 00 00 in the same lists; similarly, the problems set in Haddad 104 (i), VAT 8522 (i) and YBC 8600 are not 'different intended areas of application', although each uses a different coefficient; the numerical details of individual 'administrative and fiscal practices' are not a concern of OB maths, and are unlikely to have been as consistent over as wide a geographical area as the coefficients.

the verb is DUB, and the noun can be represented by Ì.DUB, as well as by É.UŠ.GÍD.DA[28] and ARÀH.[29] The mathematical logogram which was first read É.DUB (*MCT*: 134) is more likely to be É.KIŠIB, which is a well-attested writing for *bīt kunukkim* 'sealed room, storehouse' in non-mathematical sources. However, there seems to be no mathematical distinction between *našpakum* and É.KIŠIB; lists A and F use the logogram throughout while the other lists use the syllabic writing. The storage coefficients are referred to occasionally as 'boats',[30] presumably because boat loads were measured in capacity units in the Ur III and OB periods (Krecher 1993–7: 161; Stol 1993–7: 167).

The reading *quppum* 'vessel' in H 1 is Kilmer's (1985: 24); the reading of GÁN in BM 85194 (xxx) is uncertain (cf. Powell 1987–90: 491). The name of the coefficient in Haddad 104 (vii)–(viii), *middum*, is an otherwise unattested *pirs*-formation from the verb *madādum* 'to measure'.

The last half of D 49 is unclear, but is certainly not the meaningless but widely cited *šà-al-šu-di-im* (e.g. Nemet-Nejat 1993: 200). The published photo (*TMS*: pl. 4) shows the second of these signs to be a clear *qa*/SÌLA, without the initial vertical of the hand-copy. The following signs must comprise a verb in the genitive infinitive, with root qCd, or ᵓCd, Cᵓd or Cdᵓ respectively. It was not possible to have list D collated.

7.3 GRAIN-PILES

7.3.1 The coefficients

C 72	*e-le-pu-um* 7 12 *i-gi-gu-bu-ša*	A boat: 7 12 00 is its coefficient.
D 51	7 12 *šà na-aš-pa-ak* GUR_7	7 12 00, of the storage of a grain-pile.
H 7	7 30 *ka-ru-um*	7 30 00, a grain-pile.
J 15	7 30 [É.Ì.DUB... IGI.GUB-*ú*]	7 30 00, [storage of ...: the coefficient].

BM 95954+ BM 102366 + SÉ 93

(viii): II 4′	1 30 IGI.GUB.BA	1 30, the coefficient.
(ix): II 13′	1 30 IGI.GUB	1 30, the coefficient.
(xiii): III 5′	1 30 IGI.GUB	1 30, the coefficient.
(xiv): III 18′	1 30 IGI.GUB	1 30, the coefficient.
(xv): III 33′	40	0;00 40.
(xvi): IV 6	40	0;00 40.
(xvii): IV 19	1 30	1 30.
(xviii): IV 26	[1] 30	[1] 30.
TMS 14 (ii): II 4′	8 *na-aš-pa-ak* GUR_7	8 00 00, storage of a grain-pile.

7.3.2 Mathematical commentary

There are two surviving sets of problems on grain-piles, each of which uses a slightly different coefficient and method of solution.[31] BM 96954 + BM 102366 + SÉ 93 contains twelve or

[28] See E 23, §7.4.

[29] Thus *ur-ru-uḫ* SAG.3 (J 2) almost certainly refers to the floor-area of an equilaterally triangular granary; cf. §3.3.

[30] Not related to the geometrical figure 'barge', which is written (GÁN) (GIŠ).MÁ.GUR_8 (§3.5).

[31] They are now discussed, with comparative material from other cultures, by Friberg (1996*b*).

thirteen extant problems from an original twenty or so.[32] The final three, (xvii)–(xix), deal with relatively simple volumes: a triangular, a trapezoidal and a semicircular[33] prism respectively (Fig. 7.2). The first sixteen, nine of which are extant, concern piles in the shape of a rectangular pyramid with an elongated apex, like a pitched roof (see Figure 7.3 below).

The method of solution used corresponds to the correct algebraic formula,

$$C = \frac{khw}{3}(l + \frac{t}{2}),$$

where

C := grain capacity, 28 48 00 gur; l := length of pile, 10 nindan;
h := height of pile, 48 cubits; t := top of pile, 4 nindan;
w := width of pile, 6 nindan; k := sar_V–gur conversion, 1 30.

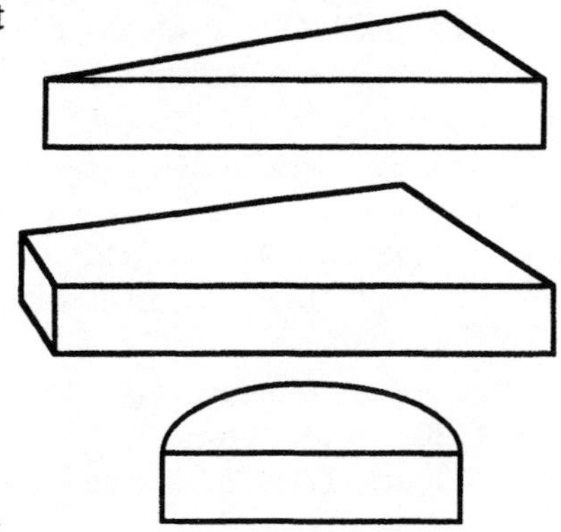

Figure 7.2: Grain-piles in BM 96954 (xvii)-(xix) (to scale)

The formula can be derived by considering the solid as a triangular prism with half a rectangular pyramid either side of it:

$$\begin{aligned} V &= \text{triangular prism} + \text{rectangular pyramid} \\ &= \frac{hwt}{2} + \frac{hw(l-t)}{3} \\ &= \frac{hwl}{3} + \frac{hwt}{6} = \frac{hw}{3}(l + \frac{t}{2}). \end{aligned}$$

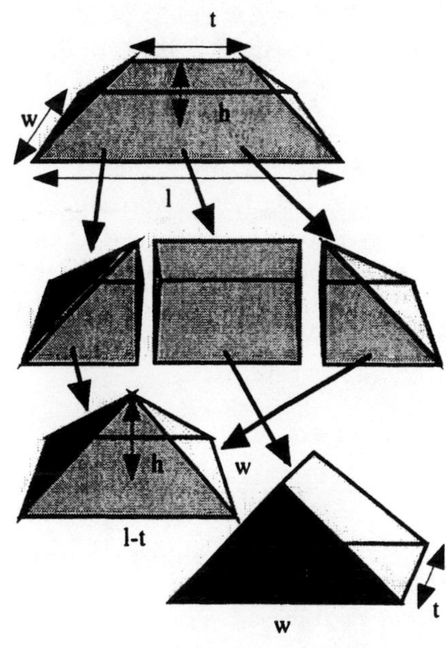

Figure 7.3: Construction of the grain-pile

The correct formula for the volume of a triangular prism is used in at least three other OB texts,[34] but problems on whole pyramids are otherwise unknown (Friberg 1987–90: 566).[35]

However, none of the extant problems use this formula explicitly, as they nearly all involve finding two parameters of the solid given a numerical relationship between them (see §A.3). In each case, the coefficient states that for a grain-pile the volume capacity equivalent 1 sar_V = 1 30 gur is used, equivalent to 1 sar_V = 7 30 00 sila, as in H 7. This is confirmed by problem (xvii): IV 19, where the capacity of the grain-pile is written as six ŠÁR signs, i.e. 6 × 3600.

TMS 14, from Susa, contains two problems about a grain-pile of similar shape but different proportions. The coefficient used is also different. Although this tablet was published over thirty years ago, it has remained uninterpreted until now. The difficulty lay in explaining both how so many parameters could be calculated from the two given and why the height of the pile

[32] See §A.3 for hand-copy, transliteration, translation and notes.

[33] See §3.2 for the semicircle coefficient used. The solution to (xvii) is corrupt and omits the triangle coefficient. See §A.3.

[34] YBC 4673 (xiii)–(xv): canal dykes with triangular cross-sections.

[35] Only one of the published problems on truncated pyramids — BM 85194 (xxvii) — uses a correct method to calculate its volume (Friberg 1987–90: 567; Nemet-Nejat 1993: 93). (The subject of the calculation is called a *ḫirītum* 'excavation', although it has a height, SUKUD.)

is cubed in lines 5–6. The sole attempt at interpretation (*TMS*: 84–8) assumed that GUR$_7$ referred to a grain *container* rather than a free-standing pile. This container was supposed to consist of a rectangular solid below ground and a triangular prism above, although there is no indication of a depth in the text itself.

(i) GUR$_7$ *a-na* 14 24 SAḪAR 3 GI *me-*⸢*lu*!⸣*-*[*ú-um*]
a-na 14 24 SAḪAR UŠ SAG *ù qà-aq-qa-du*
mi-na GAR ZA.E IGI 12 *šu-up-li pu-ṭú-úr*
5 *ta-mar* 5 *a-na* 14 24 *i-ší-ma*
1 12 *ta-mar* 3 GI *me-la-am* NIGIN 9 *ta-mar*
9 *a-na* 3 *me-le-e te-er-ma* 27 *ta-mar*
i-na 1 A.RÁ *ka-a-a-ma-ni* 20 GÍN? *šà-lu-uš-ti*
šà ka-a-a-ma-ni ⸢*iš-tu* TA X X⸣ ZA BU ZI
40 *ta-mar* 40 *aš-šum* 2 SAG? GUR$_7$ *a-na* 2 TAB.BA
1! 20 *ta-mar* 1 20 *a-na* 27 *i-ší-ma* 36 *ta-mar*
36 *i-na* [1 12] ZI 36 [*ta-mar tu*]*-úr-ma*
3 *me-la-am* NIGIN 9 *ta-*[*mar* IGI 9 *pu*]*-*⸢*ṭú*⸣*-úr*
6 40 *ta-mar* 6 40 *a-na* 36 *i-*[*ší-ma*]
4 *ta-mar* 4 *qà-aq-qa-du* 3 *me-la-*[*a-am*]
[*aš*]*-šum i-na am-ma-at am-ma-*⸢*at*⸣ [Ì.KÚ]
[*a-na* 2 TAB].BA 6 *ta-mar* 6 SAG 6 [*a-na* 4]
[*qà-aq-qa*]*-di* DAḪ 10 *ta-mar* 10 [UŠ]

A grain-pile. For a volume of 14 24 the height is 3 (reeds). For a volume of 14 24 what are put for the length, width and top? You: Take the reciprocal of 12, of depth. You will see 0;05. Multiply 0;05 by 14 24 and you will see 1 12. Square 3 (reeds), the height.[36] You will see 9. Multiply 9 by 3, the height, and you will see 27. Take from 1, *the constant factor*, 20 shekels, a third of the *constant* … . You will see 0;40. Because the grain-pile (has) two widths, double 0;40. You will see 1;20. Multiply 1;20 by 27 and you will see 36. Take 36 from [1 12. You will see] 36. Return, and square 3, the height. You will see 9. [Take the reciprocal of 9.] You will see 0;06 40. Multiply 0;06 40 by 36. You will see 4. The top is 4. Because [the slope is] 1 cubit per cubit, [double] 3, the height. You will see 6. The width is 6. Add 6 [to 4, the top]. You will see 10. The length is 10.

Top of reverse destroyed.

(ii) [ ] ⸢*i-na*⸣ [ ]
[…] 12 *a-na*] 3 *me-la-a-am* [*i-ší-ma*]
[…] 36 *ta-mar* ⸢36⸣ [*a-na* 24 *i-ší-ma*]
[14] 24 *ta-mar* SAḪAR 14 ⸢24⸣
⸢*a-na*⸣ 8 00 00 *na-aš-pa-ak* GUR$_7$ [*i-ší-ma*]
1 55 12 00 00 *ta-mar*! 23 [GUR$_7$]
ù 2-*šu* 24 GUR *še-*⸢*um*⸣ [*ki-a*]*-am* ⸢*ne*⸣*-*[*pé-šum*]

… [Multiply 12 by] 3, the height, and you will see 36. [Multiply] 36 [by 24 and] you will see [14 24], the volume. [Multiply] 14 24 by 8 00 00, the storage of a grain-pile, and you will see 1 55 12 00 00 sila. The grain is 23 [guru] and 2 24 gur. This is the method.

Although lines 7–8 on the obverse are obscure, the method of solution is clear. This problem is slightly more sophisticated than those in BM 96954, because, using the properties of a slope of gradient 1, the volume of the pile is stated solely in terms of the height h and top t. This can be derived from the BM 96954 formula by using the fact that $w/2h = 1$ and, in this case, $(l - t)/2h = 1$:

36 That is, '3 (nindan, or 6) reeds'. Reference to a metrological unit is a common way of fixing a number's order of magnitude in OB mathematics.

$$V = \frac{wh}{3}(l + t/2)$$
$$= \frac{2h^2}{3}(l + t/2) \text{ because } w = 2h,$$
$$= \frac{2h^2}{3}(2h + \frac{3}{2}t) \text{ because } l = 2h + t,$$
$$= h^2 t + \frac{4}{3}h^3$$

Thus although the problems in BM 96954 are more generalized, in that any values can be chosen for the four parameters length, width, top and height, *TMS* 14 is more sophisticated mathematically: it has refined the general solution to a specific case and has developed a more elegant method requiring half the number of parameters used in BM 96954.

Problem (i) is concerned solely with the shape of the grain-pile and does not use a capacity coefficient. After the statement of the problem — to find the length, top and width of the pile, given the volume and height — in lines 1–3, the first part of the solution is concerned with finding the top t. First the volume is divided by the height coefficient 12 (§7.1) to give a volume in nindan^3, thus avoiding the use of cubits in the calculation (ll. 3–5). Then the value of $^4/_3 h^3$ is found (ll. 5–10) and subtracted from the volume (line 11). This produces a value for $h^2 t$, which is divided by h^2 (ll. 12–14) to give the top t. Because the sides of the grain-pile slope at a gradient of 1 in 1, the width is twice the height (ll. 14–16). The length is then the sum of the top and the width (line 17).

Nothing further is preserved of this side of the tablet, but the problem has already been answered. Problem (ii), concerned with finding the volume and grain capacity of the pile, may have started directly after problem (i) on the obverse, or have begun on the top of the reverse. Only the last seven or eight lines are preserved. Presumably the two parameters given were the top and height of the pile. At the point where the extant text picks up the solution, the height is converted into cubits (line 2′; §7.1). This is multiplied by the value for $ht + {}^4/_3 h^2$, which must have been worked out and put aside in the missing part of the text. This procedure gives the volume of the grain-pile (ll. 3′–4′). The volume is multiplied by a conversion coefficient 8 00 00 to yield a capacity in sila (ll. 5′–6′), which is then given in GUR$_7$ and GUR. Traces of the phrase 'this is the method' indicate the end of the solution (ll. 6′–7′).

A further value for this grain-pile coefficient is attested in the lists. The three different equivalences are:

1 sar$_V$ = 8 00 00 sila = 1 36 gur	*TMS* 14
1 sar$_V$ = 7 30 00 sila = 1 30 gur	BM 96954; coefficient H 7
1 sar$_V$ = 7 12 00 sila = 1 26;24 gur	coefficients C72, D 51,

all of which are greater than the coefficients for volume–capacity conversion in other circumstances (§7.2). Exactly why these coefficients vary remains unclear, but it may be related to the granularity of the commodities being stored (cf §7.2.2). Piling grain is mentioned in the catalogue part of text C, but without specific details:

C 74 *še$_{20}$-e ka-ra-am* ⌈*ša-ka-na*⌉*-am*	To put grain into a pile.

A Late Babylonian problem about a grain-pile, BM 34800 (i), remains too fragmentary for interpretation (Sachs 1955: 271 no. 1647). Text J includes a broken entry J 15 which has the same value as the OB coefficients in BM 96954 and H 7, as well as two grain-pile coefficients whose significance is unclear. The numerical value of the second coefficient is an error for

4 10 37 30, a quarter of the value of the first; see §2.2.2.

J 4	16 42 30 GÁN *ka-re-e* KI.MIN (Ì.DUB) IGI.GUB-*ú*	16 42 30, the area of a grain-pile, ditto (i.e. storage): the coefficient.
J 5	4 13 07 30 GÁN *ka-*⸢X⸣ KI.MIN ⸢KI.MIN⸣ (Ì.DUB) IGI.GUB-*ú*	4 13 07 30, the area of a ..., ditto (i.e. of a grain-pile), ditto (i.e. storage): the coefficient.

For the coefficient D 40, previously read 10 *šà* SAḪAR *ka-ru*, '10, of the volume of a quay/ grain-pile' (*TMS*: 29; Nemet-Nejat 1993: 182) see §5.1.3.

7.3.3 Philological commentary

Lines 7–8 of *TMS* 14 (i), which describe how to find the multiplication factor $^4/_3$ (see above), are particularly obscure. The gist of the instructions is to subtract $^1/_3$ from 1 to get $^2/_3$, which is then doubled in the following line. Von Soden (1964: 49) suggested reading the last half of line 8, after *kayyamānum*, as SAḪAR$^{!?}$ *tu*$^!$*-uṣ*$^!$*-ṣa-bu* ZI, adding that:

> Der Satz lässt sich so, wie er kopiert ist, grammatisch nicht einwandfrei konstruieren (wovon hängt der Subjunktiv *tu-uṣ-ṣa-bu* 'was du addiert' ab); alles was zwischen 20 und ZI *usuḫ* steht, erläutert wohl die 20.

The word *kayyamānum*, which in non-mathematical texts is an adjective meaning 'permanent, constant' is also found in *TMS* 11 and 24 in broken passages, but has not been attested in mathematical texts outside Susa. In line 7 A.RÁ appears to be a noun, perhaps Akkadian *arûm*, 'product'.[37] There is no published photo of the tablet; it was not possible to have it collated.

7.4 OTHER STORAGE COEFFICIENTS

7.4.1 The coefficients

C 73	*e-ší-it-tum* ⸢3⸣ 36 *i-gi-gu-bu-ša*	A storehouse: 0;03 36 is its coefficient.
E 22	3 36 IN.NU.DA	0;03 36, straw.
F 31	3 36 *ša* É.KIŠIB IN.NU.DA	0;03 36, of the storage of straw.
F 32	3 45 *ša* É.KIŠIB SUḪ$_4$	0;03 45, of the storage of *matting*.
E 30	15 A	15, water.
F 33	15 *ša* É.KIŠIB A	15, of the storage of water.
A 39	6 IGI.GUB GI.SA	6 00 00, the coefficient of a reed-bundle.
E 23	1 15 É.UŠ.GÍD.DA IN.NU.DA	1 15, storage of straw.
A 40	7 12 IGI.GUB GI.⸢KILIB.BI$^{!?}$⸣	7 12, the coefficient of *a load of reed-bundles*.
B II 7′	50 GIŠ.MÁ.LÁ GI.ŠU.KIN	50, a cargo-boat of reeds.
B II 8′	30 GIŠ.MÁ.LÁ GI.NÍG.SA.ḪI.A	30, a cargo-boat of reed-bundles.
B II 9′	1 15 GIŠ.MÁ.LÁ GI.NÍG.SA.ḪI.A ḪA$_5$.RÁ	1 15, a cargo-boat of dried reed-bundles.
B II 10′	1 12 GIŠ.MÁ.LÁ *ḫi-im-ma-ti*	1 12, a cargo-boat of small reeds.

7.4.2 Mathematical and philological commentary

The exact meanings of the OB coefficients are unclear, although it may be relevant that

$$0;03\ 36 = {}^3/_4 \times 0;\ 04\ 48, \text{ and } 0;03\ 45 = {}^3/_4 \times 0;05,$$

[37] But certainly not *muttalliktum*, a carrying coefficient with logogram A.RÁ (see §5.1).

the two circle coefficients used for calculating the capacity of cylindrical grain containers (§3.1; §6.2). The first word of C 73 is certainly a variant of *išittum* 'granary' (Goetze 1951) rather than a writing of *esittum* 'pestle':[38] the whole of text C column IV is concerned with grain and grain storage. The reading SUḪ$_4$/*suḫḫum* 'reed matting' is due to Kilmer (1960: 295).

The four LB storage coefficients for reed products (B II 7′–10′), on the other hand, may be identified with more certainty. The OB problem texts BM 85196 (ii) and (v) are concerned with a reed-bundle GI.SA/*kiššûm* of volume 0;00 05 sar$_V$, in the shape of a tapering cylinder.[39] Using the capacity coefficient 6 00 00 (A 39) this is equivalent to a capacity of 30 sila, the value of B II 8′. The OB coefficient D 69 states that the daily rate for carrying reeds is 0;05 sar$_V$ over 60 nindan (§5.5). Dividing this by the volume of each bundle gives the total number carried in one day.[40]

Table 7.2: Reed storage coefficients of List B

Entry	*Capacity*	*Volume*	*Daily load*
B II 7′	50 sila	0;00 08 20 sar$_v$	36 bundles
B II 8′	30 sila	0;00 05 sar$_v$	60 bundles
B II 9′; E 23	1 15 sila	0;00 12 30 sar$_v$	24 bundles
B II 10′	1 12 sila	0;00 12 sar$_v$	25 bundles

E 23 is almost certainly to be identified with B II 9′. It is not apparent whether the distinction between the reed-bundles GI.SA/*kiššûm*, and the two types GI.ŠU.KIN/*šūrum*[41] and *ḫimmatum* is simply one of size. Dried reeds are presumably lighter than the other types and can thus be made into larger bundles for transporting.

The last two signs of A 40 are unclear, even after collation. The traces appear to match Kilmer's identification BUL (= LAGAB × EŠ) + IR (1960: 276), the closest meaningful reading of which is KILIB.BI$^{!?}$ for KILIB.BA, a load of 12 reed-bundles (cf §9.5.3). If so, then the coefficient may be similar in function to A 39 but, again, it is not entirely clear why different capacity conversions should be needed. Alternatively, 7 12 might be the capacity of the load measured in sila. A less likely reading for A 40 would be to see it as a badly copied GUR$_7$.

7.5 SUMMARY

There is still surprisingly little known about the exact function of storage coefficients. Although the problem texts give geometrical descriptions of storage containers — cylinders or tapering cylinders for reeds and grain, as well as free-standing pyramidal grain-piles — the motivation for choosing one conversion coefficient rather than another remains obscure. To a large extent this is because the limited terminology of storage and conversion is applied to the coefficients

[38] The translation 'mortar' (Kilmer 1960: 303; Nemet-Nejat 1993: 168) must have arisen from a misunderstanding of the dictionary entry *esittu(m)* '(Mörser-)Stößel' (*AHw*: 250).

[39] The absolute size of the bundle is fixed by BM 85196 (v): I 27, which states that 4 sar$_V$ contains 4 × 600 + 8 00 = 48 00 GI.SA.ḪI.A. See §6.4.

[40] The coefficients 5 00 00 and 6 40 00 can be excluded as they produce non-integer loads. e.g.:

1 12 ÷ 5 00 00 = 0;00 14 24 sar$_V$. 0;05 ÷ 0;00 14 24 = 20;50 bundles.
1 15 ÷ 6 40 00 = 0;00 11 15 sar$_V$. 0;05 ÷ 0;00 11 15 = 26;40 bundles.

[41] See Civil (1994: 122) for lexical evidence on the names of reeds.

with little apparent consistency, as Table 7.3 shows. This is also true in the calculations and problem texts themselves: see Table 7.4.

Table 7.3: Storage coefficients in the lists, sorted by subject matter

Description	*Entry*	*Value*	*Section*	*Description*	*Entry*	*Value*	*Section*
Boat	G 12	4 48 00	7.2	Thickness of a log	E 35	4 48 00	7.2
	E 19	5 00 00	7.2		F 58	4 48 00	7.2
	F 29	6 00 00	7.2	Vessel	H 1	6 40 00	7.2
	C 72	7 12 00	7.3	Water	E 30	15	7.4
Grain	F 61	4 48 00	7.2		F 33	15	7.4
	F 30	6 40 00	7.2		J 8	6 00 00	7.2
	J 7	6 40 00	7.2	Unclear	D 49	6 00 00	7.2
Grain-pile	D 51	7 12 00	7.3	None	G 9	6 00 00	7.2
	H 7	7 30 00	7.3		H 6	6 00 00	7.2
Millet	J 10	6 12 00	7.3		A 11	6 40 00	7.2
Oil	J 9	6 00 00	7.2		D 50	6 40 00	7.2
Reeds	A 39	6 00 00	7.2		C 73	3 36 00	7.4
	A 40	7 12 00	7.3	Broken	J 11	6 12 00	7.2
	F 32	3 45 00	7.4		J 12	6 24 00	7.2
Straw	A 37	4 48 00	7.2		J 13	6 24 00	7.2
	E 22	3 36 00	7.4		J 14	6 45 00	7.2
	F 31	3 36 00	7.4		J 15	7 30 00	7.3

The volume V of a grain-pile is considered, correctly, to be:

$$V = \tfrac{1}{3}\,\text{height} \times \text{width} \times (\text{length} + \tfrac{1}{2}\,\text{top})$$

or

$$V = \text{height}^2 \times \text{top} + \tfrac{4}{3}\,\text{height}^3$$

in the special case that the width is twice the height.

Table 7.4: Storage coefficients in calculations and problem texts, sorted by subject matter

Subject	*Problem or Calculation*	*Coefficient*	*Section*
Basket	BM 85196 (xiv)	5 00 00	7.2
Bricks in a boat	BM 85194 (xxx)	5 00 00	7.2
Grain-pile	BM 95957+102366 (vi)–(xix)	1 30	7.3
	TMS 14 (ii)	8 00 00	7.4
Logs	Haddad 104 (i)–(v)	6 00 00	7.2
Plastering	Haddad 104 (vii)–(viii)	5 00 00	7.2
Thickness of a log	YBC 8600	4 48 00	7.2
	VAT 8522 (i)	6 40 00	7.2
Unclear	*UET* 6/2 218, 247, 269, and 290	6 40 00	7.2
	UET 6/3 374 and 452	6 40 00	7.2
	UET 6/2 250	7 12 00	7.2

Conversions between units in the same metrological system are rarely shown explicitly in the problem texts or calculations, presumably because the students were trained in the use of metrological lists and tables.

8. OTHER COEFFICIENTS

The coefficients in this chapter can be grouped under five main headings:

- those which are explicable, but which do not fall into the broad categories of §§3–7 (e.g. astronomy; some metal coefficients);
- those which occur in broken and obscure contexts in problem texts, some of which have no counterparts in the coefficient lists (§8.3);
- coefficients which are legible and translatable, but have no mathematical context, as they do not occur in extant material outside the coefficient lists; these may form a sizeable body of coefficients (e.g. metals, §8.1) or be isolated cases (§8.4);
- coefficients which are complete but which are currently meaningless because they are not found in, or related to, published problem texts (§8.5);
- those which are too damaged to read (§8.5).

Not surprisingly, little work has been done on these coefficients, apart from those on metals and astronomical observation. It has mostly been restricted to comments during the initial publication of the texts, reviews of those publications (e.g. Oppenheim 1947; von Soden 1964) or collected lists of coefficients (Kilmer 1960; 1985; Nemet-Nejat 1993: 149–241 'glossary of realia'). With many of the more obscure or damaged entries it has been possible to offer little more than a summary of previous works.

In line with the aims and extent of this study outlined in §1.2.3 this chapter deals only with OB coefficients, except where later ones are also of direct relevance (as, e.g., in §8.2). The large number of obscure coefficients from LB lists are not discussed. Five of the eight OB lists contain a total of 59 entries which cannot yet be put into a meaningful mathematical context;[1] only texts D, H, and M are composed entirely of explicable coefficients.

All of the OB lists except H and M, the smallest, include metal coefficients, with 35 attestations of 15 different coefficients, yet there is only one problem text known.[2] One coefficient on astronomical observation, known from later astronomical texts, is found in lists E and F, but there is no other OB evidence of its use. One obscure coefficient, attested in four lists, is also found in a set of four problems on one tablet with a variant found in two problems on another;[3] a further two tablets contain single problems which use unclear coefficients that do not seem to be mentioned in the lists.[4]

8.1 METALS

8.1.1 The coefficients

F 28	2 08 *ru-uq-qum ša* SÙ.ÁG	2 08, a sheet of ....
F 25	2 15 *ru-uq-qum ša* KÙ.SIG$_{17}$	2 15, a sheet of gold.

[1] 16 coefficients from each of lists A and F; 13 from lists C and E; 1 from list G.
[2] YBC 4669 (B vi), about hammering out a sheet of silver.
[3] BM 85194 (x)–(xiii); BM 85210 (xviii)–(xix) — all obscure.
[4] BM 85210 (ix) — on a truncated pyramid; Str 366 (i) — obscure.

F 27	2 24 *ru-uq-qum ša* AN.NA	2 24, a sheet of tin.
F 26	4 *ru-uq-qum ša* KÙ.BABBAR	4 00, a sheet of silver.
E 33	9 KI.LÁ.BI *ša* URUDU	9 00, the (lit. its) weight of copper.
E 51	26 40 *ru-⌈qú⌉-um ša* KÙ.BABBAR	0;00 26 40, a sheet of silver.
A 18	1 12 IGI.GUB URUDU	1 12, the coefficient of copper.
C 60	[URUDU 1 12 *i-gi-gu-bu-šu* (?)]	[Copper: 1 12 is its coefficient ?]
D 54	1 12 *šà* URUDU	1 12, of copper.
E 36	1 12 ⌈*ra-ṭù*⌉*-um ša* URUDU	1 12, a mould of copper.
F 22	1 12 *ra-ṭù-um ša* URUDU	1 12, a mould of copper.
D 55	1 15 *šà lu-li-im*	1 15, of antimony.
A 19	1 20 IGI.GUB ZABAR	1 20, the coefficient of bronze.
D 56	1 20 *šà* ZABAR	1 20, of bronze.
E 34	1 20 ⌈*ra-ṭù*?*-um*⌉ ‹*ša*› KÙ.SIG$_{17}$	1 20, a mould of gold.
A 20	1 20 21 IGI.GUB AN.NA	1 20 21, the coefficient of tin.
C 64	⌈*a-na-ku*⌉*-um* 2 36 15 *i-gi-*[*gu-bu-šu*]	Tin: 2 36 15 is its coefficient.
D 57	1 24 *šà* ⌈AN⌉.NA	1 24, of tin.
A 21	⌈1⌉ 36 IGI.GUB KÙ.BABBAR	1 36, the coefficient of silver.
C 61	[KÙ.BABBAR 1 30 *i-gi-gu-bu-šu* (?)]	[Silver: 1 30 is its coefficient (?)].
D 58	1 36 *šà* KÙ.BABBAR	1 36, of silver.
F 23	1 30 *ra-ṭù-um ša* KÙ.BABBAR	1 30, a mould of silver.
A 22	1 48 IGI.GUB KÙ.SIG$_{17}$	1 48, the coefficient of gold.
C 62	KÙ.SIG$_{17}$ 1 48 [*i-gi-gu-bu-šu*]	Gold: 1 48 [is its coefficient.]
D 59	1 48 *šà* KÙ.SIG$_{17}$	1 48, of gold.
E 37	1 48 [*ra-ṭù-ùm*] ⌈*ša*⌉ [KÙ].GI	1 48, [a mould] of gold.
F 24	1 48 *ra-ṭù-um ša* KÙ.SIG$_{17}$	1 48, a mould of gold.
C 63	*a-ba-rum* 1 ⌈52⌉ 30 ⌈*i-gi*⌉-[*gu-bu-šu*]	Lead: 1 52 30 is its coefficient.
D 60	1 52 30 *šà a-ba-ri*	1 52 30, of lead.
D 61	2 *šà ḫu-lu-uḫ-ḫi*	2, of ....
D 62	2 12 *šà pa-ar-zi-il-li*	2 12, of iron.
A 23	2 24 IGI.GUB *a-*⌈*lu-lu-tum*⌉	2 24, the coefficient of ....
C 65	*a-lu-lu-tum* 2 24 *i-gi-gu-bu-*[*ša*]	...: 2 24 is its coefficient.
E 38	2 24 ⌈*a*⌉-[*lu-lu-tu*]*-um*	24, ...
G 10	[x] ⌈40⌉ IGI.GUB ŠÌTA *ra-ṭe$_4$-em*	... 40, the coefficient of a mould: a mould.

8.1.2 Mathematical commentary

The meaning of the first group of metal coefficients listed here — the *ruqqum* coefficients — is elucidated by the problem text YBC 4669 (B vi):

V 1	3 KÙŠ 1 ŠU.SI UŠ 2 KÙŠ 6 ŠU.SI SAG KI.BI EN.NAM $2^2/_3$ GÍN $20^1/_2$ ŠE	The length is 3 cubits, 1 finger. The width is 2 cubits, 6 fingers. What is the area? $2^2/_3$ shekels, $20^1/_2$ grains. 1 shekel (weight) of silver was given. They hammered out a sheet 3 fingers square. Now, how much silver was given to put (i.e. produce) an area of $2^2/_3$ shekels,
5	1 GÍN KÙ.BABBAR SUM.MA	

3 ŠU.SI.TA ÍB.SI$_8$ $20^{1}/_{2}$ grains?
ru-uq-qá-am im-ḫa-ṣú
i-na-an-na
KÙ.BABBAR EN.NAM SUM.MA
$2^{2}/_{3}$ GÍN $20^{1}/_{2}$ ŠE KI
ḪÉ.GAR.RA

Neugebauer and Sachs (*MCT*: 138–9) came very close to a complete interpretation of this problem, showing that lines 5–6 give the relationship:

1 shekel of silver measures (0;00 30 nindan)^2 = 0;00 00 15 sar$_a$,

so that 1 mina of silver measures 0;00 15 sar$_a$, and the reciprocal of this number, namely 4 00, is the coefficient found in F 26. Neugebauer and Sachs did not go on to solve the problem given, which is done simply by multiplying the area — 0;02 46 50 sar$_a$ — by this coefficient 4 00, to give a weight of 11;07 20 minas (Lewy 1949: 163). The answer is not given in the text.[5]

If the other *ruqqum* coefficients also give the weight of 1 sar$_a$ of metal, then we have:

Table 8.1: Summary of the ruqqum coefficients

Metal	*Weight of 1 sar$_a$*	*Area of 1 mina (recip. of weight)*	*Length and width of 1 shekel weight*	*Entry*
Copper	9 00 minas	0;00 06 40 sar$_a$	2 × 2 fingers	E 33
Silver	4 00 minas	0;00 15 sar$_a$	3 × 3 fingers	F 26
Gold	2 15 minas	0;00 26 40 sar$_a$	4 × 4 fingers	F 25
Tin	2 24 minas	0;00 25 sar$_a$	5 × 3 fingers	F 27
Antimony	2 08 minas	0;00 28 07 30 sar$_a$	4.5 × 3.75 fingers	F 28

It appears that E 51 lists the reciprocal of the coefficient for gold, not silver as stated in the list. Perhaps, like other mistakes in this text, this error is due to conflation or miscopying from an original source. Copper, silver, and gold all have coefficients which give a square area with integer sides, in ascending order. It seems that these were the original three coefficients and that the others have been fitted in around them with numbers that do not belong to the original pattern. The thickness of each metal sheet is not calculable from the information given.

There are no problems extant which illuminate the meaning of the remaining metal coefficients, but it is clear that they obey an underlying numerical pattern, following an arithmetical progression in steps of 12. Extra coefficients for other metals have been interpolated, in a manner which is consistent from list to list. Their absolute value is not known, and neither is it clear how — or indeed if — G 10 is related to this sequence. Table 8.2 shows to what extent each list — including the LB list K (see below) — follows the arithmetic progession, best exemplified by text D. This scheme accounts for all but 10 of 28 extant entries. A further 5 — those for antimony, bronze and lead — are straightforward interpolations whose values fall between those in the pattern. In E 34 KÙ.SIG$_{17}$ 'gold' has been written mistakenly for ZABAR 'bronze'; compare the similar conflation in E 51 discussed above.

[5] The two most recent interpretations of this text (Durand 1983: 134 n. 35; Michel 1989) are both erroneous, and arise from a misunderstanding of the function of lines 5–7. These do not refer to the piece of silver which is the subject of the problem, but present a paraphrase of the coefficient needed to solve it. Both authors, and Lewy (1949: 163 n. 1), use a modern value of the density of silver to calculate the thickness of the sheet of the metal to several decimal places. However, as both the absolute values and the degree of accuracy of the OB mina and nindan are unknown, such an exercise is inappropriate.

Table 8.2: Order of the metal coefficients in the lists

		Place in list:[6]					
Metal	*Coefficient*	*A*	*C*	*D*	*E*	*F*	*K*
Copper	1 12	1	[1]	1	1	1	–
Tin	1 24	(3)	(6)	4	–	–	(7?)
Silver	1 36	4	[2]	5	–	(2)	(2)
Gold	1 48	5	3	6	3	3	(3)
ḫuluḫḫum	2 00	–	–	8	–	–	–
Iron	2 12	–	–	9	–	–	(6)
alulūtum	2 24	6	5	–	–	–	–

The value 1 30 given for silver in F 23 arises from the fact that only three coefficients in the series are listed in this text. The value has been altered from 1 36 so that it forms a short arithmetic progression with difference 18 with the (unaltered) coefficients for copper and gold. For the same reason I have tentatively suggested the same number for the restored entry C 61.

The numbers in A 20 and C 64, both tin coefficients, are more difficult to explain. Both are sexagesimally irregular, which suggests strongly that they are erroneous. While the 1 20 21 of A 20 may arise from a miscopied 1 24, the 2 36 15 of C 64 breaks the ascending sequence exhibited in all texts but E, where the coefficients are not listed together.

The obverse of the LB text K shows the same ordering of the metals, but with antimony and tin (?) in different positions. Copper is missing at the beginning of the sequence. The left of the column — and any information in it, such as coefficient values — has been lost.

I 3	[...] ZABAR	[...] (of) bronze.
	[...] KÙ.BABBAR	[...] (of) silver.
5	[...]KÙ.SIG$_{17}$	[...] (of) gold.
	[... A].BÁR	[...] (of) lead.
	[... KÙ.]KÁM	[...] (of) antimony.
	[... AN].BAR	[...] (of) iron.
	[...*an-na*]-*ki*	[...] (of) tin$^{?}$.

Thus it appears that coefficients for metals were transmitted as a group in a fairly fixed, mathematically meaningful, order over a period of at least 1500 years.

8.1.3 Philological commentary

The identities and reading of most of the metals listed here are given by Joannès (1993–7: 97–9) and Muhly (1993–7). The exceptions are:

- SÙ.ÁG, *elmēšum* 'wohl ein kostbarer, gelbscheinender Stein' (*AHw*: 205). Still unidentified, but not the metals antimony or arsenic (Joannès 1993–7: 99; Waetzoldt and Bachmann 1984). It is almost certainly not the gold-and-silver alloy electrum (cf. Lewy 1949: 164), as this was not distinguished from gold in Mesopotamia (Moorey 1994: 217);
- *ḫuluḫḫum*, AN.ZAḪ.BABBAR 'eine helle Schmelzmasse' (*AHw*: 354);
- *alulūtum*, which von Soden (*AHw*: 39) implies may be connected to *lulûm*, antimony.

Ruqqum is the *purs* nominal form of *raqāqum* 'to be thin' (Durand 1983: 134), from which it derives its meaning 'sheet (metal)' and, by extension, refers to many objects made of this (cf. *MCT*: 139; Oppenheim 1947: 128).[7] The *rāṭum* — 'mould', 'pipe' or 'channel' (Joannès

[6] Numbers in parentheses (...) indicate values which differ from those in column two. Numbers in square brackets [...] indicate missing or broken entries in the list.

[7] It is also used to refer to various thin or narrow bodily parts (Adamson 1990: 30).

1993–7: 107) — coefficients are clearly the same as those for metals alone, as their numerical values are identical. They may refer in some way to molten metal to be poured into moulds, suggested by the title 9 IGI.GUB *šà ku-ri-im* '9 coefficients of the furnace' which precedes the text D sequence (D 53). My restorations of C 60 and 61 are tentative, but both coefficients occur in all four other OB lists of metals.

The coefficient E 38 could equally well have been placed in §4. As the coefficient is between one section on metals and another on quantity surveying, its position within text E cannot be used to decide the matter. Neugebauer and Sachs (*MCT*: 137) read the word *agurrum* 'baked brick' and interpreted it as the 'layer' coefficient of Type 11 bricks (§4.2). There are no other layer coefficients in List E. Alternatively, it could be interpreted as the brickage of the otherwise unattested brick 30 × 15 × 5 fingers, Type 8a in Powell's typology (1982*a*: 119), half the size of a Type 11 brick (§4.1).

More recently, the word has been read *abarum* 'lead' (Kilmer 1985: 21), but it is restored here as *alulūtum* because its value accords with List C for this metal. Similarly, A 23 has been understood as *alulūtum* against Kilmer's A.GAR$_5$ 'lead'. It may be, though, that lists A and E attest an alternative tradition to lists C and D, whereby A and E assign the value 2 24 to lead, while in C and D it takes 1 52 30 and *alulūtum* has the value 2 24.

8.2 ASTRONOMICAL OBSERVATIONS

8.2.1 The coefficients

E 50	12 *ma-al-ta-ak-tum*	0;12, the water-clock.
F 60	12 *ma-al-ta-ak-tum ša mu-ši-tim*	0;12, the water-clock of the night.
B III 10′	12 MÚ *ša* ^{d}30	12 00, the (first) rising of the moon.
B III 11′	12 ŠÚ *ša* ^{d}30	12 00, the (last) setting of the moon.
B III 9′	3 45 IGI.DU$_8$.A	3 45, the visibility (of the moon).
B III 12′	40 *u$_4$-mu a-na* GE$_6$ *na-pa-lu*	40, to change day into night.
B III 13′	40 GE$_6$ *a-na u$_4$-mu na-pa-lu*	40, to change night into day.
B III 14′	40 *nap-pal-tu$_4$*	40, the change.

8.2.2 Mathematical commentary

There are no extant OB problems which use these coefficients,[8] but they are found in first millennium exemplars of astronomical texts whose traditions may go back as far as OB.[9] The first was used to construct 'Table B' in the fourteenth tablet of the astronomical and astrological compendium *Enūma Anu Ellil* (Al-Rawi and George 1991–2: 56). The table lists the duration of lunar visibility for each day in an equinoctial month according to the Babylon tradition. It consists of a list of days from 1 to 30 in the left-hand column, and the corresponding periods in the right-hand column, given in terms of weights in a water-clock. The visibility on day 1 is of

[8] BM 85194 (vi)–(viii) and BM 85210 (xiv), the only extant OB problems on water-clocks, are concerned with changing water levels and do not use coefficients. The MB text HS 245 and quasi-duplicates Sm 162 and Sm 1113 do not concern astronomy but, rather, use it as a pretext for a mathematical problem on 'remarkable numbers' (Rochberg-Halton 1983; Høyrup 1993*b*).

[9] For instance, George (1991: 301) has published an NB copy of a 'forerunner' of MUL.APIN I iv 19–24 which uses OB-style equinox dates. I thank David Brown for this information.

12 shekels duration (B III 10′).[10] The durations increase by 12 shekels each day until the 15th of the month (E 50; F 60), when they start to decrease by the same amount. The visibility on the 29th day is again 12 shekels (B III 11′).[11]

It has also been suggested that this coefficient refers to the fact that at an equinox the new moon is visible for 12 00 nindan, or 0;24 danna (48 minutes) (Friberg *et al.* 1990: 497).[12] This amounts to the same thing, as in astronomical measure 1 mina = 2 danna = 4 hours, and thus 0;12 mina = 0;24 danna.

'Table A' of *Enūma Anu Ellil* XIV lists the same lunar visibilities according to the Nippur tradition (Al-Rawi and George 1991–2: 55). They are identical to those of the Babylon tradition, except for the first and last few days of the month of equinox. The moon is visible for 0;03 45 minas or 3 45 nindan — 15 minutes — on the first night (B III 9′). The length of visibility doubles each night until night 5, when it is seen for 1 mina, or 4 hours. The table is then identical to 'Table B', with lunar visibility changing by 0;12 mina a night, until night 26, when the visibility starts to halve again. The moon is seen for 0;3 45 minas on night 29 and is invisible for the last night of the month.

The third coefficient listed here occurs twice in the early first millennium astronomical text MUL.APIN II 41–2; III 15 (Hunger and Pingree 1989: 101; 107–8):

II 41–2	40 *nap-pal-ti* u_4*-mi u* GE$_6$	40, the change of day and night.
III 15	40 NINDAN *nap-pal-ti* u_4*-mi u* GE$_6$	40 nindan, the change of day and night.

Friberg *et al.* (1990: 498) have shown that this refers to the assumption that the length of the night changes uniformly throughout the year, at a rate of 40 nindan, or 0;01 20 danna (2 minutes 40 seconds), a day. The monthly change is then 30 × 0;01 20 = 0;40 danna (1 hour 20 minutes). Thus — like the brick carriage coefficients in §5.3 — the coefficient has a dual daily/monthly meaning.[13]

8.2.3 Philological commentary

The astronomical *nappaltum* is clearly related to the LB mathematical word *naplu* or *nappalu* (written *na-pal* in W 23291-x (ii)),'amount subtracted, decrease' (Friberg *et al.* 1990: 494–6). Both are derived from *napālum* 'to pay back, to convert'.

There has been some debate over whether the words *maltaktum* and *maštaqtum* are identical, and whether they refer to water-clocks or sand-clocks; see most recently Al-Rawi and George (1991–2: 57 n. 20). As the consonantal shift of *š* to *l* before dentals and sibilants appears to be a phenomenon restricted to MB and later (*GAG*: 30–1 §30f), an OB writing *ma-al-ta*-AK-*tum* suggests that the two words are to be considered separate. The first seems to be a *maprast*-formation of *latākum* 'to try, to test, to check'[14] and thus should mean 'testing instrument' or the like, while *maštaqtum* 'separator' is derived similarly from *šatāqum* 'to split off, to separate'. Either could conceivably refer to a sand- or water-clock, essentially an observational device which functions by discharging its contents uniformly over time.

[10] 6 minas of water were supposed to trickle out of the water-clock over 24 hours. Thus 12 shekels = 48 minutes (Al-Rawi and George 1991–2: 60).

[11] Actually expressed in the table as a diminution (*maššartum*) of 2;48 minas from the maximum visibility of 3 minas on day 15 (Al-Rawi and George 1991–2: 56; 63).

[12] In astronomical texts, length units are used as a measure of time. There are 12 danna ('double-hours') in a day and night, and 1 danna = 30 00 nindan as usual. Thus 1 danna = 2 hours and 1 nindan = 4 seconds.

[13] BM 17175 + BM 17284 is an OB tablet from Sippar which lists the lengths of day and night at three-monthly intervals. The same rate of change is assumed (Hunger and Pingree 1989: 163–4).

[14] Cf. OB problem BM 85196 (xiv): 9, 11, about an inaccurate measuring vessel, which 'I checked': *al-tu-uk*.

8.3 OBSCURE COEFFICIENTS IN PROBLEM TEXTS

8.3.1 The coefficients

A 17	18 IGI.GUB IM.LÁ	18, the coefficient of a ....
D 63	18 *šà im-le-em*	18, of a ....
E 14	18 IM.LÁ	18, a ....
F 40	18 *i-im-lu-ú-um*	18, a ....

BM 85194 (x): III 6	18 IM.LÁ	18, a ...
BM 85194 (xi): III 11	18 IM.‹LÁ›.BI	18, its ....
BM 85194 (xii): III 15	18	18.
BM 85194 (xiii): III 20	[1]8 IM.LÁ	18, a ...
BM 85210 (xviii): IV 4	‹1 30› DAL.BI	1;30, its diameter.
BM 85210 (xix): IV 7	1 30 *ša* DAL	1;30, of the diameter.
Str 366 (i): I 3	15 IGI.GUB.BA	0;15, the coefficient.
BM 85210 (ix): II 31–2	15 IGI.GUB $^1/_2$?	0;15, the coefficient of $^1/_2$?.

8.3.2 Mathematical and philological commentary

The mysterious IM.LÁ is the subject of six problems, yet it is still impossible to determine exactly what the word refers to. BM 85194 (x)–(xiii), two sets of essentially identical problems, use the coefficient attested in the lists. The first two read:

(x) 1 IM.LÁ GAM EN.NAM ZA.E 4 *ù* 3 NUMUN GAR.RA
III 6 4 *a-na* 3 *i-ši* 12 *ta-mar* IGI 18 IM.LÁ ‹DU_8.A›
3 20 *ta-mar* 3 20 *a-na* 12 *i-ši*
40 *ta-mar* 40 GAM
ki-a-am ne-pé-šu

The IM.LÁ is 1. What is the GAM? You: put down 4 and 3, the ratios.[15] Multiply 4 by 3. You will see 12. Take the reciprocal of 18, the IM.LÁ. You will see 0;03 20. Multiply 0;03 20 by 12. You will see 0;40. The GAM is 0;40. This is the method.

(xi) 40 GAM KI.LÁ.BI ‹EN.NAM ZA.E› 4 *a-na* 3 *i-ši* 12 *ta-mar*
III 11 IGI 12 DU_8.A 5 *ta-mar* 18 IM.‹LÁ›.BI
a-na 5 *i-ši* 1 30 *ta-mar* 1 30 *a-na* 40 GAM *i-ši* 1 *ta-mar*
ki-a-am ne-pé-šum

The GAM is 0;40. What is its KI.LÁ? You: multiply 4 by 3. You will see 12. Take the reciprocal of 12. You will see 0;05. Multiply 18, its *clay*, by 0;05. You will see 1;30. Multiply 1;30 by 0;40, the GAM. You will see 1. This is the method.

In the second pair of problems the IM.LÁ is 0;30 and the GAM 0;20, but apart from minor textual differences, the problems set and their solutions are the same as these ones. BM 85210 (xviii)–(xix) present the same subject in an even terser fashion, using a variant coefficient not attested in the lists:

(xviii) *šum-‹ma›* 1 DAGAL IM.LÁ [GAM EN.NAM] ZA.[E IGI] DAL.BI
IV 5 DU_8.A 40 *ta-mar* 40 [...] 4 *šu.si*
ne-pé-šum

[15] See §7.1.3 for a discussion of the logogram NUMUN.

If the width of the IM.LÁ is 0;01, [what is the GAM]? You: take [the reciprocal] of its diameter. You will see 0; 00 40, [...] or 4 fingers. The method.

(xix) *šum-ma* [4 ŠU].SI GAM IM.[LÁ EN].NAM ZA.E 1 30 *ša* DAL
a-na 40 *i-ši* 1 *ta-mar* 1 IM.LÁ
ne-pé-šum

If the GAM is [4] fingers, what is the IM.LÁ? You: multiply 1;30, of the diameter, by 0;00 40. You will see 0;01. The IM.LÁ is 0;01. The method.

There are several obstacles to a satisfactory interpretation, not least of which is the ambiguous terminology for the two parameters. The sign transliterated here as GAM can also be read GÚR 'circle, diameter' as well as BÙR 'depth'; it is used in both senses elsewhere on these tablets.[16] Oppenheim (1947: 126) suggested that IM.LÁ was a pseudo-logogram of *imlûm*, itself a derivative of *malûm* 'to be full'. This somehow led him to a translation 'embankment, terrace'. Other interpretations include 'ein kegelähnlicher Körper?' (*AHw*: 377) and a container or weight (Nemet-Nejat 1993: 84–5). This last is derived from the fact that in BM 85194 (xi) the subject of the problem is written KI.LÁ instead of IM.LÁ. However, KI.LÁ is widely used as a logogram for *kalakkum* 'excavation' in OB mathematics[17] as well as for *šuqultum* 'weight'.[18] The most likely possibilities, however, are summarized by Kilmer (1960: 291 n. 1), who comments that:

Its position in text D between clay and bitumen may point to a similar substance; its occurrence in *TMB* 54–7 [i.e. BM 85210 (x)–(xiii)] (where dimensions are given for it) may point to a construction, or excavation, of such a substance.

The aim each time is to find (the width of) the IM.LÁ given the GAM, or vice versa. The relationship between the two parameters seems to be:

$$\text{GAM} = (4 \times 3)/18 \text{ IM.LÁ, or GAM} = 1/(1;30) \text{ IM.LÁ}$$

in BM 85194 and BM 85210 respectively. If IM.LÁ always means the width of the body, as BM 85210 (xviii) suggests, then GAM cannot mean 'circumference', as it is shorter than the width. The method for finding the volume V of a cone of circumference c and height h is:

$$V = \frac{c^2 h}{12\pi} = \frac{\pi d^2 h}{12},$$

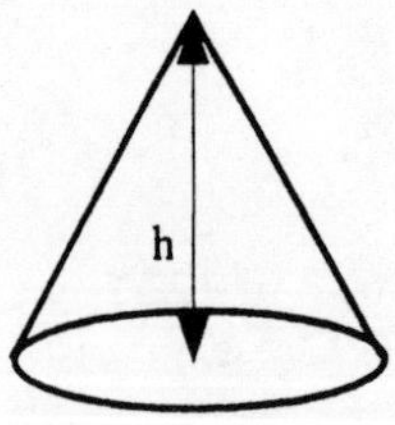

Figure 8.1: A cone

but this would only work here if IM.LÁ meant 'volume' and if the circumference of the cone never altered: the second pair of BM 85194 problems shows that the GAM is proportional to the IM.LÁ. Even if this were so, it would mean that $d^2 = 6$, an implausible occurrence in OB mathematics, which avoided the approximation of irrational numbers as much as possible. Further, this explanation does not account for the phrase '1 30 of the diagonal' in the BM 85210 problems.[19]

The embankment/excavation scenario is no more plausible. A hole whose depth was the same as the width of the bottom and half the width of the top would have an average width of

[16] GÚR: BM 85194 (iv): I 37; BM 85210 (viii): III 17. BÙR: BM 85194 (ii): I 14; BM 85210 (ix): III 27.

[17] e.g. YBC 5037 *passim*; see §6.2.

[18] e.g. YBC 7284; see §4.3.

[19] Friberg (1996*b*: 91) sees BM 95954 + BM 102366 + SÉ 93 (xx)–(xxiv) (§A.3) as problems about half-cones and cones, but his unsatisfactory interpretation depends on understanding the slope ŠÀ.GAL of the cones to be the ratio not of the base to the height, but of the square of the (semi)circumference to the square of the height. There is still no unequivocal evidence for cones in OB maths.

1;30*h* — the DAL –, from which the volume could be calculated, but this does not explain the use of the coefficient 18, nor the 'ratios' 4 and 3 in BM 85194 (x).

It was suggested in §4.3 that *imlûm* may refer to freshly made bricks or wet clay, with a density of 18 00 00 minas/sar$_v$, but if that word is to be identified with the IM.LÁ here, it still provides no obvious interpretation for the six very difficult problems under discussion. If 4 and 3 were the ratios of the length and width of the IM.LÁ they should be multiplied by (the weight of) the IM.LÁ and divided by 18 00 00 to give the depth. The problems, though, perform the inverse operations, dividing where multiplying would be expected, and vice versa. The IM.LÁ remains unexplained.

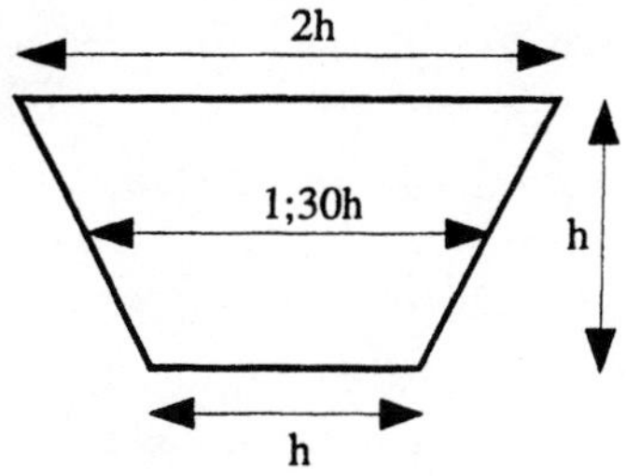

Figure 8.2: The IM.LÁ *as an excavation*

The penultimate coefficient listed in §8.3.1 is found in another very strange problem which has so far eluded all attempts at mathematical interpretation. Str 366 (i) reads:

I *ú-ma-am* 2 05 *el-qé ur-ra-am* *mi-nam e-le-qé* ZA.E KÌ.DA ZÚ!.DÈ 15 IGI.GUB.BA *a-na* 2 *u$_4$-ma-tim* NIM.MA 30 IN.SUM 30.E 4 ÍB.SI$_8$ *a-na* 2 05 NIM 8 20 *ba-ma-at* 8 20 GAZ-*ma* 4 10 *te-le-qé*	Today I took 2 05. What shall I take tomorrow? You, in your method, multiply 0;15, the coefficient, by 2, the days, and it will give you 0;30. The *root* of 0;30 is 4. Multiply by 2 05. ‹It will give you› 8 20. Break in half 8 20, and you will take 4 10.

The subject of the problem is unclear: not only what exactly is being taken 'today' but also how this implies that something different is being taken 'tomorrow'; it may be concerned with interest on a loan or investment (*MCT*: 36 n. 96d).

The seemingly nonsensical statement in line 4 that the 'square root' of 0;30 is 4 may be related to a small table text MLC 2078, which contains a list of similar 'roots', including this one in line 2. Neugebauer and Sachs (*MCT*: 35–6) saw the table as a list of 'logarithms' in base 16, so that 30.E 4 ÍB.SI$_8$ is to be interpreted as '0;30 = $\log_{16} 4$', or '$16^{0;30} = 4$', i.e. $\sqrt{16} = 4$. Friberg (1987–90: 547) has suggested that MLC 2078 was 'a help table for a class of interest rate exercises', but solely on the grounds that Str 366 might be one of those exercises.

Even if Neugebauer and Sachs' interpretation of line 4 is correct, the solution as a whole remains obscure. The function of the coefficient is unclear because it is unnamed. Put into modern notation, the method seems to be:

$$^1/_2(16^{\text{days} \times \text{unnamed coefficient}} \times \text{'today'}) = \text{'tomorrow'}.$$

Lewy (1947: 307) suggested that the problem was based on 16 mina of silver as the basic unit of investment, but was unable to give a satisfactory explanation for the peculiar solution. If the problem concerned simple interest at the usual rate of 20% (0;12) over a 360-day year (Nemet-Nejat 1993: 60–1), then the daily interest would be:

$$\text{daily interest} = \text{capital} \times 0;12/360,$$

so that a capital of 1 02;30 minas would be needed for a daily return of 2;05 shekels. If the solution required compound interest, calculated by:

$$\text{daily interest} = \text{capital} \times (1;12^{\frac{1}{360}} - 1)$$

the capital would need to be over 1 08;30 minas to earn 2;05 shekels a day. Neither method tallies with the solution given in Str 366 (i). The number 2 05 is also the starting point for the second two problems on the tablet, which are concerned with relatively straightforward problems on reciprocals (Sachs 1947; see Friberg 1987–90: 550).

The final coefficient listed in §8.3.1 constitutes the very last preserved signs of a rather damaged problem, BM 85210 (ix), on two holes in the shape of truncated pyramids. It seems that the base of the second is half the length of the first, but of the same height. Once the volume of the first is found, it is multiplied by 0;15 (i.e. $^1/_2 \times {}^1/_2$) to find the volume of the second (Thureau-Dangin 1936: 80–1).

8.4 READABLE COEFFICIENTS IN LISTS

8.4.1 The coefficients

A 44	[1] IGI.GUB HA.LA	[1], the coefficient of a share.
F 63b	1 *zi-it-tum*	1, a share.
A 45	[1] IGI.GUB GI.HAŠ.A	[1], the coefficient of a broken reed.
F 64	1 GI.HAŠ.A	1, a broken reed.
A 47a	⸢20⸣ GIŠ.ḪAB	20, madder.
E 15	11 08 ⸢IM⸣.BABBAR$_2$	11 08, gypsum.
E 27	2 13 20 *qí-it-mu-um*	2 13 20, black dye.
F 56a	⸢3⸣6 IM.BABBAR	36, gypsum.
F 56b	54 *ša ka-lu-ú-um*	54, of orpiment.
F 57b	20 *ša* NA$_4$.KUR.RA	20, of alum.
A 15	3 45 IGI.GUB NU!.LUḪ.ḪA.SAR	3 45, the coefficient of asa foetida.
A 16	4 IGI.GUB ŠE.LÚ.SAR	4, the coefficient of coriander.
A 38	4 58 IGI.GUB GI.NÍGIN.NA	4 58, the coefficient of a reed-bundle.
E 24	10 45 06 ⸢ESÍR.ḪA$_5$⸣.A *ša i-na ki-ri-im*	10, 45, 6, bitumen that is in an oven.
F 55	⸢20⸣ 48 06 *ša* GIR$_4$ AD.KE$_4$	20, 48, 6, of the reed-worker's oven.
E 52	20 Ì.UDU	20, mutton-fat.

8.4.2 Mathematical and philological commentary

Coefficients are not used in the extant OB problems on property division and measuring field with 'broken reeds'. Property (silver or land) is divided equally[20] or in arithmetic progression[21] among 'partners', 'soldiers' or 'brothers' (Friberg 1987–90: 750; Nemet-Nejat 1993: 63–7). 'Broken reed' problems[22] are based on the scenario of a field which is measured with a reed from which 1 finger's length breaks off after every measurement. The task is usually to find the original length of the reed, given the area of the field (Nemet-Nejat 1993: 72–6). A concrete link between these types of problems and the two pairs of coefficients in texts A and F has never been made.

The second group of coefficients listed here are all concerned with colourants and mordants (fixers) used in dyes and paints. They are not found in any other extant OB mathematics, and there is no clear numerical link between them. They do not form an arithmetical progression, for instance, as the metal coefficients do (§8.1). Apart from the small grouping in text F they

[20] e.g. TMS 26 (i)–(iii); VAT 7531 (i)–(iv); VAT 7621 (i)–(ii).

[21] e.g. AO 17264; Str 362 (i); *UET* 5 121 (i); VAT 6597; YBC 4608 (v); YBC 9856 (ii).

[22] AO 6770 (v); IM 53965; Str 362 (v); Str 366 (iv); Str 368; VAT 7532; VAT 7535 (i)–(iii).

occur in isolation. Madder (Akk. *ḫūratum*) is often attested in non-mathematical contexts with alum (Akk. *aban gabîm*) (*AHw*: 358) as a textile dye. Gypsum and orpiment (yellow arsenic sulphide) are used as paint and plaster colourants (Moorey 1994: 328). *Qitmum*/IM.SAḪAR.GE$_6$.KUR.RA is not yet identified; soot and other carbonated matter was used to blacken paint (Moorey 1994: 327).

None of the remaining coefficients can be interpreted mathematically either. A 38 cannot be linked numerically with other coefficients or problems on reeds and reed-bundles (§§5.5, 6.4, 7.4). Its value, which has been collated, is sexagesimally irregular, which hints at a copying error. Kilmer (1960: 291) has suggested that E 24 and F 55 are identical, with number 10 48 06, but the photos of the two tablets do not bear this out. The numbers are reminiscent of the sequences of semicircle coefficients found in the same texts (§3.2) and should perhaps be read as three separate coefficients. It has not been possible to relate them to the coefficients for spreading bitumen in §4.6. The reading of the numbers in A 15 and 16 (cf. 13 45 and 16, Kilmer (1960: 277)) were communicated to the author by John Brinkman (pers. comm. 1995); their mathematical significance is unknown.

8.5 OBSCURE AND BROKEN COEFFICIENTS IN LISTS

8.5.1 Coefficients from text A

A 9	56 IGI.GUB U+KUL X	56, the coefficient of ....
A 12	15 IGI.GUB ḪA.LA X	15, the coefficient of a share ... .
A 32	3 45 IGI.GUB SAḪAR KÙ ŠU BA.AN.TE	3 45, coefficient of receiving volume, silver?
A 33	16 IGI.GUB KÙ SAḪAR ‹ŠU› BA.AN.TE	16, coefficient of receiving silver, volume?
A 43	⸢X⸣ IGI.GUB GIŠ.NÁ GIŠ.GU.ZA	..., the coefficient of a bed and chair.
A 46	⸢X⸣ IGI.GUB ŠU.NÍGIN.NA	.... the coefficient of the total.
A 47b	1 IGI.GUB *ma-li ṣí-‹il›-pe-tum*	1, the coefficient of all rectangles?
A 48a	30 NU.KAR [X] DÈ$^{?}$	30 ...
A 48b	48 IGI.GUB SÍG$^{?}$.ḪI$^{?}$.A	48, the coefficient of wool (?).

The above readings and translations are, by and large, all as given by Kilmer (1960: 276–8; 1985). It was not always possible to get entirely satisfactory collations of these lines, as 'the physical condition of the tablet does not always permit clear readings' and 'some wedge traces are hard to distinguish from surface wear' (Brinkman, pers. comm. 1995; 1997).

The numerical value of A 9, which is clear on the tablet, is sexagesimally irregular; the final, unreadable sign is similar to ÚR. The last sign of A 12, which Kilmer reads BA, looks more like URU. A 32 and A 33 seem to refer to mutually inverse operations; they are reciprocals of each other. The numerical value of A 43 appears to be either 5 or 6, although is entirely unclear why domestic furniture should have a coefficient. A 46, which is on the last line of the reverse, may have given the total number of entries on the tablet: '... coefficients: the total'. That total may have been anything between 45 (the number of lines on obverse and reverse) and 54 (the number of coefficients), but the traces look more like ⸢1 10⸣, or 70. A 47b — whose numerical value after collation reads 1, not 10 — might be understood as the area of any rectangle, see §3.4. In A 48a the KAR may in fact be ḪI.A, while in A 48b the ḪI.A looks more like DU$_8$.A.

8.5.2 Coefficients from text C

C 31	[...] AK [ ] *i-gi-gu-bu-šu*	... is its coefficient.

C 32	[...] IG X [	]5 40 *i-gi-gu-bu-šu*	...5 40 is its coefficient.
C 33	[...] IG X [	]5 *i-gi-gu-bu-šu*	...5 is its coefficient.
C 34	[...] AŠ [	*i-gi*]*-gu-bu-šu*	... is its coefficient.
C 35	X-*ra-am* [	*i-gi*]*-gu-bu-šu*	... is its coefficient.
C 36	X-*ra-am* [	*i-gi*]*-gu-bu-šu*	... is its coefficient.
C 56	[	] *i-gi-gu-*[*bu-šu*]	... is its coefficient.
C 57	[	] *i-gi-gu-bu-*[*šu*]	... is its coefficient.
C 58	[	]16 *i-gi-gu-bu-šu*	...16 is its coefficient.
C 59	[	]11 *i-gi-gu-bu-šu*	...11 is its coefficient.
C 66	[	] ⸢*i-gi*⸣*-gu-bu-šu*	... is its coefficient.
C 67	[	*i*]*-gi-gu-bu-šu*	... is its coefficient.
C 68	[	*i*]-⸢*gi*⸣*-gu-bu-ša*	... is its coefficient.
C 47	*ma-*⸢*la* SAG⸣.KI [AN.TA ... *i-gi*]*-gu-bu-šu*		As much as the [upper] width ...: is its coefficient.

The coefficients listed here come from particularly broken parts of an already damaged text. There are no obvious parallels from the other lists which would enable C 31–6 to be restored, although brick coefficients, for instance, are missing from this text. The chances of successfully restoring the other lines are even slighter, as only the repeated word 'its coefficient' remains.

C 47 apparently starts a geometrical 'catalogue' which continues until the break at the bottom of the tablet (see §A.1). Goetze (1951: 144) described its subject matter as 'a triangle (rectangle) which is subdivided by parallels drawn to length in proportionate distances. The length of the parallels and probably the area of the various smaller triangles are sought.' However, no coefficients are attested in similar situations.[23] It may be that the signs after the break belong to the previous line (which is, however, virtually unreadable).

8.5.3 Coefficients from text E

E 7	5 ⸢24⸣ GÁN X X X	5 24, the area of a ... .
E 9	45 GÁN IN	45, the area of a ... .
E 12	4⸢5⸣ GIŠ.A.BAL	45, ...
E 16	5?4 ⸢IM.HA$_5$?⸣.A	0;54, dried earth?
E 17	6 40 IM.LAG?	6 40, earth clods?
E 18	4 16 [IM?].LAG?	4 16, earth clods?
E 28	20 [...]	20, ...
E 29	20 ⸢DUSU?⸣	0;20, a basket?
E 42	2 30 5 20 [*nu-ud-di*]*-tum*	2 30, 5 20, ...
E 43	48 01 40 01 20 ⸢*ú-ku/lu* ⸣ ...*-um*	0;48, 1;40, 1;20, ...

All these coefficients are badly preserved, and from a relatively disorganized list, which makes their identification doubly difficult. All of their values are sexagesimally regular. E 7 is discussed in §3.8. E 12, translated 'water-dipper (?)' by Neugebauer and Sachs (*MCT*: 136), in conjunction with Kilmer's reading of the number as 48 (1985: 27), might suggest that it is the irrigation coefficient 0;48 if it were not for E 26 on the same tablet (§6.4). However, it — and

[23] See Friberg (1987–90: 558–60) for a discussion of the figures he characterizes as 'striped triangles'.

presumably the virtually destroyed E 9 — may equally be a geometrical coefficient, coming as it does at the end of a geometrical section of the text (§3.8).

E 16–18 are part of an acrographic sequence in which all entries begin with the sign IM. The reading IM.KAL.A for *kalûm* 'orpiment' for E 16 is tempting, as the entry would then correspond to F 56b (§8.4), but is excluded by the traces on the tablet; the line was originally read IM.KI.A (*MCT*: 136; no translation). IM.HA$_5$.A 'dried earth/clay' is also possible, but the sun-dried bricks discussed in §4.3 are 0;48 ($^4/_5$) of their original weight, not 0;54 ($^9/_{10}$).

E 17–18 seem to have the same name, IM.LAG/*kirbānum* 'earth clods' (Kilmer 1960: 301). This is an alternative term for sun-dried clay products used in D 64 (§4.3), but once again, there is no obvious numerical connection. The last sign could perhaps be DUB (*MCT*: 136).

Little remains of E 28–9 after the numbers. Only a vertical stroke is visible to the far right of E 28, while the sole sign of E29 might be AL 'hoe', LÀL 'honey' or IL (*MCT*: 136; Kilmer 1960: 301). AL/*allum* seems to be a mathematical term in the Eshnunna region only and is never attested alone (§5.2.3). The context of the two coefficients is unhelpful.

The restoration of E 42 is conjectural. It is based on the first millennium B III 19′: 24, 5 20 *nu-ud-di-tu*$_4$, whose meaning is, however, unknown (24 and 2 30 are a reciprocal pair). E 43 should be read as '0;48, 1;40 and 1;20', i.e. $^4/_5$, $^5/_3$, and $^4/_3$. A reading *ukurrum* 'brick' is unlikely as the following line, E 44, gives the alternative spelling *agurrum*.

8.5.4 Coefficients from texts F and G

F 46 ⸢3⸣ [22] ⸢30⸣ *ša* IM	3 [22] 30, of clay.
F 57a ⸢20?⸣ *ša* ESÍR?	20, of bitumen?
F 62 24 PI-*lu-ú-tum ša* GI [(x)]	24, ... of reed?
F65 57 36 *ša eb-lu-ú-um*	0;57 36, of a rope.
G 11 [x] IGI.GUB ⸢ŠÀ⸣.GAL *ma-ša-ri*	..., the coefficient of the slope of ...

Most of the number of F 46 is broken completely away, but has been restored on the grounds that 3 22 30 is the only sexagesimally regular number which fits the traces. There are no other attested coefficients for 'clay' alone, so it is impossible to determine its purpose. It may be connected with carrying earth, as the previous entry in the list, F 45, gives the volume of a carrying-basket (§5.2). The final sign of F 57a looks like LAGAB×A, but is probably to be read ESÍR 'bitumen' (Kilmer 1960: 291). The value 0;20 does not match other known bitumen coefficients (§4.6). F 62 has remained untranslated since publication; the number of F 65 is the same as D 30 (§3.3).

G 11 occurs in a section of the list concerning storage, but seems to describe the slope of an object. The two penultimate signs could equally be read KU and TA respectively.

8.6 SUMMARY

The number of obscure or completely unexplained OB coefficients is currently around 10% of the total known. As more Old Babylonian problem texts become available, the number of 'obscure' coefficients is bound to decrease as they are put into mathematical context. It is only those from the very damaged parts of texts C and E which are probably never recoverable.

9. MATHEMATICS IN UR III AND OB ADMINISTRATION

The following two chapters of this work address the matter of why the coefficient lists were composed. Clearly they were an integral part of the mathematics taught to scribal students in the OB period, but this raises the question as to why coefficients, or indeed mathematics, were taught at all. As mentioned briefly in §1.3.6, some scholars (e.g. von Soden 1977; Walters 1970: pp. xiv, 118–19, 147–8) have already worked on the principle that the numerical values found in the school coefficient lists are those actually employed by professional quantity surveyors and accountants of the same periods. This assumption has been supported by several examinations of the mathematics and coefficients used in individual economic documents (Goetze 1962; Powell 1976*a*; Vaiman 1960), but the material presented here is the first attempt to conduct a wider survey. Those earlier studies are used here, but in general transliterations of the texts concerned are not given. However, there are also documents included in this chapter which have been published in hand-copy but which have not been analysed for their mathematical content, or even edited, before. Full transliterations are given in these cases.

Englund's recent work on Ur III accounting procedures (1988; 1991), Civil's edition of *The Farmer's Instructions* (1994), Maekawa's review of agricultural practices (1990), and Waetzoldt's studies of pot manufacture and of reeds in administrative texts from Umma (1970–1; 1992) all discuss certain quantitative aspects of Ur III bureaucracy, and have been very helpful in compiling this survey. On the other hand, little has been said of equivalent OB procedures; Lackenbacher's edition of a record of textile manufacture (1982) and Stol's study of ploughing rates (1994) are notable exceptions.

The evidence from the Ur III and OB periods is of vastly disparate character. The greater quantity, and the wider range, of sources available from Ur III mean that there is much more mathematical evidence available than for OB. This chapter thus concentrates on the earlier period, bringing in OB material where practicable. Relevant accounting and administrative practices are surveyed first, followed by a brief look at the use of mathematical tables by professional bureaucrats. Then the textual and archaeological evidence for the different types of coefficients is examined, topic by topic, following the same divisions as §§3–8: geometry; various types of quantity surveying; storage; other subjects. This examination is by no means exhaustive; instead, it focuses on single examples of the coefficients in use. Lists of Ur III documents which use coefficients found in Ur III and OB school texts are presented where appropriate in each section of the chapter. A full-scale search of Ur III and OB administrative documents was beyond the scope of this investigation, but is needed if we are to comprehend fully the extent to which coefficients were used by working scribes.

9.1 COEFFICIENTS IN ADMINISTRATIVE PRACTICE

9.1.1 Accounting for labour before Ur III

Just as mathematical education can be traced back well before the Ur III period (§10.1), so too there is evidence for early attempts to predict and manage the expenditure of labour and materials. Already in ED III workers were allotted daily lengths of canal to renovate (e.g. VAT 4851; cf. Maeda 1984). One text, *DP* 654, gives detailed measurements of completed waterworks and earthworks, without stating the number of labourers working on them. In neither case, though, is there enough information to determine whether daily work-rates, measured by volume, were used in quantity surveying at this time.

However, there is scant but clear evidence for labour-rates in use by the Akkad period. *DV* 5 64 records the dimensions of three canal banks, of total volume 14;30 sar_V. The work is said to have taken 87 days, so that it must have been carried out at a rate of 14;30 ÷ 87 = 0;10 sar_V a day (Powell 1976*b*: 100–2; Friberg 1982: 117–18). This is exactly the most commonly recorded work-rate for earth moving in both Ur III administrative records and OB school mathematics (§§6.2, 9.6.3).

9.1.2 Ur III accountancy and quantity surveying practices

There is almost an overabundance of administrative records from the Ur III period, dating for the most part from the fifty or so years between the sweeping reforms of the second half of Shulgi's reign and the rapid decline of the empire after the accession of Ibbi-Sin (*c.* 2075–2025 BC) (cf. Sigrist and Gomi 1991: 100–260). Almost every aspect of the state's activities was recorded and controlled, but of most relevance to this study are the documents with which conscripted labour was managed. Agriculture and the maintenance of waterways are the primary concerns of the extant texts, many of which are from Umma. Building work and the manufacture of goods feature to a lesser extent.

Where large-scale repair or construction work was needed, the first stage was to carry out a quantity survey, which recorded the measurements of the work to be done and noted the personnel to whom the work had been assigned.[1] The completed work was also inspected, and the workers and their wages accounted for. These records of completed work were kept and made up into yearly accounts. It is often difficult to determine whether an individual document is a preliminary survey or a record of completed work, as the type of information, and the manner in which it is recorded, can vary.[2] The text may record almost any combination of:

- the area of the field or the measurements of the construction, with the area or volume calculated from them. For constructions these measurements may be in the form of:
  - length, width, and height or depth each given individually;
  - length, and volume per nindan;[3]
- the labour needed to complete the task, with or without the daily work-rate explicitly stated;
- the workers' pay, with or without the daily pay clearly stated (cf. Krecher 1993–7: 158–9).

These are totalled and reported at the end of the text, with the place and date of the work and the responsible officials. The tablet is often sealed.

Where the rate is given, the usual formulation is to state the total volume or area to be worked, the task involved and the labour needed, as in AO 5676: VI 11–12 (cf §9.7):

4 05 sar gi ku_5-rá 10 sar-ta	4 05 sar_a of cutting reeds at 10 sar_a (a day):
á-bi u_4 24$^1/_2$	its work is 24$^1/_2$ days.

[1] e.g. BM 21335, undated, from Girsu.

[2] By contrast, in the 40-odd ED III records of irrigation work studied by Maeda (1984) three types of document could be distinguished by the verbs used:

- surveys of canals to determine where work was needed (**gíd** 'to measure');
- documents which assigned the work to overseers and their teams (**dab_5** 'to assign');
- records of completed work (**dù** 'to construct') (Maeda 1984: 33, 37).

[3] Cf. Civil (1994: 116), where these formulations are discussed, and examples given.

But to recover the work-rates used for administrative purposes it is not necessary to have those rates recorded explicitly. As long as the total volume and the number of workers are known the rate can be calculated by dividing the volume by the labourers. In MLC 2404: 1–4, for instance, it is 35;20 ÷ 3 32 = 0;10 sar_V a day:

$35^1/_3$ sar kin-saḫar	$35^1/_3$ sar_V earthworks.
á lú-ḫun-gá	The wages of the hired men
[...] sìla-ta	are ... litres each:
á-bi u_4-3 32-kam guruš u_4-1-šè	its work is 3 32 labourers for 1 day.

Equally, it is possible to calculate the workforce by dividing the total pay by the daily pay. This extract from YBC 3668: IV 64–6 gives the total volume of repairs to a dyke, the daily pay and total wage bill:

3 40 sar saḫar	3 40 sar_V earth.
á lú-ḫun-gá 6 sìla-ta	The wages of the hired men are 6 litres each:
še-bi 26 (gur) 2 (bariga)	its grain is 26 gur, 2 bariga.

The workforce is thus 2 12 00 ÷ 6 = 22 00 strong, and the work-rate of each labourer is 3 40 ÷ 22 00 = 0;10 sar_V a day (cf. §9.6). But only relatively rarely are such combinations of data given, and it is often impossible to reconstruct the work-rates.

The practice of accounting for labour in the Ur empire has been studied by Robert Englund, who summarizes his findings as follows:

> The book-keeping of expected and real labor performance achieved its most developed form during the Ur III period. The administrative texts of this period documented a continuous reckoning of all real labor performances against theoretical debit entries and imposed duties. Accounts of the activities of foreman and workers were balanced against one another. The accounting period consisted as a rule of one administrative year, that is, a year of twelve 30-day months. The posting of a balance usually resulted in a debit, since the expected labor performance was in all likelihood simply beyond the capabilities of the normal worker. Moreover, an incentive for the workers to produce more was non-existent; their remuneration consisted of no more than the minimum amount of grain and clothing to keep them able to produce. [...]
>
> Although we are able to document the performance standards and value equivalencies only through their appearance in accounting convention, there can be no doubt that such norms really existed in explicit form and that they were strictly enforced in dealing with the foremen of work gangs. The conversions, from which they can be documented, were conversions into comparable products specific to particular sectors of the economic organization. Depending on the economic sector, silver, barley, fish or workdays served as means of comparison or as measure of standardized norms, and performance expectations. Workdays were, in the debits section of Ur III accounts, the product of the number of workers multiplied by the number of days they were at the disposal of a given foreman. In the section of credits, the corresponding productive workdays were calculated through the division of real production — milled grain, delivered fish, and so on — by the amount of the product which the administration assumed could be processed by one workman in one day. A reconstruction of the conversion demonstrates that these workdays were based on stable conversion factors for the various labour performances and goods. (Englund 1991: 279–80)

It is the aim of this chapter to show that these 'stable conversion factors' are in many cases identical to, and in other cases of a similar nature to, the OB — and indeed Ur III — quantity surveying coefficients of *school* mathematics. These practical roots can also be traced through contemporary archaeological evidence. In particular, field plans and architectural features such as spiral columns demonstrate the level of geometrical knowledge in everyday use (§9.3); the measurements of complete, datable bricks are comparable to the theoretical ones of mathematical problems (§9.4.1); and contemporary house plans and modern plans of excavated works suggest that the rules of thumb used in construction were carried over into school mathematics (§9.4.2).

9.1.3 Quantitative methods in OB administrations

Although there is clear documentary evidence for quantity surveying practices after the end of the Ur III empire, the work-rates assumed are in general tantalisingly elusive because, it seems, labour was no longer accounted for so rigorously. The well-known and extensive administrative archive concerning repair work on the Larsa canal system in the early nineteenth century BC (Walters 1970) is typical in this respect. Many of the 150 or so published documents belonging to the archive give detailed numerical information on the volume of earth (to be) dug,[4] the bricks (to be) made,[5] etc., and the workers[6] and their wages,[7] but only one contains a useful combination of information (see §9.5.1). Walters had to resort to using OB mathematical coefficients to draw any conclusions about work-rates (e.g. Walters 1970: p. xix). Renger's survey of (later) OB sources on irrigation practices also concludes that the numerical information they contain is 'at best sketchy' (Renger 1990: 37). On the other hand, the OB archaeological sources such as brick sizes, pillars and house plans are just as illuminating as their Ur III counterparts (§§9.3, 9.4).

There are several technical terms associated with the calculation of work-rates in OB Akkadian, but because they usually occur in non-mathematical contexts, it is very difficult to distinguish their exact functions. For instance, *ARM* 6 7: 8–12 is an extract from a letter to Bahdi-Lim, the superintendent of the OB palace at Mari:

DUMU.MEŠ É.DUB.BI *a-da-a-am* *ú-pí-šu-ma* *e-li ši-ip-ri-im* SUMUN *a-du-ú-um ša se*[!]*-ki-ri-im*[8] 2 *li-im ṣa-bu-um qa-al*	The scribes calculated the work assignment and (found that) as well as the old work (there is) the canal-workers' work assignment. 2,000 men are insufficient.[9] (Cf. *CAD*: A I 135; Muroi 1992*a*: 173)

Akkadian *adûm* is suposedly derived from the Sumerian **á-dù** 'assignment, task, allotment', which, however, is only attested post-OB (*PSD*: A II 52). This extract confirms that something like the Ur III daily work-rates was indeed used for labour management in OB times: *adûm* may be the total labour force needed for a task, calculated in working days. Other attestations do not provide contexts specific enough to confirm or refute this interpretation. Similarly, it is not clear whether *iškarum* ever bears the very specific OB mathematical meaning of 'daily (work-)rate' (cf. §5.2.3) in non-mathematical contexts.

[4] e.g. NBC 5506 (text 88), supposedly 'a statement of labor requirements for ... seven plots assigned to ... two men [who] were responsible for seeing that the necessary earth operations were carried out on these plots', lists only the volumes of earth concerned and makes no mention of the manpower needed.

[5] e.g. NBC 9050 (text 99), which lists large quantities of (baked and sun-dried?) bricks and the overseers responsible for them.

[6] e.g. text (types) 117–18 (Walters 1970: 138–41), a total of 25 different tablets, assign to overseers various numbers — up to 83 — of labourers (ERÍN.ḪI.A), stating their occupations.

[7] e.g. NBC 5488 (text 30), which records the hire of 1800 men for 10 minas of silver, or $^1/_3$ shekel each. In NBC 5477 (text 82) 9 men receive 5 gur 120 sila barley for 3 months of (unknown) work: a rate of 2 sila a day.

[8] This word was previously read *ták-ki-ri-im* and understood as the name of a canal: *takkīrum*, meaning 'diversion-canal', from the D-stem of *nakārum* 'to ' (*AHw*: 1307). But all attestations are post-OB (Kupper 1954: 116 n. 7) except ÍD *ta-ki-ru-um* in *UET* 5 855: 14 (Groneberg 1980: 311), which is a mathematical school text (§1.3.4) and thus cannot be used as evidence for real names. On the other hand, the examples in *CAD* (A I 135) show that *adûm* is usually followed by a profession. TÁG and SI are very similar in OB script. I thank Stephanie Dalley for this interpretation.

[9] M 288 (§6.2.2) confirms that Mari scribes were taught how to compute the manpower needed for large-scale earthworks.

9.2 MATHEMATICAL TABLES

There is no unequivocal archaeological evidence that metrological and mathematical tables were used by professional administrators. Seven tablets containing mathematical tables were found in the OB palace at Mari,[10] although Margueron (1986) has since shown that all the 'archives' from which they derive are the consequence either of systematic looting after Hammurabi's conquest or of excavators' errors.[11] Further, the recent publication of the mathematical diagram M 288 (§6.2) gives rise to the possibility that they belong to a scholastic milieu.

On the other hand, although no clear school contexts for mathematical and metrological tables have been found at other archaeological sites[12] there is a good deal of internal evidence, particularly from Nippur (§10.5). However, there is simply not enough evidence to warrant the view that 'multiplication and reciprocal tables are prepared only for use in schools; they are not used by skilled mathematicians who are able to calculate by heart due to the flexibility of the sexagesimal system' (Nemet-Nejat 1993: 19; cf. Bruins 1971: 222). This statement derives from a rather romantic notion that — as the extract suggests — all those engaged in arithmetical activity were in some sense 'skilled mathematicians' rather than bureaucrats with many different levels of numeracy. It is generally agreed by modern scholars that metrological and mathematical tables must have been 'almost indispensible as tools for the multiplication of measures' (Friberg 1987–90: 543), particularly, one supposes, for forgetful or less than perfectly numerate administrators.[13]

9.3 GEOMETRY

No geometrical coefficients are attested explicitly in Ur III administrative texts, but some of the properties of triangles, trapezoids and circles must have been known. Field plans[14] show the division of irregularly shaped fields into quadrilaterals and approximately right triangles. Their

[10] *ARM* 9 299, a table of reciprocals, came from Room 5 along with a large number of administrative documents. Birot (1960: 351) suggested that it might be a school text, deposited there during the confusion of the last days of occupation of the palace, or, equally possibly, that it had been used as a calculating aid by the record keepers. A 3514, M 12306, M 6199b and M 12598 (three multiplication tables and a table of squares) were found in Room 108, from which nearly a thousand tablets, covering all sorts of subjects, were recovered. M 8613, a table of exponentials, with some decimal notation, was discovered in Room 115, packed with several thousand tablets which had been abandoned in great disorder and for the most part broken in antiquity. They included important diplomatic correspondence dating back to the reigns of the last kings of Mari. M 10681, another multiplication table, came from Room 156, the hallway of the main gate into the palace (Parrot 1958: 80–1, 102; Soubeyran 1984: 19). Soubeyran (1984: 20) said of these tablets that 'rien ne permet de penser qu'il s'agit d'exercices d'école'.

[11] According to Margueron (1986: 148) the Room 5 archive was in fact the result of a collapse in the ceiling, and had originally been stored in one or more collections on the first floor or roof. Joins in tablets from several different 'archives', including Rooms 108 and 115, indicate that many were assigned untrustworthy findspots during or immediately after excavation. Further, the tablets in Room 115 had probably been moved there by Hammurabi's followers for examination prior to the destruction of the palace (Margueron 1986: 148–9).

[12] Mathematical tables were found in two locations characterized by Woolley as 'schools' in OB Ur (Woolley and Mallowan 1976: 112, 137). Charpin (1986: 481–2) has shown, though, that the tablets found in 'No. 1 Broad Street' were used as fill during the restructuring of the house at an unknown date. In 'No. 7 Quiet Street', a private house, the archival and scholarly material was mixed together so that it is impossible to determine the function of the table texts (Charpin 1986: 432–4).

[13] For the (present) impossibility of distinguishing between Ur III and OB table texts, see §10.2.

[14] See most recently Nissen *et al.* (1990: 106–10).

areas were calculated using the so-called 'agrimensor formula' later known to the Romans[15] and the coefficient 0;30[16] respectively, and summed to give the total area of the field in question.

A few Ur III surveys of work to be done concern structures with trapezoidal cross-sections — but use difficult terminology such as **dagal ì-ḫi** 'average width',[17] and **ba-an-gi₄** 'top (of a tapered wall)'.[18] Of course, where only the cross-section is given and not the individual measurements of the width and height (see §9.1.2) it is impossible to determine the shape of the structure concerned.

Trapezoidal bricks were specially moulded for the construction of wells; indeed two OB mathematical problems concern this very subject.[19] From the measurements of baked well-head bricks given by Walker (1981: 167), the approximate size of the structures they came from can be reconstructed,[20] with essentially the same method as the OB one:

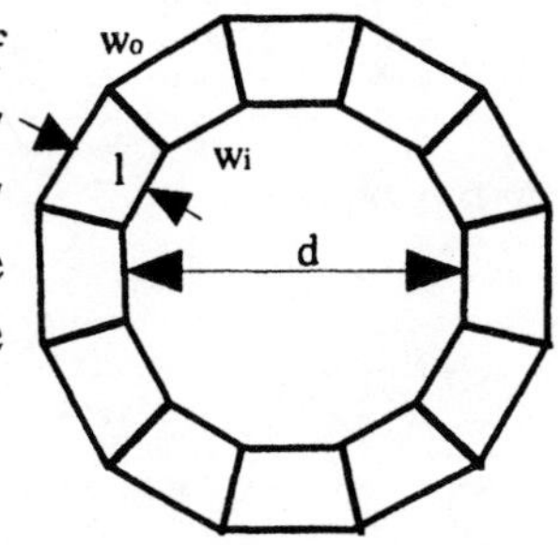

Figure 9.1: Schematic brick well-head

$$d = \frac{2w_i l}{w_o - w_i}$$

$$c = \pi d$$

$$n = \frac{c}{w_i}$$

where l = length of the bricks, w_i and w_o = inner and outer width; d = inner diameter of the well, c = inner circumference; n = number of bricks,

The first of the BM 85194 problems states the measurements of the bricks (see Table 9.1 above) and asks for the number of bricks needed per layer, and for the internal and external measurements of the well; the second gives the inner circumference of the well and the length of the bricks. If the difference between the outer and inner widths is known, what are these widths and how many bricks are needed? Neither quite conveys a realistic scenario. Under working conditions, it is much more likely that it was necessary to determine the size and number of bricks needed for a given diameter of well. In these circumstances the coefficient 3 used as an approximation to π (cf. §3.1) worked to the advantage of the quantity surveyor as it produces a slight underestimate of the circumference from the diameter. In choosing an inner width for the brick that was an integer divisor of the calculated inner circumference — and therefore ensuring a whole number of bricks for the circle — the surveyor automatically allowed for roughly 5% of the circumference to be mortar between the individual bricks.[21]

[15] i.e. that the area A is the product of the respective averages of pairs of opposite sides, l_1 and l_2, and w_1 and w_2:

$$A = \tfrac{1}{2}(l_1 + l_2) \times \tfrac{1}{2}(w_1 + w_2).$$

There are a also a few Ur III-OB school calculations and diagrams of this method, e.g. Ash 1922.168 (§A5.2), YBC 7290, YBC 11126 (§1.3.3).

[16] For instance, a triangular field **a-šà sag-dù** is mentioned in the Ur III agricultural account AO 5676: III 27. See §3.3 for the coefficient.

[17] BM 20054 (Englund 1988: 179 & n. 49); see also §9.6.3.

[18] AO 7667: see §9.4.2.

[19] BM 85194 (xvi), (xix).

[20] BM 137402, a Nebuchadnezzar well-head brick, is also mentioned by Walker but, as its inner width is not given, the size of the well it came from is not deducible.

[21] Because $(\pi - 3) \div \pi = 0.0451... \approx 0.05$.

Table 9.1: Well-head bricks

Museum number	*Ruler/provenance*	*Brick measurements*[22]		*Well measurements*[23]	
BM 90272	Nebuchadnezzar II	Length:	0.430	Inner diameter:	2.15
	(Walker 1981: no. 91)	Inner width:	0.250	Inner circumference:	6.75
		Outer width:	0.350	Number of bricks:	27
BM 90824	Nebuchadnezzar II	Length:	0.430	Inner diameter:	2.15
	(Walker 1981: no. 91)	Inner width:	0.250	Inner circumference:	6.75
		Outer width:	0.350	Number of bricks:	27
Ash 1957.183	Ashurnasirpal II	Length:	0.285	Inner diameter:	2.61
	(Walker 1981: no. 145)	Inner width:	0.275	Inner circumference:	8.21
		Outer width:	0.335	Number of bricks:	30?
BM 90780	Sennacherib	Length:	0.390	Inner diameter:	5.11
	(Walker 1981: no. 172)	Inner width:	0.295	Inner circumference:	16.06
		Outer width:	0.340	Number of bricks:	54
BM 90783	Sennacherib	Length:	0.370	Inner diameter:	6.34
	(Walker 1981: no. 172)	Inner width:	0.300	Inner circumference:	19.93
		Outer width:	0.335	Number of bricks:	66
BM 90773	Sennacherib	Length:	0.355	Inner diameter:	3.81
	(Walker 1981: no. 173)	Inner width:	0.295	Inner circumference:	11.96
		Outer width:	0.350	Number of bricks:	40
BM 85194 (xiv), (xvii)	OB mathematics	Length:	0.335	Inner diameter:	1.33
		Inner width:	0.250	Inner circumference:	4.00
		Outer width:	0.175	Number of bricks:	24

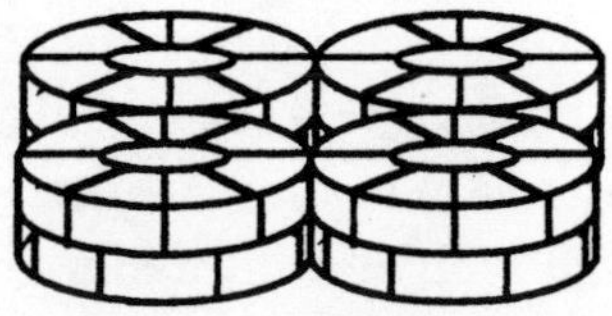

Figure 9.2: Sketch of the 'pilier des briques'

Although the examples of well-head bricks in Table 9.1 are all from the first millennium BC, Gudea's 'pilier des briques' in Girsu (de Sarzec 1884–1912: II pl. 53b; Parrot 1948: 156–8, fig. 34) shows that the technique was already established before the Ur III period. The structure comprises four adjoining cylindrical pillars,[24] each of which is constructed from circular and wedge-shaped bricks in alternating layers.

Around a central cylindrical brick roughly 250 mm (i.e. half a cubit) in diameter[25] are eight wedge-shaped bricks of the same length, which thus divide the diameter into three equal parts.[26] The layers above and below are composed of two concentric rings of eight bricks each, the central ones being half a cubit long, and the outer bricks a quarter cubit.

Figure 9.3: Alternating layers in the 'pilier'

More intricate temple columns, composed of pre-moulded bricks, are a feature of the OB period. The 'bastion of Warad-Sin' on the ziggurat terrace of Ur is an elaborately recessed and

[22] Measurements are given in metres to the nearest 5 mm. The thicknesses have been omitted as they are irrelevant to the calculations.

[23] It follows that the outer diameter = $d + 2l$, and the outer circumference is π times greater again. Measurements are in metres, to 2 decimal places.

[24] The OB problem BM 15285 (xxxvi) (§A.2) and the diagram *TSŠ* 77 from ED III Shuruppak (§10.1) both show four touching circles in the same formation as the 'pilier des briques'. Perhaps it is to be expected that certain geometrical concerns of school mathematics were reflected in architectural design — or vice versa.

[25] e.g. BM 96945, 24.5 cm in diameter and 5.5 cm thick (Walker 1981: no. 7).

[26] e.g. Ash 1930.52, outer width 29 cm and thickness 6 cm (Walker 1981: no. 7). The length and inner width are reconstructable from the inner diameter (from BM 96945) and the number of bricks in the circle.

rabbetted façade (Woolley 1939: pl. 71), at the centre of which is a doorway flanked by two columns of diameter 0.70 m:

> The columns were built of specially moulded bricks, segmental in shape and with the outer edge not only rounded to the curve of the column-shaft, but with a further boss in relief so that each set of three bricks (two in the lower and one in the upper course) produce a truncated triangle standing out from the column in low relief; these triangles (0.21 m at the base, 0.20 m high, with a flat top 0.08 m across), set in rows one above another, made of the brick shaft a very close imitation of a date-palm trunk. (Woolley 1939: 42 & pl. 30b)

Because Woolley did not illustrate this description, it is difficult to visualize how these columns were constructed; there must have been ten triangles in each row, but the shape and size of the bricks and how they fitted together is not reported.

The roughly contemporary Great Temple at Tell al-Rimah, built during the time of Shamshi-Adad I, was decorated with hundreds of engaged columns, fifty of which were constructed from pre-formed bricks to produce spiral and palm-tree effects:

> Each column, whether palm-trunk or spiral, was built with mud bricks bearing on their outer face patterns in relief which, by repetition in a standard sequence, produced the required motif. The basic shape of brick employed was a 60° sector of a circle of radius *c*.29 cm. The brick could have been made in a mould of this shape or it could have been cut from a brick made in the standard square mould. To produce the correct outline a 60° angle could easily be defined by drawing the 29 cm circle and then intersecting the circumference with an arc of the same radius. The ornament was then carved on the curved surface. Two types of palm trunk column were made. The motif for the one with a 'scale' pattern consisted of only one course, each brick bearing three scales; the 'diamond' type was four courses high and require four different patterns of brick. (Oates 1990: 294–5)

The spiral half-columns were also composed of a four-course sequence. Two of these layers comprised three or four wedge-shaped bricks dividing the column into 60° segments (shown below for a full pillar), alternating with courses of one central block surrounded by six roughly semicircular bricks continuing the spiral motif (Oates 1990: 395–6 & fig. 4).[27]

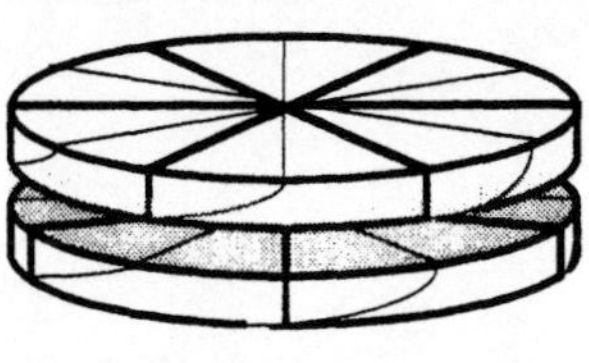

Figure 9.4: Construction of the Rimah columns

As in the case of the palm-trunk pillars, the geometry involved is on a similar level to OB maths. The cross-section of the cylindrical column is considered as a regular hexagon composed of six equilateral triangles which are themselves bisected twice. The indent with which the spiral is represented slopes between two consecutive bisections. The various coefficients for equilateral and right triangles must have been used in the design, if not the execution, of these pillars.[28, 29]

9.4 BRICKS AND BRICKWORKS

OB and Ur III school mathematics uses a complex system of brick metrology, involving a dozen or so standardized sizes of brick (see §§4.0–4.3). This section examines to what extent

[27] As Oates (1990: 396) points out, their presence was dictated not by geometrical but by structural considerations, reinforcing the bonding between the half-column and the wall behind. As can be seen from the diagram, the wedge-bricks alone are enough to ensure the continuation of the pattern.

[28] Cf. *TMS* 2 and the coefficients discussed in §3.3 and §3.6.

[29] In the 1984 season of excavations at Tell Haddad, ancient Me-Turan, an OB temple or palace complex was discovered. It comprised two rectangular courtyards whose 'surrounding walls ... were decorated with semi-engaged mudbrick columns which had date-palm and barley-sugar designs similar to those found in the OB temple at Tell al-Rimah' (Killick and Black 1985: 220). No further details have been published.

these theoretical bricks are based on real norms (§9.4.1), and whether the brickage coefficients associated with them can be traced in archaeological and documentary sources (§9.4.2). Finally, the estimates of proportions assumed in problems on building are compared to those inferred from house plans, both on tablets and in modern excavation reports (§9.4.3).

9.4.1 Baked bricks

Twelve sizes of standardized bricks are attested in OB mathematics. All are either square or two-thirds bricks (i.e. with width equal to two-thirds of the length), or half-bricks (with width equal to half of the length). There are six sizes of square bricks, with thicker variants for two of them. Two of the larger ones also come in half-sizes, and two of the smaller have two-thirds brick counterparts (see Fig. 9.5 below and §4.8).

A survey of the measurements of baked bricks excavated from various Mesopotamian sites of different periods (§A.6) suggests strongly that this system of brick metrology had a basis in reality. Although only two brick sizes are attested from the small sample of OB baked bricks, in the Ur III to OB periods taken together there are examples of all types except the 1 cubit bricks: square, two-thirds and half-bricks are all attested; very few bricks indeed are not of these shapes. In the Ur III period there is a spread of lengths between 15 and 22 cm, but the Isin, Larsa, and OB bricks all cluster around the standardized brick sizes known from OB mathematics. The thicknesses of the bricks also bunch round the theoretical 5 finger (8 cm) ideal. In no sample is the complete range of brick sizes attested, but this may be because unbaked bricks are not included in the survey: only baked, stamped bricks were chosen, so that they could be dated by their inscriptions.

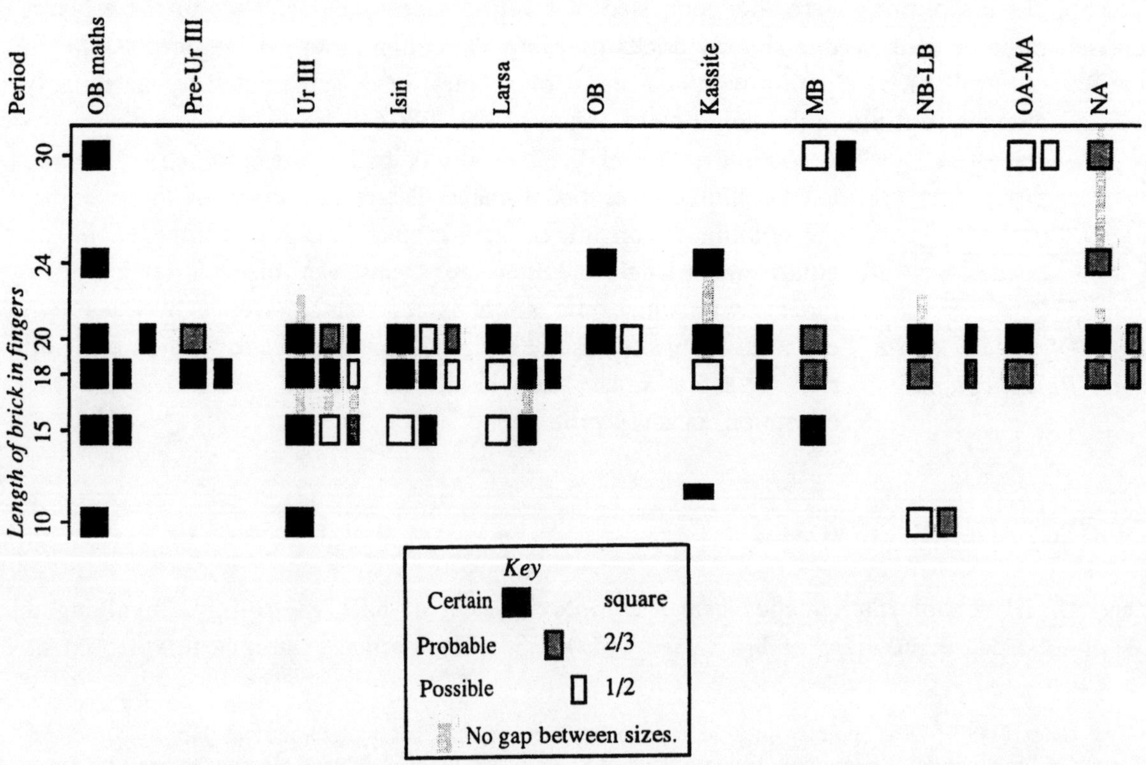

Figure 9.5: Sizes of baked bricks in the British Museum compared to those in OB maths

It is more pertinent that for the late third–early second millennium there is only one *certain* attestation of a brick size not found in OB maths but which might be expected to have been used (i.e. the 18 finger half-brick from the Larsa period). In later periods the square brick approximately $^2/_3$ cubit long remained an important vehicle for royal inscriptions, although it seems that to a large extent other sizes were not used for this purpose — if indeed they were used at all. The favourite size in the early first millennium, though, seems to have been the half-cubit square brick. The survey shows that Assyrian baked bricks are moulded in similar proportions to the south Mesopotamian ones, as squares and half-squares. In general, though, they are rather thicker, at around 10 cm, and there is less differentiation between different sizes. There seem instead to be two ranges of brick lengths: 30–35 cm and 40–55 cm. None of the Assyrian baked bricks are less than 30 cm square. Baked bricks whose width is three-quarters of the length are found only in the Neo-Assyrian period, it seems.

9.4.2 Ur III records of building work

VAT 7038, from Ur III Umma, is either a preliminary survey of work to be done (cf. §9.1.2) or a school practice text.[30] It bears several anomalous characteristics, being undated, anonymous and rather inconsistent in terminology. There are also unusually many results of intermediate calculations.[31]

Table 9.2: Summary of VAT 7038

		Length (in nindan) of walls of different widths (in cubits)					
Lines	*Description*	*5*	*4*	*3;20*	*3*	*2*	*0;40*
1–4	**iz-zi gu-la**	12;35					3;45[32]
5–7	**ús-sa bàd-da**				9;10	1;22 30	
8–11	**á-1-e ki-tuš**			2;00	6;40	2;32 30	
12–15	**zà é ugula gìr-šè**			2;05	3;50	0;30	
16–20	**iz-zi ì-sá-šè**??		1;20	1;00	7;15	1;20	
					7;30		
20–5	Total length (nindan)	12;35	1;20	5;05	34;05	5;00	3;45
26–31	Total area (sar_a)	5;14 35	‹0;26 40›	1;24 43 20	8;36 15	0;50	0;12 30

The first twenty lines give the lengths of numerous walls of different widths and short descriptions of small groups of these walls. Lines 21–6 give the total lengths of wall for each width, in descending order. The following five lines list the areas of those walls (the product of the width and total length in each case). The area of the 4-cubit thick wall is missing; the remaining results are not listed in the expected order. Finally, line 33 correctly gives the total floor-space of wall as 16;44 43 20 sar_a,[33] and a number of bricks: 2 ešè, $2^1/_2$ iku, or 24 10 sar_b. The height of the constructions is not stated on the tablet, but if it were 1 nindan, or 12 cubits, the volume would be 16;44 43 30 × 12 = 3 20;56 sar_v. The number of Type 2 bricks in this volume is 3 20;56 × 7;12 = 24 06;43 12 sar_b, just short of the rounded figure 24 10 given in the text.

[30] The fact that there is no obvious mathematical order to the entries suggests that it is a survey rather than a model document, written line by line as each structure was measured in turn.

[31] It is impossible to determine the physical appearance of the tablet from Schneider's copy (1930: no. 482). For collations see Waetzoldt (1978: 54).

[32] Listed as **9 dub-lá 5 kùš-ta** '9 pilasters of 5 cubits each' (line 2). For the translation 'pilaster' see George (1995: 182–6). Thus we are dealing with a wall 5 cubits wide with shallow pilasters at regular intervals along it, to produce the 'niche and projection' façade typical of Mesopotamian temple architecture. Coincidentally, these pilasters are almost exactly the same size as those on the north wall of the central courtyard of the Esagil temple in late first millennium Babylon — around 1500 years later (George 1995: 184 fig. 4).

[33] Excluding the area omitted from the previous section.

AO 7667, also from Ur III Umma, summarizes records of completed work of the sort that the model document YBC 9819 (§4.5) is based on. The dimensions of ten pairs of constructions (marked **a-rá 1-kam** and **a-rá 2-kam**) are given with the numbers of bricks needed for each pair. Underneath these figures the official responsible for the original tablet is named (**dub** PN).[34] There is a subtotal near the top of column III, for work under the authority of (**gìr**) **Lugal-má-...**. The second subtotal, in the fourth and last column, is for work done under **Lú-dingir-ra**. The tablet ends:

⌜šu-nigin⌝ 2 22 sar 12 gín sig$_4$ ù-ku-ru-um	Total: 2 22;12 sar **ukurrum**-bricks
[šu-nigin 1]⌜3⌝ sar sig$_4$ za-rí-in	Total: 13 sar **zarin**-bricks
[...] šà ga-eš^{ki}	... in Gaesh
[mu en]-⌜maḫ⌝-gal-an-na [en ^{d}]nanna ba-ḫun	Amar-Suen year 4 (2043 BC)

The numerical information contained in the text is summarized in Table 9.3 below, with data not on the tablet, but calculable from it, given in italics. Dividing the total volume of each pair of structures by the number of bricks associated with it gives the brickage of each structure (§4.1). In each case this is 2;42: the brickage for Type 8 bricks ($^2/_3$ cubit square).[35] These bricks are called *agurrum* in OB and later texts but **ukurrum** in YBC 9819 and here. This confirms their identification in YBC 9819, which is obscured by a consistent calculation error (§4.5). The total given for **zarin** bricks is **[1]⌜3⌝ sar**. This must be composed of the entries given as **[...]2 40 sig$_4$** in II 21 (part of the subtotal **11 sar sig$_4$** in III 8) and **2 sar sig$_4$** in III 22.[36] Because no details of the constructions are given it is impossible to determine the size of these bricks but, analogously, we would expect them to be the same as the bricks called **zarin** in YBC 9819. These are Type 2 bricks, often simply called SIG$_4$ in OB mathematics (cf. §4.0): this is also their designation throughout AO 7667 except in the final total.

In short, the two Ur III administrative records AO 7667 and VAT 7038 concern the only brick sizes — Types 2 and 8 — attested in the contemporary school mathematical texts AOT 304 (§4.4) and YBC 9819 (§4.5).

9.4.3 Ur III and OB house plans, ancient and modern

Two Ur III period house plans retain enough numerical data to allow their areas to be calculated, and compared to the OB mathematical estimate that a third of the floor-space of a house is occupied by its walls (§4.5).

VAT 7031, reproduced here in Fig. 9.6, is a house plan from Ur III Umma. The measurements in parentheses, and marked out with fine lines, are not shown on the tablet, but can be calculated from the existing lengths. It has been assumed that all walls are 2 cubits thick, and that the doorways are 2 cubits wide.

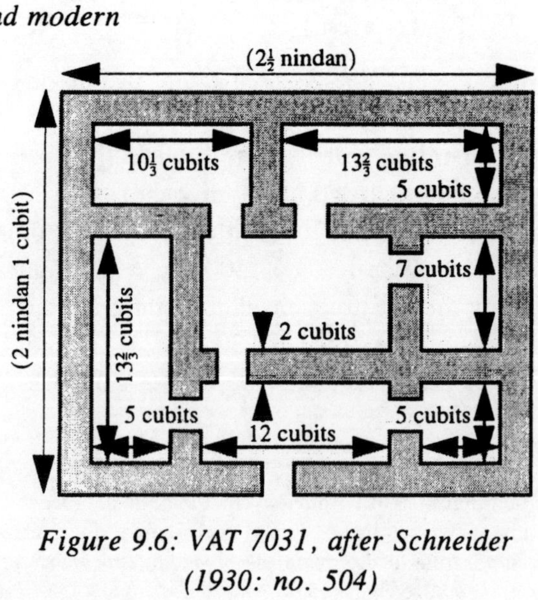

Figure 9.6: VAT 7031, after Schneider (1930: no. 504)

34 Summaries of two tablets together are denoted **dub-bi 2-àm**.

35 Established already in *MCT*: 95–6, except for those entries with the phrase **ba-an-gi$_4$**.

36 This is repeated in Lu-dingira's subtotal, perhaps because it is the only entry for **zarin** bricks under his supervision.

Table 9.3: Summary of AO 7667

	Line nos.	*Length (nindan)*	*Width (nindan)*	*Height (cubits)*	*Volume (sar_v)*	*Bricks (sar_b)*	*Brickage*
1)	I 1–9	broken			*6;55?*	[18;40]	*2;42*
2)	I 10–11	2	0;31 40	2	*2;06 40*		
	I 12–13	2	0;20	4	*2;40*		
	I 14				*4;46 40*	12;54	*2;42*
3)	I 17–18	1;30	0;31 40	2	*1;35*		
	I 19–20	1;30	0;20	4	*2;00*		
	I 21				*3;35*		*2;42*
						9;40[37]	
4)	II 1–2	1	0;31 40	2	*1;03 20*		
	II 3–4	1	0;20	4	*1;20*		
	II 5				*2;23 20*	6;27	*2;42*
5)	II 9–10	1	0;31 40	2	*1;03 20*		
	II 11–12	1	0;20	4	*1;20*		
	II 13				*2;23 20*	6;27	*2;42*
6)	II 14–16	3	0;30	2	*3;00*		
	II 17–18	3	0;17 30[38]	4	*3;30*		
	II 19				*6;30*		*2;42*
						17;30[39]	
7)	II 24–5	0;30	0;31 40	2	*0;31 40*		
	III 1–2	0;30	0;20	4	*0;40*		
	III 3				*1;11 40*		*2;42*
						3;13[40]	
	III 6			Subtotal:		1 14;5[1]	
8)	III 10–11	2;30	0;4[2 30]	3	*5;18 45*		
	III 12–13	2;30	0;23 45[41]	5;20	*5;16 40*		
	III 15				*10;35 25*		*2;42*
						28;30[42]	
9)	III 17–18	0;30	0;42 30	3	*1;03 45*		
	III 19–20	0;30	0;23 45[43]	5;20	*1;03 20*		
	III 21–2				*2;07 05*		*2;42*
						5;41[44]	
10)	IV 3–4	1;30	0;42 30	3	*3;11 15*		
	IV 5–6	1;30	0;30	1;45[45]	*1;18 45*		
	IV 8				*4;30*	12;09	*2;42*
	IV 11					21;00	
	IV 14			Subtotal:		1 07;21	
	IV 17			Total:		2 22;12	

[37] Rounded figure; correct value 9;40 30.

[38] Average of 0;30, the base, and 0;05, the top (written **ba-an-gi$_4$** because the wall tapers, or 'goes back', **gi$_4$**).

[39] Rounded figure; correct value 17;33.

[40] Rounded figure; correct value 3;13 30.

[41] Average of 0;42 30, the base, and 0;05, the top, written **ba-an-gi$_4$**.

[42] Rounded figure; correct value 28;35 37 30.

[43] Average of 0;42 30, the base, and 0;05, the top, written **ba-an-gi$_4$**.

[44] Rounded figure; correct value 5;43 07 30.

[45] Average of 2;30 and 1, written **ba-an-gi$_4$**.

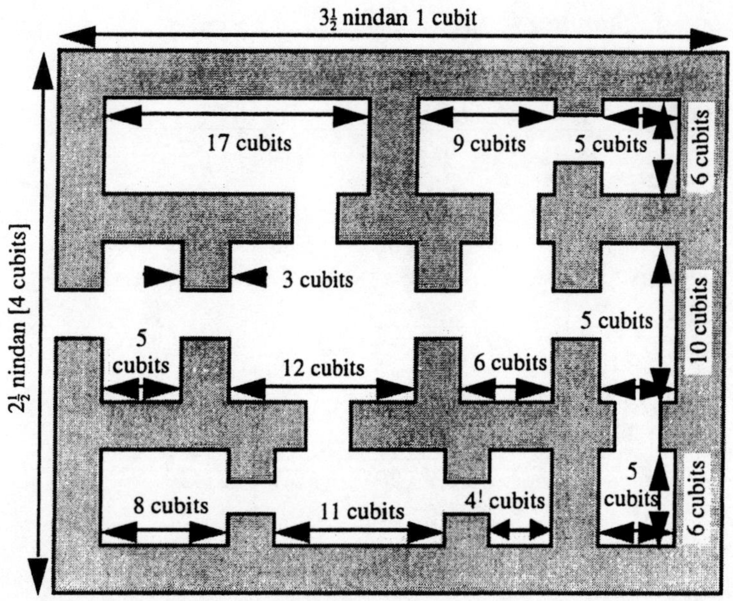

Figure 9.7: House plan from the John Rylands Library, after Donald (1962)

The plan illustrated in Fig. 9.7, in the John Rylands Library, Manchester, is of unknown provenance, but may be of Ur III or Isin period date (Heinrich and Seidl 1967: 26). The width of one of the lower rooms must be modified from 7 to 4 cubits (as shown on the plan) if the measurements are to make sense.

There is only one house plan of the OB period which is complete enough to calculate from: *PRAK* 2 D 30, a lenticular school text from Kish, depicted in Fig. 9.8. Two room-dividers, shown here, are marked on the tablet as as single lines. They are not included in the following calculations.

External measurements are also given on the reverse of *PRAK* 2 O 107, another school house plan from OB Kish: 0;50 nindan by 1;05 nindan. The measurements of the courtyard and five surrounding rooms are not stated, but around the edge of the tablet the numbers 10 35 and 33 20 are written. On the obverse, under a slightly different sketch, is SIG$_4$ 45 55. It is not clear to the author how these numbers relate to the house plans on the tablet.

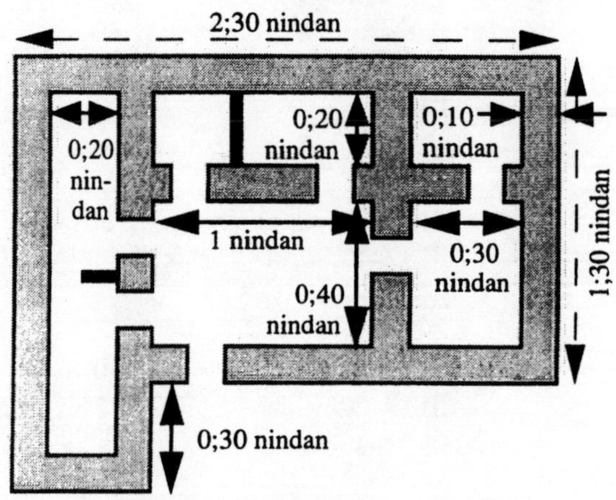

Figure 9.8: PRAK *2 D 30, after de Genouillac (1924–5: II pl. 39)*

Note the difference in character of the Ur III documents and the OB exercise: whereas the measurements in the former are accurate to the nearest cubit or third of a cubit, the rooms in the latter are all 'natural fractions' (Friberg 1987–90: 535) of the nindan. In other words, they are deliberately chosen figures, and not conditioned by practical considerations such as the area available to build on.

Elizabeth Stone (1981: 20) has suggested that in house sale contracts from Old Babylonian Nippur the meaning of É.DÙ.A is 'roofed floor-space', excluding the area of the walls and central courtyard. However, in quantity surveys it is precisely the area of the walls that is under consideration, so É.DÙ.A must have a broader sense here. When a built house is sold the most important consideration is the amount of living space it provides; the area of the walls is largely irrelevant, not to mention very difficult to measure. With an empty building plot, on the other hand, the surveyor is concerned with the site as a whole. The proportion of walls to *roofed* floor-space would be meaningless in this context. However, it often happened, especially in urban sites, that the external walls were party walls with another property, and may have already been standing (cf. Van De Mieroop 1992*a*: 124). It may be, then, that the floor-space coefficient refers not to the proportion of all walls to the external area of the house, but the proportion of the internal walls to the internal area.

Table 9.4 below shows the measurements in sar_a of the houses depicted on the three house plans, and gives three ratios of walls to area for each. AW/EA is the proportion of all walls to the external area of the house; IW/IA is the ratio of the internal walls to the internal area, and EW/EA is the fraction of the whole house occupied by the external walls alone.

Table 9.4: Proportions of wall to total area in three Ur III–OB house plans

Plan	*External area (EA)*	*All walls (AW)*	*External walls (EW)*	*AW/EA*	*IW/IA*	*EW/EA*
Fig. 9.6	5;12 30	2;17 21 40	1;25	0.44	0.23	0.27
Fig. 9.7	10;09 10	5;02 30	2;57;30	0.50	0.29	0.29
Fig. 9.8	4;05	2;10 40	1;21 40	0.53	0.30	0.33

Even though two of these houses are approximately 5 sar in area, the size of the house in the OB problems, none of these ratios exactly reflects the ideal fraction $^1/_3$. The proportion of all walls to external area is certainly to be excluded as the model for this rule of thumb.[46]

Extracting similar information from modern plans of excavated houses from the same periods can be fraught with difficulties, not least because they have often been of secondary importance to excavators in comparison to large-scale public constructions.[47] However, because in general 'les emprises au sol des édifices sont toujours du même ordre et le principe architectural proche' in whatever city they are built, and 'ne dépassent, en règle générale, 200 m^2 [≈ 5;30 sar_a]' (Calvet 1994: 125), a preliminary survey of just a few dwellings will for the moment serve to test the reliability of the ancient house plans.

Many domestic residences from the OB 'new town' of Haradum on the Middle Euphrates were excavated, and have been published with plans and measurements of the external areas and the sizes of the individual rooms in them (Kepinski-Lecomte 1992: 71–150). The author chose twelve of the most complete houses[48] and used the excavators' figures for each to calculate the proportion of walls in the whole house (AW/EA). The buildings were typically much smaller than those on the ancient plans, at around 100 m^2 (≈ 2;48 sar_a).[49] The proportion of walls in the houses was independent of the size of the house,[50] between 33% (Building 8) and 58% (Building 28), with mean 44.6%, slightly less than the average of 49% for the ancient house plans.

46 Defining É.DÙ.A as the *roofed* floor-space only (Stone 1981: 20) would give:

Plan	*Area of roofed floor-space (sar_a)*	*Proportion of walls to roofed floor-space*
Fig. 9.6	2;30;50	0.91
Fig. 9.7	4;03 45	1.22
Fig. 9.8	1;23 20	1.46

That is, the walls occupy about as much room as the roofed floor-space.

47 Cf. Van De Mieroop (1992*a*: 124) on OB Ur: 'Unfortunately, the published plans of the domestic quarters are of such a reduced scale that accurate measurements are impossible, and it seems that the scale of the plans is not always accurate either'. The plans of OB and Ur III private dwellings in Nippur have been published at a similarly small scale (see McCown *et al.* 1967: pl. 55) and could not be used in this study.

48 Namely, buildings 2 (p 111), 3 (p 137), 4 (p 82), 5 (p 88), 6 (p 74), 7 (p 116), 8 (p 85), 9 (p 92), 11 (p 95), 14 (p 102), 28 (p 150), and 34 (p 71).

49 Largest: building 3 (213 m^2); smallest: building 11 (34 m^2).

50 Correlation between the total floor-space of the house and the percentage occupied by walls: –0.127.

Few contemporary private dwellings from south and central Mesopotamia have been published in such detail. The OB 'bâtiment centrale' of the 1978 excavations at Tell ed-Dēr (Sippar Amnanum) is one exception (De Meyer 1978: pl. 5). Its external area is approximately 222.75 m^2 (≈ 6;10 sar_a), with external walls of 49 m^2, and total wall area 85 m^2. Using the abbreviations of Table 9.4, the proportions of the house are: AW/EA = 0.38; IW/IA = 0.21; EW/EA = 0.22. In this case, the ratio of all the walls to the whole house (AW/EA) is closest to the OB mathematical ideal of 0.33.

These few examples show that there is not yet an obvious candidate for a professionally implemented precursor to the OB mathematical coefficient. In particular, the internal and external walls together appear to take up nearer a half than a third of the floor-space of the houses surveyed here, including those on the ancient plans. However, one would need a much larger sample of large-scale plans from several different sites to determine whether either of the other ratios discussed here was on average 0.33. Equally, it is at present virtually impossible to determine the exact proportions of brick to mortar in published photos and plans of buildings, but 5:1, the ratio implicit in OB mathematics, appears to be plausible, at least.

9.5 MAKING BRICKS; CARRYING MATERIALS

Neither brick-making nor carrying appear to have been monitored particularly closely by the Ur III administration. Typically only the working days taken to perform the task were recorded, without quantifying the bricks produced, the goods carried, or the distance over which those goods were transported. This extract from AO 5676: IX 13–14 is a representative example:

33 guruš u_4-1-šè	33 workers for 1 day,
gi ÍL-gá	carrying reeds.

There are occasional instances of reeds being carried in fixed numbers of bundles; these are discussed in the third part of this section. Before that is a discussion of an unusually detailed Ur III record of building materials and labour published by Farber (1989). The first document discussed here, though, is concerned not with expected work-rates but with the actual output of several teams of brick-makers working on the extensive irrigation project of early OB Larsa.

9.5.1 An OB record of making bricks

NBC 5506 is part of the OB irrigation archive from Larsa, and dates to 1881 BC. It is a list of bricks for a reservoir on the Isin canal and the number of labourers employed to make (?) them under each overseer. Some of the brick quantities have been rounded up, as would be expected, but the total numbers of both bricks and workers are incorrect. The daily work-rate of each team can be calculated by dividing the number of bricks made by the size of the workforce, as shown in Table 9.5.[51] This is clearly a record of actual production: assuming that the bricks are all of the same type, the teams are working at very different rates, with larger teams tending to be slower.[52] For instance, the two-man team in line 8 produces twice as many bricks per head as the six-man team in line 4. Presumably this reflects particular working conditions which are not mentioned in the document.

[51] The archive was published fourteen years before Haddad 104, which contains the only known mathematical problems on making bricks (§§5.1–2). Walters did not attempt to draw conclusions about work-rates from it.

[52] The correlation between the team size and the daily rate is –0.455.

Table 9.5: Brick-making work-rates in NBC 5506

Lines	*Bricks in* sar_b		*Workers*	*Daily rate in* sar_b		*Comments*
1	2;20	(1680 bricks)	7	0;20	(240 bricks)	Number of bricks written 2 SAR 4[53]
2–3	3;40	(2640 bricks)	13	0;16 55	(203 bricks)	Actually 2639 bricks?
4	1;30	(1080 bricks)	6	0;15	(180 bricks)	
5–6	2;40	(1920 bricks)	7	0;22 55	(275 bricks)	Actually 1925 bricks?
7	6	(4320 bricks)	21	0;17 05	(205 bricks)	Actually 4305 bricks?
8	1	(720 bricks)	2	0;30	(360 bricks)	
9	2[;15]	(1620 bricks)	6	0;22 30	(270 bricks)	
10	2;30	(1800 bricks)	6	0;25	(300 bricks)	
11	2;15	(1620 bricks)	6	0;22 30	(270 bricks)	Number of bricks written 2 SAR 3
12	3	(2160 bricks)	10	0;18	(216 bricks)	
13–14	1;50	(1320 bricks)	6	0;18 20	(220 bricks)	
15	0;50	(600 bricks)	3	0;16 40	(200 bricks)	
16–17	29	(20880 bricks)	85	≈ 0;20 30	(246 bricks)	Totals on tablet
Total	29;50	(21480 bricks)	93	≈ 0;19 15	(231 bricks)	Actual totals

According to OB mathematics, one worker can excavate 0;05 sar_v of earth and make it into bricks in one day[54] (§5.1), rather less if the earth has to be carried to the brick-making site. If indeed NBC 5506 records this same excavation and brick-making process, and we assume that the bricks concerned are Type 2, then the average daily output of (approximately) 0;20 sar_b is equivalent to 0;20 ÷ 7;12 = 0;02 46 40 sar_v, rather more than half the ideal, but still of the same order of magnitude.[55]

Put differently, it appears that the OB brick-making coefficients were based on realistic data. There are no such work-rates explicitly attested from Ur III, but an identical output to that of NBC 5506 can be inferred from the document discussed below.

9.5.2 An Ur III record of wages for carrying and construction

A 2976 is an Ur III economic text dating from the 46th year of king Shulgi's reign (2049 BC), of unknown provenance. The text lists wages, administered by one Shu-Adad, for workers building a storehouse (**é-kišib-ba**, line 15) for a calf shed.[56] The operations involved are described in unusual detail. In order of appearance on the tablet they are:

24 men to make 8 sar of bricks	(lines 1–2)
16 men to carry 8 bricks each over 60 nindan	(lines 3–5)
8 men to mix mortar	(line 6)
9 men to carry 90 reed-bundles for mats	(lines 7–8)
3 men to mix mortar	(line 9)
18 men to carry 180 reed-bundles for putting into the brickwork	(lines 10–11)
6 men to mix mortar	(line 13)
Total: 84 man-days	(line 14)

The tasks fall into three groups of related activities, all of which conclude with mixing mortar to incorporate the materials into the structure. There is no explicit record of the construction

[53] The 4 stands for 4 00, not 4 GÍN (*pace* Walters 1970: 128).

[54] This is equivalent to 432 Type 2 unbaked bricks, 324 Type 3 half-bricks, 300 Type 4 two-thirds bricks or 162 Type 8 square baked bricks (cf. §4.8).

[55] The rate is equivalent to that for a theoretical situation in which the earth is carried 16 nindan before the bricks are manufactured (cf. §5.2 and §9.5.2 below).

[56] Farber (1989: 139) reads this line as **é-dub-ba šà é-amar-ra-ka dù-a**, translating it 'die Schule, die im "Kälberhaus" gebaut worden ist'. However, the word 'school' is always written **é-dub-ba-a**, and a storehouse is a much more probable appendage to a cattle shed than an educational establishment.

tasks themselves. It may be that these were executed by skilled workmen and not by the unskilled labourers (**guruš**) whose wages are documented here; alternatively it is suggested below that the bricklaying is subsumed under the expression 'mixing mortar' (**al-tar**) and that the two operations are performed by the same team of labourers.

Starting with the first group of tasks: at first sight it appears that we have a clear attestation of the OB coefficient for filling brick moulds (§5.1):

$$8 \div 24 = 0;20 \text{ sar}_V \text{ a day,}$$

but this would lead to an impossibly high rate for carrying those same bricks, as is attested in the following line.[57]

If we assume, on the other hand, that the 'sar' measurement refers to a number of bricks rather than a unit of volume, then we are dealing with:

$$8 \times 720 = 5760 \text{ bricks.}$$

Taking the carrying first, the number of trips made by each of the 16 men carrying 8 bricks a time is:

$$5760 \div (16 \times 8) = 45.$$

45 trips over a distance of 60 nindan is a daily distance of 45 × 60 = 2700 nindan, exactly the value of the *muttalliktum* or 'daily distance' coefficients given in the OB lists (§5.2). Then the 'carriage' or *nazbalum* for the bricks is (§5.3):

$$8 \times 45\ 00 = 6\ 00\ 00, \text{ or } 21{,}600,$$

and their 'brickage' or *nalbanum* is (§4.1):

$$6 \times 1;36 = 9;36.$$

These two constants do not correspond to the coefficients of any of the standard brick sizes of OB maths (cf. §4.8). If the preceding calculations are valid, the size of the bricks must be:

$$\frac{1}{9;36 \times 1\ 12} \text{ sar}_V = 0;00\ 00\ 31\ 15 \text{ sar}_V,$$

or 562.5 cubic fingers. If they are 5 fingers thick, then the area of the top surface is 112.5 square fingers.

On the other hand, it may be that the text concerns the common Type 2 unbaked brick, carried at a higher rate of 8 at a time instead of the 6 attested in OB mathematics (§5.4). In that case:

$$8 \text{ sar}_b = 1;06\ 40 \text{ sar of volume,}$$

and the men make the bricks at a rate of:

$$1;06\ 40 \div 24 = 0;02\ 46\ 40 \text{ sar}_V \text{ a day;}$$

or, conversely, it takes 21.6 days to make 1 sar_V of these bricks.[58] If we assume from OB maths that the manufacture of 1 sar_V bricks takes 12 days (§5.1), this leaves 21.6 − 12 = 9.6 days for

[57] For instance, this volume would be equivalent to 7;12 × 8 = 57;36 sar_b Type 2 bricks (which have a brickage of 7;12: see §4.1). If, as the next line says, 16 men carried them 8 at a time over 60 nindan, they would each have to make: 57;36 × 12 00 ÷ (16 × 8) = 5 24, or 324 trips a day, compared to the 45 trips a day over this distance expected from OB maths (§5.3) — a total distance of ≈116.6 km as opposed to ≈16.2 km!

[58] Compare this rate with the average inferred from NBC 5506 (§9.5.1), a record of brick manufacture from OB Larsa.

carrying the earth for them, so that it is brought from a distance of (§5.2):

$$9;36 \times 1;40 = 16 \text{ nindan.}$$ [59]

Eight men are needed to mix mortar for the bricks, or one for every sar_b of bricks. If we assume that the volume of mortar needed is a fifth of the volume of the bricks,[60] then this quantity is:

$$1;06\ 40 \div 5 = 0;13\ 20 \text{ sar}_V,$$

and the total volume of the wall will be:

$$1;06\ 40 = 0;13\ 20 = 1;20 \text{ sar}_V.$$ [61]

Each of the 8 men must mix:

$$0;13\ 20 \div 8 = 0;01\ 40 \text{ sar}_V.$$

This is $^1/_{36}$ sar_V, or 108,000 cubic fingers. If the mortar is 1 finger thick, then it will cover 0;50 sar_a, or about 30 square metres. This is 0;50, or $^5/_6$, of the quantity needed for a full day's plastering or mortaring work according to one OB coefficient (§4.6). Could it be that the activity **al tar** describes not only mixing the mortar (Farber 1989) but laying the bricks with it as well? In that case $^1/_6$ of each labourer's day would be spent mixing the mortar at a daily rate of 0;10 sar_V,[62] and the remaining $^5/_6$ day in laying 1 sar_b of Type 2 bricks with the prepared mortar.[63]

The second and third groups of activities both concern carrying reed-bundles and mixing the plaster for them. In both instances 10 bundles are carried a day:

$$1\ 30 \div 9 = 10$$
$$3\ 00 \div 18 = 10.$$

If, as in OB maths, the bundles each have a volume of 0;00 05 sar_V,[64] and the daily distance is 45 00 nindan (§5.5), then the bundles have been carried over:

$$45\ 00 \times 0;00\ 06\ 40 \div 0;00\ 50 = 6\ 00 \text{ nindan,}$$

or just over 2 km. The total volume of the reeds would be 0;07 30 sar_V and 0;15 sar_V respectively.

If we assume that in both cases the work-rate for mixing plaster is the same as before, then the 6 and 3 man-days produce respectively:

59 If, on the other hand, we continue to assume a brickage of 9;36, then the 8 sar_b = 0;50 sar_V, and the bricks are made at a rate of 0;50 ÷ 24 = 0;02 05 sar_V a day. The distance the earth is carried, following the same method as above, works out to be 28 nindan.

60 Cf. BM 96957 + VAT 6598, in which five-sixths of a brick wall are assumed to be bricks (§4.5).

61 Thus, according to the OB coefficient for the proportion of walls in a house (§4.5), the total volume of the building would be 1;20 ÷ 0;20 = 4 sar_V. If the walls were 4 cubits, approximately 2 m, high the floor-space would be 1 sar_a, an area roughly 6 m square.

62 Identical to the OB coefficients for mixing earth for bricks: much the same activity (§5.1).

63 But this daily rate — 0;08 20 sar_V — is exactly twice that attested in the one known OB problem on building a brick-pile (§5.6).

64 According to AO 6036 XII 25–35, an Ur III labour account from Umma 1 reed mat with a **ki-lá** of 1 sar_x is made of 36 reed-bundles (**sa-gi**) and takes 6 days to make (i.e. at a daily work-rate of 0;10 sar_x) (Goetze 1948*b*; Civil 1964: 80; Englund 1988: 170 n. 43). Thus each bundle is equivalent to 0;01 40 sar_x. If, as seems sensible for a flat object such as a mat, the metrological unit in question is the area measure sar_a, then this is consistent with the volume of the bundles being 0;00 05 sar_V each. They would then have a thickness of 0;00 05 ÷ 0;01 40 = 0;03 cubits, or 1.5 fingers (≈ 2.5 cm).

$$3 \times 0;01\ 40 = 0;05\ \text{sar}_V$$
$$6 \times 0;01\ 40 = 0;10\ \text{sar}_V$$

and so to secure each reed-bundle into the brickwork the volume of mortar needed is:

$$0;10 \div 3\ 00 = 0;00\ 03\ 20\ \text{sar}_V$$
$$0;05 \div 1\ 30 = 0;00\ 03\ 20\ \text{sar}_V,$$

or 3600 cubic fingers, which covers an area of 0;01 40 sar_a, or a square of length 2 cubits, to a depth of 1 finger. This is exactly the area of a mat 1;30 fingers thick made from 1 reed-bundle of volume 0;00 05 sar_V which was deduced in the note above. The mortarers would then each be responsible for laying 0;02 30 sar_V of reeds in the remaining $^5/_6$ of a day, at a rate of 0;03 sar_V.

After so much detail, it seems appropriate to summarize the many conclusions that can be drawn from this one small document.

- Type 2 bricks are carried 8 at a time, a rate 1;20 greater than that attested in OB school maths, but over the same distance in a day, namely 45 00 nindan.
- It appears that the OB mathematical rates for making bricks were also used, although this is not conclusively proved. At least, the output is very similar to that attested in one OB administrative text.
- The data is also commensurate with OB estimates that walls were composed of bricks and mortar in the ratio 5:1 and that roughly a third of the volume of a building was occupied by the walls.
- The daily rate of mixing mortar is 0;10 sar_V, but this task is combined with bricklaying or laying reed matting, at daily rates of 0;08 20 sar_V and 0;03 sar_V respectively. The first of these rates is identical with OB coefficients, the second twice as large as expected, and the third otherwise unattested; the OB 'combined coefficient' method (§5.1) is used to calculate the work-rates of labourers doing different tasks in a day.
- Reed-bundles are the same size as those attested in OB maths (0;00 05 sar_V), and are used to make matting 1;30 fingers thick, of the sort documented widely in Ur III administrative records. They are laid in a 1 finger depth of mortar.

In short, even where the data presented in A 2976 does not conclusively match the coefficients of OB maths they are of a similar whole-number character: there is nothing in this text to contradict the hypothesis that OB maths, and in particular OB-type quantity surveying coefficients, were already used in administrative documents of the Ur III period.

9.5.3 Other evidence from Ur III administrative texts

Carrying is rarely quantified in Ur III records; the texts usually state the number of working days needed to transport an unspecified quantity of goods.[65] Occasionally the number of reed-bundles is specified, but never the distance over which they are carried, so it is impossible to tell whether the OB work norms were in use. However, where the size of one load is given one

[65] Goetze (1962) read O 56-7/17: 4–5 as:

5 30 sar ⌜ú-ÍL-gá⌝ 10 sar-ta	15 30 sar, carrying grass at 10 sar_v each.
á-bi u_4-1 33-kam	Its wages are 1 33 days.

If correct, this would be a unique occurrence of an Ur III carrying work-rate expressed as a volume. It is much more likely that the very damaged signs in line 4 refer to a ground-clearing operation like that which immediately precedes it on the tablet. Both the value and the format of the work-rate are more consonant with this interpretation. See also §9.8.1.

can at least make rough comparisons of those distances.

For instance, three documents record carrying reeds in loads (**kilib-ba**) of 12 bundles each. Table 9.6 summarizes their contents, with data deducible from the texts given in italics:

Table 9.6: Work-rates for carrying reed-bundles

Text	*Bundles*	*Size of load*	*Loads*	*Work-days*	*Daily rate*
AO 5675: IX 28–32	11 50	12 bundles	59;10	23;40	2;30 loads
Ist P 1051: 1–6	46 12	12 bundles	3 51	30	*7;42 loads*
NBC 1332: 197–204	36! 00	12 bundles	3 00	1 30	2 loads

If we assume that these bundles are all approximately the same size and that the same working conditions pertain, it would appear that the reeds were carried over the longest distance in NBC 1332, and that the journeys recorded in AO 5675 and Ist P 1051 are roughly $^4/_5$ and $^1/_4$ of that distance respectively.

If each bundle has volume 0;00 05 sar_V, as in OB maths (§6.4) and A 2976 discussed above, then the volume of each load is 12 × 0;00 05 = 0;01 sar_V, nine times larger than the theoretical volume (§5.5). Clearly, then, we are dealing with bundles and loads of quite a different order from those in the mathematical texts. If, on the other hand, we assume that each 'load' is indeed what a worker can carry at any one time, then the daily rates shown in the table above are identical to the number of trips made in a day. According to OB maths, the total daily distance for carrying is 45 00 nindan (§5.1). If this were also the basis on which these Ur III texts were calculated then the journeys would be:

45 00 ÷ 2;30 = 18 00 nindan (approx. 6.5 km)
45 00 ÷ 7;42 ≈ 5 50;40 nindan (approx. 2 km)
45 00 ÷ 2;00 = 22 30 nindan (approx. 8 km)

respectively. These are at least plausible distances. See Waetzoldt (1992) and Van De Mieroop (1992*b*) for more on reeds in Ur III and OB administrative records.

9.6 EARTHWORKS AND WATERWORKS

According to *The Farmer's Instructions*,[66] the repair and construction of irrigation canals and their dykes was an integral part of the annual agricultural routine (ll. 2–3, Civil 1994: 28–9). Indeed, large agricultural accounts from Ur III Umma typically include a section at the end recording the labour expended on such tasks.[67] There are also many dozens of individual surveys and records of completed work which give the daily rates, implicitly or explicitly. Such data cannot be deduced from the OB irrigation archive from Larsa, however (Walters 1970).

Whereas irrigation dykes were constructed by simply heaping up the earth excavated from the canals they served, earthen walls (IM.DÙ.A/*pitiqtum*) that sloped at an angle of less than 45° to the vertical required a more involved method of production. Braidwood and Howe (1960: 40) describe the modern process as follows:

66 A widely copied OB school text in Sumerian (Civil 1994: 3), which purports to instruct the budding farmer on the routine of the agricultural year. Provenances include Ur, Nippur, Sippar and Me-Turan, all of which have also produced OB mathematics. Leaving aside the question of the authenticity of the procedures it describes, the order in which it presents them serves here to structure this discussion.

67 e.g. AO 5676 IX 15–XI 10; NBC 1332: 171–210.

In Iraq, *tauf* walls are still used, generally as garden or courtyard walls, and forms [i.e. moulds] are not employed ... A *tauf* wall is built of a mud mix of sufficient fluidity so that the lowest 'course' may be molded with the builder's hands, with a vertical face on either side, to a height of about 3 or 4 inches without slumping. Having laid the first 'course' the builder simply waits a day or so for complete sun-drying before he adds the second 'course', after which he must wait again, and so on. The mud mix contains straw or grass to prevent cracking.

The first two texts discussed in this section concern the construction and demolition of such walls respectively; they are followed by an overview of the work-rates attested in documents on irrigation earthworks.

9.6.1 An Ur III survey of work on an orchard

CBS 8162 surveys repairs to the walls and irrigation system of an orchard. The first entry reads:

I 1 **$1^1/_2$ nindan 5 kùš gíd $^1/_2$ kùš dagal 3 kùš sukud**
$1^1/_2$ nindan 3 kùš gíd $^1/_2$ kùš dagal 2 kùš sukud
5 nindan gíd $^1/_2$ kùš dagal 3 kùš sukud
21 nindan 3 kùš gíd $^2/_3$ kùš dagal 3 <kùš> sukud
5 **im-du$_8$-a-bi $4^2/_3$ sar $^1/_2$ gín $22^1/_2$ še**
guruš-bi 1 $14^1/_3$ u$_4$-1-šè
⸢še⸣-bi 1 gur 4 pi 5 bán $8^1/_3$ sìla gur
á šà-gal 8 [sìla]

Length $1^1/_2$ nindan, 5 cubits; width $^1/_2$ cubit; height 3 cubits.
Length $1^1/_2$ nindan 3 cubits; width $^1/_2$ cubit; height 2 cubits.
Length 5 nindan; width $^1/_2$ cubit; height 3 cubits.
Length 21 nindan 3 cubits; width $^2/_3$ cubit; height 3 ‹cubits›.
Its earth wall is $4^2/_3$ sar $^1/_2$ shekel 22 $^1/_2$ grains.
Its labourers are 1 $14^1/_3$ for 1 day.
Its grain is $598^1/_3$ litres. Wages and fodder? 8 [litres] (each).

Lines 1–4 give the measurements of four constructions, in nindan and cubits, which are summarized in Table 9.7 below. Lines 5–8 give further data; as before, information not explicitly recorded on the tablet is set in italics:

Table 9.7: Numerical data in CBS 8162

Line	*Length (nindan)*	*Width (nindan)*	*Height (cubits)*	*Volume (sar$_v$)*	*Work-rate (sar$_v$/day)*	*Work-days*
I 1	1;55	0;02 30	3	*0;14 22 30*		
2	1;45	0;02 30	2	*0;08 45*		
3	5;00	0;02 30	3	*0;37 30*		
4	21;15	0;03 20	3	*3;32 30*		
5–6			Total:	4;40 37 30	× *0;03 45* =	1 14;50

The total volume of the earth walls is in fact 4;33 07 30 sar$_V$. The total given in the text is 0;07 30 sar$_V$ too large. This excess is exactly equal to the volume of a wall measuring 1 nindan × 0;02 30 nindan × 3 cubits, so that the error could have come about by misrepresenting the length in either line 1 or line 3 (as 2;55 or 6 respectively) when calculating the individual volumes of the walls. The number of man-days needed for work on these walls is $74^5/_6$. Dividing the total volume (as given in the text) by the number of work-days gives the average daily work-rate of the labourers: 0;03 45 sar$_V$. This is the number given as the IM.DÙ.A or *pitiqtum* coefficient in the OB coefficient lists (§6.1).

The rest of the first column is broken. There is some information remaining on the second column of the obverse, but the first of the reverse (column III) is completely missing. The last part of column IV reads:

6 nindan gíd 1 [kùš dagal 3 kùš sukud] — Length 6 nindan; width 1 [cubit; height 3 cubits].
sahar-bi $1\frac{1}{2}$ sar — Its earth is $1\frac{1}{2}$ sar.
guruš-1-e 6-gín-ta — 1 labourer at 6 shekels each.
guruš-bi 15 u$_4$-1-šè — Its labourers are 15 for 1 day.
kù-bi 4 ma-na — Its silver is 4 minas.
kiri$_6$ bàd še-ga — Making an orchard (and) wall satisfactory.
dug$_4$-ga um-mi-a — Craftsman: Dugga.

There is no dividing line between this section and the preceding one, perhaps because they concern different jobs on the same piece of land or under the same overseer. The single construction is wider than the others on the tablet, although it is the same height and of the same order of length. The wages are paid in silver and the daily work-load is smaller, at 0;06 sar$_V$.[68] The labourers are paid 16 shekels of silver each, which suggests that they are performing skilled work which earns a higher wage than constructing earth walls.

At the end of column IV (ll. 11′–14′), and on the left edge, is a summary of the tablet. It records 521 man-days of work for which $4{,}168\frac{1}{3}$ sila of grain were paid out. This is equivalent to 8 sila per day, with $\frac{1}{3}$ sila remaining. Fifteen days were paid for with 4 minas of silver. This refers to the work in the immediately preceding section.

šu-nigín á-bi 8 41 guruš u$_4$-1-šè — Total work: 8 41 labourers for 1 day.
še-bi 13 gur 4 pi 2 bán $8\frac{1}{3}$ silà [gur] — Its grain: $4168\frac{1}{3}$ litres.
šu-nigín 15 guruš u$_4$-1-šè — Total: 15 labourers for 1 day.
kù-bi 4 ma-[na] — Its silver: 4 minas.
l.e. **im-du$_8$-a ak ù** — Making an earth wall and
pa$_5$ ba-al-la — digging an irrigation ditch
kiri$_6$ lugal-kù-zu — (at) Lugal-kuzu's orchard.
gìr ur-^{d}IM — Under the authority of Ur-Ishkur,
ur-me-me ugula — foreman Ur-meme.

The extant entries in the tablet account for least 344 work-days paid in grain. The missing parts must then record 177 days of work, equivalent to around 11 sar$_V$ of wall being built at a rate of 0;03 45 sar$_V$ a day. The tablet is not dated.

Three further documents are known which record work at this same rate (Table 9.8). It has been assumed that they all concern building earth walls, although only one says so explicitly and another refers to a brick wall, **iz-zi**. This work-rate — which is stated clearly in each of these cases — was one of the first to be recognized as a precursor to an OB mathematical coefficient (Vaiman 1960: 9).

Table 9.8: Ur III records of building mud walls

Daily rate	*Task*		*Text*	*Date*	*Comments*
0;03 45 sar$_v$	**iz-zi**	'brick wall'	*UET* 3 1386	Ibbi-Sin 24	From Ur.
0;03 45 sar$_v$	**im-du$_8$-a**	'earth wall'	NBC 3581	Shu-Sin 6	From Umma.
0;03 45 sar$_v$	–	–	BM 20054[69]	–	No provenance.

[68] Cf. OB coefficient A 27: 6 IGI.GUB ‹ÉŠ›.GÀR LÚ.1.KAM (§6.6).

[69] It is unclear from the published transliteration whether this is the whole text or just an extract from it. If it is complete, we may be dealing with a school text: the text as Englund (1988: 179) presents it is not dated or authorized, and the task is not designated. Further, there is an unusual amount of detail in the calculation.

9.6.2 An Ur III record of demolition work

Ist Ni 236 is a record of completed work from Nippur, which dates to 2035 BC.[70] It reads:

22 nindan gíd iz-zi gul-la	The length of demolished wall is 22 nindan
2 kùš dagal-ḫi-a	The *average*[71] width is 2 cubits.
sukud-bi $^1/_2$ nindan 2 kùš	Its height is $^1/_2$ nindan, 2 cubits.
iz-zi-bi $28^1/_3$ sar	Its wall (i.e. volume) is $28^1/_3$ sar (*sic*).
á guruš 1-a $^1/_3$ sar-ta	The wages per workers are $^1/_3$ sar_V each.
guruš-bi 1 28 u_4-1-šè	Its workers are 1 28 for 1 day.
íb-gul	It was demolished.
iz-zi til	The wall is finished.
é ^{d}nè-iri$_{11}$-<gal> X-BAR-ka	House of *Nergal*-...
ki ur-^{d}ištaran	From Ur-Ishtaran.
itu še-kin-kud	Month 12,
u_4-5-ba-zal	day 5,
mu si-ma-num^{ki} ba-ḫul-a	Shu-Sin year 3.

This is the only example known to me of a detailed record of demolition work. The measurements of the brick wall are given and the area calculated from them.[72] From the explictly stated daily work-rate of 0;20 sar_V — which is identical to that attested in one OB problem text (§5.6) — the labour force needed is calculated at 88 men. The phraseology and end-matter are typical of Ur III archival records, and there is nothing to suggest that this may be a school text. In other words, this document provides unequivocal evidence that the daily rate of demolition used in Ur III quantity surveying calculations was later used in OB scribal training.[73]

9.6.3 Other Ur III records of earthworks and waterworks

Civil (1994: 109–40) has studied the terminology and dimensions of irrigation structures mentioned in Ur III administrative records, and the types of work involved in their construction and maintenance. The rates at which those tasks were performed are touched on only incidentally (cf. Civil 1994: 134). Earlier examinations of Mesopotamian waterworks — apart from publications of individual texts, such as Lafont (1980) — do not as a rule deal with their quantitative management either.[74]

The many extant records of completed work from the Ur III period use a bewildering array of terminology to describe the tasks undertaken on canals (**pa$_5$**) and levees (**ég** = E). Civil (1994: 115) distinguishes two different categories of work: moving earth (**KIN-saḫar**), and reinforcing the structures with reeds (**ég-e ú-sag$_{11}$-a**). Under the first heading fall such activities as digging canals and ditches, and piling up dykes. Often these two activities comprise one and the same operation, as the earth is moved from below to above ground level. Many texts record the combined procedure, under the title **KIN ú saḫar-ba** (Civil 1994: 124). Table

[70] Lines 1–4 were published in transliteration as 'Const 236' by Allotte de la Fuÿe (1907: 77–8); cf. Nemet-Nejat (1993: 50, 130). Allotte de la Fuÿe refers to a hand-copy published by Scheil, but I have not been able to find it.

[71] Literally 'mixed width'. Cf. **dagal ì-ḫi** 'average width' in BM 20054 (Englund 1988: 179 & n. 49).

[72] The correct volume is 29;20 sar_V but, as this figure was used to calculate the labour, the $28^1/_3$ of the hand-copy may be a modern copyist's error.

[73] Albeit with one other rate — 0;15 sar_V (§5.6) — which, however, may have been used under different circumstances.

[74] See, most recently, Nissen (1976–80); Stol (1976–80); Maeda (1984); Hruška (1988); Hunt (1988); Pemberton *et al.* (1988); Steinkeller (1988); Renger (1990) and Waetzoldt (1990).

9.9 shows that the various tasks are carried out at three different work-rates, apparently independently of the type of work done.[75]

Table 9.9: Earthworking rates attested in Ur III administrative records

Daily rate	*Task*		*Tablet*	*Date*[76]	*Comments*
0;05 20 sar_V	**ég**	'dykes'	U 3	SS 2	Implicit rate.
	ég si-ga	'piling dykes'	Ist Um 1904	SS 2	Implicit rate.
0;07 30 sar_V	**KIN ú saḫar-ba**	(moving earth and reinforcing)	Syr 385	AS 8	
			Ist Um 1904	SS 2	
			Syr 445	Sh 48	
	nag-ku$_5$ si-ga	'piling (earth from) reservoirs'	AO 5676	SS 2	
	pa$_5$ ba-al-la	'digging ditches'	Syr 271	AS 6	
	saḫar zi-ga	'removing earth'	CB 6	AS 2	From Puzrish-Dagan.
0;10 sar_V	**ég**	'dykes'	YBC 3668	SS 7	Implicit rate.
	ég ba-a-si	'piling dykes'	CMAA 83.101	SS 4	
			DV 5 165	SS 5	
	íd ba-al-la	'digging canals'	NBC 1332	AS 3	
	KIN saḫar	(moving earth)	MLC 2404	Sh 46	Implicit rate.
	KIN saḫar al-e	(moving earth)	AO 5670	Sh 48	
	KIN ú saḫar-ba	(moving earth and reinforcing)	Syr 393	AS 9	
	nag-ku$_5$ si-ga	'piling (earth from) reservoirs'	AO 5676	SS 2	
	saḫar zi-ga	'removing earth'	Platt	AS 3	Implicit rate.
			MLC 2407	Sh 46	
	šu-luḫ ak	'cleaning'	CMAA 83.101	SS 4	

None of these rates is for dyke reinforcement alone, although there are two instances of mixed work. It may be that reed-work is included, but not mentioned, in other cases too. W 47a, an Ur III model document, uses a work-rate of 0;10 sar_V a day to calculate the labour needed first for constructing a levee and then for reinforcing it with reeds (§6.4). If the tasks were indeed performed at the same rate — and by the same workers — they would not need to be recorded separately in labour accounts. This same school text suggests that reeds comprise a quarter of the volume of the earth, or a fifth of the total volume of the dyke. This is in the middle of the range of proportions attested in the Ur III documents, where the earth-to-reed ratio varies from 3:1 to 5:1 (Civil 1994: 121). The OB mathematical problems about structures composed partially of reeds all concern dams, for which there are no relevant Ur III records of construction. The lower two-thirds of the dam is supposed to be made of reed-bundles, which cannot be washed away by the current as easily as the earth comprising the upper third (§6.4).

Only CB 6 records the (added) height of the levees being repaired, all of which are 0;30 to 1 cubit (25–50 cm) (Lafont 1980). It is thus impossible to tell whether the work-rate increases according to the depth or height of the work done, as is the case in OB mathematics. However, it is striking that 0;10 sar_V and 0;07 30 sar_V are both widely attested OB coefficients for excavations and earthworks, both in problems and lists (§6.2–3). Further, 0;05 20 sar_V is stated to be the daily rate of reinforcing dykes in coefficient list E and may also be implicit in one mathematical problem (§6.4).

[75] For places of publication see the index of tablets, §I.1. All tablets are from Umma unless otherwise stated.
[76] Sh: Shulgi; SS: Shu-Sin; AS: Amar-Suen.

9.7 STORAGE AND METROLOGY

The metrological system developed in the Old Akkadian–early Ur III period, called the 'standard (scientific) system' by Powell (1987–90: 458), was used throughout the Ur III and OB periods, in both administrative and mathematical documents. Coefficients for conversions between metrological units are attested widely in OB mathematical problems and, to a lesser extent, in the coefficient lists (§7.1–3). One would not expect to find evidence for their use (as opposed to, for instance, tables of metrological conversions) in Ur III economic records as, by their very nature, those records do not display the calculations performed in the course of formulating them. At best, one might hope to discover a draft document from which the auxiliary calculations have not been erased,[77] in which a metrological conversion is explicitly performed. However, in the Ur III school calculation AOT 304 (§4.4) the data are converted into various length units with no indication as to how it was done. This might lead one to expect that similar procedures obtained in contemporary working documents too. It is unlikely that we shall ever be able to determine whether conversion coefficients were preferred over metrological tables in everyday practice.

Civil (1994: 96–8) has shown that the Ur III and OB phrases **še giš è-a** 'grain coming out from the stick' and **gidru-šè ná** / *ina/ana ḫaṭṭim šunūlum* 'to lay under the stick' refer to the practice of making a first estimate of the volume of newly winnowed grain:

> Once the grain was completely clean, the conical piles of grain were approximately measured by vertically inserting a stick in the center all the way to the bottom. The covered stick length gave the approximate amount of grain in the pile. Fractions of this length could be used to measure shares. Where this system is in use, it is claimed that experienced farmers can estimate the grain amounts with a stick, often the handle of a tool, with a very small margin of error. (Civil 1994: 96)

The grain was then measured more accurately before being taken to more permanent storage. Is this the backgound to the OB mathematical problems on grain-piles (§7.4)? Those problems describe piles with rectangular, not circular, bases, but it appears on current evidence that the geometry of cones was not part of the OB mathematical repertoire.[78] Using the method given in *TMS* 14 one can (correctly) work out the amount of grain in a pile solely from its height and the length of the apex, assuming that it slopes at a 45° angle. If the base is square the calculation is simpler, as the apex does not have to be taken into account.

Two OB share-cropping contracts refer to dividing grain with a stick so that the tenant receives two-thirds and the owner a third (Civil 1994: 97). Such a scenario may be the pretext for OB mathematical problems in which the volumes of upper or lower fractions of a grain-pile or container must be found.[79]

[77] YBC 1793 (cf. §10.2) is currently the only Ur III draft document known to the author which shows both the text to be recorded and the auxiliary calculations. Metrological conversions are not involved.

[78] Friberg (1987–90: 566). See further §8.3.2.

[79] e.g. BM 96954 + 102366 + SÉ 93 (vi), in which the volume of the upper half of a rectangular-based grain-pile must be found (§A.3), and Haddad 104 (iii), which concerns a grain container in the shape of an elongated truncated cone. Assuming that the whole contents are worth 1 talent of silver, the problem is to find the length of container containing 1 mina's worth of grain (i.e. $^1/_{60}$ of the volume), measuring successively from the top and bottom of the container.

9.8 WORK-RATES NOT ATTESTED IN OB MATHS

9.8.1 Agricultural tasks

Maekawa (1990) and Civil (1994) have both surveyed the work-rates attested in Ur III administrative documents, predominantly from Umma, for various ground preparation processes prior to cultivation. The following paragraphs summarize their findings and discuss some further, less widely attested, agricultural tasks and their work-rates. All the daily rates are measured in sar_a and iku_a.

According to *The Farmer's Instructions* (ll. 5–8, Civil 1994: 28–9), the cultivator's first task after irrigating the fields is to cut back the weed growth. The Ur III administrative texts use a variety of expressions to describe this task:

> The verb **zé** = *baqāmu* ['to pluck'] is one of several terms used in Ur III texts to designate the removal of plants and weeds, the others being **ku$_5$(r)**, **bù-(r)** and **SIG$_7$-a**. (Civil 1994: 70)

Whatever the phraseology, though, the work-rates — which are always stated explicitly — are all in the range 10–30 sar_a a day, as shown in Table 9.10.

Hoeing, **al (ak)**, was the next stage in field preparation (*Farmer's Instructions* ll. 11–12, Civil 1994: 28–9). Civil (1994: 79) has collected 128 mentions of hoeing in Ur III records, at rates from 1;30 sar_a to 10 sar_a a day. Over two-thirds of those attestations are for rates of 4, 5, and 6 sar_a. One Ur III school exercise, W 47a, mentions hoeing, at 3 sar_a and 2;30 sar_a a day — a rate which is not found in the working records (§6.4).

Table 9.10: Weeding work-rates in Ur III records

Daily rate	*Task*		*Example*[80]
10 sar_a	**gi ku$_5$-rá/kud-a**	'cutting reeds'[81]	A 5975
	giš-dìḫ ku$_5$-rá/kud-a	'cutting camel-thorn'[82]	O 56–7/17
	giš gul-ak	'pulling up trees' ?	A 4408
	SIG$_7$-a	'tearing out'	Duke 4
12 sar_a	**gi ku$_5$-rá/kud-a**	'cutting reeds'	NBC 1332
	gi SIG$_7$-a	'tearing out reeds'	A 2536
15 sar_a	**gi ku$_5$-rá/kud-a**	'cutting reeds'	A 4408
	ú-kiši$_{16}$ ku$_5$-rá/kud-a	'cutting false carob'	FLP 501
	giš-ma-nu bù-ra	'pulling up dogwood'	AO 5676
	ú-HAR-an bù-ra	'pulling up ... plants'	A 4408
20 sar_a	**gi SIG$_7$-a**	'tearing out reeds'	AO 5676
	giš-dìḫ ku$_5$-rá/kud-a	'cutting camel-thorn'	A 4408
	ú-kiši$_{16}$ ku$_5$rá/kud-a	'cutting false carob'	A 22183
	ú zé-a	'tearing up plants'	UM 55-21-134
30 sar_a	**gi ku$_5$-rá/kud-a**	'cutting reeds'	A 2963
	gi SIG$_7$-a	'tearing out reeds'	AO 5676
	giš-dìḫ ku$_5$-rá/kud-a	'cutting camel-thorn'	A 2536
	ú-kiši$_{16}$ ku$_5$-rá/kud-a	'cutting false carob'	AO 5676
	ú-HAR-an bù-ra	'pulling up ... plants'	A 2981

[80] For place of publication, see §I.1.

[81] Not to be identified with the OB coefficients '1, a cut reed' A 45 and F 64 (§8.4); but cf. D 68 '10, cutting reeds' (i.e. at a rate of 10 sar_a?) (see §6.5).

[82] See Civil (1987: 47) and Maekawa (1990: 123) respectively for the identification of **giš-dìḫ**/*baltum* with camel-thorn and **ú-kiši$_{16}$** (=GÍR-*gunû*)/*ašāgum* with 'false carob'.

The related activity of clod-breaking, **níg gul** (Maekawa 1990: 124), is performed at a much faster rate, between 12 sar_a and 20 sar_a daily.[83] Perhaps this is the same activity as the otherwise unexplained **al dù**, which is carried out on an area of 10–30 sar_a a day (Civil 1994: 80).

Whereas the agricultural tasks discussed so far could be performed by any number of individuals, ploughing and harrowing were carried out by teams of three labourers (de Genouillac 1924–5: 49), and seeding by teams of four (Maekawa 1990: 124). According to the *Farmer's Instructions* (ll. 26–8) and an OB contract, ploughs were expected to have a useful life of 180 iku — or 90 to 240 working days (see work-rates below). The same machine could be used for both deep ploughing and seeding, by using different shares for each (Civil 1994: 75–6). Maekawa (1990: 119, 126) has identified four different cultivation regimes in operation simultaneously in the Umma region, in which the type and frequency of ploughing and harrowing varied according to the condition of the land.

Stol (1994) has carried out an analysis of OB ploughing contracts, in which oxen and their drivers are hired out for 15 days at a time. This hire period is called an *inītum*, and costs $^1/_2$ shekel of silver; in addition the ox eats 4 litres of barley a day. The daily work-rates that Stol (1994: 233–5) has deduced are comparable with those attested for Ur III (Maekawa 1990: 138–43; Civil 1994: 76–7), as Table 9.11 shows.

Although the labour required for harvesting does not seem to have been quantified, threshing rates of 80, 240 and 300 sila a day are attested in three Ur III administrative documents (Civil 1994: 95). The existence of OB coefficients for harvesting and winnowing is signalled by two entries in the mathematical 'catalogue' C 75–6 (§6.6), although what their numerical values might have been remains unknown.

Table 9.11: Comparison of attested ploughing rates

Task	*Ur III name*	*Ur III rate*	*OB name*	*OB rate*
Deep ploughing	**túg-še-gur$_{10}$**[84]	0;45–1 iku	*mayyāram maḫāṣum*	1 iku
Loosening the soil	**túg-sig$_{18}$**	0;45–1 iku	*pašārum*	
Harrowing	**giš-ùr-ra**	4;30–6 iku	*šakākum*	3 iku
Breaking the soil			*šebērum*[85]	6 iku
Seed-ploughing[86]	**numun-na**	2 iku	*erēšum*	≈ 2;30 iku

9.8.2 *Manufacturing goods*

The OB coefficient lists appear to contain no reference to rates of manufacture, and there is only one very terse problem on weaving (§6.6). The Ur III bureaucracy, on the other hand, seems to have taken a very close interest in the output of their craft workers. For instance, two production records from Ur III Umma (Waetzoldt 1970–1: 12) record over 40 different work-rates for making pots. The rates are proportional to the size and style of the container, with the simplest 1 sila bowl being made in $^1/_{15}$ of a working day, and a 1 gur (300 sila) vessel needing 10 days to execute.

83 UM 72-25-10: 12 and 15 sar_a; Noor I (cf. Maekawa 1990: 142): 15, 16, 20 sar_a.

84 For a discussion of the reading of Ur III plough names, see Civil (1994: 167–70).

85 It may be that this activity was subsumed under the title **giš-ùr-ra** 'harrowing' in the Ur III records.

86 Sowing could also be done by hand: the Ur III document A 2981 records sowing (**ú-numun**) at a daily rate of 12 sar_a.

Another Umma labour record details the construction of 3 sieves (**^{gi}ma-an-si-im**) over a total of 15 days, each from $^1/_4$ sila of bitumen, 10 shekels weight of grass, $^1/_2$ cubit length of wood, 1 palm heart and $^1/_6$ palm rib (AO 6036: rev. I 9–14). According to the same text (rev. III 30–5), 6 standard reed-bundles could be woven into a mat of area $^1/_6$ sar_a ($6m^2$) over the course of a working day.[87] A further document concerns the construction of 10 reed baskets (**^{gi}pisan-im-sar**), each of which takes 2 days to complete (Oberhuber 1960: no. 28).

Flour-milling was also subject to output quotas in the Ur III period, of between 6;40 and 20 litres per female miller per day, depending on the coarseness of the flour (Nissen *et al.* 1990: 130; Englund 1991: 270–2). The various stages in textile production were monitored too, in both Ur III and OB, as *UET* 3 1432 and AO 7026 attest (Waetzoldt 1972: 155; Lackenbacher 1982).[88]

9.9 SUMMARY

The beginnings of quantitative labour management can be traced back at least to ED III, with the first documented work-rate attested in the Akkad period. During the early Ur III period a sophisticated accounting procedure was developed for some sectors of the state economy, whereby the output of work-gangs could be monitored and compared to a theoretical maximum productivity. This overly bureaucratic system did not survive the end of the Ur III empire, and records of production from the OB period reveal a truer picture of variable efficiency. Nevertheless, the rates of output deducible from the OB documents appear to cluster around the Ur III ideals. A few of the Ur III work quotas appear in contemporary school practice texts, and many more seem to be precursors to the OB quantity surveying coefficients. Table 9.12 compares some of the work-rates assumed in Ur III accounting practices with the actual outputs attested in OB work records and with Ur III/OB mathematical coefficients.

Table 9.12: Coefficients and work-rates compared

Task	*Ur III work-rate*	*OB output*	*Ur III/OB coefficient*	
Brick-making	0;20 sar_b	0;15–0;30 sar_b	0;05 sar_v	(§5.1)
Bricklaying	0;08 20 sar_v	–	0;04 10 sar_v	(§5.6)
Canal repairs	0;05 20–0;10 sar_v	–	0;05 20–0;20 sar_v	(§6.3)
Carrying distance	45 00 nindan	–	45 00 nindan	(§5.1)
Demolishing walls	0;20 sar_v	–	0;15–0;20 sar_v	(§5.6)
Hoeing	1;30–10 sar_a	–	2;30–3 sar_a	(§6.4)
Making earth walls	0;03 45 sar_v	–	0;03 45 sar_v	(§6.1)
Mixing mortar	0;10 sar_v	–	0;10 sar_v	(§4.6)

The Old Akkadian–early Ur III period also saw wide-ranging metrological systemization (Powell 1987–90: 458), giving rise to a complex brick metrology (Powell 1987–90: 490), which can be traced through Ur III and OB not only in the administrative and school texts of

[87] Goetze (1948*b*); Englund (1988: 170 n. 43); see also §9.5.2 above.

[88] AO 7026 is unlikely to be 'un exercice de scribes économiques ou mathématiques' (*pace* Lackenbacher 1982: 147): it is exceptionally long and detailed, with up to 13 separate activities involved in the making of 10 different products. Two BAR.DIB ÚS garments both weighing 6 minas are put through 12 of those 13 processes at different rates. This suggests that the document is a record of actual productivity rather than one based on theoretical norms, whether it be a school text or a working estimate.

the time but also in the actual bricks excavated from contemporary sites. This is summarized in Table 9.13. (See §4.7 for an overview of these brick types.)

Table 9.13: Bricks in mathematics, administration, and the archaeological record

	Brick type 1	2	3	4	5	[6]	7	8	[8a]	9	10	11	[12]
Ur III administration		•						•					
Ur III school		•						•					
Ur III bricks	•		•	•	•		•	•					
OB administration	(No clear statements of brick size)												
OB school	•	•	•	•	•		•	•		•	•	•	
OB bricks		•		•		•	•	•			•		

Further architectural conventions and proportions of the late third millennium left their mark on OB mathematics, from the geometry of circles and triangles inherent in the construction of decorated pillars, to rough guidelines on the ratio of walls to floor-space in a house and bricks to mortar in a wall.

10. MATHEMATICS IN UR III AND OB SCRIBAL EDUCATION

This chapter looks at the educational context of coefficients and coefficient lists. The material presented here is necessarily little more than an overview although, like the topic of administrative mathematics, it is worthy of an in-depth study in its own right. Each contains areas which could profitably be explored more closely, or which remain obscure for want of further data.

Until now, little attention has been paid to the scholarly milieu of Mesopotamian mathematics.[1] The primary sources of information have in the past been the so-called 'eduba texts', which present an idealized picture of life in the OB 'school' (Sjöberg 1976; Civil 1985; cf. Volk 1996: 178–83).[2] Nemet-Nejat (1993: 5–10) collected several references to mathematics teaching in these and other texts, and Friberg's survey of Mesopotamian mathematics (1987–90: 582–3) refers briefly to the circumstances in which it might have been written. Høyrup has covered the issue most extensively (1985; 1990*a*: 326 n. 144, 337; 1994), but the present study is the first to use not only the 'eduba texts' but also the mathematical texts themselves and other types of school exercises.[3]

This investigation is concerned with what scribal students learned and how they were taught, rather than the environment in which they learned it, or who they and their teachers might have been.[4] The exact connotations of the word **é-dub-ba-a**, usually translated 'school', remain unclear;[5] it is enough for the moment to assume that scribal skills were passed on in some sort of 'stream of tradition' (cf. Oppenheim 1977: 13). This chapter sets out to answer, at least in part, the following questions:

- What evidence is there for mathematics education in the Ur III period, and indeed earlier? What role, if any, did coefficients play?
- How was mathematical education affected by the end of the Ur III empire?
- How did mathematics fit into OB scribal training?
- What can be deduced about OB methods of teaching mathematics from the coefficient lists and other extant mathematical texts?

[1] Compare Kraus on Sumerian literature (1959–62: 16): 'Es ist möglich, dass die gesamte oder fast die gesamte eigentliche Literatur der alt-babylonischen Zeit entweder aus Schulbüchern besteht oder als Schulbücher benutzt wurde, aber wir wissen und begreifen zu wenig vom Wesen des "Tafelhauses" ... um uns darüber ein begründetes Urteil bilden zu können. Jedenfalls gilt die mannigfache Beachtung, welche die Literatur der altbabylonischen Zeit in der Assyriologie findet, ihr selbst und nicht der vielleicht hinter ihr stehenden Schule.'

[2] '[...] Une formule introductive stéréotypée [...] place l'action à une époque lointaine où les écoles étaient plus près de l'idéal qu'à présent. Les textes décrivent une situation, non telle qu'elle est, mais telle qu'elle devrait être' (Civil 1980: 229).

[3] Ideally, archaeological evidence should also be used (e.g. Charpin 1986: 419–86), but even where school tablets come from controlled excavations data on their findspots are often frustratingly meagre, inaccessible or ambiguous.

[4] So that, e.g., Michalowski's justified concern that 'we cannot establish how much training in practical matters was provided by the schools and how much was left for on-the-job training' (1983: 226) is not at issue here.

[5] For a summary and critique of the view of scribal schools presented in 'eduba texts' and accepted uncritically by earlier Assyriologists, see Charpin (1986: 420–3, 485–6).

- Why were scribal students taught mathematics? Were the educational aims of the Ur III and OB periods the same?

10.1 THIRD MILLENNIUM MATHEMATICS

The few extant mathematical texts dating from before the Ur III period have been surveyed by Friberg (1987–90: 539–41). Already in late fourth millennium Uruk model economic documents such as W 20517, 1 were written, which can be distinguished from working records by their lack of end-matter (cf. §1.3.4) and by the very large or small number units they deal with, which are often also 'conspicuously round' (Friberg 1987–90: 539).

The earliest known problem texts are from ED III Shuruppak[6] and from Ebla,[7] in each of which a single problem is stated and the numerical answer given below (Powell 1976*a*: 431–4; Archi 1989). There is no working shown on the tablets, but Powell has shown that *TSŠ* 50 and *TSŠ* 671 are more than simple practical exercises, as each worker receives 7 sila of rations:

> [...] having read thousands of Sumerian and Akkadian texts from the third millennium, I cannot recall a single case where 7 functions as a divisor. Thus, the choice of 7 can hardly be motivated by any other cause than that it is an irregular number [i.e. one without a terminating reciprocal]. Moreover, the two different answers [on TSŠ 50 and 671 respectively] suggest that the object is an exercise in using the reciprocal of an irregular number (Powell 1976a: 433).

In other words, these are precursors of a common OB class of problems, which use a practical pretext to explore the division properties of the so-called 'remarkable numbers' 7, 11, 13, 17 and 19, which are both irregular and prime (cf. Høyrup 1993*b*). *TSŠ* 77 is a geometrical diagram on a lenticular tablet from mid-third millennium Shuruppak. It too appears to anticipate the mathematical concerns of the OB period, being identical to one of the illustrations on BM 15285 (§§9.3, A.2) (Powell 1976*a*: 431 fig. 2). Further, two ED tables of squares from Shuruppak and Adab display 'consciously sexagesimal' characteristics (Powell 1976*a*: 431), such as the use of gín in the general sense of $^{1}/_{60}$.

There are eight known tablets containing mathematical problems from the Old Akkadian period, from Girsu and Nippur.[8] Whiting (1984: 59 n. 2) has described them as follows:

> The exercises are characterized by the occurrence of the expression **ba-pà** (variant **pa**) from the Sumerian verbal root **pà(d)** 'to find', 'to reveal' [...]. The exercises occur in two types:
> (1) Given the area and one side of a rectangle, find the other side [...];
> (2) Given the side of a square, find its area.

The texts are of two types: those that consist of the statement of a single problem and its numerical answer, which are the ancestors of OB-style problem texts; and those in which two stated problems are allocated to named students. In these cases the answers are not given, and they may have more to do with OB-style 'catalogues' than problem texts in that they appear to have been written by an instructor in preparation for teaching (Whiting 1984: 65). Indeed, one of these assigned problems has a solved counterpart amongst the problem texts (Whiting 1984: 66). Once again, there is evidence that some elements of the sexagesimal place system were in

[6] In particular, *TSŠ* 50 and *TSŠ* 671, variants on a single problem on distributing a large quantity of grain equally among thousands of people in portions of 7 sila each; *TSŠ* 188, an exercise on squaring. For a full list, see Powell (1976*a*: 436 n. 19).

[7] TM.75.G.1392 and TM.75.G.2346, which are similar to *TSŠ* 50 and *TSŠ* 671.

[8] From Girsu: Limet 1973: nos. 36–9; Gelb 1970: no. 112; A 5443 and A 5446 (Whiting 1984); from Nippur: Pohl 1935: no. 65.

use, as indicated by errors such as the writings 0;36 15 for 0;30 06 15, and 38 38 for 38 20 (Limet 1973: nos. 36, 37; Whiting 1984: 62).

IM 58045, a lenticular tablet from Old Akkadian period Nippur, shows a mathematical diagram which displays a concern with the construction of problems to produce integer solutions, later typical of OB mathematics. It is described by Friberg as follows:

> There is drawn on [the tablet] a trapezoid with a transversal line parallel to the base, dividing the trapezoid into two parts of equal area. The lengths of the sides are chosen in such a way that the length of the transversal line can be expressed in whole numbers (no square roots). (Friberg 1987–90: 541)

No mathematical tables are known from this period, but model documents of various kinds have been identified, including a practice account from Eshnunna (Westenholz 1974–7: 100 no. 11) and several land surveys and building plans (Foster 1982: 239–40).

In short, many of the foundations of the sexagesimal place value system had been laid by the twenty-third century BC. Further building-blocks for OB mathematics were also in place, including the manipulation of reciprocals, the formulation of problems to give integer answers, the use of practical pretexts for problems primarily concerned with the properties of numbers. Equally importantly, all types of mathematical text attested in OB (§1.3), or their prototypes, were in use before Ur III. The sole exception appears to be the coefficient lists — but coefficient-like work-rates with OB-style values were already being put to administrative use (§9.1.1).

10.2 MATHEMATICS IN UR III SCRIBAL EDUCATION

The Ur III empire began to expand rapidly towards the east in the second half of Shulgi's reign (*c.* 2070–2047 BC). At its widest extent it stretched to the foothills of the Zagros mountains, encompassing the cities of Urbilum, Assur, Eshnunna, and Susa. To cope with the upkeep of these new territories and the vastly increased taxation revenues they brought in, large-scale administrative and economic reforms were executed over the same period, producing 'a highly centralized bureaucratic state, with virtually every aspect of its economic life subordinated to the overriding objective of the maximization of gains' (Steinkeller 1987: 20). These administrative innovations apparently included:

> - the creation of a unified administrative system for southern and northern Babylonia;
> - the creation of an enormous bureaucratic apparatus, as well as of a system of scribal schools that provided *highly uniform* scribal and administrative training for the prospective members of the bureaucracy;
> - the radical reform of the writing system;
> - the introduction of new accounting and recording procedures and of new types of archival records;
> - the reorganization of the systems of weights and measures;
> - the introduction of a new calendar, [...] which became the official calendar throughout the Ur III state. (Steinkeller 1987: 21; italics added)

Although little is known of Ur III scribal education — no contemporary eduba has yet been excavated — a high degree of uniformity must have been essential to produce such wholesale standardization in the bureaucratic system. Students must have been trained in the new writing, accounting and administrative methods in use throughout the empire. Shulgi is supposed to have established scribal schools at Nippur and Ur, 'in order to indoctrinate future bureaucrats with the propaganda of empire and the charismatic force of the king' (Michalowski 1987: 63). Two hymns which purport to be written by Shulgi extol the virtues of mathematical ability, in well-known passages on literacy and education:[9]

nam-dub-sar-ra ki-nam-kù-zu-ba lú im-me-re$_6$-re$_6$
zi-zi-i gá-gá šid níg-ka$_9$-dè zà im-mi-til-til (Shulgi B: 16–17, Castellino 1972: 32)

(Where) one goes to the place of wisdom of the scribal art;
I have completed subtraction, addition, counting and accounting.

šid níg-ka$_9$ giš-ḫur-kalam-ma-ka
igi-gál-sum-mu-bi á-bi-šè in-ga-zu (Shulgi C: 46–7, Castellino 1972: 253)

Counting, and accounting, the plans of the nation:
I have also mastered their given wisdom.

As yet only a few school mathematical texts can be dated with any certainty to the Ur III period, but between them they reveal a good deal about contemporary educational practice. There are two serious obstacles to the confident identification of school texts from the Ur III period. Firstly, it is often difficult to distinguish between competently written model documents and genuine archival texts (see §1.3.4). Indeed, YBC 9819, which is now firmly identified as a model mathematical document, was for many years assumed to be an administrative text (Legrain 1935; Dunham 1980: 422–30).[10] There must be many more unrecognized school texts which have been published as archival material.[11]

Secondly, when texts are neither dated nor excavated from well-defined findspots, palaeographic criteria must be used to assign a period to them. In many cases it is matter of dispute whether a text is from the late third millennium or was written using archaizing script in the early second millennium.[12] Numerals such as 19 written as 20.LÁ.1 might be one such archaizing indicator. In particular, it was long thought that the sexagesimal place system, which represents numerals using just tens and units signs, was an innovation of the OB period (cf. Powell 1976*a*: 417), so that any text using that notation was assumed to date from the early second millennium or later. However, on the basis of the draft economic document YBC 1793, Powell (1976*a*: 420–1)[13] established that it was already being used by around 2050 BC — and that the conceptual framework for it had been under construction for several hundred years

[9] Almost all surviving exemplars of the Shulgi hymns date to OB but, as Klein (1981: 24 n. 13) notes, 'their general authenticity cannot be questioned. It follows not only from the accurate historical and biographical references that these hymns contain, but also from the appearance of some of them in literary catalogues from the Ur III period proper', with the proviso that 'there are indications that the fixation of their final form was a process which sometimes continued up to the Old-Babylonian period.' According to Hallo (1966: 139), royal praise poems of this type were composed annually to commemorate each regnal year, and were often thematically linked with the official year name. For instance, Frayne (1983: 96) ascribes Shulgi C to regnal year 21, or 2074 BC.

[10] The fact that the text is not dated or signed later led Dunham (1982) to conclude that it was a school exercise. That there are serious calculation errors in it too (§4.5), which arise from misusing two coefficients, lends further credence to this hypothesis.

[11] As, for instance, the two other Ur III model surveys discussed in this work: AOT 304 (§4.4) and W 74a (§6.3). On the other hand, BM 20054 (§9.6.1) and VAT 7038 (§9.4.2), which have been provisionally classified as draft documents here, may turn out to be school exercises. AO 5499, which has not been discussed in this study, chronicles the (remarkably regular) growth of a herd of cattle over a decade, from Shulgi 40 to Shulgi 49. Friberg (1987–90: 542) has described it as 'either a school text ... or a theoretical, *ideal* model intended for practical use in the central planning of the administration of herds of cattle.' This latter interpretation is most likely, as the tablet is signed by a sanga of the temple of the god Ishtaran, and is probably from Puzrish-Dagan, the Ur III state's livestock control centre (Nissen *et al.* 1990: 139).

[12] See, most recently, Civil (1994: 9–10).

[13] See also Powell (1976*a*: 435 nn. 6, 7) for some further Ur III administrative texts which use a limited amount of sexagesimal place notation.

(see §10.1 above). Crucially, though, he also concluded that:

> calculations in sexagesimal notation were made on temporary tablets which were then moistened and erased for reuse after the calculation had been transferred to an archival document *in standard notation*. (Powell 1976*a*: 421; italics added)

This procedure is succinctly illustrated by AOT 304 (§4.4), a lenticular model document from Ur III Girsu. On one side of the tablet is a (slightly incorrect) model entry from a quantity survey, giving the dimensions of a wall and the number of bricks in it. The measurements of the wall are given in standard metrological units, but were (mis-)copied on to the reverse in sexagesimal notation. The volume of the wall, and the number of bricks in it, were then worked out using the sexagesimal numeration, and converted back into standard volume and area measure, in which systems they were written on the obverse of the tablet. These conversions were presumably facilitated by the use of metrological tables similar to the many hundreds of OB exemplars known.[14] In other words, scribal students in the Ur III period were already taught to perform their calculations — in sexagesimal notation — on tablets separate from the model documents to which they pertained, which were written in the ubiquitous mixed system of notation. It is hardly surprising, then, that pre-OB examples of sexagesimal notation are so rare.

YBC 9819 (§4.5) is a more extensive model document, listing, according to the colophon, bricks needed for (repairs to?) the temples of Shara and Ninurra in Umma. This tablet provides clear evidence that scribes were being trained in how to carry out quantity surveys, and to lay out their measurements and derived figures correctly, but also that the same brick coefficients that are widely in evidence in OB school mathematics were already in place. Not only are the bricks counted the same way (*MCT*: 95–6), but indeed, the calculation is based on the same standardized brick sizes and the estimate that five-sixths of the wall was made of bricks. The fact that there are no intermediate calculations on the tablet suggests that these were done on a separate tablet.

Less certainly of Ur III date are Ist T 7375 and HS 201, both tables of reciprocals, from Girsu and Nippur respectively.[15] They differ from OB examples of the genre in that they list all integers from 1 to 60, stating their reciprocals where calculable, and **nu** 'no' otherwise.[16] OB reciprocal tables, on the other hand, typically list all integers between 1 and 81 which have finite reciprocals, and omit those, such as 7 and 11, which do not. CBS 10201, a list of successively doubled and halved pairs of reciprocals from Nippur, may also be Ur III, or at least Isin, in date (Friberg 1987–90: 549).

In short, even with such little evidence, we can say that practice calculations, model documents, and perhaps tables were used to train scribes in the late third millennium. Further, both coefficients and the sexagesimal place value system were already firmly established. Despite the paucity of contemporary mathematical material in our possession, the arithmetical skills revealed by the vast quantities of contemporary economic documents would suggest that mathematics was a corner-stone of Ur III scribal training.[17] Exactly what that scribal training entailed must wait for the identification of a larger body of school exercises.

[14] Indeed, some of these tables may turn out to be of Ur III date.

[15] 'Schrift etwa Shulgi' (*MKT*: I 10); 'not conclusively Ur III' (Powell 1976*a*: 435 n. 3). Ist T 7375 apparently belongs to the large group of Ur III economic documents found by de Sarzec at Tello in 1898–1900 (Delaporte 1911: 131). For the date of *MDP* 27 61, see §10.5.

[16] Cf. BM 96957 + VAT 6598 (xx), which uses the logogram NU as a statement of its insolubility (§A.4). This is the only known OB mathematical text of its kind.

[17] Cf. Friberg (1987–90: 542): 'the conspicuous absence of preserved mathematical cuneiform texts from the Ur III period does not necessarily mean that mathematics did not exist as an important item in the curriculum of the Ur

10.3 AFTER UR III: DIVERSITY IN OB MATHEMATICS

The empire of the Third Dynasty of Ur began to disintegrate early in the reign of Ibbi-Sin (*c*. 2025 BC), as first Susa and Eshnunna asserted their independence, followed one after another by cities closer to the heartland of the kingdom. Within 20 years Ur itself had been conquered and its domination was in effect over (Michalowski 1989: 1–3). The resulting political fragmentation, in which a dozen or so city states, including those of Isin, Larsa, Eshnunna, Susa, and Mari, vied for domination and control of greater Mesopotamia, lasted until Hammurabi of Babylon briefly reunited them in the mid-eighteenth century BC.

The period of the Ur III empire appears, on present evidence, to be the time at which a more or less uniform mathematical culture was spread throughout southern Mesopotamia and the surrounding area. Not only is there a prime candidate for the means of transmission in the form of a highly centralized and standardized bureaucracy, but there are also the coefficients as a *terminus post quem*. It has been shown in Chapter 9 that standard work-rates and building constants were intrinsic to the management of the Ur III state, and that they were adopted more or less wholesale as coefficients by OB school mathematics at a time when they were no longer in widespread use. The coefficients, and their associated mathematics, must have travelled with the expanding empire, and remained in outposts such as Susa and Eshnunna long after their independence from Ur.

Under these circumstances, without a centralizing authority, we should expect a degree of diversity in educational materials and techniques, perhaps deriving from a common source in the Ur III period or earlier. That is not to say that communication and transmission of ideas and texts did not occur on an individual scale, simply that official diffusion of them did not — could not — take place once the mechanism for that transfer was no longer in place. The innovatory — and geographically diverse — character of the early second millennium lexical lists is a recognized phenomenon; in particular many new lists were created 'avec des traditions très diverses entre les localités, surtout pour les listes thématiques', and others were supplied with Akkadian translations for the first time. Yet 'malgré ces innovations du IIe mill. la rupture avec la période précédente n'est peut-être pas si brutale qu'on peut l'imaginer' (Cavigneaux 1980–3: 616). This is in many ways the same situation that we find with the school mathematical texts.

Goetze (1945) used putative regional characteristics in orthography to assign the mathematical texts then known to northern (Sippar) and southern groups (Larsa and Uruk). However, with the subsequent identification and publication of groups of mathematical material from well-documented findspots (Susa, Shaduppum and the Eshnunna region, Ur, Sippar, Mari) it is now possible to look at the issue from a different angle: given the established provenances of different groups of mathematical tablets, how do they differ? A few preliminary remarks are made here, but for an in-depth study see now Høyrup (1996*b*).

Two examples of regional terminology have already been mentioned. In §§5.2.3 and 6.1.3 the words *iškarum* and *allum* were examined. In most mathematical texts *iškarum* denotes a daily work-load defined in terms of sar$_{v}$; in those from the Eshnunna region it refers to the output resulting from that daily rate, expressed in the number of bricks made, or the length of wall built. There is no equivalent term in other mathematical texts. In the Shaduppum problems *igigubbûm* has taken over the meaning 'daily rate (in sar$_{v}$)', whereas *allum* has this function in Haddad 104, from Me-Turan, and in the Shaduppum coefficient list C. Secondly, in §7.1.3 it was noted that, on present evidence, the use of NUMUN as a quasi-synonym of IGI.GUB is

III scribe schools'.

apparently restricted to the Sippar area. Other geographically restricted features include the introductory phrase *šumma kīam išālka umma šūma* 'if someone asks you, saying as follows: ...' in Shaduppum,[18] and the predominantly southern use of the second person present instead of the imperative in the solutions of problems (e.g. *teḫeppema* 'you halve, and ...' instead of *ḫepema* 'halve, and ...').[19] The solutions to Sippar and Susa problems all begin ZA.E 'you: ...', while those from the Eshnunna region start *atta ina epēšika* 'you, in your working: ...'.[20] On the other hand, the Sippar instructions end with *nēpešum* or *kīam nēpešum* '(this is) the method', while in Susa and Shaduppum they usually finish without a formal marker.[21] There is also the well-known phrase in the Susa problem text *TMS* 9 (ii): 15 which describes the solution of a problem as *kīam akkadû* 'this is the Akkadian (method)'.

On the other hand, not all terminological variation is a regional phenomenon. For instance, the problems on the ten tablets excavated together from Shaduppum use either *tammar* 'you will see' or *illi* 'will come up' to announce results, and either *kī maṣi* or *minûm* for 'how much?', in any combination.[22] Usage is consistent within each problem, though, and so may have been a matter of individual scribal preference.

A comparison of the widespread mathematical material on grain-piles (§7.3) reveals four ways in which the coefficients for a single mathematical topic may vary from text to text, even those from the same findspot, as summarized in Table 10.1.

Table 10.1: Data on grain-piles in OB mathematics

Value	*Name*	*Unit*	*Text*	*Provenance*
7 12 00	*ša našpak* GUR$_7$	sila	D 51	Susa
7 30 00	*karûm*	sila	H 7	Shaduppum
1 30	IGI.GUB	gur	BM 96954 + BM 102366 + SÉ 93	Sippar
8 00 00	*našpak* GUR$_7$	sila	*TMS* 14	Susa

First, not only does each text use a slightly different description of the coefficient — which might be most fully desribed as *IGI.GUB *ša našpak* GUR$_7$ 'the coefficient of the storage of a grain-pile' — but, second and third, three different numerical values are attested, one of which is expressed in two different units of measure (H 7, BM 96954 + ...). One might expect minor variations in wording: even within single problems the name of a coefficient is often abbreviated (see §2.3.2). It is the numerical differences which are more significant. Two of the three values are attested at Susa and the other is found at both Sippar and Shaduppum, albeit in different guises.

Fourth, it has already been pointed out in §7.3.2 that the Susa method for solving grain-pile problems is an elegant simplification of that used in the Sippar tablet, resulting from the assumption that the grain-pile slopes with a gradient of 1. An examination of the values taken by the parameters in each of the problems confirms this conclusion, as shown in Table 10.2.

[18] e.g. IM 53961; IM 54011 (i), (ii) (§6.1). Note the variant *šumma ṣiliptam išālka umma šūma* 'if someone asks you about a rectangle, saying as follows: ...' in Db$_2$-146 from Tell Dhiba'i. Haddad 104, from ancient Me-Turan, does not use such an introductory formula.

[19] e.g. AO 8862 (i): I 12 *te-ḫe-ep-pe-e-ma*, supposedly from Larsa; BM 96954 (ix): II 16 *ḫe-pé*, from Sippar (§A.3). Almost all the texts assigned by Goetze (1945: 147–9) to the southern groups 1 and 2 (using other criteria) exhibit this usage, as do around half of those belonging to the 'Uruk' group 3. It is very rare in the Sippar problem texts and in those from the Eshnunna region and Susa — but see *TMS* 2 (ii), §3.6.2.

[20] Sippar and Eshnunna examples as above; Susa: *TMS* 14 (§7.3).

[21] Sippar and Eshnunna examples as above; Susa: *TMS* 25 (i) (§6.5). *TMS* 14 (ii) exceptionally ends with *kīam nēpešum*.

[22] e.g. *tammar* and *kī maṣi*: IM 54478; *illi* and *minûm*: IM 53953; *tammar* and *minûm*: IM 54011.

The length, width and top of the two piles are identical. The crucial change to the *TMS* 14 data is the reduction of the height from 48 cubits to 36 resulting from the 1 in 1 slope. This in turn affects the volume of the pile and its capacity (although the reasons for using a different capacity conversion coefficient are not apparent). Thus it appears that in some sense the BM 96954 + ... problems are 'forerunners' of those on *TMS* 14, whose parameters were retained in transmission and altered as necessary to accommodate refinements in the method of solution.

Table 10.2: Dimensions of the grain-piles in BM 96954 + ... and TMS *14*

Text	*Provenance*	*Length*	*Width*	*Top*	*Height*	*Volume*	*Capacity*
BM 96954 + ...	Sippar	10 nindan	6 nindan	4 nindan	48 cubits	19 12 sar_V	28 48 00 gur
TMS 14	Susa	10 nindan	6 nindan	4 nindan	36 cubits	14 24 sar_V	23 02 24 gur

The broad homogeneity of Old Babylonian mathematics hardly needs to be emphasized: it is the assumption on which all studies on the subject, including this one, have been founded. Nevertheless, the brief observations made here indicate that there are also a number of ways in which the material may vary, both in the mathematical techniques used and in the language with which they are expressed. A much larger study is needed to determine to what extent this diversity can be attributed to geographical and chronological factors, or to the result of individual scribal preference. The overall picture, though, is one of textual transmission through relatively informal means, resulting in the widespread diffusion of problem types, and even of appropriate sets of data for those problems,[23] but very few word-for-word copies (see §10.4 below). Transmission was, to a large extent, conditioned by the way in which mathematics was taught.

10.4 TEACHING METHODS: THE ROLES OF THE SIX TEXT TYPES

Civil identified four characteristic types of tablet on which school copies of lexical texts were inscribed at Nippur:

Type I Large tablets of eight columns or more and large prisms with the full text of the series or a sizeable extract of it and nothing else.

Type II/1 Obverse of medium-to-large tablets divided into two columns: the left one is the instructor's model, the copy by the apprentice scribe is on the right column. The right side is reused, after erasing the exercise, and in most of the tablets recovered this side is now blank. After repeated use of such tablets, the right side tended to become too thin and was cut away, and the left side was preserved, presumably to serve again as a model. [...]

Type II/2 Relatively long extracts of at least forty lines, on the reverse of tablets of the type just described. [...]

Type III One column tablets with a short extract of the series continued on the reverse, and nothing else.

Type IV Lentil-shaped exercise tablets with a couple of lines, sometimes repeated; the more convex side is normally uninscribed. (Civil 1969: 27–8)

This has become the standard physical typology of OB elementary school texts in general (cf. Veldhuis 1997: §2.3.3). But how well does it fit the generic analysis of mathematical texts outlined in §1.3, and does it help our understanding of how the three strands of mathematical

[23] i.e., those which produce integer answers to the problems.

skills — arithmetical proficiency, problem-solving, and their implementation in the preparation of working documents — were taught to OB scribal students?

Arithmetic skills were encouraged by the rote learning of metro-mathematical lists and tables.[24] Let us take multiplication and reciprocal tables as exemplary. Neugebauer (*MCT*: 20–33, with previous lit.) categorized multiplication tables as single, where they were the sole contents of the tablet, and combined, where there were two or more on the same tablet, almost always in a standard order.[25] Many tablets with single tables on them end with colophons describing them as IM.GÍD.DA,[26] which is also the name given to single-column tablets of 30–40 line extracts from lexical lists (Civil 1979: 5). Like them, they typically end with a double ruling, and may also have a colophon and/or a catch-line consisting of the first line of the next table in the series.[27] The single tables correspond, then, to Civil's Type III tablets,[28] but the combined tables present a more complex picture. While only two Type I tablets and prisms must have originally contained the whole series,[29] many more contain substantial extracts, of half or a third of the total length.[30] There are also, however, numerous four- or six-column tablets too small to be Type Is but too big to be Type IIIs, which contain extracts from the series.[31] Types I and III, and the middle-sized tablets, are found both at Nippur and elsewhere.

Type II multiplication tables, on the other hand, are known only from Nippur. Of particular interest are Type II/1 tablets, with multiplications in teacher–student copies on the obverse, listed in Table 10.3.

On the other hand, there are no Type IV multiplication or reciprocal tables, or any other type of mathematical text, from Nippur; perhaps because the tables were treated as inseparable units, two- or three-line extracts were not considered appropriate. Indeed *MDP* 27 61 from Susa is the only Type IV multiplication table known to me.[32]

[24] Cf. 'Eduba text D' lines 40–2, which describe copying tables 'from 1 gur to 600 gur' and 'from 1 shekel to 10 minas' (Civil 1985: 72).

[25] The order of the whole series is a reciprocal table, followed by multiplication tables (by 1–20, 30, 40, 50) for:

50	40	24	16 40	12	8	6 40	4	2 30	1 40
48	36	22 30	16	10	7 30	6	3 45	2 24	1 30
45	30	20	15	9	7 12	5	3 20	2 15	1 20
44 26 40	25	18	12 30	8 20	7	4 30	3	2	1 15,

sometimes ending with a table of squares (if squares have not been given in the course of the series).

The only non-standard sequences known to me are: Ist Ni 2937, a surface flake from Nippur which lists a 50 and an 18 times table followed by reciprocals (*MKT*: II Taf. 67); and the unpublished IM 55292, which, according to Bruins (1954), contains multiplication tables for 18, 44 26 40, 16, 25, 28, 22 30, and 20 in that order. No other table for 28 is known.

[26] Not to be confused with the unexplained mathematical term IM.GÍD.DA (e.g. text C: II 43, text J: II 4 §A.1).

[27] Example with a double ruling: 2 times table HS 222a, from Sippar or Nippur; example with a catch-line: UM 29-15-482, reciprocal table from Nippur with catch-line [50] A.RÁ 1 ⌜50⌝ '50 times 1 is 50'; example with a colophon: YBC 6739, unprovenanced 8 times table:

IM.GÍD.DA *u-bar-rum*	Ubarrum's 'long tablet'
ITI KIN.^{d}INANA U_4.20.KAM BA.ZAL	Month 6, day 20.

[28] 144 single multiplications and 12 reciprocal tables are known to the author, and 76 combined tables, of which 28 preserve some or all of the reciprocal table. For the most recent listings, see the notes to §1.3.5.

[29] A 7897, a 13-column cylinder from the Diyala region, and Ist Ni 2733, a 20-column tablet from Nippur.

[30] Examples of half the series: BM 80150, an unprovenanced 8-column tablet, with a reciprocal table, multiplications for 50 to 9, and a table of halved and doubled reciprocals, and HS 202a, a multi-columned tablet from Nippur with a reciprocal table and multiplications for 50 to 8.

Examples of a third of the series: AOS 15375, a 10-column tablet from Susa containing a reciprocal table and multiplications for 50 to 16 40, and HS 203, from Nippur, with multiplications for 6 40 to 1 15.

[31] For instance, HS 207, a 4-column tablet from Nippur, with multiplications for 24 to 16 40, and the unprovenanced BM 78267, the upper half of a 4-column tablet containing multiplications for 2 24 and 2.

[32] On the reverse of a two-line teacher-and-student copy of a lexical extract. It is not clear whether the numerous copying errors are the ancient scribe's or the modern copyist's.

In short, multiplication and reciprocal tables fit Civil's physical typology well. But this should not be particularly surprising, as they share several characteristics with the lexical lists. They are ordered in a long, standard series and, being susceptible to mastery by prodigious memory if not by arithmetical skill, were apparently taught by repeated copying.[33] On the other hand, the tablets containing other genres of mathematical texts cannot all be so easily categorized.

Table 10.3: Type II/1 multiplication tables

Tablet	*Provenance*	*Obverse*	*Reverse*
CBS 6095	Nippur	3 45 times table, teacher's copy only.	Multiplications for 50 to 12 30 preserved.
CBS 8226	Nippur	2 times table, teacher's and student's copies.	Multiplications for 40 to 24 preserved.
CBS 11340 + CBS 11402	Nippur	45 times table, teacher's and student's copies (cf. Hilprecht 1906: pl. IV).	Reciprocals and 50 times table, copied three times over four columns.
CBS 11368	Nippur	50 times table, teacher's copy only (cf. *MKT*: II Taf. 62).	Reciprocals and 50 times table, copied three times over four columns.
HS 203	Nippur	16 40 times table, teacher's copy only (cf. *MKT*: II Taf. 64).	Reciprocals, multiplications for 50 to 16 40, reciprocals again, over seven columns.
Ist Ni 2726	Nippur	44 26 40 times table, teacher's copy only (cf. *MKT*: II Taf. 66).	Multiplications for 50 to 30, over three? columns.
UM 29-15-474	Nippur	7 30 times table, teacher's copy only.	Three columns of badly preserved multiplication tables.
UM 29-15-478	Nippur	15 times table, teacher's and student's copies, badly preserved, with catch-line for 12 30 times table.	Badly preserved multiplications for 1 40 to 1 20.

At Ur and elsewhere, arithmetic was practised on the small round Type IV tablets, which show that there were rules on the layout of sums (see §A.5). The Ur tablets give us the added insight that five or more students may have been set the same problem to solve, each with different parameters — chosen by the teacher from a 'catalogue' (§A.5.1.4)? These erasable, round tablets were called IM.ŠU (Civil 1985: 75; George 1993: 70–1). The calculations on IM.ŠU were probably carried out in the course of composing model documents or solving mathematical problems, and explain why such texts rarely contain any evidence of arithmetical workings. YBC 4657 sets 31 problems on KI.LÁ excavations (cf. §6.2), which are described at the end as 31 IM.ŠU KI.LÁ '31 exercises on excavations'. YBC 4663 and YBC 4662 give the step-by-step solutions to problems (i)–(viii) and (xix)–(xxviii) of YBC 4657. Neither contains a colophon of any sort, although one might expect an identifying phrase similar to that on the first tablet.[34] Both tablets are long and rectangular, measuring roughly 75 × 190 mm (*MCT*: pl. 32–3), and so cannot be IM.ŠU in the sense of round tablets. If neither the objects YBC 4663 and YBC 4662 nor the text on them are IM.ŠU, then the word must refer solely to the contents of YBC 4657: namely, the problems to be solved, but not the pertinent procedures. So, for mathematical texts at least, IM.ŠU refers not only to the Type IV tablets themselves but also to what was written on them, namely the computations.

[33] For erroneous writings of A.RÁ and 20.LÁ.1, see Hilprecht (1906: 23). Omitted lines and numerical copying errors abound; see, for instance, CBS 11340 + CBS 11402 rev. (Hilprecht 1906: fig. 20).

[34] Solutions of this type are called *kibsātum* 'steps' (a plural *pirist* formation from *kabāsum* 'to step on', 'to tread') in colophons of tablets from Sippar: e.g. *ki-ib-‹sa›-tum* BM 85210. See Nemet-Nejat (1993: 184) for a full list.

In Nippur, however, arithmetic was practised on square, not round tablets. Some examples can be seen in §A.5.2, and more are listed in Tables 1.1 and 1.2 in §1.3.3. Veldhuis (1997: §2.3.3.4) contrasts his experience of well-preserved, visually appealing Nippur lenticular tablets with Faivre's characterization of them as showing erasures and signs of re-use (1995: 60). These views can be reconciled simply by noting that the Nippur tablets exhibit a rather different physical typology from those from elsewhere: round tablets are reserved for student–teacher copies while the roughly square ones are kept for rough work; by contrast in Ur, for instance, the Type IV tablets served both purposes.

Civil's tablet typology cannot be applied to the remaining genres of mathematical text. None are large enough to be considered Type I tablets, while those small enough to be thought of as Type III show none of the formal characteristics outlined above. In general it is the compilations of several problems which conclude with colophons and catch-lines: the only exception known to the author is the Susa 'catalogue' *TMS* 5, which describes itself as an IM.GÍD.DA.[35] Type II tablets are restricted to Nippur tables, while Type IVs and the related square tablets are used for rough work and a few model documents (cf. §1.3.4). Why this might be is addressed in §10.5.

While the emphasis in many scribal subjects, including arithmetic, seems to have been on memorization of texts, with most existing in several duplicates and variants, this was not always the case for mathematics. The calculations mentioned above provide good evidence that maths was actively practised in the classroom, and a few tablets contain problems with partially erased scribbled calculations which seem to have been performed while the text was being written.[36] On the other hand, some texts show clear signs of being copies. A line of text may have been omitted,[37] or the scribe may have confused a sign or group of signs,[38] producing a piece of work which does not make mathematical sense. Two word-for-word duplicate sets of problem texts are known, but these copies reproduce only incomplete procedures, which suggests they were copied for non-mathematical purposes, such as handwriting practice.[39]

The otherwise unattested words in the coefficient lists, as well as some unknown mathematical terms in lexical lists, show that we have recovered only a small proportion of OB mathematics. But to what extent does this reflect a predominantly oral tradition of teaching? Problem texts are structured in the form of a series of instructions in answer to a question. Indeed, several problem texts from Shaduppum preface this question with the phrase, 'If a man asks you', which seems to reflect an oral pattern of teaching. Further, *TMS* 16 'looks very much as a written documentation of a sort of didactical explanation which would normally be given orally' (Høyrup 1990*a*: 337), and in *TMS* 9 (i), at least according to Muroi (1994: 231), 'a teacher blames a pupil for not following his instructions to solve the problem, and he again teaches the pupil step by step'.

However, the 'step-by-step' approach of the problem texts is not only a useful pedagogical tool: it is intrinsic to the very conception of OB mathematics. Knuth (1972: 672–3) characterized OB mathematical solutions as 'genuine procedures for solving a whole class of problems [... which] explain an algorithm by example as the algorithm itself [is] being

[35] For a full discussion and typology of colophons on OB mathematical texts, see Robson (fc.).

[36] See the list in §1.3.3 and Friberg (1981*b*) on the mathematical 'catalogue' BM 80209 (§1.3.2).

[37] e.g. the systematic 'simplification' of some of the solutions in BM 96954 (§A.3) which renders them nonsensical. See also Powell's discussion (1982*b*) of the corrupt problem BM 85194 (iv).

[38] e.g. the writing ZA.E TUK.ZÚ.DÈ for ZA.E KÍD.DA.ZU.DÈ 'you, in your working' in IM 52301: I 3, 19.

[39] YBC 9852 is a partial copy of YBC 4675; YBC 10522 seems to be a partial duplicate of an unknown text. As discussed in §2.1, it appears that coefficient lists E, F and G were copies of earlier originals, but as coefficient lists are essentially reference works, this fact does not bear on the teaching of mathematical *methods*.

delivered', by using the numerical values of the parameters as generic examples of any values.[40] However, because the followers of Old Babylonian mathematical procedures were human beings and not computers, those procedures could often be subverted or simplified according to the conditions of the problem.[41] This quasi-algorithmic aspect of OB mathematics is now confirmed beyond doubt by BM 96957 + VAT 6598 (xx) (§A.4). Underneath the statement of the problem is the single word NU 'no'. It is soluble by algebraic means, but not by the 'algorithm' being taught in the text (Robson forthcoming).

Problem texts are found in three different arrangements on the tablets:

- one problem (and sometimes a second which is just a variant of the first),[42] with the full solution given;
- several thematically linked problems,
  - with the full solutions given;
  - with no solutions but sometimes with numerical answers.[43]

We have already seen that this third type of tablet can be provided with sets of solutions in the form of the second type,[44] but it remains unclear whether this was always the case. Nevertheless, the ways in which the problems are grouped within these collections provide useful information on didactic techniques. At one extreme there are tablets such as YBC 4708 which comprise extensive explorations of a single topic, while at the other, tablets like BM 85194 contain collections of problems on diverse subjects, with themed subgroupings apparent.[45] These last have the appearance of compilations from single-subject collections,[46] which are structured in the following way, as typified by BM 96954 + 102366 + SÉ 93 (§A.3):

- the same set of parameters is used throughout the sequence;
- each problem concerns a different parameter in turn, continuing the sequence with problems on two parameters once they have all been dealt with;
- the problems are carefully arranged in order of difficulty, with the easiest first.[47]

Clearly, the primary function of these collections cannot be to teach arithmetic: at that level all problems after the first are trivial, in that the values of all the parameters are known. They are

[40] Hence the inclusion of apparently trivial operations such as multiplication by 1, where 1 is a generic example of a parameter that could take any (positive, rational) value (e.g. IM 53961: 6, §6.1)

[41] Høyrup (pers. comm. 1995). See. for instance, BM 96957 + VAT 6598 (vi)–(vii) (§4.5.2).

[42] e.g. IM 54011 (i) and (ii), §6.1; *TMS* 14 (i) and (ii), §7.3.

[43] Using Nemet-Nejat's lists of mathematical tablets (1993: 113–48) one can pick out a few exceptions to this dichotomy. These are: AO 8862 (p. 144), in which problem (iv) is not solved. This is the last of a group of 'algebraic' problems, while the rest of the text concerns carrying bricks; Str 362 (p. 146), in which only the first problem is solved; YBC 4669 (p. 148), in which only one problem is solved; YBC 4673 (p. 140), in which only problems (xvii)–(xxi), on building a dam, are solved (§6.4). To be excluded from this list of exceptions are: *UET* 5 859 (p. 122), in which the second problem is unsolved because the text is unfinished; VAT 8522 (p. 147), in which all four problems are solved — although the third is obscure; YBC 9874 (p. 142), in which '#1' is in fact a diagram, not a problem, and so is neither stated nor solved.

[44] See YBC 4657, YBC 4662, and YBC 4663, discussed above.

[45] For instance, among the groupings on BM 85194 are: (i)–(v): earthworks with a daily construction rate of 0;10 sar_v (cf. §6.3), including one about a circular city (see below); (vi)–(viii): water-clocks; (x)–(xiii): IM.LÁ (cf. §8.3); (xiv)–(xv): reed-bundles (cf. §6.4); (xxv)–(xxvi): circular cities, without construction rates, as well as other groupings and individual problems.

[46] e.g. BM 96957 + VAT 6598, with problems on building with bricks and on right-angled triangles, explicitly states that it has been copied (§A.4).

[47] Cf. Vanstiphout's characterization of OB teaching of Sumerian grammar from the text 'Lipit-Eshtar's praise in the Eduba', as practical, exemplary, contrastive and progressive (1978; 1979: 125).

concerned, rather, with the exhaustive presentation of mathematical procedures, both in the enumeration of all steps of individual solutions, including the arithmetically trivial, and in the attempt to develop as many different problems as possible from a single mathematical scenario.[48]

We can now put forward a tentative summary of the educative roles of the various types of mathematical text attested in OB:

- Large collections of the statements of mathematical problems (such as BM 15285, §A.2) functioned as 'textbooks' in that the students were set problems to do from these. They might contain the numerical answers, but never the procedure by which they were obtained. Teachers may have copied them from from earlier exemplars or composed them anew.
- There were accompanying sets of model solutions, designed to be applicable to any set of values, which the student might learn by copying them out (hence the partial copies of solutions mentioned in §10.3 above).
- The actual mathematical exercises were performed on small square or round tablets (IM.ŠU), after which the answer could be checked against the 'textbook'. The students laid out their arithmetic in certain standard formats, and might also use diagrams to help them.
- Students were expected choose the necessary coefficients themselves, as these were not given when the problems were set. Coefficient lists may have served as reference documents.
- Arithmetical and metrological relationships were memorized through the repeated copying of lists and tables.
- Teachers kept lists of appropriate values for problems, in the texts now known as 'catalogues' (see §1.3.2), to provide the students with extra practice in a particular problem, if needed, or to assign identical problems with different parameters to a class of students.

10.5 THE PLACE OF MATHEMATICS IN OB SCRIBAL TRAINING

The existence of OB tablets containing a variety of different school subjects was noted as early as 1916 (Chiera 1916–19: I 16–17), but since then the individual sections have been used for their separate contents, not read together in context,[49] and the coexistence of different genres of text on the same tablet has to a large extent been underexploited.

While the fact that coefficient list B is on the same tablet as non-mathematical material has been cause for much comment (see §2.2.1), this is a far from unique phenomenon, and it is restricted neither to Nippur nor to post-OB mathematics. Published OB period tablets of mixed mathematical and non-mathematical content include those given in Table 10.4, but this list is by no means exhaustive.

We can add to this list seventeen Type II and two Type IV tablets containing Sumerian Proverb Collections 1 and 2 and various metro-mathematical tables (Gordon 1959: 31–8, 161–72, 492, 519), and forty-six lenticular tablets from Ur containing proverbs and calculations

[48] Høyrup (pers. comm. 1995) adds that 'one should not overlook [...] that the problems in question may still serve to train the use of the place value system for multiplication, finding of the IGI, and finding square roots in the table. The known end result would then serve as an automatic check, but would not give you the intermediate results directly.'

[49] Cf. the remark that Sumerian proverbs 'can better be described in the context not of scribal education but of religious literature and higher learning' (Hallo and Simpson 1971: 158).

published in §A.5. But there are as yet no confirmed instances of mixed-subject tablets containing *problems*: all the identified mathematical texts are tables and calculations.[50] Why might this be?

Table 10.4: OB mixed-subject school tablets

Tablet	*Provenance*	*Type*	*Obverse*	*Description*
CBS 8153	Nippur	II/2	'Left-hand side of a single multipl. table and remains of a vocabulary' (*MCT*: 28).	Reciprocal table and multiplications for 50 to 18.
CBS 9044	Umma	—	Sumerian PNs.	Model account.
CBS 11097	Nippur	II/2	'Vokabular' (*MKT*: I 12).	Reciprocal table and multiplications for 50 to 24 preserved; originally contained tables to 18? over six? columns (cf. *MKT*: II Taf. 62).
CBS 11397	Nippur	II/2	'Vokabular' (*MKT*: I 12).	Multiplications for 50 to 7 30 preserved; originally contained entire series? over ten columns (cf. *MKT*: II Taf. 63).
CBS 14233	Nippur	II/2	'Bruchstücke eines literar. Textes' (*MKT*: I 69).	Table of squares preserved — end of the standard multiplication series?
CBS 19790	Nippur	II/2	'Vokabular' (*MKT*: I 55).	Multiplication tables for 22 30 to 7 12 preserved; originally contained tables from 30 to 7? over seven columns (cf. *MKT*: II Taf. 63).
Erm 15063	unknown	IV	Three-line lexical extract.	Metrological list (capacity measures).
FLP 1283	Uruk	IV	Proverb.	Calculation (pair of reciprocals).
Ist O 3833 + Ist O 3848	Kish	—	—	'Nicht mathem.(?) Text-abschnitt' (*MKT*: I 12), followed by a reciprocal table and multiplications for 50 to 15.
Ist Si 441	Sippar	IV	Opening of model letter.	'Carelessly written numbers' (Kraus 1959–62: 18).
MDP 27 61	Susa	IV	Teacher's and student's copy of a two-line extract from a lexical text.	3 times table with numerous scribal errors.[51]
PTS 259	unknown	—	—	2 30 times table, followed by practice list of names (*MCT*: 23 n. 83g).
UM 29-16-90	Nippur	II/2	'Remains of a vocabulary' (*MCT*: 29)	Reciprocal table and multiplications for 50 to 20.

Veldhuis (1997: §2.4.1) has examined nearly 800 Type II tablets from Nippur, numerically analysing the contents of their obverses and reverses. He has convincingly demonstrated a curricular order for a 'Phase 1' of scribal education at Nippur, in which metro-mathematical lists are taught in 'cluster 3: advanced lists' (Veldhuis 1997: §2.4.1.5). Although no problem texts are known from Nippur, mathematics must have been taught as something more than a list tradition: how else are we to explain the square tablets containing mathematical rough work? Veldhuis (1997: §2.4.2) considers 'Phase 2' of the Nippur curriculum to have consisted entirely of Sumerian literature but, in my opinion, we should locate mathematics in this phase

50 This does not preclude the possibility that some of those labelled 'mathematical' in the published catalogues may turn out to be problems.

51 Neugebauer (*MKT*: III 50) dated it to Ur III on the basis of the script, but the preface and introduction to *TMS* states that the *MDP* 27 and *TMS* tablets, which have been firmly dated to the late OB period (Tanret 1986), were from the same locus.

too. Just as Civil's physical typology enabled table texts to be assigned a place in 'Phase 1' of the Nippur curriculum, a similar comprehensive analysis of literary tablets would help enormously in comparing formal characteristics of the remaining mathematical corpus with other school material. While this is lacking, it suffices to note that, like mathematics, literary compositions and extracts are found most commonly on single- and double-columned tablets, but very rarely on Type II and IV tablets (Veldhuis 1997: §2.4.2).

10.6 THE AIMS OF OB MATHEMATICS EDUCATION

We return finally to the issue first raised at the end of §1.3.6: why was mathematics taught? Should we think of it merely as a practical training for future overseers, accountants and surveyors? This is the picture drawn by the widely cited 'Examenstext A', line 27. It asks:

a-rá igi igi-ba igi-[gub-ba] níg-ka$_9$ kuru$_7$ šid-dù ga-lá á-dù-à dù-a-bi dù-a ḫa-la ḫa-la-bi a-šà si-ge-dè ì-zu-ù

a-ra-a i-gi i-gi-ba-a i-gi-gu-ba-a nik-kas-si pi-qit-tam pa-qa-a-da a-de-e ka-la e-pe-šú zi-it-tam za-a-zu eq-la pa-la-ku ti-de-e

Do you know multiplication, reciprocals, coefficients, balancing of accounts, administrative accounting, how to make all kinds of pay allotments, divide property and delimit shares of fields? (Sjöberg 1975: 144)

Although this is a post-OB exemplar of a genre concerned more with presenting an idealized scribal heritage than with accurately recording educational practice,[52] all the subjects it mentions belong to OB mathematics — and indeed to the school texts of the third millennium too (see §10.1). Yet it does not present the whole picture, for which we must turn to the evidence of the mathematical texts themselves.

Once again we take the topic of grain-piles (§7.3) as a starting point. In the first sixteen problems of BM 96954 + ... the measurements of the grain-pile remain the same, while each parameter is calculated in turn. The first few problems are missing, but judging from other texts we would expect them to be on finding the length, then the width, height etc. (cf. §A.3). The first preserved problem concerns finding the volume of the top half of the pile. One could imagine how such techniques might be useful to a surveyor making the first estimate of the capacity of a grain-pile after harvest (cf. §9.7). However, then things start to get complicated. The remaining problems give data such as the sum of the length and top, or the difference between the length and the thickness, or even the statement that the width is equal to half of the length plus 1. It is hardly likely that an agricultural overseer would ever find himself needing to solve this sort of a problem in the course of a working day.

Similarly, although the mathematical grain-pile is a realistic shape, even simply calculating its volume involves some rather sophisticated three-dimensional geometry, at the cutting edge of OB mathematics as we know it (cf. §7.3.2).[53] Further, it appears that at some point the scenario was further refined to enable mathematically more elegant solutions to be used in *TMS* 14 (§10.3 above). In both sets of problems the pile is 10 nindan (60 m) long and 36–48 cubits

[52] Although surviving exemplars all date to the Neo-Assyrian period, their content — as this extract illustrates nicely — mirrors the concerns of the schools depicted in the OB 'eduba texts'. Cf. the introduction to this chapter and notes there.

[53] The solutions to some of the more advanced problems of BM 96954 + ... were beyond the comprehension of the ancient copyist (cf. §A.3).

(18–24 m) high. It is difficult to imagine how a grain-pile this big could ever be constructed, let alone measured with a stick (cf. §9.7).

The accurate mathematical modelling of the real world was not a priority of OB mathematics; rather it was concerned with approximations to it that were both *good enough* and mathematically pleasing. Høyrup (1994: 24) describes it as follows:

> They would [...] calculate as we do, knowing that their results were not precise but apparently without any definite idea about the nature and size of the errors. Presumably, they would see no decisive difference between the imprecision of manpower calculations and those of area determinations.

This is apparent, for instance, in the eight problems about right-angled triangles on BM 96957 + VAT 6598 (§A.4). Each of the three methods used produces a different value for the diagonal of a 2 × 8 cubit rectangle. Where the length of the diagonal constitutes part of the data given at the beginning of the problem, it is not the most accurate value which is given, but that which produces an integer answer with the method of solution used.

Clearly, then, one cannot draw a clear distinction between 'pure' and 'applied' OB mathematics, let alone define 'practical' mathematics purely by its terminology (*pace* Nemet-Nejat 1988: 289). Such a dichotomy is inappropriate to OB mathematics, as it functions at two levels. It was undoubtedly taught primarily as a practical skill, with the emphasis on knowing how to use mathematical techniques and tables in appropriate situations,[54] and to document the results appropriately. On the other hand, many of those methods no longer had real-life applications.[55] For instance, although the coefficients for bricks and basic geometry were still used by working scribes (§9.3–4), work-rates were no longer part of administrative practice (§9.1.3) — yet all three types of coefficients were listed and used in OB scribal education.

However, with such a dearth of sources from the second half of the third millennium it is as yet impossible to determine exactly when Mesopotamian mathematics evolved into the sophisticated, multi-faceted manifestation of the OB period.

10.7 SUMMARY

Many of the characteristic features of Old Babylonian mathematics were already discernible in the mid-third millennium, as Powell (1976*a, b*) and Whiting (1984) have convincingly demonstrated. Further, each of the text types known from OB was already in use, at least in prototype form (see Table 10.5 below). The only one for which direct evidence is lacking is the coefficient list, but coefficients themselves are firmly attested for Ur III.

The fact that many of the OB coefficients have unequivocal Ur III antecedents suggests that the tail end of the third millennium saw the diffusion of mathematics across the Near East, as centrally trained numerate bureaucrats were dispersed across the Ur III empire. That is not to say that there was no second millennium transmission of mathematics or that it ceased to develop. On the contrary, it seems that, without a formal network for the dissemination of

[54] Cf. Høyrup (1994: 24): 'The technical table constituted the nexus between mathematical computation and administrative and engineering reality. Mathematics was taught in schools because scribes should be able to calculate the areas of fields and the volume of canals to be dug out and siege ramps to be built (and, not least, the manpower needed for that).'

[55] Some of the more sophisticated geometry, in particular techniques utilizing irregular coefficients, had clearly never been practically motivated. Their technical names often derive from realia by means of simple visual analogy, but otherwise the mathematics of 'bows' and 'barges' are no more useful than the geometry of 'moons' (semicircles) (cf. §3). Nevertheless, the entire repertoire of OB two-dimensional geometry is derivable from squares, triangles and circles, all of which *were* utilized in the practical mathematics of the third millennium.

scribal knowledge, local traditions and variants grew up, ranging from the choice of tablet-types, and the use of technical terms and the grammatical forms in which they were couched, to mathematical refinements of problems and their solutions, and the adoption of new values for coefficients. Nevertheless, the shared character of this mathematics — its subject matter and style, and the documents in which it was written — persisted. With such a paucity of Ur III mathematical material at our disposal, though, it is as yet impossible to ascertain the point at which mathematics began to serve more than immediately practical purposes.

Table 10.5: The chronological spread of mathematical texts

Period	*Uruk III-IV*	*ED/Ebla*	*Akkadian*	*Ur III*	*OB*	*post-OB*
Problems		•	•		•	•
'Catalogues'			(•)		•	
Calculations			•	•	•	
Diagrams		•	•		•	
Tables		•		•	•	•
Coefficient lists				(•)	•	•
Model documents	•		•	•	•	•

Each type of mathematical text served a separate pedagogical function. The tables were written out while committing to memory metrological and arithmetical relationships (in 'Phase 1' of the Nippur curriculum). Calculations were carried out, in formal layouts, on small square or round IM.ŠU tablets in the course of solving problems from 'textbooks' consisting of a series of (often minimally different) problems and their numerical answers. Perhaps five or more students were set similar problems at once, with parameters chosen from mathematical 'catalogues'. Model solutions, in the form of quasi-algorithmic instructions, were not only didactically similar to other types of educational text, but were also intrinsic to the very way mathematics was conceived. Although many ostensibly practical scenarios were used as a pretext for setting non-utilitarian problems, they had little concern with accurate mathematical modelling. Although the bottom line for OB education must have been to produce literate and numerate scribes, those students were also instilled with the aesthetic pleasure of mathematics for its own sake.

11. CONCLUSIONS

This final chapter draws together the various conclusions reached in this study. They concern both the mathematical techniques of the OB and Ur III periods and the place of mathematics within the broader history and culture of Mesopotamia. Generally speaking, the conclusions are drawn from three sources: the study of coefficient lists which makes up §§2–8 of this work; the new texts and hand-copies from the British Museum presented in the Appendices (§§A.1–5); and the mathematical and non-mathematical material already available which is referred to throughout.

For ease of reference, and at the risk of some repetition, the conclusions summarised here are divided into four sections:

- how the coefficient lists, both OB and LB, are related to each other, and what they can tell us about Mesopotamian mathematical techniques;
- how the previously unpublished mathematical texts presented here contribute to our understanding of OB maths and its cultural context;
- how mathematics fits into Mesopotamian history, with emphasis on scribal education and administrative practices;
- the practical implications of this study for other historians of mathematics, Assyriologists and archaeologists.

11.1 COEFFICIENTS AND COEFFICIENT LISTS

11.1.1 The lists

The documents with which mathematics was taught and learnt in ancient Mesopotamia can usefully be divided into six types of text, each of which served a specific educational purpose (see §11.3 below). Problems and mathematical tables have long been treated separately from each other, while 'catalogues', calculations and diagrams, and model documents are presented as individual categories for the first time in this study. But it is the coefficient lists which are the primary focus of this work. Coefficient lists are reference documents, but differ from other sorts of mathematical tables in that they list not long series of numerical relationships (e.g. squares, reciprocals, etc.) but thematically grouped numerical constants (coefficients), each of which is given a name or short description. Unless a particular coefficient is used in a problem, these descriptions — which often vary from list to list — are the sole means of identifying their mathematical function and absolute numerical value.

Eight OB coefficient lists have been published, two of which are from Shaduppum, two from the Larsa area, and one each from Sippar, Susa, and Ur, while one list is unprovenanced. While they all share the same basic tabular structure, with numerical values listed to the left of the descriptions, each has very individual characteristics. Not only do the lengths of the lists range from 3 to 77 coefficients, but the way in which those coefficients are described varies from list to list, both in the names used and in the way in which the names are written. Further, even where the same thematic clusterings of coefficients are evident, they may not be in the same sequence, or follow the same internal arrangement within those groupings. Occasionally coefficients are organized acrographically or in numerical order. Two lists, F and H, are compilations, whose format varies from subject to subject, while C and E both contain copying errors. There is no evidence for a common 'Urtext' from which these lists derive, although clearly they all belong to a common mathematical culture.

Also presented here are two fragments of coefficient lists from LB Babylon (see §11.2.1), and a re-examination of the two post-OB lists already published (see §11.3).

11.1.2 The coefficients

In Mesopotamian mathematics a 'coefficient' IGI.GUB/*igigubbûm* is a numerical constant considered intrinsic to a particular type of problem and independent of the parameters chosen in setting individual problems of that type. Thus coefficients may be recognized as numbers introduced where appropriate into the solution of a problem but which are not mentioned in the statement. A coefficient may be designated by the word IGI.GUB/*igigubbûm*, or given a descriptive name, or both, or neither. They are multiplied using the verb *našûm*. Even where a text consists solely of the statement of a problem and its numerical answer the value of the coefficient used to find that answer can often be inferred. Coefficients are sometimes a component of the data given in an illustration to a problem.

There are around 220 coefficients attested in OB mathematics, counting coefficients as separate if they have identical or similar names but different numerical values (except for mutual reciprocals). One hundred and eighty-seven of these are found in lists of coefficients, and 61 are attested in problems and diagrams. Only 22, or 10%, of the known coefficients are found in problems but not in the lists. Many of these occur in more than one document, so that there are over 500 attestations of them altogether, a little under two-thirds of which are in lists. All but 26 of the coefficients have sexagesimally regular numerical values. Fifteen of those exceptions are for complex geometry and two are for metals; the remaining nine are almost certainly erroneous. All the coefficients used in problems and calculations are sexagesimally regular. The mathematical purpose of around 40 coefficients — or rather less than 20% of the whole — remains unclear for the time being. The remainder can be divided roughly into four subject groupings: geometry; quantity surveying; metrology and storage; and metals, which are dealt with in §§11.1.3–8 below.[1]

Coefficients were not an innovation of the OB period. Three Ur III model documents implicitly attest six different quantity surveying coefficients, four of which have identical OB counterparts. Coefficients continued to play an important role in mathematics after the early second millennium too, as four later lists attest. Of the 131 more or less complete entries in those lists, 55 (42%) are directly descended from OB coefficients.[2] Even where their numerical values are no longer preserved, the coefficients often keep the same names and continue to be grouped in similar ways.

11.1.3 Geometry

The geometrical coefficients are perhaps the most fully understood of OB coefficients. Constants for circles, semicircles and triangles are a feature of all the OB lists, and were used in

[1] For numerical data, see §§2.1.6, 2.3.2.

[2] Geometry: 1; quantity surveying: 24; storage and metrology: 17; metals: 7; astronomy: 6. The remainder are for the most part readable but without established mathematical contexts. A further 30 entries in lists B and K survive only as traces of numerals.

the construction of architectural features such as wells and pillars from at least the Ur III period onwards. Triangles were considered to be either (almost) right or equilateral; one of the diagrams on BM 15285 shows that figures with one or more curved sides were also classed as triangles. The words for circle, rectangle, and square were also the names for their defining components: the circumference, diagonal, and side respectively.

More complicated figures, especially those inscribed in circles or composed of circles, are attested most fully in list D and in the set of geometrical exercises BM 15285. Two distinct shapes are given names of bows: one describes a rhombus inscribed in a third of a circle, while the other refers to a B-like figure based on two touching semicircles. 'Grain(-field)s' and 'barges' are different designations of the same petal-shaped figure, formed from two quarter-circle arcs. Regular 5-, 6-, and 7-sided polygons are constructed from triangle geometry and do not presuppose a knowledge of angles.

The *apsamikkum* remains the most intriguing of the OB geometrical figures. The illustrations on BM 15285 show that the (GÁN) GEŠTÚ ZÀ.MÍ is a concave square, which should be identified with the *apsamikkum* described by the coefficients in most of the lists; a different figure based on a rectangle, but preserved only in one rather damaged illustration, is called GÁN GIŠ.ZÀ.MÍ. The 'three-sided *apsamikkum*' is a concave triangle. The literal meaning of the word remains obscure, but it now seems unlikely that it is a reference to the 'cow's nose' on a lyre, as the animals depicted there are bulls, GU_4. Many geometrical figures are named after realia, but do not purport to be life-like representations of those objects.

All geometrical coefficients refer to the area of a figure with respect to its defining component, unless the name explicitly states otherwise.

11.1.4 Bricks and brickworks

Ten types of bricks were used for mathematical calculations in the Ur III and OB periods: six sizes of square brick ranging from $^1/_4$ to 1 cubit in length, and two each of associated half- and two-thirds bricks. Baked bricks of all but two of these sizes are known from the same periods; further it seems unlikely that other measurements were used, at least for baked bricks. Bricks continued to be manufactured in these same proportions, and approximately the same sizes, throughout Babylonia and Assyria right until the second half of the first millennium BC. Indeed, four of the OB brick sizes are attested in three of the four known post-OB coefficient lists.

The metrology of bricks has been understood for half a century now, but text L, an LB coefficient list newly published here, adds further detail. It now transpires that the only previously known OB coefficient for the density of bricks refers to their weight when oven baked; sun-dried bricks are $1^1/_5$ times as heavy, and a full $1^1/_2$ times heavier when freshly made and still full of water.

Brick metrology is one of the few mathematical subjects for which we have Ur III school exercises, in the form of practice quantity surveys: two of them give the dimensions of various brick-piles and calculate the number of bricks in them, using brickage coefficients already well known from OB coefficient lists and problems. A third text is in a similar format, yet concerns not brick-piles but mortared walls. The composition of the walls is usually considered to be $^5/_6$ brick and $^1/_6$ mortar, as shown conclusively in the new set of OB problems on BM 96957 + VAT 6598. However, in the Ur III model document YBC 9819 a scribal miscalculation produces the result that $^6/_5$ of the wall is made of bricks! The (correct) proportions suggest that there was a 1 finger layer of mortar for every 5 finger course of bricks, and this is indeed the depth to which plaster or mortar is spread in two further OB problems (at a rate of 1 sar_a, or

0;02 sar_V, a day). Two Ur III administrative surveys for brickworks are known, with coefficients for brick-piles rather than for mortared walls.

BM 96957 + VAT 6598 and one OB coefficient also state that the floor plan of a house is considered to be $^1/_3$ wall. This is a rather lower estimate than the 45%–50% given by both ancient and modern plans of OB and Ur III domestic dwellings. It may be that this rule of thumb assumes that the external walls of the house, shared with the neighbouring plot, are already standing, as according to those same house plans the internal walls are 25%–30% of the internal area as a whole; this second interpretation fits better with the established legal meaning of the phrase É.DÙ.A.

Although bricks and brickworks were clearly a widespread subject in Ur III and OB maths, there is no common group of brick coefficients found in every list, nor a standardized set of names for the different brick sizes.

11.1.5 Making bricks and carrying goods

The set of three work-rate coefficients for making bricks, known from two OB problems, can now also be identified in list D. The harmonic mean of these rates is called *neqrebum* 'combined coefficient' and gives the output of one man who undertakes all three tasks in one day. This 'combined coefficient' method is used to solve two further OB problems on performing different tasks; the proportion of the day spent on each is called IGI.TE.EN/*igitennûm*. The daily rates for destroying walls and building brick-piles can thus be elucidated for the first time; both have identical precursors in Ur III administrative texts.

The mathematics of carrying earth and other materials is based on two constants, which are the same for all goods carried, including bricks: the total distance travelled in a day, and the weight of one load of goods. The volume of a load will thus vary according to its density. The corrupt carrying coefficients from list C are shown to result from a simple copying error, and one rather broken OB problem turns out to be about carrying home sheaves of grain laid out equidistantly in a line 1 danna long. Two corrupt brick carriage coefficients result from sexagesimal place errors in their calculation from the corresponding brickages, but one rather garbled problem on carrying bricks remains obscure.

One Ur III document lists the labour needed for various carrying and construction tasks in some detail, although it is not always clear whether the work-rates used are the same as the OB coefficients. The outputs deducible from an OB inventory of bricks manufactured by different teams are comparable to the OB coefficient for brick-making.

11.1.6 Earthworks, waterworks and other work-rates

It is relatively well known that the OB coefficients for building earth walls and for performing various earth-moving activities are also frequently attested, with the same values, in Ur III administrative records. In OB maths the same work-rates come to be used for any earthwork, not necessarily connected with canals. The 'combined coefficient' method is used in several OB problems on digging canals, as the work proceeds at different rates according to the depth of the excavation. In the Ur III documents several different phrases are used to describe a single task: (re-)digging a canal and piling up the earth on the banks to form or repair the associated levees. The additional activity of reinforcing those levees with reeds may sometimes be implied in those records, but is always calculated separately in school documents.

According to OB problems, dams are composed of $^1/_3$ earth and $^2/_3$ plant material; on the other hand Ur III administrative texts record that dykes were topped with $^1/_5$ to $^1/_3$ vegetable matter. According to one Ur III exercise, which also represents the only known instance of a hoeing rate in a school text, reeds comprise $^1/_4$ of a levee.

Irrigation is the topic of three OB problems, all of which assume that $^4/_5$ of the water in a watercourse is available for use. Four coefficients from the late list B are closely related.

Two isolated OB coefficients — on cutting reeds and weaving — are the sole remnants of further large areas of concern of Ur III bureaucracy: agricultural work-rates, measured in area units a day, and manufacturing rates for objects such as pots, tools and garments, usually expressed in the number of days needed to complete one item. The agricultural work-rates for ploughing and harrowing, although they have not yet been found in OB school mathematics, were none the less used in contemporary administrative calculations. It may yet turn out that other work-rates were also used by working scribes after the Ur III period.

11.1.7 Storage and metrology

Coefficients for storage and for conversion between metrological units are not explicitly attested in the Ur III period, although they must have been used in auxiliary calculations of the type which are so rarely found. Similarly, metrological conversions are relatively rarely explicitly signalled in OB problems. Linear measurements in nindan are transformed into cubits to facilitate the computation of volumes; and volume measures are themselves expressed in capacity measure by means of a range of coefficients whose exact import remains obscure. The terminology of these coefficients is not consistent from list to list, and there are not yet enough clear examples of their use in problem texts to enable them to be untangled. Many of them appear in the LB list J, with a set of coefficients which appear to relate to different types of container.

The new set of OB problems BM 96954 + BM 102366 + SÉ 93 shows that grain-piles were conventionally depicted as symmetric pyramids on a rectangular base, with an elongated apex. *TMS* 21 considers a special case in which the width is twice the height. In both texts, the grain-pile is taken as a pretext for the setting of unprecedentedly complex problems. Agricultural custom in both Ur III and OB seems to have been to store newly threshed grain in such piles before distribution. The height of the pile was measured by vertically inserting a stick into the apex as a means of estimating the volume; the sophisticated OB mathematics of grain-piles must have had its origins in such practices.

11.1.8 Metals and other topics

There are many OB coefficients which remain open to interpretation, as they still lack firm mathematical contexts. A large group of coefficients for molten metals forms an arithmetic progression with constant difference 12 but, without problems in which at least one of them appears, they remain obscure. On the other hand, *ruqqum* coefficients turn out to be an index of metallic density and ductility: they give the maximum area to which 1 shekel weight can be hammered out. Apparently neither type of metal coefficient is attested in Ur III economic documents.

Two OB coefficients record the constant difference in the visibility of the moon from night to night of an equinoctial month; this forms part of the astronomical tradition embodied later in *Enūma Anu Ellil* and MUL.APIN; related coefficients also occur in the post-OB list B.

Four difficult coefficients found in highly abbreviated or broken OB problems, and several dozen which are attested in the lists alone, cannot be interpreted until such time as clear mathematical contexts are found for them in problem texts.

11.2 EVIDENCE FROM THE NEW TEXTS

This section highlights the significant contributions to our understanding of OB mathematics made by the new texts and copies presented in §§A.1–5 of this study.

11.2.1 BM 36776 and BM 37096

These two small fragments of LB coefficient lists (texts K and L, §A.1) provide further vital evidence for an 'unbroken tradition' (Friberg *et al.* 1990: 548) of mathematics from OB to LB. Not only are their layouts identical to those of the OB lists, but they also share similar concerns: the preserved entries concern metals and bricks, both common OB topics. The entries are listed in a similar order to the OB lists, and the brick coefficients in text L concern four sizes of bricks common in OB. Indeed, the full listing of brick weights at different stages in the drying process elucidates several previously obscure OB coefficients.

11.2.2 BM 15285

The new hand-copy of this large geometrical text (§A.2) reveals, in a way that previously published photographs have not, the technical drawing skills of its ancient author. The scale of the geometrical shapes is remarkably accurate, given the highly convex surface of the reverse of the tablet, and the lack of a concept of angle. Auxiliary lines drawn to aid the author in the construction of the diagrams are still visible (*pace* Friberg 1987–90: 565), as are the marks left by the fixed compass used to draw arcs onto the tablet. It is also possible to distinguish which lines have been drawn freehand.

11.2.3 BM 96954 + BM 102366 + SÉ 93

This three-column tablet contains a compilation of OB problem texts (§A.3). The subject of the first extant sections — and of *TMS* 21 and a number of coefficients — is a grain-pile in the shape of a rectangular pyramid with an elongated apex. Comparison of the two sets of problems, and the lists, reveals four ways in which the coefficients for a single mathematical topic may vary from text to text:

- the numerical value of the coefficient;
- the name of the coefficient;
- the unit in which the coefficient is expressed;
- the mathematical 'formula' which the coefficient represents.

We also get an interesting insight into the mathematical understanding of the scribe of this collection: some of the solutions to the problems are 'simplified' by (incorrectly) omitting stages which give identical numerical results to the immediately preceding step. This is strong evidence both that the text was copied from an earlier exemplar, and that not all scribes of mathematical texts can be considered as 'mathematicians': their mathematical understanding is not necessarily deep enough to preserve the meaning of the text in transmission.

The second and third sets of problems on this tablet concern unclear three-dimensional objects, called *berūtum* and *mazarum*. They also use coefficients, perhaps for a circle.

11.2.4 BM 96957 + VAT 6598

The first set of problems illuminates previously obscure coefficients, by showing that a brick wall was considered to be composed of bricks and mortar in the ratio 5:1. Houses were assumed to have a ground plan composed of floor and wall in the ratio 2:1. These problems also elucidate the meaning of YBC 9819, an Ur III model document. It is no longer necessary to hypothesize the existence of further standard brick sizes, if it is assumed that the author of the text misapplied these same brick wall coefficients.

The second set of problems is perhaps even more important for our understanding of OB mathematics. It presents three different methods of solution for a single set of problems on right triangles. This is a unique and remarkable occurrence in itself, but at least one of those problems is stated to be 'insoluble'. This unprecedented comment on an OB problem text confirms the quasi-algorithmic thinking which underlies Mesopotamian mathematics, as the answer is in fact obtainable by non-algorithmic methods.

Equally crucially, this is the only known OB problem text which states explicitly that it has been checked against an original. (There is indirect evidence for copying from other mathematical texts.) The colophon names the copyist of this text — and thereby also all the other Sippar texts written in the same hand: BM 85194, BM 85196, BM 85200 + VAT 6599, BM 85210, BM 96954 + BM 102366 + SÉ 93, and VAT 6597. All other OB problem texts are unsigned.

11.2.5 Mathematical rough work from Ur and elsewhere

A group of 46 round tablets from OB Ur, which are now in the British Museum, turn out to have computations on their reverses, as well as the known, published Sumerian proverbs on their obverses. This discovery has led to the definition of calculations as a genre of mathematical texts, the identification of a further 60 or so previously published examples of this text type, and the edition in this work of another seven such tablets. They show that, in OB Ur at least, scribal students were taught to lay out their calculations and other intermediate work such as diagrams on a separate tablet from the problem text or model document on which they were working. This explains why calculations are rarely found on economic and administrative texts written by working scribes.

Standard formats were used, some with horizontal or vertical rulings, for each type of calculation. Multiplications, squares and reciprocals, as well some as yet obscure calculations, were carried out. Two groups of around half a dozen multiplications can be identified, which use different parameters but the same coefficients. This suggests that several students were set similar problems to solve simultaneously. One group may concern the grain capacities of rectangular containers; the other may be about finding the time needed, in months and days, to construct various earthworks. One calculation in the first group apparently uses the wrong coefficient, but still comes up with the right answer in the end!

Similar standard layouts were also used in Nippur, where, however, more or less square tablets were used instead of round ones for mathematical rough work. Unprovenanced examples of calculations and diagrams are found on both round and square tablets.

11.3 THE HISTORICAL DEVELOPMENT OF MESOPOTAMIAN MATHEMATICS

Metrological and mathematical exercises are known to date back to at least ED III, but the widespread dissemination of mathematical knowledge was aided enormously by Shulgi's bureaucratic reforms of the early Ur III period. To cope with the administrative demands of a rapidly expanding empire, many aspects of scribal education were improved and systematized, a vital part of which were the numerical and metrological skills needed for accountancy and surveying. In this way a core body of mathematical techniques was transmitted in a more or less uniform manner throughout Sumer and Akkad, and at least as far as Susa, the Diyala region and Ashur. Of particular relevance to this study was the practice of keeping yearly accounts of labouring teams. Records were kept of the team's output over the year, which was then compared to their theoretical, ideal output, calculated on the basis of work-rate coefficients: the volume of earth shifted, or reeds carried, or bricks made, that could optimally be expected of a single worker in one day. Teams which had not fulfilled their work quotas were expected to catch up the following year.

There is little direct evidence for methods of scribal mathematical education at the end of the third millennium. Model documents of all kinds were certainly used to teach students the format of the texts they would be writing as working scribes. Tables of important numerical relationships were probably constructed. Calculations in the sexagesimal place system were performed on separate tablets and the results then inserted into the model documents. Work-rate coefficients, and those for standardized brick sizes, were learned: both were necessary for life as a quantity surveyor. Mathematical procedures seem to have been taught orally; at any rate, no texts have yet been discovered from this period which give instructions for solving mathematical problems.

Once the Ur III empire, and its concomitant bureaucratic network, began to disintegrate mathematical knowledge was no longer systematically disseminated throughout the region. Nevertheless, the techniques were kept alive, and even developed, by the scribal schools of the successor states. Because there was little contact on an official level, mathematics began to develop independent strands in different places. Within the Eshnunna kingdom, for instance, there was a shift in terminological usage not found in other areas. Susa too explicitly distinguished some of its own mathematical methods from those of the 'Akkadians'. Despite the procedural differences which are manifested, there remained a common concern with particular mathematical themes and topics. Coefficients continued to be an integral part of both Old Babylonian mathematics and administration, and are attested in all areas which have revealed mathematical activity at this period.[3] The subject matters of the Ur III work-rates and the OB coefficients are not identical: some topics, such as agricultural activities, were dropped from OB scribal studies, while others, such as carrying goods, were treated in much more detail than is generally found in the earlier administrative records. Nevertheless, where continuity can be proven, whether into OB education (e.g. canal construction) or into OB administrative practice (e.g. ploughing and harrowing), the numerical values of the coefficients are in general identical to those found in the Ur III documents.

Scribal schooling in mathematics seems to have continued along more or less the same lines as in the Ur III period, but with the addition of problem texts to teach procedure. Arithmetic was learnt by rote through mathematical tables, during 'Phase 1' of schooling at Nippur. Exercises from the problem texts were calculated on separate 'rough-work' tablets, and model documents were written to practise the correct layout of working documents. 'Catalogues' may have provided scribal teachers with extra exercises for students; coefficients were organized into lists arranged thematically. The subject matter of OB maths, though, seems to have been more wide-ranging and at times more sophisticated and abstract than in the Ur III period. However, one cannot distinguish between a 'pure' or 'applied' mathematics: many topics of practical concern were the starting point for an exploration of deeper mathematical properties. OB maths was not concerned with accurate mathematical modelling of the figures it purported to represent; even if the procedure was a correct one, realistic numbers were not necessarily

[3] Too little OB mathematics has been discovered yet at Mari to tell whether it goes back to Ur III influence, or was a scribal import into Mari from Eshnunna during OB times.

inserted into that procedure. Simple values, which often gave integer answers, enabled the student to concentrate on the method without getting snarled up in difficult arithmetic.

It also appears that the methods used to teach mathematics and Sumerian were in some ways similar to each other, being in Vanstiphout's terminology (1979: 125) practical, exemplary, contrastive, and progressive. That different school subjects are found on the same tablets indicates that scribal students studied a whole range of topics and did not, as a rule, specialize in one particular area or another.

There is still only sporadic documentation of mathematical activity after the early second millennium. The present study, though, adds to the small but growing body of evidence for a connection between OB and later mathematics. List B, still of uncertain date, is shown to be part of a compendious school tablet, on which the scribal student wrote a number of different 'lessons': the musical section which has attracted so much scholarly attention is unrelated to the coefficient list, but forms a separate subject on the same tablet. Eighteen of the extant coefficients in list B belong to the OB tradition. Nine of the storage coefficients of the 'esoteric' list J also derive from OB antecedents; after collation, the remainder are shown to be numerically related to each other. The previously unpublished fragments K and L provide the strongest links with the second millennium lists, with the same layout, subject matter, order of entries and numerical values as their OB precursors. In short, first millennium mathematics is not necessarily corrupt or exclusively esoteric, but shares many of the concerns, techniques and conventions of OB mathematics.

11.4 COEFFICIENTS AND THE STUDY OF ANCIENT MESOPOTAMIA

Some mathematical coefficients known from OB — those for work-rates and brick sizes in particular — were already used by the Ur III administrators in accounting for large-scale construction, irrigation and agricultural activities. It seems reasonable, then, that the OB coefficients with equivalent Ur III work-rates can give us a rough approximation of the amount of labour, time and expense involved in constructing anything from a pot to a palace. But how far can we take this? If the methods taught for brick metrology and manpower calculations were firmly based on *realia* we might be able to assume tentatively that other aspects of quantity surveying were too. Thus we would be able to take into account the time and cost of manufacturing the bricks for, and transporting the raw materials to, our putative palace.

However, although we can almost certainly take the *coefficients* as an accurate reflection of reality (or at least of reality as perceived by Ur III bureaucrats), other numerical data from mathematical texts should be treated with much more caution. In OB maths the method was paramount, not the accuracy or appropriateness of the measurements of the object under discussion. It is important to remember when using the more practical texts as evidence of everyday custom that they do contain useful, albeit sometimes simplified, information about the sorts of shapes such constructions might take, but not about their size.

The subject of coefficient lists in ancient Mesopotamia is far from being exhausted. As indicated throughout this book, further research is needed, not only into the mathematical techniques of the late third and early second millennium, but also into the educational and administrative contexts in which they were taught and used. Only thereby will we develop a clear picture of the role of mathematics in general, and coefficients in particular, in the working lives of scribes and administrators, and thus their broader impact on ancient Mesopotamian society.

APPENDIX 1: TRANSLITERATIONS OF THE COEFFICIENT LISTS

This first appendix serves a double purpose: it gives a transliteration of the full texts of all twelve coefficient lists discussed here, and also acts as an index for the individual coefficients by noting the section of this work in which each coefficient is discussed. Hand-copies of the two previously unpublished LB fragments, texts K and L, also appear here. Throughout Appendix 1 new readings are marked with a dot •.

Four of the remaining appendices comprise presentations of cuneiform mathematical texts. BM 15285 and BM 96957 + VAT 6598 + SÉ 93 have been published elsewhere; the remaining texts are presented here for the first time. All copies are larger than the originals, except those of BM 15285, which are approximately the same size as the tablets themselves. The last of the six appendices summarizes information on the sizes of the Mesopotamian baked bricks now in the British Museum and the Ashmolean Museum, Oxford.

A.1.1 LIST A

				INDEX
Obv	30 *ù* 1	IGI.GUB SAG.DÙ		§3.3
	5 *ù* 6	IGI.GUB KA.KÉŠ		§3.1
	30	IGI.GUB *ṣí-li-ip-tum*		§3.4
	10 15 36	IGI.GUB GÁN U$_4$.SAKAR		§3.2
	12 30	IGI.GUB GÁN GIŠ.PAN	•	§3.5
	26 40	IGI.GUB GEŠTÚ ZÀ.MÍ		§3.7
	⸢45⸣	IGI.GUB GÁN MÁ.GUR$_8$	•	§3.5
	3 45	IGI.GUB IM.DÙ.A		§6.1
	56	IGI.GUB U+KUL X		§8.5
	20	IGI.GUB É.DÙ.A		§4.5
	6 40	IGI.GUB É.KIŠIB	•	§7.2
	15	IGI.GUB ḪA.LA.X		§8.5
	15	IGI.GUB ESÍR.HA$_5$.RÁ		§4.6
	16	IGI.GUB ESÍR		§4.6
	3 45	IGI.GUB NU$^{!}$.LUḪ.ḪA.SAR	•	§8.4
	4	IGI.GUB ŠE.LÚ.SAR	•	§8.4
	18	IGI.GUB IM.LÁ		§4.3, §8.3
	1 12	IGI.GUB URUDU		§8.1
	1 20	IGI.GUB ZABAR		§8.1
	1 20 21	IGI.GUB AN.NA		§8.1
Edge	⸢1⸣ 36	IGI.GUB KÙ.BABBAR		§8.1
Rev	1 48	IGI.GUB KÙ.GI		§8.1
	2 24	IGI.GUB *a-⸢lu-lu-tum⸣*	•	§8.1
	7 12	IGI.GUB SIG$_4$.ANŠE		§4.4
	5 24	IGI.GUB PEŠ SIG$_4$.AL.ÙR.RA		§4.1

	2! 42!	IGI.GUB SIG$_4$.AL.ÙR.RA		§4.1
	6	IGI.GUB ‹ÉŠ›.GÀR LÚ.1.KAM		§6.6
	45	IGI.GUB A.RÁ		§5.2
	4 30	IGI.GUB SIG$_4$ ÍL.ÍL		§5.3
	3 22 30	IGI.GUB SIG$_4$ IN.⌜x⌝		§5.3
	1 41 15	IGI.GUB SIG$_4$ ÍB.⌜SI$_8$⌝	•	§5.3
	3 45	IGI.GUB SAḪAR KÙ ŠU BA.AN.TE		§8.5
	16	IGI.GUB KÙ SAḪAR ‹ŠU› BA.AN.TE		§8.5
	11 30	IGI.GUB SAḪAR KI.LÁ	•	§5.6
	1 20	IGI.GUB SAḪAR ÍL.ÍL SIG$_4$.DÙ	•	§5.6
	11 06 40	IGI.GUB SAḪAR ÍL.ÍL IM.DÙ.A	•	§5.6
	4 48	IGI.GUB É.KIŠIB IN.NU.⌜DA⌝		§7.2
	4 58	IGI.GUB GI.NÍGIN.NA		§8.4
	6	IGI.GUB GI.SA		§7.4
	7 12	IGI.GUB GI.KILIB.BI!?	•	§7.4
	6	IGI.GUB IZ.ZI		§4.5
	1 21	IGI.GUB *ṣí-li-ip-tum*		§3.4
	⌜x⌝	IGI.GUB GIŠ.NÁ GIŠ.GU.ZA		§8.5
Edge	[1]	IGI.GUB ḪA.LA		§8.4
	[1]	IGI.GUB GI ḪAŠ.⌜A⌝		§8.4
	⌜x⌝	IGI.GUB ŠU.NIGIN$_2$.NA		§8.5
Edge	⌜20⌝	GIŠ.ḪAB 1 IGI.GUB *ma-li ṣí-‹il›-pe-tum*	•	§8.4; §8.5
	30	NU.KAR [x] DÈ? 48 IGI.GUB SÍG?.ḪI?.A		§8.5; §8.5

A.1.2 LIST B

INDEX

The top of the column is broken.

I	⌜2 4⌝	⌜SA *ti-tur* MURUB$_4$⌝-*tu*$_4$
	4 3	SA *kit-mu*
	3 6	SA *ti-tur i-šar-tu*$_4$
	7 4	SA *pi-tu*$_4$
	4 6	SA *ṣer-du*
	SA *qud-mu-ú*	*ù* SA 5-*šú* 1 5 SA *niš ga-‹ba›-ri*
	SA 3 *uḫ-ri*	*ù* SA 5-*šú* 7 5 SA *še-e-ru*
	SA *ša*-GE$_6$	*ù* SA 4 *uḫ-ri* 2 6 SA *i-šar-tu*$_4$
	SA *qud-mu-ú*	*ù* SA 4 *uḫ-ri* 1 6 SA *šal-šá-tu*$_4$
	SA 3!-*šú* SIG	*ù* SA 3-*šú uḫ-ri* 3 4 SA *em-bu-bu*
	SA *ša*-GE$_6$	*ù* SA 3-*šú uḫ-ri* 2 4 SA 4-*tu*
	SA ^{d}*é-a*-DÙ	*ù* SA *qud-mu-ú* 4 1 SA ŠUB.MURUB$_4$
	SA *qud-mu-ú*	*ù* SA 3-*šú* SIG 1 3 SA GIŠ.ŠUB.BA
	SA MÍN 3-*šú*	*ù* SA *ša*-GE$_6$ 5 2 SA MURUB$_4$-*tu*
	SA *ša*-GE$_6$	*ù* SA ^{d}*é-a*-DÙ 2 4 SA *ti-tur* MURUB$_4$-*tu*$_4$
	SA 4 *uḫ-ri*	*ù* SA 3-*šú* SIG 6 3 ⌜SA⌝ [*kit-mu*]
	SA 3-*šú* SIG	*ù* SA 5-*šú* ⌜3⌝ [5 SA *ti-tur i-šar-tu*$_4$]
	SA 3-*šú uḫ-ri*	*ù* [SA ^{d}*é-a*-DÙ 7 4 SA *pi-tu*$_4$]
	SA ^{d}*é-a*-DÙ	[*ù* SA 4 *uḫ-ri* 4 6 SA *ṣer-du*]

The end of column I and the beginning of column II are missing.

II 3′	[6	GIŠ.MÁ.LÁ]	SIG$_4$	• §4.3, §7.2
	[4] ⌜30⌝	GIŠ.MÁ.LÁ	SIG$_4$.ÁB	• §4.3, §7.2
	[2] 15	GIŠ.MÁ.LÁ	SIG$_4$.AL.ÙR.RA	• §4.3, §7.2
	4 10$^!$	GIŠ.MÁ.LÁ	SIG$_4$.$^2/_3$-*ti*	• §4.3, §7.2
	50	GIŠ.MÁ.LÁ	GI.ŠU.KIN	§7.4
	30	GIŠ.MÁ.LÁ	GI.NÍG.SA.ḪI.A	§7.4
	1 15	GIŠ.MÁ.LÁ	GI.NÍG.SA.ḪI.A ḪA$_5$.RÁ	§7.4
	1 12	GIŠ.MÁ.LÁ	*ḫi-im-ma-ti*	§7.4
	1	IGI.GUB.E	1 GUR	
	50	IGI.GUB.E	2 (PI) 2 BÁN	
	45	GIŠ.«BA».BA.RÍ.GA 1 (PI) 2 BÁN		
	40	GIŠ.BA.RÍ.GA 1 PI		
	30	GIŠ BA.RÍ.GA 2 BÁN		
	20	GIŠ.BA.RÍ.GA 3 BÁN : 20 GIŠ.BÁN		
	12	GIŠ.BÁN.LUGAL		
	13 03 SÌLA : 12 02 SÌLA : 10 ⌜11⌝ SÌLA			
	7 30 $^2/_3$? SÌLA: ⌜6 40⌝ $^1/_2$ SÌLA : 6 $^1/_3$ SÌLA			
	5 15 GÍN : ⌜4⌝ 10 GÍN : 3 45 2$^!$ GÍN			
	[	...] 1 GÍN : 2 $^1/_2$ GÍN	6 X-*šú-nu*	
	[	]DAL GIŠ.‹BA›.RÍ.GA	1 GUR	
	[	]⌜DAL⌝ GIŠ.‹BA›.RÍ.GA	2 (PI) 2 BÁN	
	[	DAL] ⌜GIŠ⌝.‹BA›.RÍ.GA	1 (PI) 2 BÁN	
	[	DAL GIŠ.‹BA›.RÍ].⌜GA⌝	1 PI	
	[		] 3 BÁN	
	[		] 2 BÁN	
	[		GIŠ].⌜BÁN⌝.LUGAL	
	[		] ⌜2⌝ SÌLA	
	[		] SÌLA	
	[		] ⌜X⌝	
	[		] ⌜X⌝	

The end of column II and the beginning of column III are missing.

III 1′	[...]	IGI.GUB.E *ra-ḫi-ṣi*	
	⌜3⌝ 20	*zi-qip* ŠE.GIŠ.Ì	
	2 30	*zi-qip* ŠE.GIŠ.Ì	
	1 20	*gu-ru-un* ŠE.GIŠ.Ì	
	20	KI.MIN	
	1 20	KI.MIN	
	1	KI.MIN	
	7 30	ŠE.GIŠ.Ì	•
	3 45	IGI.DU$_8$.A	§8.2
	12	MÚ *ša* ^{d}30	§8.2
	12	ŠÚ *ša* ^{d}30	§8.2
	40	*u$_4$-mu a-na* GE$_6$ *na-pa-lu*	§8.2
	40	GE$_6$ *a-na u$_4$-mu na-pa-lu*	§8.2
	40	*nap-pal-tu$_4$*	§8.2

19 30	gišNU.ÚR.MA		
17 30	gišNU.ÚR.MA		
15 30	gišNU.ÚR.MA *ši-na-nu*	•	
13 30	gišNU.ÚR.MA *ši-na-nu*	•	
24 05 20	*nu-ud-di-tu*$_4$	•	
3 20	*ši-na-nu*	•	
1 36	A.MEŠ *mi-lu-ti*		§6.5
1 12	A.MEŠ *zi-rí-qum*		§6.5
1 48	A.MEŠ *pat-ti*		§6.5
48	A.MEŠ *pat-ti*		§6.5
9 39	A.MEŠ GIŠ.Á.LÁ		
1 30	A.ŠÀ A.MEŠ *ša-qu-ú*		
18	A.ŠÀ A.MEŠ *ša-qu-ú* KI.MIN		
2 46 40	A.ŠÀ A.MEŠ *ša-qu-ú* KI.MIN		
3 BÁN	ŠE GÚN A GAR ŠA RA	•	
1 (GUR) 1 PI	*ḫa-a-mu* GÚN GIŠ.MAR.GÍD.DA		
4 30	*za-ba-lu* 6 (GUR) IN.NU GÚN KI.[MIN]		
12 30	*za-ba-lu* 1 (GUR) IN.NU GÚN [KI.MIN]		
3 45	*za-ba-lu* 2 46 40 GÚN [...]		
[...]3	*za-ba-lu* 1 PI [...]		

The end of column III, all of columns IV and V, and the beginning of column VI are missing.

VI 1′	10	[...]
	33 20	[...]
	26 40	[...]
	40	SAG [...]
	30	SAG [...]
	25	UŠ [...]
	1	UŠ [...]
	15	UŠ [...]
	5	GAM [...]
	1	BAL [...]
	20	DAL [...]
	[...	]
	⌈30⌉	[...]
	⌈3⌉	[...]
	⌈4⌉	[...]

The rest of the column is broken.

A.1.3 LIST C[1] INDEX

Obv. 1′ ⌈GU?⌉ [...]
a-⌈*na*⌉ [A.ŠÀ LAGAB ...]
a-⌈*na*⌉ [A.ŠÀ LAGAB *eš-re-et* UŠ-*ia wa-ṣa-ba-am*]

[1] Extensive restorations of the obverse of the text were made using the quasi-duplicate IM 52304 + 52685 (after Goetze 1951). The top and bottom of the tablet are missing.

a-[*na* A.ŠÀ] LAGAB [*ti-še-at* UŠ-*ia wa-ṣa-ba-am*]
[*a-na* A.ŠÀ LAGAB *sa-ma-na-at* UŠ-*ia wa-ṣa-ba-am*]
[*a-na* A.ŠÀ LAGAB *se-be-et* UŠ-*ia wa-ṣa-ba-am*]
[*a-na* A.ŠÀ LAGAB *ši-še-et* UŠ-*ia wa-ṣa-ba-am*]
[*a-na* A.ŠÀ LAGAB *ḫa-am-ša-at* UŠ-*ia wa-ṣa-ba-am*]
[*a-na* A.ŠÀ LAGAB *er-be-et* UŠ-*ia wa-ṣa-ba-am*]
[*a-na* A.ŠÀ LAGAB *ša-la*]-⌜*ša*⌝-*at* [UŠ-*ia wa-ṣa-ba-am*]
[*a-na* A.ŠÀ LAGAB *ši*]-*ne*-[*pa-at* UŠ-*ia wa-ṣa-ba-am*]
[*a-na* A.ŠÀ LAGAB *iš*]-*te-en* [*ši-di-i wa-ṣa-ba-am*]
[*a-na* A.ŠÀ LAGAB *iš-te*]-⌜*en*⌝ *ši-di-i ù* ⌜*ba-ma-at* UŠ-*ia*⌝ [*wa-ṣa-ba-am*]
[*a-na* A.ŠÀ LAGAB *iš-te-en ši*]-*di-i ù* ‹*ši-in-ne*›-*pa-tu* UŠ-*ia* ⌜*wa-ṣa*⌝-[*ba-am*]
[*a-na* A.ŠÀ LAGAB *iš-te-en ši*]-⌜*di-i*⌝ *ù ša-li*-‹*iš*›-*ti* UŠ-*ia wa-ṣa*-⌜*ba-am*⌝
[*a-na* A.ŠÀ LAGAB *iš-te-en ši-di-i*] ⌜*ù*⌝ *ra-ba-at* UŠ-*ia wa-ṣa-ba*-⌜*am*⌝
[*a-na* A.ŠÀ LAGAB *iš-te-en ši-di-i ù*] ⌜*ḫa*⌝-*am-ša-at* UŠ-*ia wa-ṣa*-⌜*ba*⌝-*am*
[*a-na* A.ŠÀ LAGAB *iš-te-en ši-di-i ù še*]-⌜*ša*⌝-*at* UŠ-*ia wa-ṣa-ba-am*
[*a-na* A.ŠÀ LAGAB *iš-te-en ši-di-i ù se-ba*]-*at* UŠ-*ia wa-ṣa-ba-am*
⌜*a*⌝-[*na* A.ŠÀ] ⌜LAGAB *iš*⌝-[*te-en ši-di-i ù ša-am*]-⌜*na*⌝-*at* UŠ-*ia wa-ṣa-ba-am*
a-⌜*na*⌝ [A.ŠÀ] ⌜LAGAB *iš*⌝-[*te-en ši-di-i ù ti-ša-at*] ⌜UŠ-*ia wa-ṣa*⌝-*ba-am*
a-na [A.ŠÀ LAGAB] ⌜*iš*⌝-[*te-en ši-di-i ù eš-re-et* UŠ-*ia wa-ṣa*]-*ba-am*
a-na [A.ŠÀ LAGAB *ši-na ši-di-i* UŠ-*ia wa-ṣa-ba-am*]
a-na [A.ŠÀ LAGAB *ši*]-⌜*na*⌝ *ši*-⌜*di*⌝-[*i ù ba-ma-at* UŠ-*ia wa-ṣa-ba-am*]
a-⌜*na*⌝ [A.ŠÀ LAGAB *ši*]-*na ši-di-i* [*ù ši-in-ne-pa-at* UŠ-*ia wa-ṣa-ba-am*]
a-⌜*na*⌝ [A.ŠÀ] ⌜LAGAB⌝ *ši*-⌜*na*⌝ *ši-di-i* [*ù ša-li-iš-ti* UŠ-*ia wa-ṣa-ba-am*]
⌜*a*⌝-[*na* A.ŠÀ] LAGAB *ša-la-ša-at* [*ši-di-i ù* ... UŠ-*ia wa-ṣa-ba-am*]
[*a-na* A.ŠÀ] LAGAB *er-bé-et* [*ši-di-i ù* ... UŠ-*ia wa-ṣa-ba-am*]
a-na A.ŠÀ LAGAB *ḫa-am-ša*-⌜*at*⌝ [*ši-di-i ù* ... UŠ-*ia wa-ṣa-ba-am*]
a-na A.ŠÀ LAGAB *ši-še-et* [*ši-di-i ù* ... UŠ-*ia wa-ṣa-ba-am*]
i-na A.ŠÀ LAGAB *ši-ša-at* [UŠ-*ia a-su-uḫ-ma*]
i-na A.ŠÀ LAGAB ⌜*ḫa-am-ša*⌝-[*at* UŠ-*ia a-su-uḫ-ma*]
i-na A.ŠÀ LAGAB *ši-in*-[*ne-pa-at* UŠ-*ia a-su-uḫ-ma*]
i-na A.ŠÀ LAGAB *iš*-[*te-en* UŠ-*ia a-su-uḫ*]-⌜*ma*⌝
i-na A.ŠÀ LAGAB *ši*-[*ne* UŠ-*ia a-su*]-⌜*uḫ-ma*⌝
i-na A.ŠÀ LAGAB ⌜*ša*⌝-[*li-iš-ti* UŠ]-⌜*ia*⌝ *a-su-uḫ-ma*
i-na A.ŠÀ LAGAB *er*-[*be-et* UŠ]-⌜*ia*⌝ *a-su-uḫ-ma*
i-⌜*na*⌝ A.ŠÀ LAGAB *ši*-⌜*na*⌝ [UŠ]-*ia a-su-uḫ-ma*
i-⌜*na*⌝ A.ŠÀ LAGAB *ḫa*-[*am-ša-at* UŠ]-*ia a-su-uḫ-ma*
i-⌜*na* A⌝.ŠÀ LAGAB [*ši-še-et* UŠ]-*ia a-su-uḫ-ma*
i-⌜*na*⌝ [A.ŠÀ LAGAB *se-be-et* UŠ]-*ia a-su-uḫ-ma*
i-⌜*na*⌝ [A.ŠÀ LAGAB *sa-ma-na-at* UŠ-*ia*] *a-su-uḫ-ma*
i-⌜*na*⌝ [A.ŠÀ LAGAB *ti-ša-at* UŠ]-*ia a-su-uḫ-ma*
i-⌜*na*⌝ [A.ŠÀ LAGAB *eš-re-et* UŠ-*ia*] *a-su-uḫ-ma*
i-⌜*na*⌝ [A.ŠÀ LAGAB *iš-te-en ši-di-i ù ba-ma-at*] UŠ-*ia a-su-uḫ-ma*
i-⌜*na*⌝ [A.ŠÀ LAGAB *iš-te-en ši-di-i ù pa-tu*] ⌜UŠ⌝-*ia a-su-uḫ-ma*
[*i*]-⌜*na*⌝ [A.ŠÀ LAGAB *iš-te-en ši-di-i ù ša-li-iš*]-⌜*ti*⌝ UŠ-*ia a-su-uḫ*-[*ma*]
[*i-na* A.ŠÀ LAGAB *iš-te-en ši-di-i ù ra-ba-at*] ⌜UŠ⌝-*ia a-su-uḫ*-⌜*ma*⌝
[*i-na* A.ŠÀ LAGAB *iš-te-en ši-di-i ù ha-am-ša-at*] ⌜UŠ⌝-*ia a*-⌜*su*⌝-[*uḫ-ma*]
[*i-na* A.ŠÀ LAGAB *iš-te-en ši-di-i ù ši-še-et* UŠ-*ia*] *a-su*-⌜*uḫ*⌝-[*ma*]

Rev [...] ⌜X X X⌝[...]
[...] ŠE EḪ RI [...]
[...] A MA RA [...]
⌜*ki*⌝-*pa-ta-am* ⌜5⌝ [*i-gi-gu-bu-ša*] §3.1
ta-al-li ki-pa-[*tim* ... *i*]-⌜*gi-gu*⌝-[*bu-šu*] §3.1
SAG.DÙ-*kum i-*[...] *i-gi-gu-*[*bu-šu*] §3.3
SAG.DÙ-*kum ša sa-am-na-*[*tu na*]-*ás-ḫa* 26 15 *i-*[*gi-gu-bu-šu*] • §3.3
ta-al-li SAG.DÙ-*ki* 52 30 *i-*[*gi-bu-bu-šu*] §3.3
a-ap-sà-mi-kum 26 15 *i-*[*gi-gu-bu-šu*] §3.7
ta-al-‹li› a-ap-sà-mi-ki 48 *i-gi-*⌜*gu*⌝-[*bu-šu*] §3.7
ṣí-li-ip-ti a-ap-sà-mi-ki 1 20 *i-gi-*⌜*gu*⌝-[*bu-šu*] §3.7
i-in al-pi-im 16 52 30 *i-*⌜*gi*⌝-[*gu-bu-šu*] §3.5
ús-ka-rum 2 30 *i-gi-gu-bu-šu ta-al-li ús-ka-rum* [... *i-gi-gu-bu-šu*] §3.2
[*na*]-*al-ba-tam i-na li-bu na-al-ba-tim e-pé-ša-am*
na-ra-am na-ar ta-ra-hi-i e-pé-⌜*ša*⌝-*am*
⌜*na*⌝-*ar ša-ka-na-ka-tim na-ra-am šu-zu-ba-am*
⌜*pa*⌝-*ta-am e-pé-ša-am*
ki-pa-tam i-na li-bu ki-pa-tim e-pé-ša-am
ki-pa-⌜*tam*⌝ [*i*]-*na ši-na at-hi za-za-am e-pé-sa-am*
i-na li-bu na-al-ba-tim ki-pa-ta-am
i-na li-bu ki-pa-tim na-al-ba-ta-am
na-az-ba-al SIG$_4$ 45 *i-gi-gu-bu* §5.3
na-az-ba-al SAHAR 1 ⌜40⌝ *i-gi-gu-bu* §5.2
na-az-ba-al ŠE.IN.NU.DA 3 20 *i-gi-gu-bu* §5.5
a-na 40 NINDAN *a-za-bi-il* 2 13 20 *al-lu-um* §5.2
a-na 20 NINDAN *a-za-bi-il* 4 *al-lu-um* • §5.2
a-na 15 NINDAN [*a-za-bi-il* 5] *al-lu-um* • §5.2
[*a-na*] ⌜10⌝ NINDAN *a-*⌜*za*⌝-[*bi-il*] 6 ⌜40⌝ *al-lu-um* §5.2
[*a-na* 5] NINDAN *a-za-*[*bi-il*] 10 *al-lu-um* §5.2
[*na*]-*al-*[*ba*]-*at-ta-*[*am i-na i-in*] *al-pí-im e-pé-ša-am* •
[...] AK [...] *i-gi-gu-bu-šu* §8.5
[...] IG X [...]5 40 *i-gi-gu-bu-šu* §8.5
[...] IG X [...]5 *i-gi-gu-bu-šu* §8.5
[...] AŠ [... *i-gi*]-*gu-bu-šu* §8.5
X-*ra-am* [... *i-gi*]-*gu-bu-šu* §8.5
X-*ra-am* [... *i-gi*]-*gu-bu-šu* §8.5
IŠ [...]-*mu ša re-ša-am* •
[...]-*am*
še-še-et [...]
⌜*ṣí*⌝-*li-*⌜*ip*⌝-[...]-⌜*ar*⌝
⌜*ṣí*⌝-*li-*⌜*ip*⌝-[...]-*ar*
ṣí-li-⌜*ip*⌝-[...]TA X ZA
IM.GÍD.DA [...]TE 10
áš-še-ša-am $^{??}$[...]-*ar*
KU TA AB ⌜SAG?⌝[...] X
ÁŠ X SI GU [...] X
ma-⌜*la* SAG⌝.KI [AN.TA ... *i-gi*]-*gu-bu-šu* §8.5

	sa-ba-at SAG.KI [AN.TA i-na UŠ wa-ra]-da-am	
	⌜ši⌝-ša-at SAG.KI AN.TA [i-na UŠ wa]-⌜ra-da⌝-am	
	⌜ḫa⌝-am-ša-[at SAG.KI AN].TA i-na UŠ wa-ra-[da]-am	
	[ra]-ba-at [SAG.KI AN.TA] i-na UŠ wa-[ra]-⌜da-am⌝	
	[ša-li-iš-ti SAG.KI.AN.TA i-na UŠ] ⌜wa-ra⌝-[da-am]	
	[ba-ma-at SAG.KI AN.TA i-na] UŠ wa-ra-[da-am]	
	ši-[ni-pa-at SAG.KI AN.TA i]-na UŠ wa-ra-[da-am]	
	ti-[... SAG.KI AN.TA i-na UŠ wa]-ra-⌜da⌝-[am]	
Edge I	[ ] i-gi-gu-[bu-šu]	§8.5
	[ ] i-gi-gu-bu-[šu]	§8.5
	[ ]16 i-gi-gu-bu-šu	§8.5
	[ ]11 i-gi-gu-bu-šu	§8.5
II	[URUDU 1 12 i-gi-gu-bu-šu (?)]	§8.1
	[KÙ.BABBAR 1 30 i-gi-gu-bu-šu (?)]	§8.1
	KÙ.GI 1 48 [i-gi-gu-bu-šu]	§8.1
	a-ba-rum 1 ⌜52⌝ 30 ⌜i-gi⌝-[gu-bu-šu]	§8.1
	a-na-ku-um 2 36 15 i-gi-[gu-bu-šu]	§8.1
	a-lu-lu-tum 2 24 i-gi-gu-bu-[ša]	§8.1
III	[ ] ⌜i-gi⌝-gu-bu-šu	§8.5
	[i]-gi-gu-bu-šu	§8.5
	[i]-⌜gi⌝-gu-bu-ša	§8.5
	[ ]	
	[ ]	
	[ ]	
IV	e-le-pu-um 7 12 i-gi-gu-bu-ša	§7.3
	e-ší-it-tum ⌜3⌝ 36 i-gi-gu-bu-ša	§7.4
	še$_{20}$-e ka-ra-am ⌜ša-ka-na⌝-am	§7.3
	[iš]-ka-ar e-⌜ṣe⌝-[di] ⌜e-pé-ša⌝-am	§6.6
	[iš]-ka-ar za-ri-i ša še-im e-pé-⌜ša-am⌝	§6.6
	[iš-ka-ar ...] e-pé-ša-⌜am⌝	§6.6
V	iš-ka-ar [... e-pé-ša-am]	§6.6
	qa-na-am [ ]	
	qa-na-am [ ]	
	qa-na-am 25 [ ]	
	qa-na-am el-qé mi-id-di-X [...]	
	i-na ša-am-mu-ti-šu 1 iš-rum-[...]	

A.1.4 LIST D

			INDEX
Obv	IGI.GUB	šà mi-im-ma ka$^{!}$-li-šu	§2.1.3
	5	IGI.GUB šà GÚR	§3.1
	20	DAL šà GÚR	§3.1
	10	[pi]-ir-ku šà GÚR	§3.1
	16	[IGI].GUB šà GÚR šà 2 ŠE i-na ŠÀ GÚR GAR	§3.8
	16 26 46 40	IGI.GUB šà GÚR šà 3 ŠE i-na ŠÀ GÚR GAR	§3.8
	15	IGI.GUB šà ús-ka$_4$-ri	§3.2

	40	DAL	*šà ús-ka*$_4$*-ri*		§3.2
	20	*pi-ir-ku*	*šà ús-ka*$_4$*-ri*		§3.2
	6 3⸢3⸣ 45	IGI.GUB	*šà* GÁN *pa-na-ak-ki*		§3.5
	52 30	DAL	*šà* GÁN *pa-na-ak-ki*		§3.5
	15	⸢*pi-ir-ku*⸣	*šà* GÁN *pa-na-ak-ki*		§3.5
	13 ⸢07⸣ 30	IGI.GUB	*šà* GÁN ILLAR		§3.5
	52 30	DAL	*šà* GÁN ILLAR		§3.5
	30	⸢*pi*⸣*-ir-ku*	*šà* GÁN ILLAR		§3.5
	1⸢3⸣ 20	IGI.GUB	*šà* A.ŠÀ ŠE		§3.5
	5⸢6⸣ 40	DAL	*šà* A.ŠÀ ŠE		§3.5
	2⸢3⸣ 20	[*pi*]-ir-ku	*šà* A.⸢ŠÀ⸣ ŠE		§3.5
	1⸢6⸣ 52 30	IGI.GUB	*šà* IGI.GU$_4$		§3.5
	52 30	DAL	*šà* IGI.GU$_4$		§3.5
	30	*pi-ir-ku*	*šà* IGI.GU$_4$		§3.5
	26 40	IGI.GUB	*šà a-pu-sà-am-mi-ki*		§3.7
	1 20	BAR.TA	*šà a-pu-sà-mi-ki*		§3.7
	3⸢3⸣ 20	*pi-ir-ku*	*šà a-pu-sà-mi-ki*		§3.7
	15	IGI.GUB *šà a-pu-sà-mi-ik-ki šà* 3			§3.7
	1 40	IGI.GUB	*šà* SAG.5		§3.6
	2 37 30	IGI.GUB	*šà* SAG.6		§3.6
	3 41	IGI.GUB	*šà* SAG.7		§3.6
	52 30	IGI.GUB	*šà* SAG.DÙ		§3.3
	57 36	IGI.GUB	*šà* UB	•	§3.3
	1 25	IGI.GUB	*šà* BAR.TA *šà* NIGIN		§3.4
	1 15	IGI.GUB	*šà* BAR.TA UŠ *ù* SAG		§3.4
	48	IGI.GUB	*šà* PA$_5$.SIG		§6.5
Edge	1 12	⸢IGI.GUB⸣ *šà šu-pu-ul-ti*		•	§6.5
	30	⸢IGI⸣.[GUB] *šà pu-ut* ÍD[!?] *ḫe-ri-im*		•	§6.5
	45	IGI.GUB *šà* ⸢*na-az-ba-li*⸣*-im*		•	§5.2
	2 13 20	[IGI.GUB *šà tu-up-ši-ki-im*]		•	§5.2
Rev	20	⸢*šà*⸣ [ ]		•	§5.1
	20	⸢*šà*⸣ ⸢SAḪAR⸣ [*ḫa-ba*]*-ti*		•	§5.1
	10	*šà* SAḪAR	*ba-la-lu*	•	§5.1
	2⸢4?⸣	*šà* SIG$_4$	*šà-ḫa-a-ṭì*	•	§5.1
	20	*šà* SAḪAR	*za-ba-li*		§5.2
	3 45	*šà*	*pí-ti-iq-ti*		§6.1
	15	*šà*	*pí-ti-iq-ti*		§5.6
	5	*šà*	*na-al-ba-ni*		§4.1
	10	*šà* SAḪAR *za-ba-li* ÉŠ.GÀR LÚ 1			§5.2
	5	*šà*	SAḪAR		§5.2
	5	*šà*	*a-ma-ri-im*		§4.4
	6	*šà na-aš-pa-ak šà* SÌLA X*-di-im*		•	§7.2
	6 40	*šà*	*na-aš-pa-ki-im*		§7.2
	7 12	*šà*	*na-aš-pa-ak* GUR$_7$		§7.3
	10	*šà* 1 2 *ù* 1			§7.1
	9	IGI.GUB	*šà ku-ri-im*		§2.1.3
	1 12	*šà*	URUDU		§8.1

1 15	*šà*	*lu-li-im*		§8.1
1 20	*šà*	ZABAR		§8.1
1 24	*šà*	⸢AN⸣.NA		§8.1
1 36	*šà*	KÙ.BABBAR		§8.1
1 48	*šà*	KÙ.GI		§8.1
1 52 30	*šà*	*a-ba-ri*		§8.1
2	*šà*	*ḫu-lu-uḫ-ḫi*		§8.1
2 12	*šà*	*pa-ar-zi-il-li*		§8.1
18	*šà*	*im-le-em*		§4.3, §8.3
14 24	*šà*	*ki-ir-ba-ni*		§4.3
12	*šà*	*ḫa-aṣ-bi*		§4.3
15	*šà*	ESÍR		§4.6
12	*šà*	*ku-up-ri*		§4.6
10	*šà*	*šu-ri ka-sà-mi*		§6.4
5	*šà*	*šu-ri za-ba-li*		§5.5
⸢1⸣ SAR	*a-na qa-ti sé-ki-ri iš-šà-ka-[an]*		•	§6.3
⸢7⸣ 30	SAḪAR *i-za-ab-bi-il*			

A.1.5 LIST E — INDEX

Obv. 5	KA.[KÉŠ]		§3.1
15	GÁN [U$_{4}$.SAKAR$^{?}$]	•	§3.2
10	GÁN ⸢U$_{4}$⸣.[SAKAR KI.2$^{?}$]	•	§3.2
⸢2⸣6 40	GÁN ⸢ZÀ.MÍ⸣		§3.7
12 30	GÁN PAN	•	§3.5
⸢13 20⸣	GÁN ⸢MÁ⸣.GUR$_{8}$	•	§3.5
5 ⸢24⸣	GÁN x x x		§3.8
15 36 10	GÁN ⸢U$_{4}$.SAKAR KI$^{?}$⸣.IN.ŠUB		§3.2
⸢45$^{?}$⸣	GÁN IN		§3.8, §8.5
1 ⸢24 51 10⸣	⸢*ṣi-li*⸣*-ip-tum* ‹*ša*› ÍB.SI$_{8}$		§3.4
1	*ṣi-li-ip-tum*		§3.4
4⸢5⸣	GIŠ.A.BAL		§3.8, §8.5
3 45	IM.DÙ.A		§6.1
18	IM.LÁ		§4.3, §8.3
11 08	⸢IM⸣.BÁBBAR		§8.4
5$^{?}$4	⸢IM⸣.ḪA$_{5}$$^{?}$.A		§8.5
6 40	IM.LAG$^{?}$		§8.5
4 16	[IM$^{?}$].LAG$^{?}$		§8.5
5	⸢GIŠ⸣.MÁ		§7.2
10	⸢ÉŠ⸣.GÀR ‹KI$^{?}$›.LÁ		§6.3
20	⸢PA$_{5}$⸣ *ši-*⸢*lu*⸣*-tum*		§6.3
3 36	IN.NU.DA		§7.4
1 15	É.UŠ.GÍD.DA IN.NU.DA		§7.4
10 45 06	⸢ESÍR.UD⸣.A *ša i-na ki-ri-im*		§8.5
15	⸢*ni-qí*⸣*-ir-tum*	•	§5.6
48	*me-e* A.ŠÀ		§6.5

Rev	2 13 20	*qí-it-mu-um*		§8.4
	20	[ ]		§8.5
	20	⸢DUSU?⸣	•	§8.5
	15	A		§7.4
	15	⸢ESÍR.ḪI⸣.A		§4.6
	6 40	*na-az-ba-al* SAḪAR		§5.2
	9	KI.LÁ.BI *ša* URUDU		§8.1
	1 20	⸢*ra-ṭù?-um*⸣ ‹*ša*› KÙ.GI		§8.1
	4 48	*ku-*⸢*bu-ur i-ṣi-im*⸣		§7.2
	1 12	⸢*ra-ṭù*⸣*-um ša* URUDU		§8.1
	1 48	[*ra-ṭù-um*] ⸢*ša*⸣ [KÙ].GI		§8.1
	2 24	⸢*a*⸣-[*lu-lu-tu*]*-um*	•	§8.1
	1 12	[SIG$_4$ 1].⸢KÙŠ⸣	•	§4.1
	7 12	⸢SIG$_4$.ANŠE⸣		§4.4
	4 30	*na-*⸢*áz*⸣-[*ba*]-⸢*lu*⸣*-um*		§5.3
	2 30 5 20	[*nu-ud-di*]*-tum*	•	§8.5
	48 1 40 1 20	⸢*ú-ku/lu*⸣ ...*-um*		§8.5
	⸢2⸣ 42	⸢SIG$_4$ *a-gu*⸣*-ru-um*		§4.1
	1 41 15	*ša* [SIG$_4$] ÍB.‹SI$_8$›	•	§5.3
	45	*ta-la-ak-tum*		§5.2
	3	*ša* $^1/_2$ KÙŠ ‹SIG$_4$›	•	§5.3
	5 20	Ú.SAG$_{11}$	•	§6.4
	1 15	*na-az-ba-lu-um*		§5.4
	12	*ma-al-ta-ak-tum*		§8.2
	26 40	*ru-*⸢*qú*⸣*-um ša* KÙ.BABBAR		§8.1
	20	Ì.UDU		§8.4

A.1.6 LIST F

				INDEX
	IGI.GUB.BA	*ša né-pe-eš$_{20}$-tum*		§2.1.4
	4 30	*na-az-ba-al* SIG$_4$		§5.3
	7 12	*na-al-ba-an-ša*	•	§4.1
	5 24	*na-al-ba-an* SIG$_4$.ÁB		§4.1
	3 22 30	*na-az-ba-al-ša*		§5.3
	2 42	*na-al-ba-an* SIG$_4$.AL.ÙR.RA		§4.1
	1 41 15	*na-az-ba-al-ša*		§5.3
	1 40	*na-az-ba-al* SAḪAR		§5.2
	1 12	*na-al-ba-an* 1 KÙŠ SIG$_4$		§4.1
	45	*na-az-ba-al-ša*		§5.3
	4 48	*na-al-ba-an* $^1/_2$ KÙŠ SIG$_4$		§4.1
	4 23 36	*na-az-ba-al-ša*		§5.3
	9	*na-al-ba-an* $^1/_3$ KÙŠ SIG$_4$		§4.1
	8 20	*na-al-ba-an* IM.DUGUD		§4.1
	7 38	*na-az-ba-al-ša*		§5.3
	7 12	*ša* SIG$_4$.ANŠE		§4.4
	3 45	*ša* IM.DÙ.A		§6.1

Line	Number	Text		Section
	15	*ša* ESÍR.HA$_5$.RÁ.A		§4.6
	30	*ša* SAG.DÙ		§3.3
20	⌈5?⌉	*ša* GÁN GÚR		§3.1
	⌈1⌉5 *ù* 10	*ša* GÁN U$_4$.SAKAR		§3.2
	1 12	*ra-ṭù-um ša* URUDU		§8.1
	1 30	*ra-ṭù-um ša* KÙ.BABBAR		§8.1
	1 48	*ra-ṭù-um ša* KÙ.GI		§8.1
25	2 15	*ru-uq-qum ša* KÙ.GI		§8.1
	4	*ru-uq-qum ša* KÙ.BABBAR		§8.1
	2 24	*ru-uq-qum ša* AN.NA		§8.1
	2 08	*ru-uq-qum ša* SÙ.ÁG		§8.1
	6	*ša* GIŠ.MÁ.LÁ		§7.2
30	6 40	*ša* É.KIŠIB ŠE		§7.2
	3 36	*ša* É.KIŠIB IN.NU.DA		§7.4
	3 45	*ša* É.KIŠIB SUH$_4$		§7.4
	15	*ša* É.KIŠIB A		§7.4
Rev	3 20	*ša* SIG$_4$.AL.ÙR.RA		§4.1
35	1 52 30	*ša* SIG$_4$.AL.ÙR.RA KI.2		§4.1
	‹1›4 24	*ma-aš-qá-al-tum ša* SIG$_4$	•	§4.3
	«1» 12	*ša-ḫa-al-lu-ú-um*	•	§4.3
	14 24	*ta-ad-di-tum ša* SIG$_4$		§4.2
	45	*mu-ut-ta-al-li-ik-tum*		§5.2
40	18	*i-im-lu-ú-um*		§4.3, §8.3
	6	*ma-aš-⌈šu-ú-um⌉ [ša]* ⌈SIG$_4$⌉		§5.4
	1 40	*ma-aš-šu-ú-um ša* SAḪAR		§5.4
	10 GÍN	ÉŠ.GÀR LÚ.1		§6.5
	1 20 *ù* ⌈15⌉	*na-az-ba-al* A		§5.5
45	2 13 ⌈20 *ù* 1⌉ 15 *ša* GIŠ «*tup*» DUSU		•	§5.2, §5.4
	⌈3⌉ [22] ⌈30⌉	*ša* IM	•	§8.5
	[13]⌈20⌉	*ša* GÁN MÁ.GUR$_8$	•	§3.6
	⌈53⌉ 20	*ša* GÁN ZÀ.MÍ	•	§3.7
	⌈5⌉2 30 *ù* 12 30 *ša* GÁN PAN		•	§3.5
50	26 40	*ša* GEŠTÚ ZÀ.MÍ		§3.7
	3 30	*ša* GÁN ⌈PAN⌉ KI.2	•	§3.5
	4 20	*ša* GÁN U$_4$.SAKAR KI.2		§3.2
	4 36 10	*ša* GÁN GÚR KI.2		§3.2
	45	*ša* GÁN U$_4$.SAKAR KI.3		§3.2
55	⌈20⌉ 48 06	*ša* GIR$_4$ AD.KE$_4$		§8.4
56a, b	⌈3⌉6	IM.BABBAR 54 *ša ka-lu-ú-um*		§8.4; §8.4
57a, b	⌈20?⌉	*ša* ESÍR? 20 *ša* NA$_4$.KUR.RA		§8.5; §8.4
	⌈4⌉ 48	*ša ku-bu-ur-re-e* GIŠ		§7.2
	40	*ša* DÙ GIŠ.*šu-rum*		§6.4
60	12	*ma-al-ta-ak-tum ša mu-ši-tim*		§8.2
	4 48	É.KIŠIB ŠE		§7.2
	24	PI-*lu-ú-tum ša* GI [(X)]		§8.5
63a, b	1	*ṣí-li-ip-tum* 1 *zi-it-tum*		§3.4; §8.4
	1	GI ḪAŠ.A		§8.4

65	57 36	*ša eb-lu-ú-um*	• §8.5
		^{d}NISABA	§2.1.4

A.1.7 LIST G

					INDEX
1	5	IGI.GUB	○	*ki-pa-tim*	§3.1
	7 30	IGI.GUB	▷	SAG-*ga-kum*	§3.3
	2 13 20	IGI.GUB	GIŠ.DUSU	*šu-up-ši-ki-im*	§5.2
	1 40	IGI.GUB	SAḪAR	*na-az-ba-al* SAḪAR	§5.2
5	4 30	IGI.GUB	SIG$_4$	*na-az-ba-al* SIG$_4$	§5.3
	7 12	IGI.GUB	.BI	*a*!*-ma-ri-im*	§4.4
	6	IGI.GUB		*i-ga-ri-im*	§4.5
	5	IGI.GUB	*mi-nu-ut*	*li-bi-tim*	§4.3
	6	IGI.GUB	É.ÙR!.RA	*na-aš-pa-ku-um*	§7.2
Rev 10	[...] ⸢40⸣	IGI.GUB	ŠÌTA	*ra-ṭe$_4$-em*	§8.1
	[...]	IGI.GUB	⸢ŠÀ⸣.GAL	*ma*?*-ša*?*-ri*	• §8.5
	4 [4?]8	IGI.GUB	GIŠ.MÁ	*e-le-pí-im*	§7.2
	3 45	IGI.GUB	«*pí-iq*»	*pí-*⸢*ti*⸣*-iq-tim*	• §6.1
	[1] 30	IGI.GUB	*na-az-ba-al*	*a-gu-ri-im*	• §5.3
15	3	IGI.GUB	*na-*[*az*]*-ba-al*	*ar-ḫu-um*	§5.3
	45	IGI.GUB	*mu-ta-*	*li-ik-tim*	§5.2

A.1.8 LIST H[2]

			INDEX
1	6 40	*i-gi-gu-ub-bi-im qù-up-pí-im*	§7.2
(col. I)	3 45	*pí-ti-iq-tum*	§6.1
	4 10	*ša* SIG$_4$	• §5.6
	5	*ša ki-pa-ti*	§3.1
5	30	*ša sa*?*-ta*?*-*⸢*ki*?⸣	• §3.3
(col. II)	6	*ša na-aš-pa-kum*	§7.2
	7 30	*ka-ru-um*	§7.3

A.1.9 LIST J

					INDEX
Obv 1	4 26 40	GÁN AM.AM.MA	⸢É⸣.Ì.DUB	IGI.GUB-*ú*	• §2.2
	26 15	GÁN UR.RU.UḪ SAG.3	KI.MIN	IGI.GUB-*ú*	• §2.2, §3.3
	33 45	GÁN *ke-er-ṣi*	KI.MIN	IGI.GUB-*ú*	• §2.2
	16 42 30	GÁN *ka-re-e*	KI.MIN	IGI.GUB-*ú*	• §2.2, §7.3
5	4 13 07 30	GÁN *ka-*⸢X⸣ KI.MIN	⸢KI.MIN⸣	IGI.GUB-*ú*	• §2.2, §7.3
	12 36 22 30	GÁN GAB-*ri*	KI.MIN	IGI.GUB-*ú*	• §2.2
	6 40	É.Ì.DUB *še-im*		IGI.GUB-*ú*	§7.2
	6	É.Ì.DUB *me-e*		IGI.GUB-*ú*	§7.2

[2] Because the list is on the reverse of the tablet (after three unrelated problem texts), the order of the columnar groupings within it is from right to left after the first entry, which uses the whole width of the tablet.

6	É.Ì.DUB ⌈Ì?⌉	IGI.GUB-*ú*	•	§7.2
6 12	É.Ì.DUB ŠE *duḫ-nu*	IGI.GUB-[*ú*]	•	§7.2
6 12	É.Ì.DUB ⌈ŠE⌉ [...] X	⌈IGI.GUB⌉-[*ú*]		§7.2
6 24	É.Ì.[DUB ...	IGI.GUB-*ú*]		§7.2
6 24	⌈É⌉.[Ì.DUB ...	IGI.GUB-*ú*]		§7.2
6 45	[É.Ì.DUB ...	IGI.GUB-*ú*]		§7.2
7 30	[É.Ì.DUB ...	IGI.GUB-*ú*]		§7.3

The rest of the column is missing.

Rev. The beginning of the column is missing.

[ ] HI DU [ ]
[ ] ⌈2⌉ 40 IGI.GUB.BA [ ]
[ ]2 40 A.RÁ 6 16 16 DUG [ ]

⌈GAM *ša*!⌉ É.KUR «5 20 IM» 5 20 IM.GÍD.DA [ ]
UR.A.MEŠ *tal-qé-ma eš-re-e-*[( )]*-ti gi-mir-* [ ]
kul-lat né-me-di sa-ni-qu mit-ḫar- [ ]
ša IGI.GUB-*bé-e šá* A.ŠÀ.MEŠ *aḫ-zu* X-[ ]
ka-a-a-nam-ma ... ... *ta-nam-*[ ]

---

ina UR.A *me-tì uš-ziz* A.ŠÀ *šá* 45 *ta-* X-[ ]
13 30 ŠE.NUMUN *me-tì ša* UR.A 45-*šu* [ ]
ás-ḫur ur-dam-ma mit-ḫar-ta GÚR-*up šá* UR.A 12

I.NAM.GIŠ.ḪUR.AN.KI.A *ḫur-ḫur-ra ki* KA DUB GABA.RI *la-bi-ri-šú*
ša-ṭir-ma ba-ri DUB ^{m}EN-*na-din*-DUMU.UŠ
LÚ.GÍR.LÁ ^{d}AMAR.UTU LÚ *šà-suk-ku*
pa-liḫ ^{d}AMAR.UTU NU.TÙM ITI.ZÍZ U$_{4}$.12.KAM
MU.33.KAM ^{m}*da-a-ri-ia-uš*$_{11}$ LUGAL ŠÚ
ša-ṭir

A.1.10 LIST K[3]

INDEX

Obv. I	[...] BAL	
	[...]⌈ESÍR⌉.ḪI.A	
	[...] ZABAR	§8.1
	[...] KÙ.BABBAR	§8.1
	[...] KÙ.GI	§8.1
	[...] ⌈A⌉.BÁR	§8.1
	[... KÙ.]KÁM	§8.1
	[... AN].BAR	§8.1
	[... *an-na*]-*ki*	§8.1

The rest of the column is missing.

Obv. II 7 ⌈30⌉ [...]
9 [...]
10 [...]

[3] Fragment from top of tablet; text below obverse and above reverse missing. Hand-copy in Fig. A.1.1 below.

11$^+$ [...]
16$^?$ [...]
10 [...]
10 [...]
1$^+$ [...]

The rest of the column is missing.

Rev I The start of the column is missing.

10$^+$ [...]
10$^+$ [...]
11$^+$ [...]
20$^+$ [...]
15 [...]

Rev II The start of the column is missing.

[ ] ⌜X X⌝	
[ ] SIG$_4$	§4.3
[ ]⌜SIG$_4$⌝.ÁB	§4.3
[SIG$_4$].⌜AL⌝.ÙR.RA	§4.3
[SIG$_4$].$^2/_3$-*ti*	§4.3
[ ]⌜X⌝ ḪU SU TU	

A.1.11 LIST L[4]

INDEX

The beginning of the column is missing.

[7 12 SIG$_4$ ...]		§4.1
[5 24 SIG$_4$.ÁB ...]		§4.1
[2] ⌜42 SIG$_4$⌝.[AL.ÙR.RA ...]		§4.1
5 SIG$_4$.$^2/_3$-*ti* [...]		§4.1
12 30	KI.LÁ SIG$_4$ [...]	§4.3
16 40	KI.LÁ SIG$_4$.⌜ÁB⌝ [...]	§4.3
33 20	KI.LÁ SIG$_4$.AL.⌜ÙR⌝.[RA ...]	§4.3
18	KI.LÁ SIG$_4$.$^2/_3$-*ti* [...]	§4.3
10	KI.LÁ SIG$_4$ ḪA$_5$.⌜RÁ⌝	§4.3
13 20	KI.LÁ SIG$_4$.ÁB ḪA$_5$.RÁ	§4.3
26 40	KI.LÁ SIG$_4$.AL.ÙR.RA ḪA$_5$.RÁ	§4.3
14 24	KI.LÁ SIG$_4$.$^2/_3$-*ti* ḪA$_5$.RÁ	§4.3
[8] 20	KI.LÁ SIG$_4$ *ṣa-rip-ti*	§4.3
[11 06 40	KI].LÁ SIG$_4$.ÁB *ṣa-rip-ti*	§4.3
[22 13 20	KI.LÁ SIG$_4$].AL.ÙR.RA KI.MIN	§4.3
[12	KI.LÁ SIG$_4$.$^2/_3$]-*ti* KI.MIN	§4.3

The rest of the column is missing.

4 First column of obverse; beginning and end of column missing. Hand-copy in Fig. A.1.2 below.

A.1.12 LIST M

		INDEX
1	6 SIG$_4$ SI.SÁ GÚ.UN LÚ.1.E	§5.4
	1 SIG$_4$ 1 KÙŠ ÍB.SI$_6$ GÚ.UN LÚ.1.E	§5.4
3	2 15 SIG$_4$ KÙŠ.TA ÍB.SI$_8$ GÚ.UN LÚ.1.E	§5.4

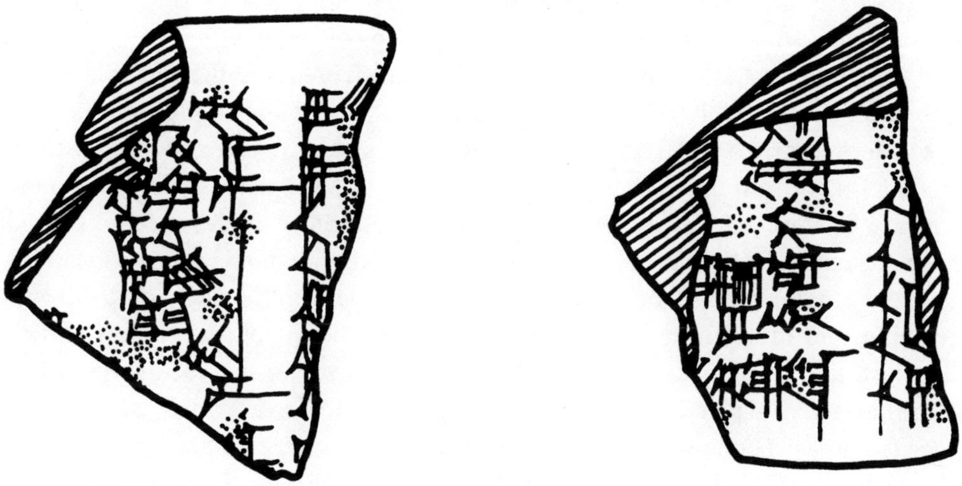

Figure A.1.1: Hand-copy of BM 37096 obverse and reverse (List K)

Figure A.1.2: Hand-copy of BM 36776 obverse (List L)

APPENDIX 2: BM 15285

This text has previously been published only in parts. Gadd (1922) published a full — though, it must be said, rather inaccurate — copy of BM 15285, namely the top right of the obverse/bottom right of the reverse, which was then treated in *MKT* (I 137–42) and *TMB* (53–7). Saggs (1960) published hand-copies of the other, unnumbered fragment with a photograph of the extant portions of the tablet. The complete tablet must have measured somewhat more than 30 × 50 cm, and is shown at actual size in the following hand-copy.

A complete copy has never been made until now, although the contents are very well known. Substantial parts are still missing, but it must have originally contained around 40 problems, of which 31 are now completely or partially preserved. Neither Gadd's drawings nor Saggs' photograph show clearly the partially erased construction lines and compass marks[1] on the tablet which the scribe used in making the drawings. An attempt has been made to distinguish guide-lines from the outlines of figures by using a thinner pen to draw them. All the large squares are 48 mm (*c*.3 fingers) in length, often with more or less accurate subdivisions into halves and quarters. The circles are all of radius 11 mm, which means that they do not always touch the other figures where they should. Considering the lack of a Mesopotamian concept for angle, and the convex surface of the reverse of the tablet, it is hardly surprising that many of the illustrations seem slightly skew to the modern eye. In the following transliteration and translation, and in Chapter 3, idealized figures represent the drawings; reconstructions are shown by shaded areas. Guide-lines, surface damage and inaccuracies are not marked, but can all be seen clearly on the hand-copy. Numbering of the problems does not follow Gadd or Saggs but runs consecutively.

Column I

(i)

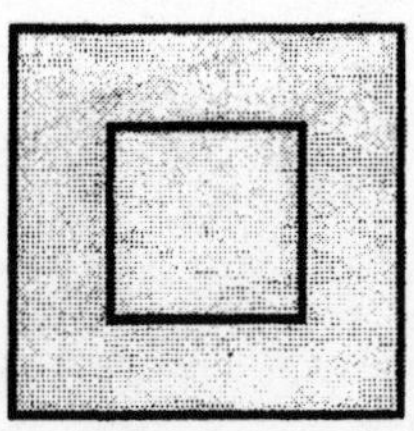

[1 UŠ *mi-it-ḫa-ar-tum*]
[PAD.TA.ÀM *ad-ku-uš-ma*]
[*mi-it-ḫa-ar-tam ša-ni-tam ad-di*]
[A.ŠÀ.BI EN.NAM]

[The side of the square is 1 UŠ.[2] I made a border each side and I drew a second square. What is its area?]

(ii)

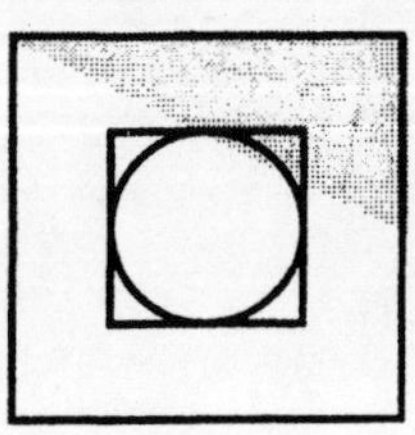

1 UŠ *mi-it-ḫa-ar-tum*
PAD.TA.ÀM *ad-ku-ša-ma*
⸢*mi*⸣-*it-ḫa-ar*-⸢*tam*⸣ *ša-ni-tam*
ad-di i-na li-ib-bi
ÍB.SI$_8$ *ki-ip-pa-tam ad-di*
A.ŠÀ.BI EN.NAM

The side of the square is 1 UŠ. I made a border each side and I drew a second square. Inside the square I drew a circle. What are their areas?

[1] Also visible on the illustration to BM 85194 (iv) (*MKT*: II pl. 5).
[2] After Høyrup (1996*a*: 54–5).

(iii)

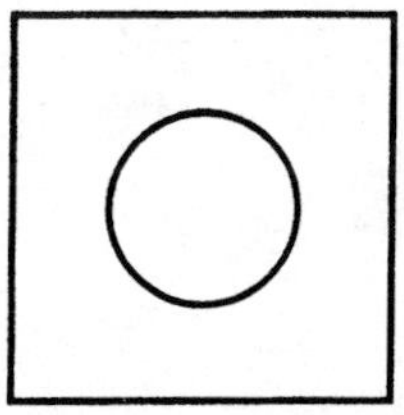

1 ⌈UŠ *mi-it-ḫa-ar-tum*⌉
PAD.TA.ÀM *ad-ku-uš-ma*
ki-ip-pa-tam ad-di
A.ŠÀ.BI EN.NAM

The side of the square is 1 UŠ. I made a border each side and I drew a circle. What is its area?

(iv)

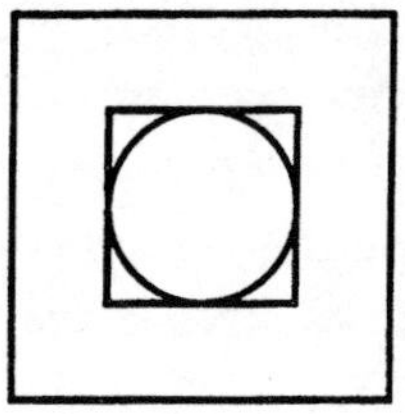

1 UŠ *mi-it-ḫa-ar-tum*
ŠÀ.BA *mi-it-ḫa-ar-tim*
ki-ip-pa-tam ad-‹di›
ki-ip-pá-tum
ša ad-du-ú mi-it-ḫa-ar-tam
i-mi-id A.ŠÀ.BI EN.NAM

The side of the square is 1 UŠ. Inside it I drew a square and a circle. The circle that I drew touched the square. What are their areas?

Column II

(v)

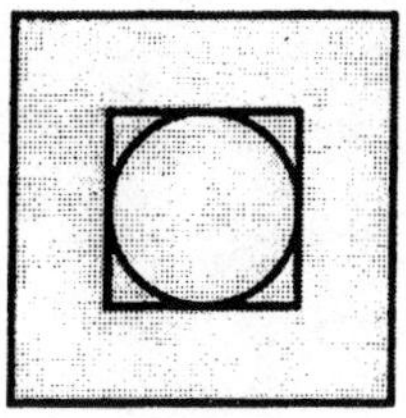

[1 UŠ *mi-it-ḫa-ar-tum*]
[*mi-it-ḫa-ar-tam*] *ša-ni-tam*
[ŠÀ.BA] [*mi-it-ḫa*]*-ar-tim*
[*ša-ni-tim*] 4 SAG.DÙ 1
ki-ip-pá-tum
[*ad-di*] A.ŠÀ.BI EN.NAM

[The side of the square is 1 UŠ.] ‹I drew› a second [square. Inside the second square I drew] 4 triangles and 1 circle. What are their areas?

(vi)

[1 UŠ *mi-it-ḫa-ar*]-⌈*tum*⌉
[*mi-it-ḫa-ar-tam ša-ni-tam*]
[ŠÀ.BA *mi-it-ḫa-ar-tim*]
[*ša-ni-tim* 4 ÍB.SI$_8$ 1 *ki-ip-pa-tam*]
⌈*ad-di*⌉ A.ŠÀ.BI EN.NAM

[The side of the square is 1 UŠ. ‹I drew› a second square. Inside the second square I drew 4 squares and 1 circle.] What are their areas?

(vii)

1 UŠ *mi-it-ḫa-ar-tum*
ŠÀ.BA *mi-it-ḫa-ar-tam*
ša-ni-tam ad-di ÍB.SI$_8$
ša ad-du-ú mi-it-ḫa-‹ar›-tam
ki-di-tam i-mi-id
A.ŠÀ.⌈BI EN⌉.NAM

The side of the square is 1 UŠ. Inside it I drew a second square. The square that I drew touches the outer square. What is its area?

(viii)

1 UŠ *mi-it-ḫa-ar-tum*
ŠÀ.BA 4 SAG.DÙ
1 ÍB.SI$_8$ *mi-it-ḫa-ar-tum*
ša ad-du-ú mi-it-ḫa-ar-ta
ša-ni-tam i-mi-id
A.ŠÀ.BI EN.NAM

The side of the square is 1 UŠ. Inside it are 4 triangles and 1 square. The square that I drew touches the second square. What is its area?

Column III

(ix)

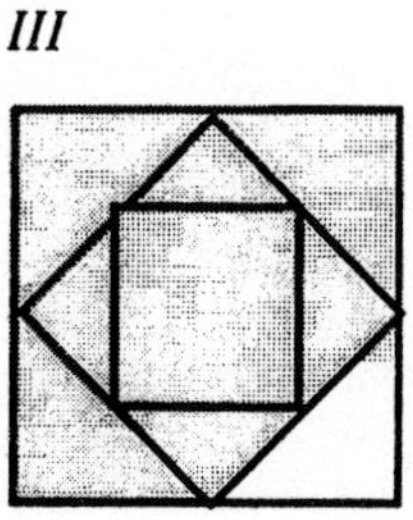

⌈1 UŠ⌉ ÍB.SI$_8$ ŠÀ.BA ÍB.SI$_8$ *ad-di*
⌈ÍB.SI$_8$⌉
ša ad-du-ú ÍB.SI$_8$ *i-mi-id*
i-na ŠÀ ÍB.SI$_8$ *ša-ni-tim* ÍB.SI$_8$
ša-lu-uš-tam ad-⌈*di*⌉ *ša ad-du-ú*
mi-it-ḫa-ar-tam i-mi-id
A.ŠÀ.BI EN.NAM

The side of the square is 1 UŠ. In it I drew a square. The square that I drew touches the ‹first› square. Inside the second square I drew a third square. ‹The square› that I drew touches the ‹second› square. What is its area?

(x)

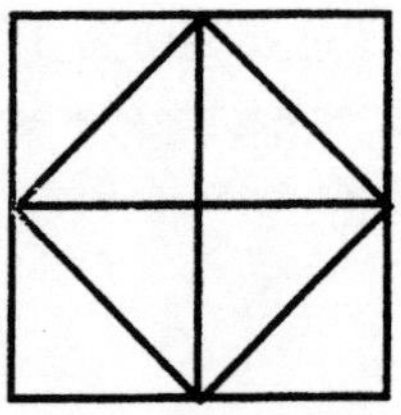

1 UŠ *mi-it-ḫa-ar-tum*
⸢ŠÀ.BA 8⸣ SAG.DÙ *ad-⸢di⸣*
[A.ŠÀ].BI EN.NAM

The side of the square is 1 UŠ. Inside it I drew 8 triangles. What are their [areas] ?

(xi)

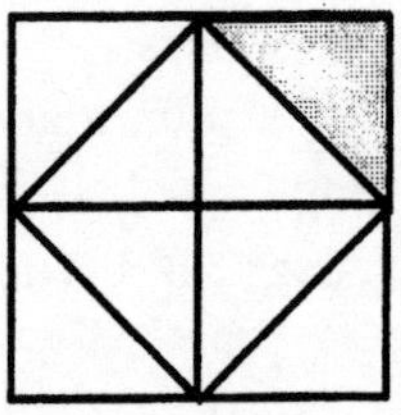

1 UŠ *mi-it-ḫa-ar-tum*
ŠÀ.BA *mi-it-ha-ar-‹tam› ad-di*
mi-it-‹ḫa-ar-tum› ša ad-du-ú
ÍB.SI$_8$ *i-mi-id* ŠÀ.BA ÍB.SI$_8$
4 SAG.DÙ *ad-di*
‹A.ŠÀ.BI EN.NAM›

The side of the square is 1 UŠ. Inside it I drew a square. The square that I drew touches the ‹first› square. Inside the ‹second› square I drew 4 triangles. ‹What are their areas?›

(xii)

1 UŠ *mi-it-[ḫa-ar-tum]*
ŠÀ.BA 16 ⸢SAG⸣.[DÙ *ad-di*]
A.ŠÀ.BI EN.[NAM]

The side of the square is 1 UŠ. Inside it [I drew] 16 triangles. What are their areas?

Column IV

(xiii)

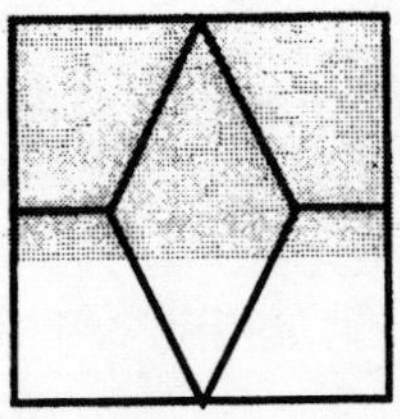

1 UŠ *mi-it-ḫa-ar-tum*
⸢ŠÀ⸣.BA 4 SAG.KI.GUD
2 SAG.DÙ *ad-di*
A.ŠÀ.BI EN.NAM

The side of the square is 1 UŠ. Inside it I drew 4 trapezoids and 2 triangles. What are their areas?

(xiv)

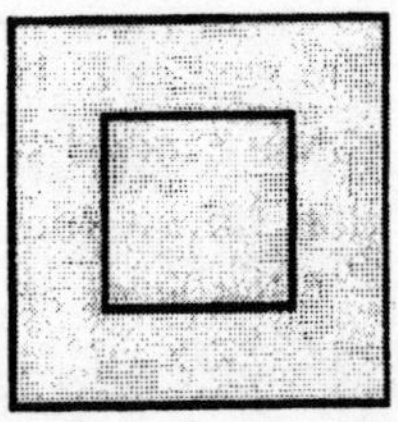

1 UŠ *mi-it-ḫa-ar-tum*
BA.TA.‹ÀM› *ad-ku-uš-ma* ÍB.SI$_8$
ad-di
i-na ŠÀ ÍB.SI$_8$ *ša-ni-tum*
ÍB.SI$_8$ *ša-lu-uš-tam ad-di*
A.ŠÀ.BI EN.NAM

The side of the square is 1 UŠ. I made a half-(border)[3] each side and I drew a square. Inside the second square I drew a third square. What is its area?

Problems (xv)–(xvi) are missing.

Column V

(xvii)

1 UŠ *mi-it-ḫa-ar-tum*
12 SAG.DÙ 4 ÍB.SI$_8$ *ad-di*
A.ŠÀ.BI EN.NAM

The side of the square is 1 UŠ. ‹Inside it› I drew 12 triangles and 4 squares. What are their areas?

[3] Written for 'half' — perhaps as an abbreviation of *bāmtum*, 'the half resulting from a "breaking" operation' (Høyrup 1990*a*: 52 & n65), i.e. half of *something*, as opposed to *mišlum*, the number $\frac{1}{2}$ = 0;30. This writing is also found in AO 8862 (i): I 12, (ii): II 19; (iii): III 13–14; BM 13901 (v): I 3. BA.TA in this text was previously read IGI *ša* (Gadd 1922; *MKT*: I 137) and ⸢ŠÀ⸣ *ša* (*TMB*: 55 no. 108).

(xviii)

1 UŠ *mi-it-ḫa-ar-tum*
ŠÀ.BA 4 SAG.DÙ *ad-di*
A.ŠÀ.BI EN.NAM

The side of the square is 1 UŠ. Inside it I drew 4 triangles. What are their areas?

Problems (xix)–(xx) are missing.

Column VI

Problems (xxi)–(xxii) are missing.

(xxiii)

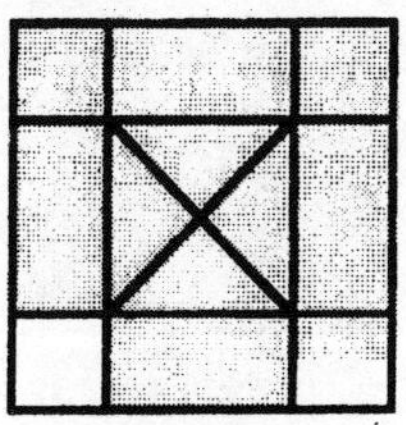

1 UŠ ÍB.SI$_8$ ŠÀ 4 ÍB.SI$_8$
4 *ṣi-il-pa-tum* 4 SAG.DÙ
A.ŠÀ.BI EN.[NAM]

The side of the square is 1 UŠ. Inside it are 4 squares, 4 rectangles and 4 triangles. What are their areas?

(xxiv)

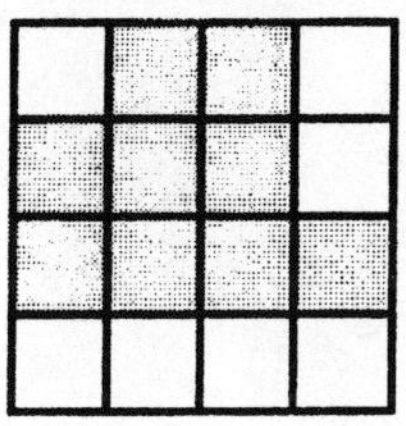

[1 UŠ *mi*]*-it-ḫa-ar-tum*
ŠÀ.BA 16 *mi-it-ḫa-ra-tim*
ad-di A.ŠÀ.BI EN.NAM

The side of the square [is 1 UŠ]. Inside it I drew 16 squares. What are their areas?

(xxv)

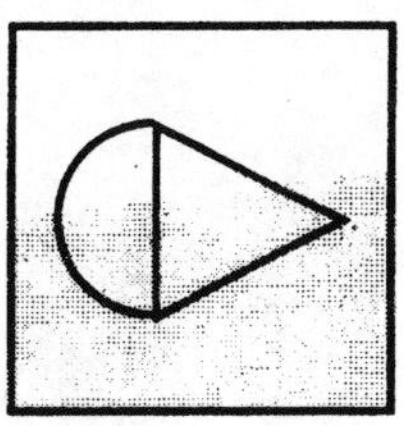

1 UŠ X [...] X SAG

The ... is 1 UŠ. ... the width.

Column VII

Problems (xxvi)–(xxvii) are missing.

(xxviii)

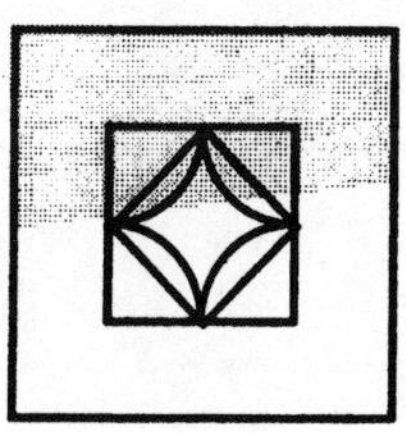

1 [UŠ ÍB].⌜SI$_8$⌝ PAD.TA
ad-ku-uš-ma
⌜ÍB⌝.[SI$_8$] ⌜*ad*⌝*-di* ŠÀ.BA ÍB.SI$_8$
[*ša ad-du*]*-ú* 1 GÁN ⌜GEŠTÚ
ZÀ.MÍ⌝
[A.ŠÀ.BI] ⌜EN⌝.NAM

The side of the square is 1 UŠ. I made a border each side and I drew a square. Inside the square that I drew is 1 concave square. What is its area?

(xxix)

[Text missing.]

Column VIII

(xxx)

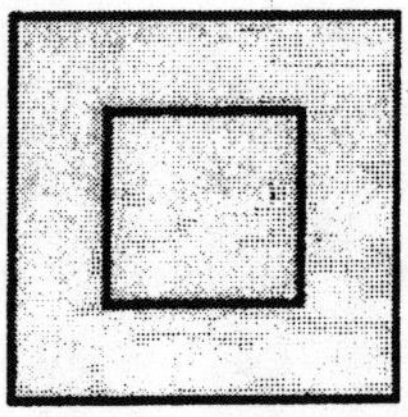

1 UŠ ÍB.SI$_8$ PAD.TA *ad-ku-uš-*⌜*ma*⌝
GÁN GIŠ.ZÀ.MÍ *ad-*⌜*di*⌝
A.ŠÀ.BI EN.NAM

The side of the square is 1 UŠ. I made a border each side and I drew the area of a lyre/harp. What is its area?

(xxxi)

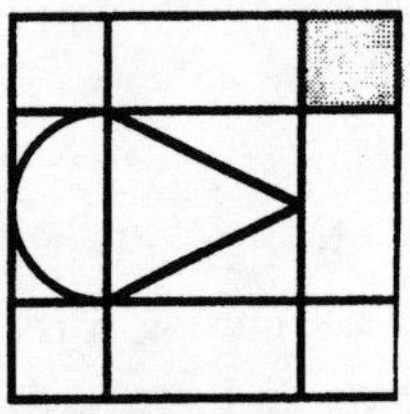

⌜UŠ⌝ 1 *mi-it-ha-ar-tum*
[ŠÀ.BA] 2 GÁN U$_4$.SAKAR
[1] SAG.DÙ 1 GÁN
U$_4$.KAK$^?$.SAKAR
1 *ṣi-li-ip-tum* 4 ÍB.SI$_8$.MEÍ
A.ŠÀ.BI EN.NAM

The side of the square is 1 UŠ. [Inside it] are 2 semi-circles, [1] triangle, 1 cone$^?$, 1 rectangle and 4 squares. What are their areas?

(xxxii)

[1] ⌜UŠ *mi-it-ḫa-ar*⌝-[*tum*]
⌜ŠÀ$^!$⌝.[BA 2] ⌜*ṣi-il*$^!$⌝-[*pa-tum*]
[1] GÁN [... 4] ⌜ÍB.SI$_8$⌝
⌜A.ŠÀ⌝.[BI] ⌜EN⌝.[NAM]

The side of the square is [1] UŠ. Inside [it are 2] rectangles, [1] oval$^?$ and 4 squares. What are their areas?

(xxxiii)

[Text missing.]

Column IX

(xxxiv)

1 UŠ *mi-it-ḫa-ar-tum*
⌜ŠÀ⌝.BA 3 GÁN GIŠ.PAN
1 *ṣi-li-ip-*⌜*tum*⌝
A.ŠÀ$^!$.BI EN.⌜NAM⌝

The side of the square is 1 UŠ. Inside it are 3 bows and 1 rectangle. What are their areas?

(xxxv)

1 UŠ ÍB.SI$_8$ ŠÀ.BA 2 GÁN GIŠ.PAN
1 GÁN GIŠ.[...].TE 4 SAG.KI.GU$_4$
[......]-*ar-tam*
[A.ŠÀ].BI EN.NAM

The side of the square is 1 UŠ. Inside it are 2 bows, 1 diamond$^?$ and 4 trapezoids. ... What are their [areas]?

(xxxvi)

[Text missing.]

Problem (xxxvii) is missing.

Column X

(xxxviii)

1 UŠ *mi-it-ḫa-ar-tum*
ŠÀ.BA 1 *ki-ip-pa-tum* 6 GÁN U$_4$.SAKAR
A.ŠÀ.BI EN.NAM

The side of the square is 1 UŠ. Inside it are 1 circle and 6 semicircles. What are their areas?

(xxxix)

1 UŠ ÍB.SI$_8$ 2 KA.[KÉŠ]
2! GÁN U$_4$.SAKAR 4 ⌜ÍB⌝.[SI$_8$]
[A].ŠÀ.BI ⌜EN⌝.[NAM]

The side of the square is 1 UŠ. ‹Inside it are› 2 circles, 2 semicircles and 4 squares. What are their areas?

(xl)

⌜1 UŠ ÍB.SI$_8$⌝ ŠÀ.BA 4 SAG.DÙ
⌜16⌝ GÁN GIŠ.MÁ.GUR$_8$
⌜5⌝ GÁN GEŠTÚ.ZÀ.MÍ
A.ŠÀ.BI EN.[NAM]

The side of the square is 1 UŠ. ‹Inside it are› 4 triangles, 16 barges, 5 concave squares. What are their areas?

One problem missing?

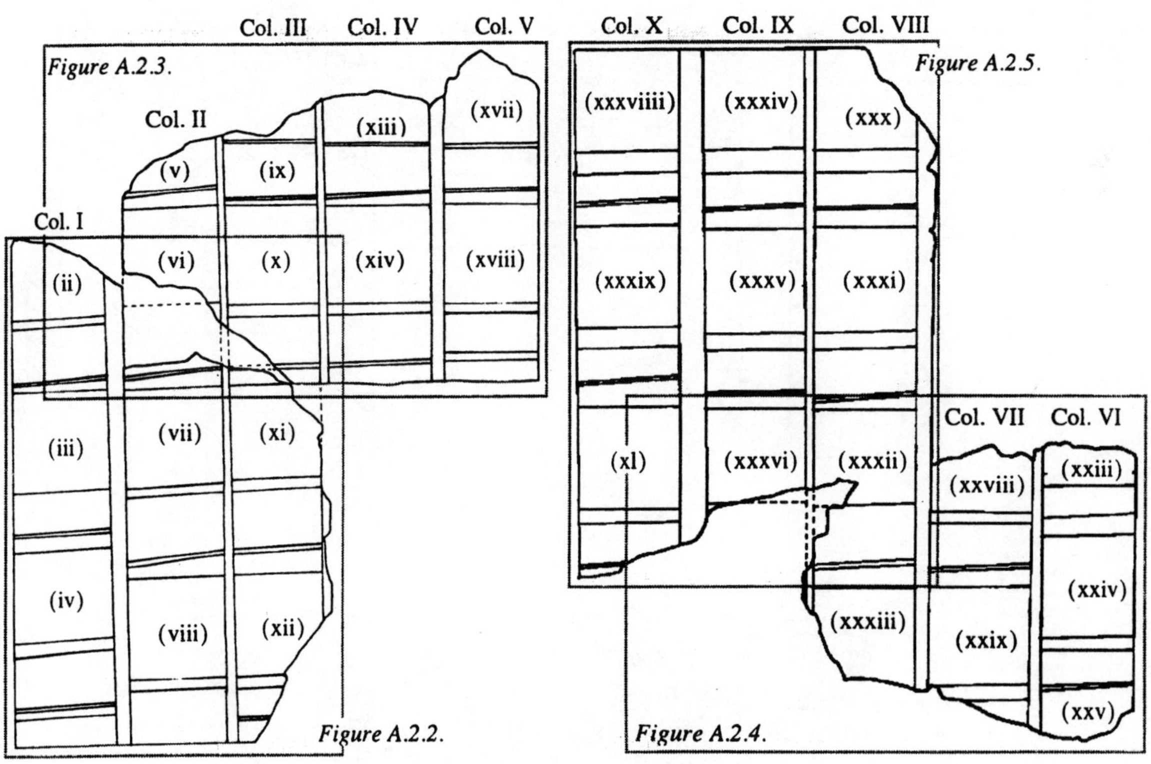

Figure A.2.1: Arrangement of problems on BM 15285 and their distribution over the following pages

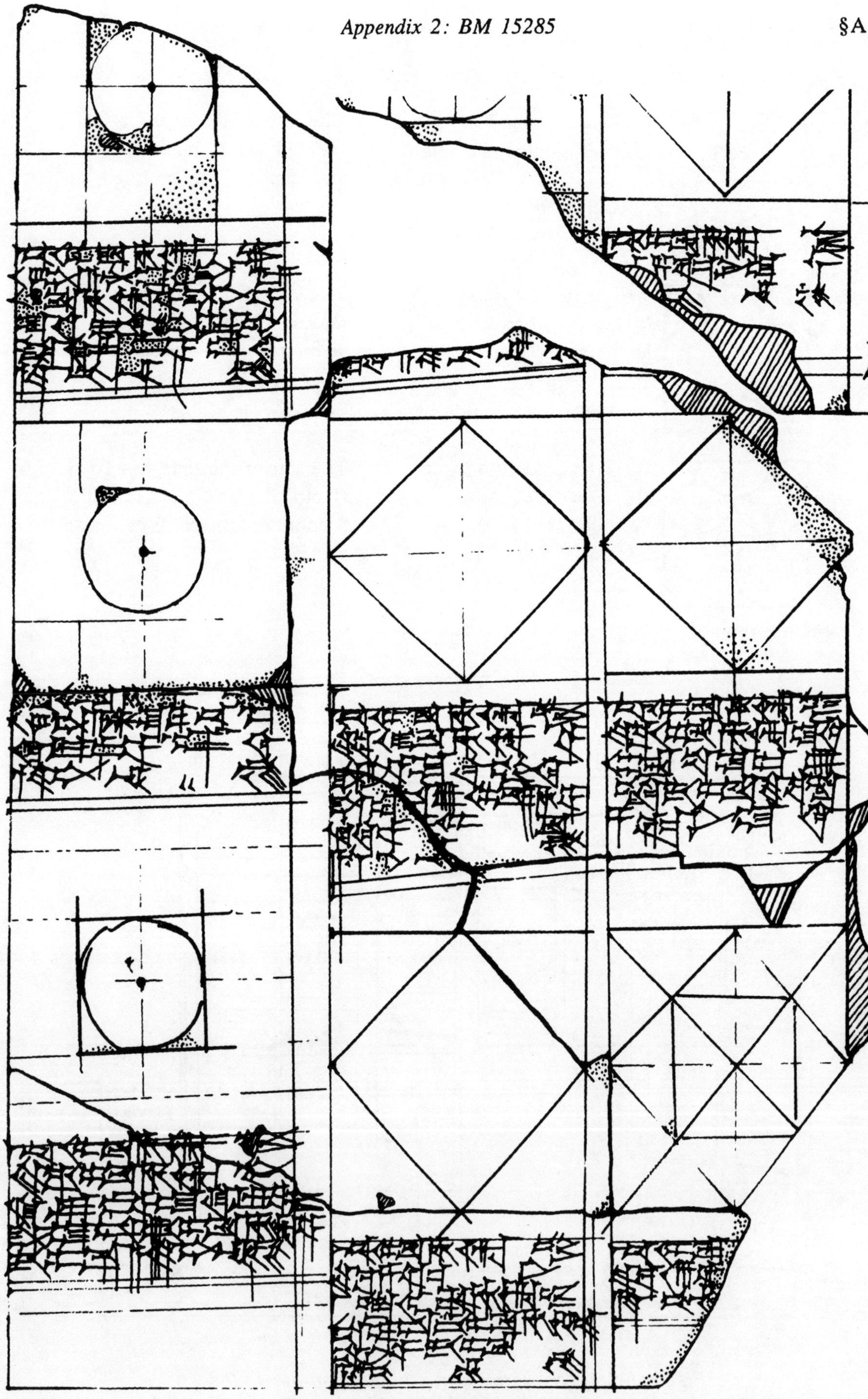

Figure A.2.2: BM 15285 obverse top

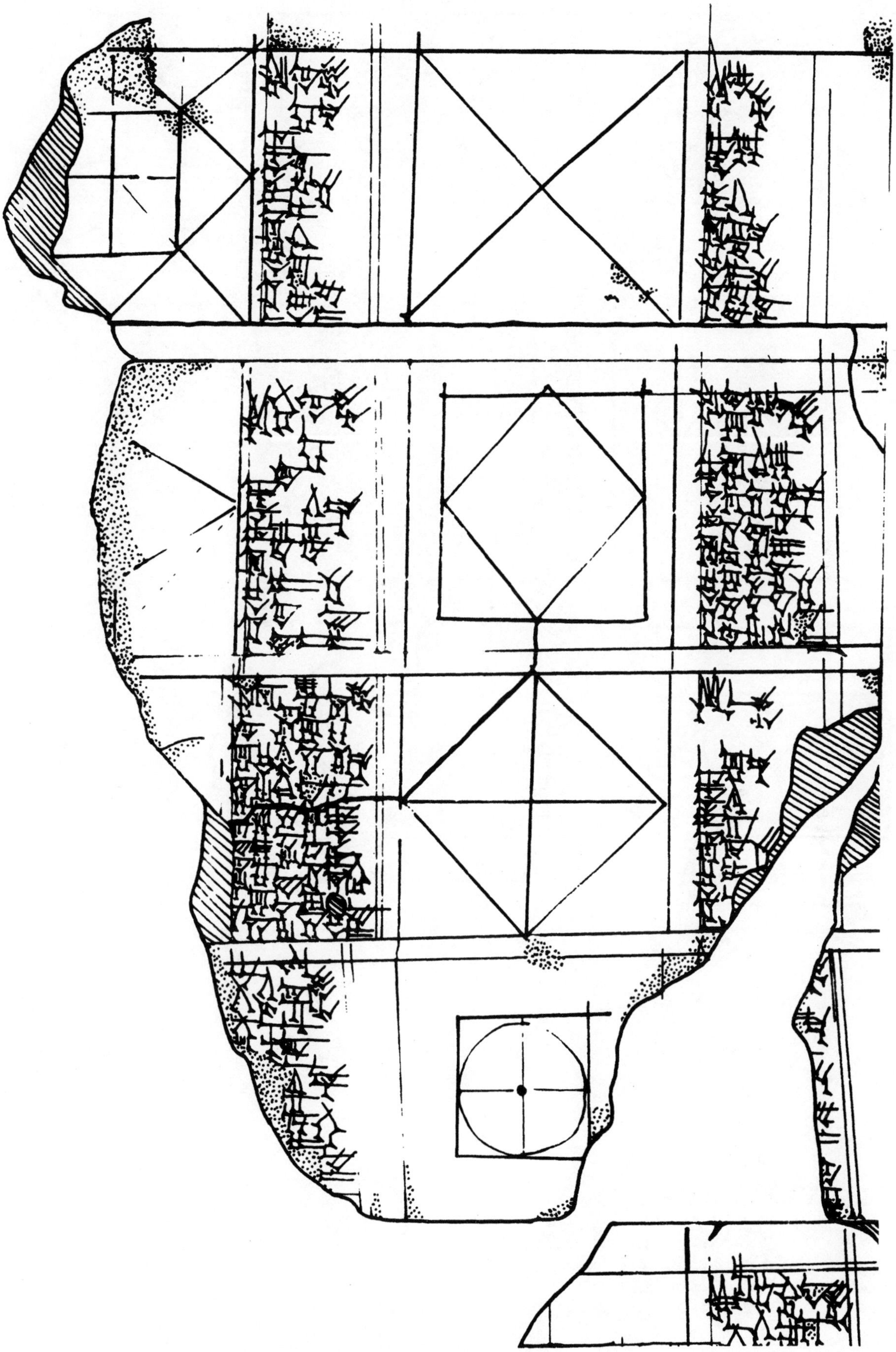

Figure A.2.3: BM 15285 obverse bottom

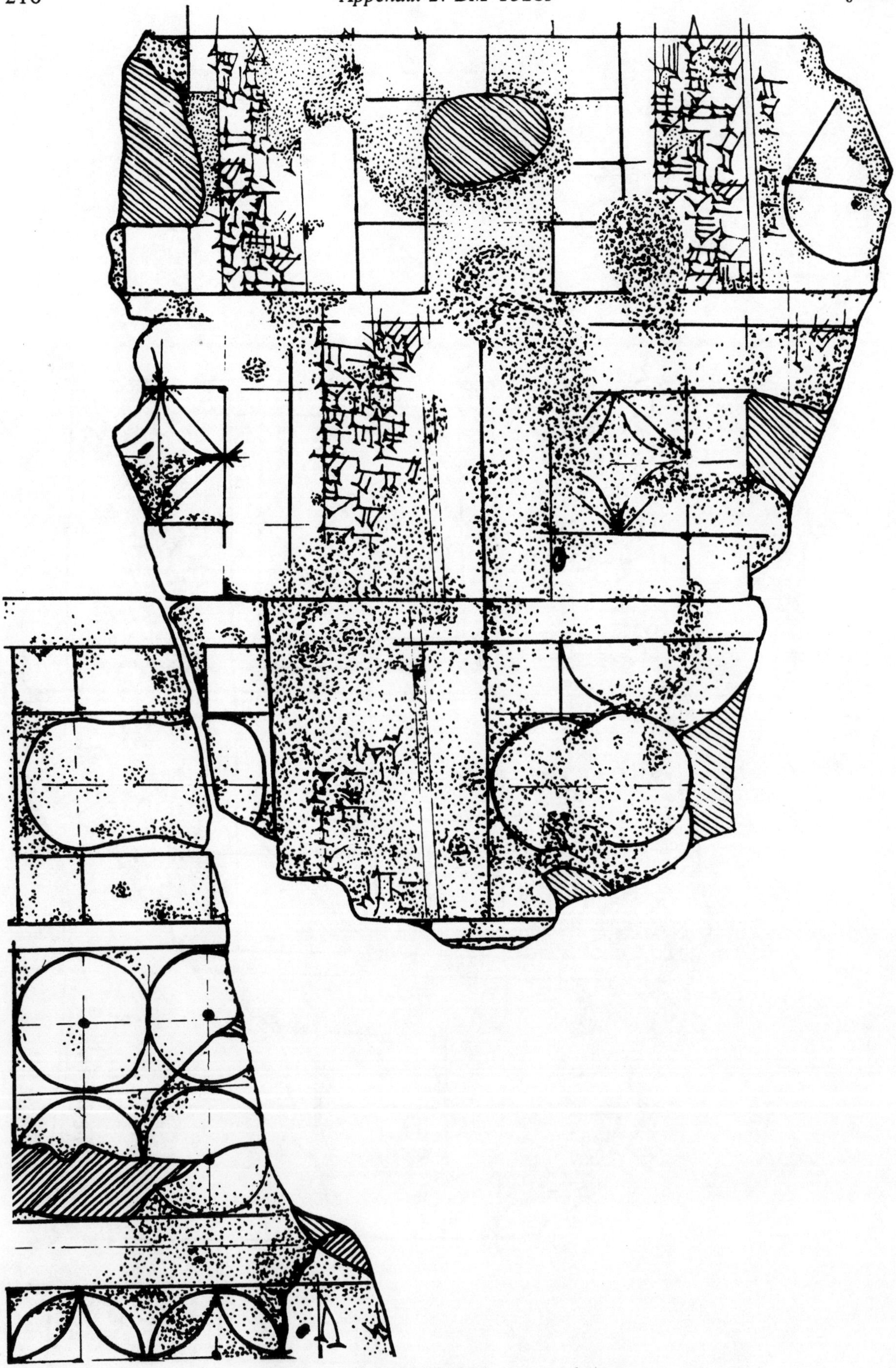

Figure A.2.4: BM 15285 reverse bottom

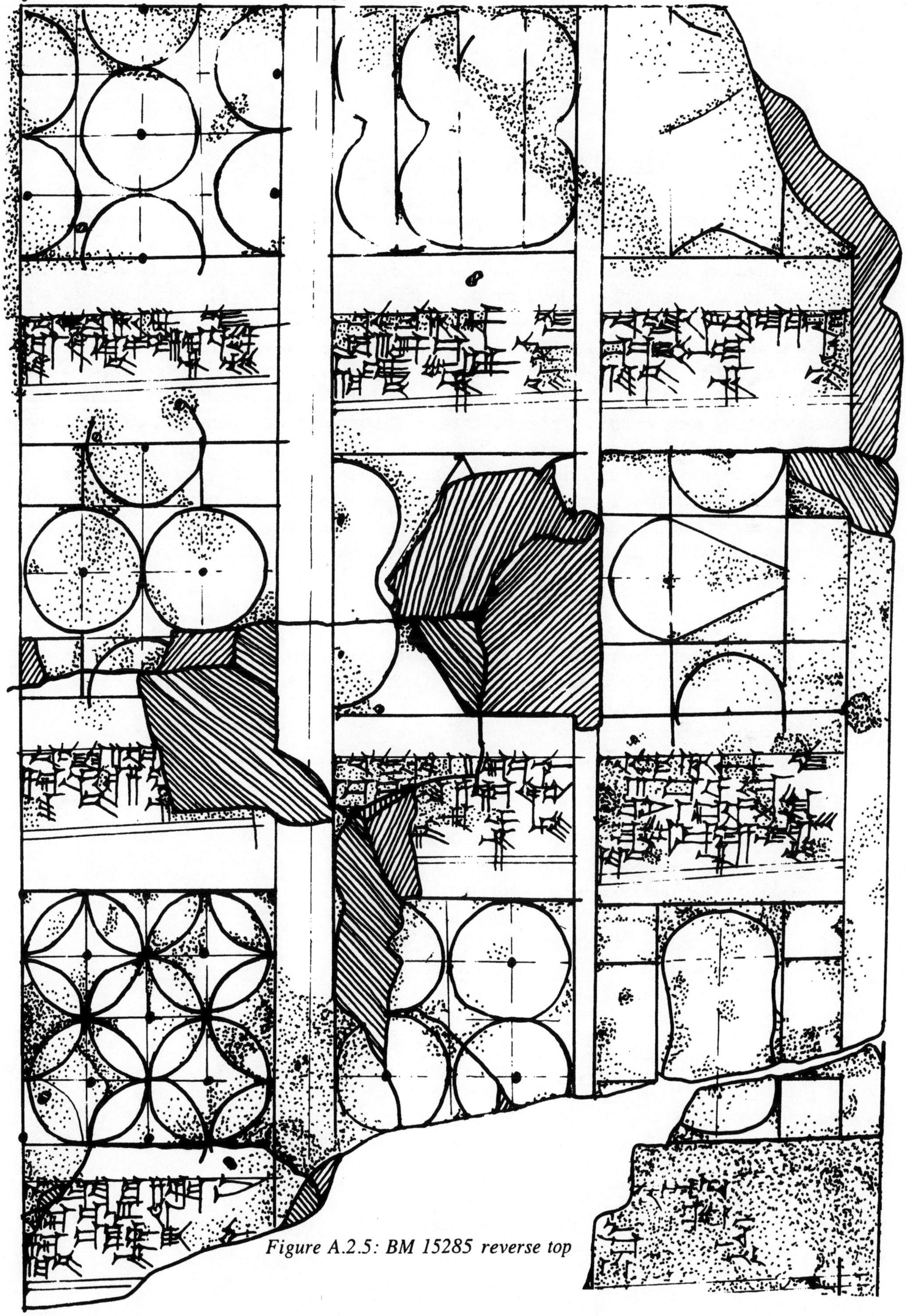

Figure A.2.5: BM 15285 reverse top

APPENDIX 3: BM 96954 + BM 102366 + SÉ 93

Together these three pieces form part of a three-column tablet, probably from Sippar. The first two were identified by Christopher Walker of the British Museum, who very generously passed his preliminary transliteration of them on to me.[1] SÉ 93, from the collection of the Couvent Saint-Étienne, was published by Jursa and Radner (1995–6: 103–5), and drawn to my attention by Stephanie Dalley. BM 96954 is a substantial piece of the lower half of the original tablet, whereas BM 102366 and SÉ 93 jointly form a smaller fragment from the top edge. The complete tablet must have measured about 155 × 220 mm, approximately the same size as BM 85194 and BM 96957 + VAT 6598 (§A.4), three-column tablets from the same archive. The exact distance between the two pieces is unclear. The problems are numbered consecutively, including probable missing ones.[2]

The text, which concerns the most sophisticated three-dimensional geometrical figure yet attested for the OB period, exhibits peculiar copying errors, which seem to be the result of steps in the solutions being omitted. The copyist noticed the arithmetical discrepancies and altered them, while failing to realise that the solutions were themselves nonsensical. See notes to the translation of problems (viii), (xv), and (xvii).

Column I

Missing. Presumably contained five simple problems, on finding:
(i) grain capacity C = 28 48 00 gur;
(ii) length l = 10 nindan;
(iii) width $\grave{w}$ = 6 nindan;
(iv) height h = 48 cubits;
(v) top t = 4 nindan;
of a grain-pile, using coefficient k = 1 30. Problems (vi)–(xvi) use the same data.[3]

Column II

(vi) 1 GUR$_7$ 10 UŠ 6 SAG 4 SAGŠU 28 48 ‹ŠE.GUR›[4]
48 SUKUD 24 *ur-dam* DAL *ù še-um* EN.NAM
⸢ZA.E⸣ IGI 48 SUKUD DU$_8$.A 1 15 *ta-mar* 1 15 *a-na*
[6 ...] TA UŠ UGU KA[!] [5] *i-ši* 7 30 *ta-mar*
[7 30 *a-na*] ⸢24⸣ *i-ši* 3 *ta-mar* 3 *i-na* 10 UŠ
[BA.ZI 7 *ta-mar* 7] ⸢DAL⸣ 3 *i-na* 6 SAG BA.ZI
[3 *ta-mar*]⸢X X⸣ *i-ši* 1 *ta-mar*
[ ] ⸢X X⸣

A grain-pile. The length is 10, the width 6, the top 4, the ‹grain capacity› 28 48 00, the height

[1] I thank Farouk Al-Rawi for suggesting improved readings of these fragments, which are discussed briefly by Friberg (1996*b*: 91–2).

[2] See §7.3 for the coefficients in (i)–(xix); §§3.1–.3 for (xvii), (xix) and (xxii)–(xxiv); §8.5.4 for (xxiii).

[3] Compare the order of problems on BM 96957 + VAT 6598, especially columns III–IV.

[4] 8 is often written defectively in this text, with one or two strokes missing in the middle row.

[5] Expected: 6 *ša* UŠ UGU SAGŠU DIRIG.

48. It went down 24. What are the cross-width and the grain?[6] You: take the reciprocal of 48, the height. You will see 0;01 15. Multiply 0;01 15 by [6] *by which the length exceeds the top*. You will see 0;07 30.[7] Multiply [0;07 30 by] 24. You will see 3.[8] [Take] 3 from 10, the length. [You will see 7.[9] The cross-width is 7]. Take 3 from 6, the width. [You will see 3. ...] multiply. You will see 1. ...[10]

(vii) Probably one missing problem. Rest of column missing, until:

(viii) [GUR$_7$ 6 SAG 48 SUKUD 28 48 ŠE GUR] UŠ SAGŠU![11]
[UL.GAR 14 UŠ SAGŠU] ⸢EN⸣.NAM ZA.[E] IGI 48
[DU$_8$.A 1 15 *ta*]-⸢*mar*⸣ 1 15 *a-na* 28 48 ŠE
[*i-ši*] ⸢36⸣ *ta-mar* IGI 1 30 IGI.GUB.BA DU$_8$.A
[40 *a*]-⸢*na*⸣ 36 *i-ši* 24 *ta-mar* IGI 6 SAG
[DU$_8$].⸢A⸣ 10 *ta-mar* 24 *a-na* 10 *i-ši* 4 *ta-mar*[12]
[4] ⸢*i*⸣-*na* UL.‹GAR› BA.ZI 10 *ta-mar* 10 UŠ 4 SAGŠU!
[*ki*]-⸢*a-am*⸣ *ne-pé-šum*

[A grain-pile. The width is 6, the height 48, the grain capacity 28 48 00. The sum of] the length and top [is 14.] What are [the length and top]? You: [take] the reciprocal of 48. [You will see 0;01 15. Multiply] 0;01 15 by 28 48 00, the grain. You will see 36 00. Take the reciprocal of 1 30, the coefficient. Multiply [by] 36 00. You will see 24.[13] [Take] the reciprocal of 6, the width. You will see 0;10. Multiply 24 by 0;10. You will see 4.[14] Take [4] from the sum. You will see 10. The length is 10; the top is 4. This is the method.[15]

(ix) [GUR$_7$] ⸢6 SAG 48 SUKUD 28⸣ [4]⸢8⸣ ŠE GUR
[$^1/_2$] ⸢SAGŠU⸣ *ki-ma* IGI.5.GÁL UŠ UŠ SAGŠU! ⸢EN.NAM⸣
⸢5⸣ NUMUN UŠ 2 NUMUN SAGŠU GAR.RA NIGÍN.NA IGI 48
DU$_8$.A 1 15 *ta-mar* 1 15 *a-na* 28 48
i-ši 36 *ta-mar* IGI 1 30 IGI.GUB DU$_8$.A
40 *ta-mar* 40 *a-na* 36 *i-ši* 24 *ta-mar*
IGI 6 SAG DU$_8$.A 10 *ta-mar* 10 *a-na* 24 *i-ši*
4 *ta-mar* $^1/_2$ 4 *ḫe-pé* 2 *ta-mar* GAR.RA

[6] i.e., the length half-way down, and the grain capacity of the top half of the pile. Compare Haddad 104 (iii), in which grain is removed first from the top, then from the bottom, of a container in the shape of a truncated cone.

[7] The combined slope of the pile, or $(l-t)/h = w/h$.

[8] $l - {}^1/_2(l-t) = {}^1/_2 l + {}^1/_2 t = t + {}^1/_2(l-t) = l'$, the length half-way down.

[9] $(l-t)/h \times {}^1/_2 h = {}^1/_2(l-t)$.

[10] The rest of the solution is missing, but it presumably went on to calculate the grain capacity of the top half of the pile, using the method ${}^1/_3 kh'w'(l' + {}^1/_2 t)$, where h' is the half-height and w' is the width half-way down. The answer is 5 24 00 gur.

[11] SAG written for SAGŠU (= U + SAG); also in lines 7 and 10 of this column.

[12] Following line missing. Should probably read: 4 *šu-ul-li-iš* 12 *ta-mar* 12 TAB.BA 24 *ta-mar*, with the first half of line 7 reading: 14 *i-na* 24 BA.ZI 10 *ta-mar* instead of [4] ⸢*i*⸣-*na* UL.‹GAR›. See notes to translation.

[13] Problems (viii)–(xvi) all start with this procedure, which divides the capacity $C = {}^1/_3 kwh(l + {}^1/_2 t)$ by k and h to establish that ${}^1/_3 w(l + {}^1/_2 t) = 24$. The problems are each concerned with the numerical relationships between two of l, w, and t, so the values of k and h are essentially irrelevant.

[14] This gives ${}^1/_3(l + {}^1/_2 t) = 4$.

[15] The preserved method, calculating that $l + t - {}^1/_3(l + {}^1/_2 t) = l$, is a meaningless 'simplification' of the correct solution, as it is clearly not true for all values of l. It seems that a line of solution was omitted by the scribe, probably because it ended with same word as the previous line. The following line was then altered to provide a continuous, if nonsensical solution. The missing text probably went: 'Triple 4. You will see 12. Double 12. You will see 24 (i.e. $2l + t = 24$)'. The following line would have begun: 'Take the sum from 24. You will see 10' (i.e. $2l + t - (l + t) = l = 10$).

$^1/_2$ 2 NUMUN SAGŠU *ḫe-pé* 1 *ta-mar a-na* 5 NUMUN UŠ
DAḪ.ḪA 6 *ta-mar* IGI.3.GÁL BA.ZI 2 *ta-mar*
5 *a-na* 2 *i-ši* 10 *ta-mar* 2 *a-na* 2 NUMUN SAGŠU
i-ši 4 *ta-mar* SAGŠU *ne-pé-šum*

[A grain-pile.] The width is 6, the height 48, the grain capacity 28 48 00. [$^1/_2$] the top is like a fifth of the length. What are the length and the top? ‹You:› put down 5, the ratio of the length; 2, the ratio of the top.[16] Return. Take the reciprocal of 48. You will see 0;01 15. Multiply 0;01 15 by 28 48 00. You will see 36 00. Take the reciprocal of 1 30, the coefficient. You will see 0;00 40. Multiply 0;00 40 by 36 00. You will see 24. Take the reciprocal of 6, the width. You will see 0;10. Multiply 0;10 by 24. You will see 4. Break 4 in half. You will see 2. Put it down. Break 2, the ratio of the top, in half. You will see 1. Add to 5, the ratio of the length. You will see 6. Take a third. You will see 2. Multiply 5 by 2.[17] You will see 10. Multiply 2 by 2, the ratio of the top. You will see 4, the top. The method.

(x) GUR$_7$ 48 SUKUD 28 48 ‹ŠE GUR› 10 UŠ $^2/_3$ SAG
[*ki-ma* SAGŠU] ⌈SAG *ù* SAGŠU EN.NAM ZA.E⌉ IGI 48
[DU$_8$.A 1 15 *ta-mar* 1 15 *a-na* 28 48]
[*i-ši* 36 *ta-mar* IGI 1 30 IGI.GUB DU$_8$.A]
[40 *ta-mar* 40 *a-na* 36 *i-ši* 24 *ta-mar*][18]
[ ]-⌈*na*⌉ 10 UŠ
[ ] ⌈*i*⌉-*ši*

A grain-pile. The height is 48, the grain capacity 28 48 00, the length 10. $^2/_3$ of the width [is like the top]. What are the width and top? You: [take] the reciprocal of 48. [You will see 0;01 15. Multiply 0;01 15 by 28 48 00. You will see 36 00. Take the reciprocal of 1 30, the coefficient. You will see 0;00 40. Multiply 0;00 40 by 36 00. You will see 24. Take a third of 24. You will see 8.][19]

Two or three lines missing at the end of the column.

Column III

(xi) ⌈3⌉ 20 *ta-mar* $^1/_2$ 3 20 *ḫe-pé* 1 40 *ta-mar*
1 40 NIGIN 2 46 40 *ta-mar* 2 46 40 *a-na* 2 40 DAḪ$^!$.ḪA
5 26 40 *ta-mar* EN.NAM ÍB.SI$_8$ 2 20 ÍB.SI$_8$
1 40 *i-na* 2 20 BA.ZI 40 *ta-mar* $^1/_2$ 40 *ḫe-pé* 20 *ta-mar*
20 NIGIN 6 40 *ta-mar* IGI 6 40 DU$_8$.A 9 *ta-mar*
40 *a-na* 9 *i-ši* 6 *ta-mar* SAG 40 *a-na* 6 *i-ši*
4 *ta-mar* ⌈4⌉ SAGŠU
ki-a-am ⌈*ne*⌉-*pé-šum*

You will see 3;20. Break 3;20 in half. You will see 1;40. Square 1;40. You will see 2;46 40. Add 2;46 40 to 2;40. You will see 5;26 40. What is the square root? 2;20 is the square root.

[16] Because this is the smallest integer solution to $l/5 = t/2$. The problem is to find n, where $l = nl'$ and $t = nt'$. This method is common in OB mathematics, and is sometimes known as the 'rule of false value' (because $l' = 5$ and $t' = 2$ are 'false values' of l and t) (Friberg 1987–90: 574–5).

[17] Not the 2 just seen (equal to the 2 kept), but the 2 by which the 4 was halved in line 16′. For the solution to be convincing, lines 15′ – 16′ should come after lines 17′–18′. Lines 17′–18′ calculate $^1/_3(l' + {}^1/_2t')$, while 15′ – 16′ establish that $^1/_3(l' + {}^1/_2t') = {}^1/_6(l + {}^1/_2t)$, i.e. that l and t are twice l' and t' (calculated in lines 19′–20′).

[18] Restoration and line division based on (ix): II 12′–14′. Rest of solution uncertain.

[19] The solution so far has simply reduced the problem to $t = {}^2/_3w$ and $^1/_2w(l + {}^1/_2t) = 24$.

Take 1;40 from 2;20. You will see 0;40. Break 0;40 in half. You will see 0;20. Square 0;20. You will see 0;06 40. Take the reciprocal of 0;06 40. You will see 9. Multiply 0;40 by 9. You will see 6, the width. Multiply 0;40 by 6. You will see 4. The top is 4. This is the method.[20]

(xii) ⌜GUR$_7$⌝ [ ] ⌜ÀM?⌝ *im-ta-ḫar* 6 SUKUD *še-um* EN.NAM
[ ] ⌜5?⌝ *ta-mar* 10 *a-na* 5 *i-ši* 50 *ta-mar*
[ ] ⌜X⌝ 3 45 *ta-mar* X
[ ] ⌜30 *ta-mar*⌝ SAḪAR.ḪI.A
[ ] ⌜*ta-mar*⌝

A grain-pile. [...][21] *The square sides are each* [...]. The height is 6. What is the grain? [You: ...] You will see (...) 5?. Multiply 10 by 5. You will see 50. [...] You will see (...) 3 45 [...] You will see (...) 30, the volume. [...] You will see [...]

Rest of column lost until:

(xiii) [GUR$_7$ 48 SUKUD 4 SAGŠU] ⌜28 48⌝ ŠE GUR
[UŠ *ù* SAG UL.GAR]⌜16⌝ UŠ SAG EN.NAM
[ZA.E IGI] ⌜48⌝ SUKUD DU$_8$.A 1 15 *ta-mar*
[1 15] *a-na* 28 48 *i-ši* 36 *ta-mar*
⌜IGI⌝ 1 30 IGI.GUB DU$_8$.A 40 *ta-mar* 40 *a-na* 36 *i-ši*
24 *ta-mar* IGI.3.GÁL 24 BA.ZI 8 *ta-mar*
$^1/_2$ 4 SAGŠU *ḫe-pé* 2 *ta-mar* $^1/_3$ 2 BA.ZI
40 *ta-mar* $^1/_3$ 16! BA.ZI 5 ‹20› *ta-mar* 40 *a-na* 5 20 DAḪ.ḪA
6 *ta-mar* $^1/_2$ 6 *ḫe-pé* 3 *ta-mar* NIGIN 9 *ta-mar*
8 *i-na* 9 BA.ZI ‹1› *ta-mar* 1 EN.NAM «1» ÍB.SI$_8$
1 ÍB.SI$_8$ 1 *a-na* 3 DAḪ.ḪA 4 *ta-mar* 1 *i-na* 3 BA.ZI
2 *ta-mar* 4 *šu-ul-li-iš* 12 *ta-mar*
2 *i-na* 12 BA.ZI 10 *ta-mar* UŠ *i-na* 16 UL.GAR ‹BA.ZI›
6 *ta-mar* 6 SAG *ne-pé-šum*

[A grain-pile. The height is 48, the top 4], the grain capacity 28 48 00. [The sum of the length and width is] 16. What are the length and width? [You:] take [the reciprocal of] 48, the height. You will see 0;01 15. Multiply [0;01 15] by 28 48 00. You will see 36 00. Take the reciprocal of 1 30, the coefficient. You will see 0;00 40. Multiply 0;00 40 by 36 00. You will see 24. Take a third of 24. You will see 8.[22] Break 4, the top, in half. You will see 2. Take $^1/_3$ of 2. You will see 0;40. Take $^1/_3$ of 16. You will see 5;‹20›. Add 0;40 to 5;20. You will see 6. Break 6 in half. You will see 3.[23] Square. You will see 9. Take 8 from 9. You will see ‹1›.[24] What is the square root of «1»? The square root is 1.[25] Add 1 to 3. You will see 4. Take 1 from 3. You will see 2.[26] Triple 4. You will see 12. Take 2 from 12.[27] You will see 10, the length. Take from 16, the sum. You will see 6. The width is 6. The method.

(xiv) GUR$_7$ 4 SAGŠU 28 48 ‹ŠE GUR› 48 SUKUD $^1/_2$ UŠ *ù* 1 SAG

[20] The traces suggest that the problem was to find the top and width of the pile, given their sum, $t + w = 10$.

[21] This may in fact be the beginning of problem (xiii).

[22] This division by 3, like those in the rest of the solution, is simply a numerical simplification to keep the arithmetic as simple as possible. It contributes nothing to the method of solution.

[23] This gives $^1/_6(l + w + {}^1/_2t) = 3$.

[24] This gives $\{(l + {}^1/_2t)^2 + 2w(l + {}^1/_2t) + w^2\}/36 - w(l + {}^1/_2t)/9 = \{(l + {}^1/_2t)^2 - 2w(l + {}^1/_2t) + w^2\}/36 = 1$

[25] This gives $^1/_6(l + {}^1/_2t - w) = 1$

[26] This gives $\{(l + {}^1/_2t + w)\pm(l + {}^1/_2t - w)\}/6 = {}^1/_3(l + {}^1/_2t) = 4$, and $^1/_3w = 2$.

[27] i.e. $l + {}^1/_2t - {}^1/_2t$ (calculated in line 7′). The value $^1/_3w$ just produced is not used.

UŠ SAG EN.NAM ZA.E IGI 48 SUKUD DU$_8$.A!
1 15 *ta-mar a-na* 28 48 *i-ši* 36 *ta-mar*
IGI 1 30 IGI.GUB DU$_8$.A 40 *ta-mar* 40 *a-na* 36 *i-ši*
24 *ta-mar aš-šum* $^1/_2$ UŠ *ù* 1 NUMUN SAG DUG$_4$.GA
1 *ù* 30 GAR.RA NIGÍN.NA $^1/_2$ 4 SAGŠU *ḫe-pé* 2 *ta-mar*
$^1/_3$ 2 BA.ZI 40 *ta-mar* $^1/_3$ 1 BA.ZI 20 *ta-mar*
20 *a-na* 30 *i-ši* 10 *ta-mar* 24 *a-na* 10 *i-ši* 4 *ta-mar*
4 EN.NAM ÍB.SI$_8$ 2 ÍB.SI$_8$ IGI 30 DU$_8$.A 2 *ta-mar*
2 *a-na* 2 *i-ši* 4 *ta-mar* 4 *šu-ul-li-iš*
12 *ta-mar* 2 *i-na* 12 BA.ZI 10 *ta-mar* UŠ
4 SAGŠU *i-na* 10 UŠ BA.ZI 6 *ta-mar* SAG
ki-a-am ne-pé-šum

A grain-pile. The top is 4, ‹the grain capacity› 28 48, the height 48. $^1/_2$ the length plus 1 equals the width. What are the length and width? You: take the reciprocal of 48, the height. You will see 0;01 15. Multiply by 28 48 00. You will see 36 00. Take the reciprocal of 1 30, the coefficient. You will see 0;00 40. Multiply 0;00 40 by 36 00. You will see 24. Because they said $^1/_2$ the length plus 1 equals the ratio of the width, put down 1 and 0;30. Return. Break in half 4, the top. You will see 2. Take $^1/_3$ of 2. You will see 0;40. Take $^1/_3$ of 1. You will see 0;20. Multiply 0;20 by 0;30. You will see 0;10. Multiply 24 by 0;10. You will see 4. What is the square root of 4? The square root is 2. Take the reciprocal of 0;30. You will see 2. Multiply 2 by 2. You will see 4. Triple 4. You will see 12. Take 2 from 12. You will see 10, the length. Take 4, the top, from 10, the length. You will see 6, the width. This is the method.[28]

(xv) GUR$_7$ 4 SAGŠU 28 48 ŠE GUR 48 SUKUD *i-na* 1 KÙŠ
⌜7 30 KÙŠ⌝ ŠÀ.GAL UŠ *ù* SAG EN.NAM ZA.E
[7 30 ŠÀ.GAL] *a-na* 48 SUKUD *i-ši* 6 *ta-mar*!
a-na 4 SAGŠU DAḪ.ḪA 10 *ta-mar* UŠ
IGI 48 DU$_8$.A 1 15 *ta-mar a-na* 28 48
i-ši 36 ‹*ta-mar*› 40 *a-na* 36 *i-ši* ⌜24⌝ [29]
⌜*ta*⌝*-mar* $^1/_2$ 24 *ḫe-pé* 12 *ta-mar* ⌜IGI.3.GÁL⌝ ‹BA.ZI›
[4] *ta-mar* IGI 4 DU$_8$.A 15 *ta-mar*

Column IV

24 *a-na* 15 *i-ši* 6 *ta-mar* SAG
ki-a-am ne-pé-šum

A grain-pile. The top is 4, the grain capacity 28 48 00, the height 48. In 1 cubit the slope is 1;30 cubits. What are the length and width? You: multiply [0;07 30, the slope,] by 48, the height. You will see 6. Add to 4, the top.[30] You will see 10, the length. Take the reciprocal of 48. You will see 0;01 15. Multiply by 28 48 00. You will see 36 00. Multiply 0;00 40 by 36 00. You will see 24.[31] Break 24 in half. You will see 12. ‹Take› a third. You will see [4].[32]

[28] This solution is corrupt; the original, correct procedure has not been established.

[29] Following line missing; it should probably read: *ta-mar* 10 TAB.BA 20 *a-na* 4 DAḪ.ḪA 24. See note 31.

[30] i.e. $h \times (l - t)/h = l - t = 6$.

[31] There is a line missing here, which probably went: Double 10, the length. You will see 20. Add to 4, the top. You will see 24. It was omitted either accidentally because it ended in 24 like the previous one, or deliberately because it was thought to be redundant. This calculated $2l + t = 24$.

[32] This gives $^1/_3(l + {}^1/_2 t) = 4$

Take the reciprocal of 4. You will see 0;15. Multiply 24 by 0;15. You will see 6, the width.[33] This is the method.

(xvi) GUR$_7$ 10 UŠ 48 SUKUD 28 48 ŠE GUR
ma-la SAGŠU *ù* $^1/_3$ SAG UL.GAR 6 SAGŠU SAG ‹EN.NAM›
ZA.E IGI 48 SUKUD DU$_8$.‹A› 1 15 *ta-mar* 1 15 *a-na*
28 48 *i-ši* 36 *ta-mar* 40 *a-na*
36 *i-ši* 24 *ta-mar* 1 30 *ù* 30
NUMUN GAR.RA IGI.3.GÁL 10 UŠ BA.ZI 3 20 *ta-mar*
$^1/_3$ 1 30 BA.ZI 30 *ta-mar* IGI.3.GÁL 30 BA.ZI
10 *ta-mar* 1 30 *a-na* 10 *i-ši* 15 *ta-mar*
1 30 *a-na* 3 20 *i-ši* 5 *ta-mar* 24
a-na 15 *i-ši* 6 *ta-mar* 5 *i-na* 6 BA.ZI
1 *ta-mar* IGI 10 DU$_8$.A 6 *ta-mar a-na* 1 *i-ši*
6 *ta-mar* SAG 40 *a-na* 6 SAG *i-ši* 4 *mu-ḫu*!
ki-a-am ne-pé-šum

A grain-pile. The length is 10, the height 48, the grain capacity 28 48 00. The sum of all the top and $^1/_3$ of the width is 6. What are the top and width? You: take the reciprocal of 48, the height. You will see 0;01 15. Multiply 0;01 15 by 28 48 00. You will see 36 00. Multiply 0;00 40 by 36 00. You will see 24. Put down 1;30 and 0;30, the ratios. Take a third of 10, the length. You will see 3;20. Take $^1/_3$ of 1;30. You will see 0;30. Take a third of 0;30. You will see 0;10. Multiply 1;30 by 0;10. You will see 0;15. Multiply 1;30 by 3;20. You will see 5. Multiply 24 by 0;15. You will see 6. Take 5 from 6. You will see 1. Take the reciprocal of 0;10. You will see 6. Multiply by 1. You will see 6, the width. Multiply 0;40 by 6. The top is 4. This is the method.[34]

(xvii) GUR$_7$ SAG.DÙ 30 UŠ 10 SAG 48 SUKUD
še-um EN.NAM ZA.E 30 UŠ *a-na* 10 SAG ‹*i-ši*› *ta-mar* (sic)
5 *ta-mar a-na* 48 SUKUD *i-ši* 4 *ta-mar*
1 30 *a-na* 4 *i-ši* 6 *ta-mar* 6 × 3600 ŠE GUR
ki-a-am ne-pé-šum

A triangular grain-pile. The length is 30, the width 10, the height 48. What is the grain? You: ‹multiply› 30, the length, by 10, the width. You will see 5 00. Multiply by 48, the height. You will see 4 00 00. Multiply 1 30 by 4 00 00. You will see 6 00 00 00. The grain capacity is 6 00 00 00 gur. This is the method.[35]

(xviii) GUR$_7$ 30 UŠ 20 SAG AN.NA 10 SAG KI.TA
48 SUKUD SAḪAR.ḪI.A *ù še-um* ‹EN.NAM› ZA.E SAG AN.NA
ù SAG KI.TA UL.GAR 30 *ta-mar* $^1/_2$ 30 *ḫe-pé*
15 *ta-mar* 15 *a-na* 30 *i-ši* 7 30 *ta-mar*
7 30 *a-na* 48 SUKUD *i-ši* 6 *ta-mar* SAḪAR.ḪI.A
[1] 30 *a-na* 6 *i-ši* 9 *ta-mar še-um*
[*ki*]-⌈*a-am*⌉ *ne-pé-šum*

[33] This gives $^1/_3 w(l + ^1/_2 t) \div ^1/_3(l + ^1/_2 t) = w$

[34] This solution is corrupt; I have not yet been able to establish the original, correct procedure.

[35] This method gives the volume of a rectangular block. The product of length and width should be divided by 2. Perhaps the original triangle had length 30 and width 20, as in A 24194 and A 24195, but was then modified to harmonize with the miscopied solution.

A grain-pile. The length is 30, the upper width 20, lower width 10, the height 48. ‹What are› the volume and grain? You: add the upper width and the lower width. You will see 30. Break 30 in half. You will see 15. Multiply 15 by 30. You will see 7 30. Multiply 7 30 by 48, the height. You will see 6 00 00, the volume. Multiply [1] 30 by 6 00 00. You will see 9 00 00 00, the grain. This is the method.[36]

(xix) [GUR$_7$ GÁN U$_4$.SAKAR] 30 GÚR 20 DAL 48 SUKUD
[SAḪAR.ḪI.A *ù še-um* EN.NAM ZA].⌜E⌝ 30 UŠ *a-na* 20 DAL
[*i-ši* 10 *ta-mar a-na* 15 IGI].GUB.BA GÁN U$_4$.SAKAR
[*i-ši* 2 30 *ta-mar* 2 30 *a-na* 48] ⌜SUKUD⌝
[*i-ši* 2 *ta-mar* 2 SAḪAR.ḪI.]A[37]
[2 *a-na* 1 30 *i-ši* 3 *ta-mar še-um ne-pé*]-⌜*šum*⌝

[A semicircular grain-pile.] The semicircumference is 30, the diameter 20, the height 48. [What are the volume and grain? You: multiply] 30, the length, by 20, the diameter. [You will see 10 00. Multiply by 0;15,] the coefficient of a semicircle. [You will see 2 30. Multiply by 48,] the height. [You will see 2 00 00, the volume. Multiply 2 00 00 by 1 30. You will see 3 00 00 00, the grain. This is the method.][38]

(xx) [*bé-ru-tum* 30 GÚR 1 SUKUD] *i-na* 1 KÙŠ
[ŠÀ.GAL EN.NAM ZA].⌜E IGI⌝ 1 SUKUD DU$_8$.A 1 *ta-mar*
[1 *a-na* 30 GÚR *i*]-⌜*ši*⌝ 30 ⌜*ta-mar*⌝ $^1/_2$ 30 *ḫe-pé*
[15 *ta*]-⌜*mar*⌝ 15 *i-na* 1 [KÙŠ] ⌜ŠÀ⌝.GAL
[*ki-a*]-⌜*am*⌝ *ne-pé-šum*

[A heap.[39] The circumference is 0;30, the height 1.] In 1 cubit, [what is the slope? You:] Take the reciprocal of 1, the height. You will see 1. [Multiply 1 by 0;30, the circumference.] You will see 0;30. Break 0;30 in half. You will see [0;15]. The slope is 0;15 in 1 [cubit. This is] the method.

(xxi) ⌜*bé*⌝-*ru-tum* 30 GÚR *i-na* 1 KÙŠ 15 ŠÀ.GAL
SUKUD EN.NAM ZA.E 15 ŠÀ.GAL TAB.BA 30 *ta-mar*
IGI 30 DU$_8$ 2 *ta-mar* 30 GÚR *a-na* 2 *i-ši*
1 *ta-mar* SUKUD *ne-pé*-⌜*šum*⌝

A heap. The circumference is 30. In 1 cubit the slope is 0;15. What is the height? You: Double 0;15, the slope. You will see 0;30. Take the reciprocal of 0;30. You will see 2. Multiply 0;30, the circumference, by 2. You will see 1, the height. The method.

Column V

(xxii) [*bé-ru-tum* ... *ta*]-⌜*mar*?⌝
[1 15!] ⌜GAR⌝.RA
[IGI 5 IGI.GUB DU$_8$.A 12 *ta-mar a-na* 1] ⌜15⌝ *i-ši* 15 *ta-mar*
[15 ÍB.SI$_8$ EN.NAM 30] ÍB.SI$_8$ 30 NINDAN GÚR
[*ki-a-am*] ⌜*ne*⌝-*pé-šum*

36 Correct solution.

37 This is the beginning of the fragment SÉ 93; there may in fact be several lines missing here.

38 Correct solution. See §3.2 for the semicircle coefficient.

39 For a discussion of the word *berūtum* see Jursa and Radner (1995–6: 104).

[A heap. ...] you will see [...] Put down [0;01 15!. Take the reciprocal of 0;05, the coefficient. You will see 12]. Multiply [by 0;01] 15. You will see 0;15. [What is the square root of 0;15?] The square root is [0;30]. The circumference is 0;30 nindan. [This is] the method.[40]

(xxiii) [*bé-ru*]-⌜*tum*⌝ 25 SAḪAR.ḪI.A MA ZA RUM[41] *ù di-ik-šum*
⌜EN.NAM⌝ ZA.E 25 *šu-ul-li-iš* 1 15 *ta-mar*
IGI 5 IGI.GUB DU$_8$.A 12 *ta-mar* 12 *a-na* 1 15 *i-ši*
15 *ta-mar* GAR.RA NIGIN.NA IGI 15 ŠÀ.GAL DU$_8$.A 4 *ta-mar*
15 *a-na* 4 *i-ši* 1 *ta-mar* 1 EN.NAM ÍB.SI$_8$ 1 ÍB.SI$_8$
SUKUD 15 ŠÀ.GAL TAB.BA 30 *ta-mar* 30 *a-na* 1 SUKUD *i-ši*
⌜30⌝ *ta-mar* MA ZA RUM *ne-pé-šum*

[*A heap.*] The volume is 25. What are the *mazarum* and *dikšum*?[42] You: triple 25. You will see 1 15. Take the reciprocal of 0;05, the coefficient. You will see 12. Multiply 12 by 1 15. You will see 15 00. Put it down. Return. Take the reciprocal of 0;15, the slope.[43] You will see 4. Multiply 15 00 by 4. You will see 1 00 00. What is the square root of 1 00 00? The square root is 1 00, the height. Double 0;15, the slope. You will see 0;30. Multiply 0;30 by 1 00, the height. You will see 30, the *mazarum*. The method.[44]

(xxiv) [*bé-ru-tum*] 30 MA ZA RUM 1 SUKUD 25 SAḪAR.ḪI.A 2½ NINDAN
[...] X SAḪAR.ḪI.A EN.NAM ZA.E IGI 1 SUKUD DU$_8$.A
[1 *ta-mar* 1 *a*]-⌜*na*⌝ 30 MA ZA RUM *i-ši* 30 *ta-mar* 30 *a-na*
[30 ... *i*]-⌜*ši*⌝ 15 *ta-mar* 15 *i-na* 30 MA ZA RUM[45]
[BA.ZI 15 ...] ⌜SAḪAR⌝.ḪI.A KI.TA! 30 MA ZA RUM KI.TA
[15 ... 15 *a-na* 5 IGI].GUB.BA *i-ši* 1 15 *ta-mar*
[1 15 *šu-ul-li-iš* 3 4]5 *ta-mar* 3 45 *a-na*
[ ] 15 *ù* 18 45
[ ] ⌜12⌝ 30 *ta-mar*
[½ ...] 45 *ḫe-pé*
[ ] ⌜6 15⌝

[*A heap.*] The *mazarum* is 30, the height 1 00, the volume 25. [*It went down*] 2½ nindan. [...] What is the volume?[46] You: take the reciprocal of 1 00 , the height. [You will see 0;01.] Multiply [0;01 by] 30, the *mazarum*. You will see 0;30. Multiply 0;30 by [30, *by which it went down*]. You will see 15. [Take] 15 from 30, the *mazarum*. [*You will see 15, the upper mazarum.*][47] Tell the volume. 30, The lower *mazarum* is 30, [*the upper mazarum*? *is 15.*] Multiply [15 by 0;05] the coefficient. You will see 1 15. [... Triple 1 15. You will see 3 4]5.

[40] These last few lines of solution to an unknown problem calculate the circumference of a circle of area 0;01 15 sar$_a$. See §3.1 for the coefficient.

[41] Previously unattested word; reading and meaning unknown.

[42] *Mazarum* (a conventional transcription, for convenience) is otherwise unattested and of obscure origin. *Dikšum* means something like 'border', but has only been attested in connection with two-dimensional figures until now (Friberg 1987–90: 564). The *mazarum*'s dimensions seem to be completely determined by its volume.

[43] Because this information is not given in the statement of the problem, it must be common to all of this class of problems. This makes it a coefficient; compare G 11, '[...], the coefficient of the slope of a *mašarum*' (§8.5.4).

[44] It has not yet been possible to establish the reasoning behind this procedure.

[45] One would expect *ma-za-ri-im* if this were in syllabically written Akkadian. It is presumably a logogram or pseudo-logogram like ZA.ZUM 'base' (e.g. BM 85194 (iv): I 45; II 14).

[46] This problem seems to be like (xi): the volume of the whole solid is given, and the aim is to determine the volume of the bottom half if the top is cut off, and the *mazarum* at this half-way point.

[47] This part of the calculation uses the fact that the solid slopes uniformly, so that $m/h = m/2 \div h/2$, i.e. that the *mazarum* is proportional to its distance from the top of the figure.

[*Multiply*] 3 45 by [...] 15 and 18 45 [...] You will see 12 30 [...] Break [...] 45 [in half. ...] 6 15 ...[48]

Rest of column lost until:

(xxvi) ⌜29 51 40⌝ [*ta-mar* A.ŠÀ 29 51 40]
a-na 1 SUKUD *i-ši* 29 51 40 ⌜*ta-mar* SAḪAR⌝.[ḪI.A]

[... You will see] 29 51 40, [the area]. Multiply [29 51 40] by 1, the height. You will see 29 51 40, [the volume].[49]

Column VI

(xxvii) One problem missing; 7 or 8 lines lost.

(xxviii) [ ] 29! 51 40 SAḪAR.ḪI.<A>
[ ] ⌜X⌝ 12 30

[...] 29 51 40, the volume [...] 12 30 [...]

Rest of column missing: perhaps three problems (xxix)–(xxxi) lost.

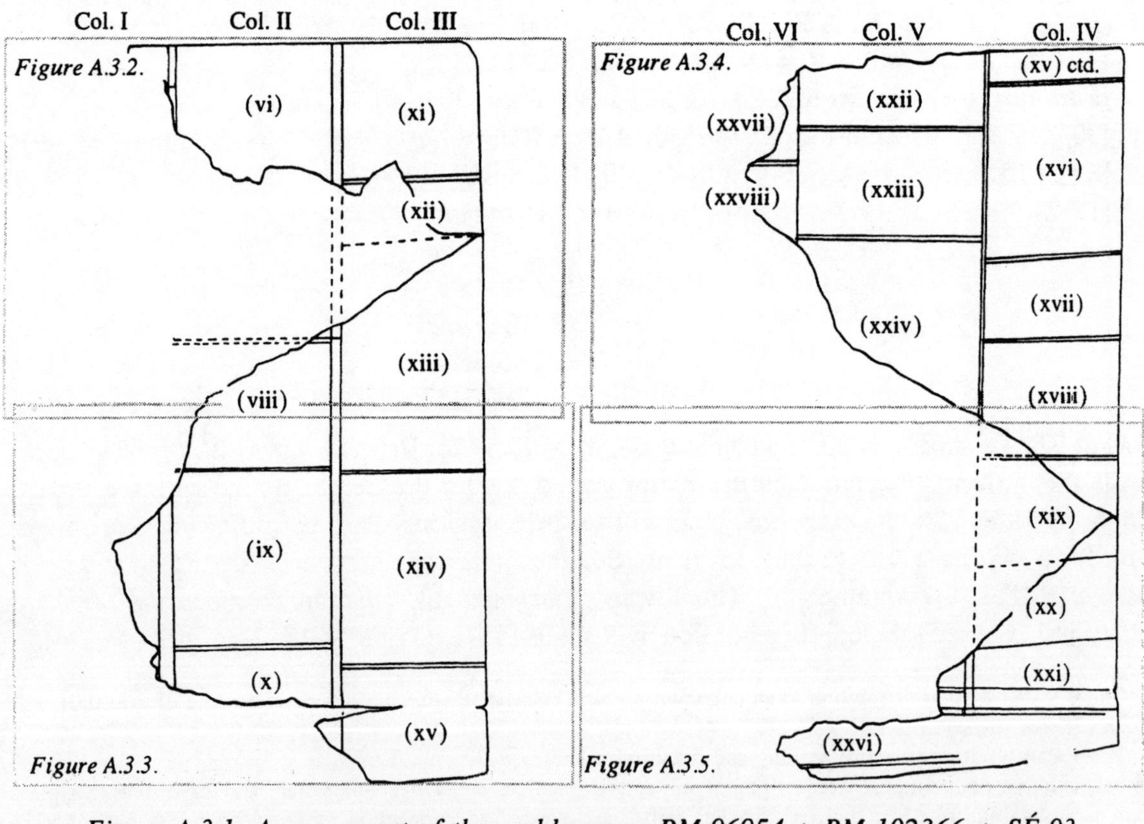

Figure A.3.1: Arrangement of the problems on BM 96954 + BM 102366 + SÉ 93, and an overview of the following hand-copies

48 This part of the solution is not entirely clear, but almost certainly involves calculating the area of the upper face and the base. These would be added together and halved, to give an 'average' cross-section, which would be multiplied by the half-height to give the volume of the truncated figure. Compare the methods of Haddad 104 (iii) and BM 85196 (xi).

49 It is clearly impossible to determine the subject of this problem and (xxvii) from the extant information, although they undoubtedly concern a single solid figure.

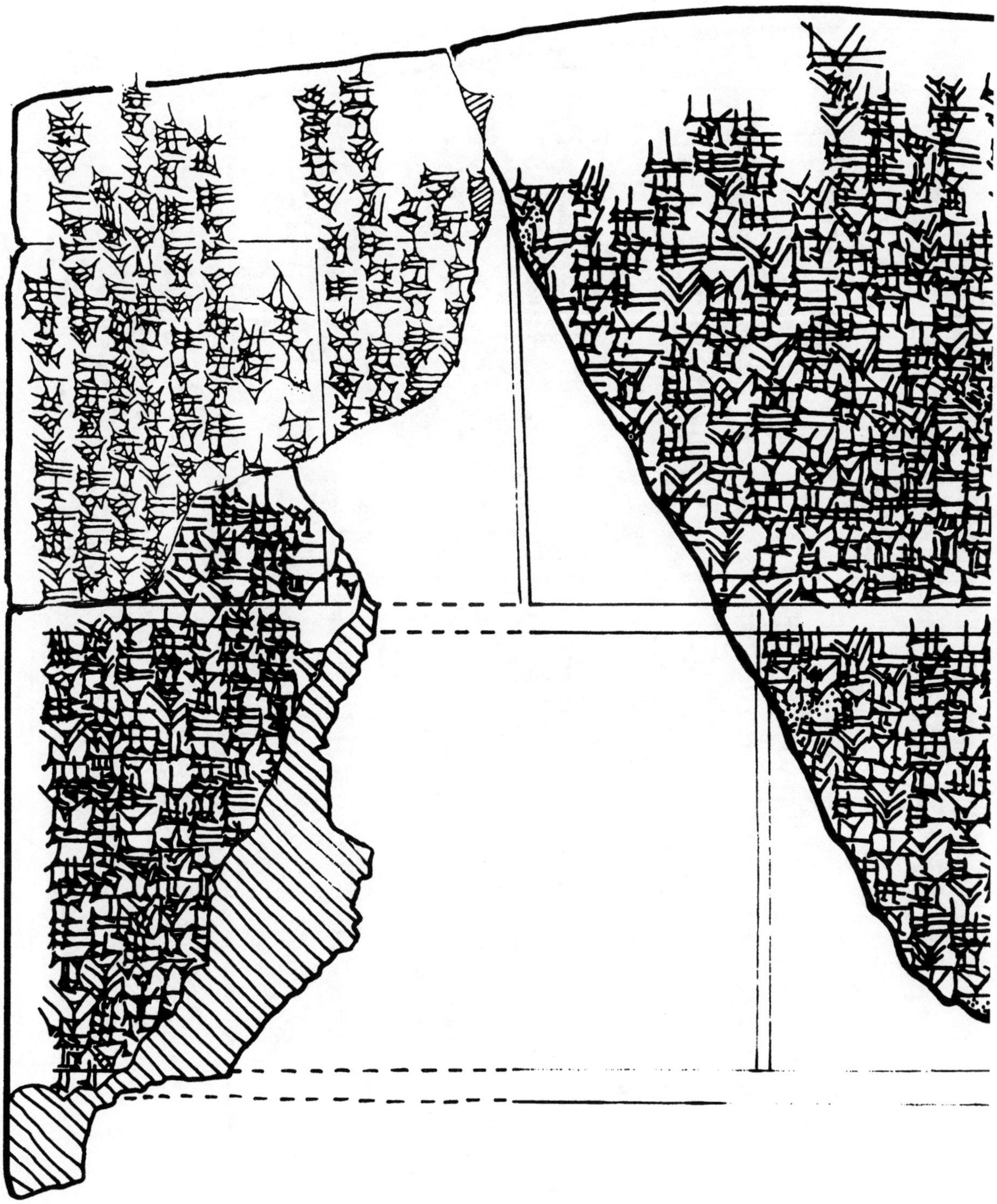

Figure A.3.2: BM 96954 + BM 102366 + SÉ 93 obverse top. Copy of SÉ 93 (top left) by Jursa and Radner (1995–6: 104)

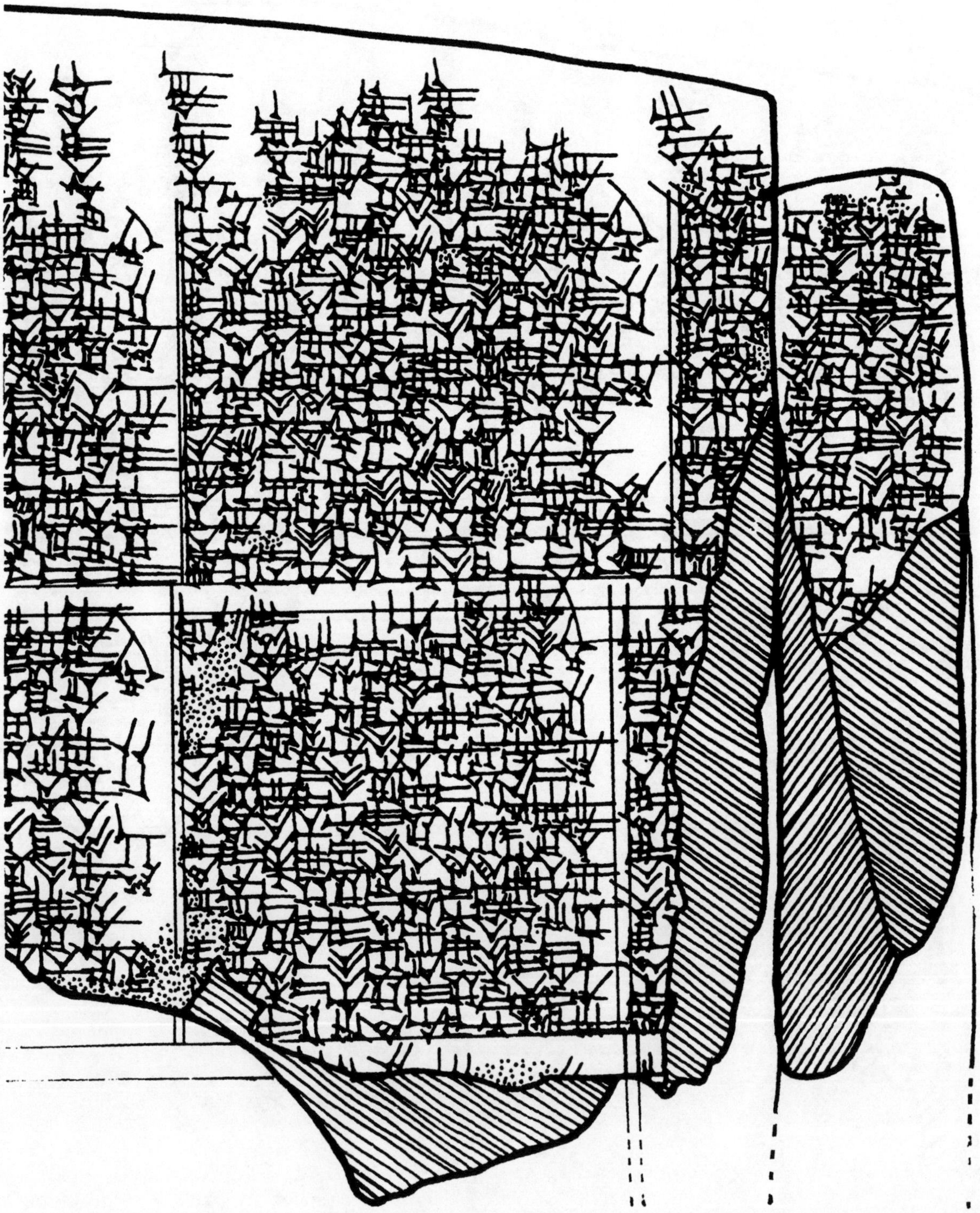

Figure A.3.3: BM 96954 + BM 102366 + SÉ 93 obverse bottom

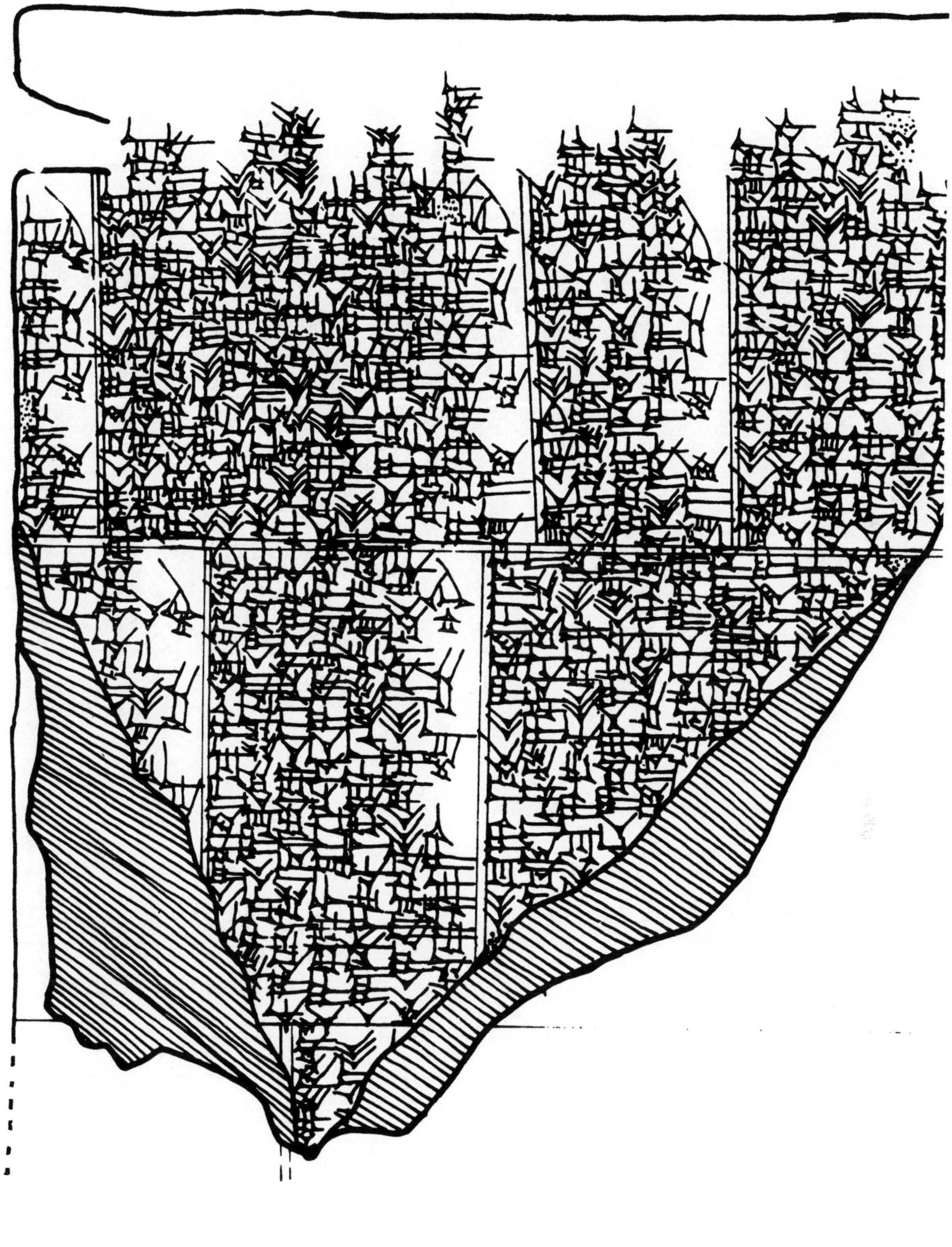

Figure A.3.4: BM 96954 + BM 102366 + SÉ 93 reverse top

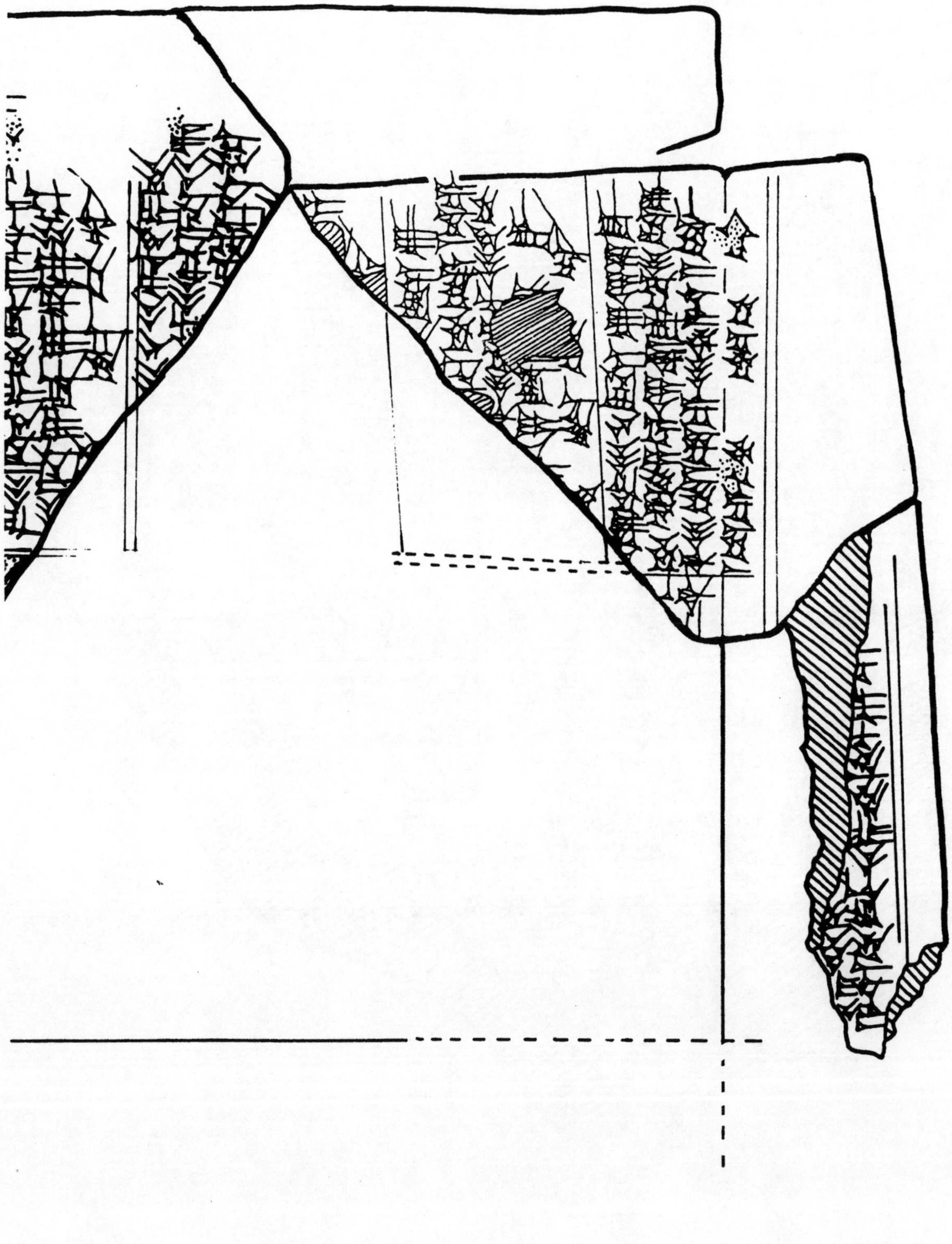

Figure A.3.5: BM 96954 + BM 102366 + SÉ 93 reverse bottom. Copy of SÉ 93 (top right) by Jursa and Radner (1995–6: 104)

APPENDIX 4: BM 96957 + VAT 6598

These two pieces comprise rather more than half of a three-column tablet, probably from Sippar. The larger fragment, VAT 6598, was published in part by Weidner (1916) and later by Neugebauer (1928; 1931–2); the smaller, BM 96957, was identified as a join to it by Christopher Walker of the British Museum. Mr Walker very generously made available to me his preliminary work on the BM tablet as well as a hand-copy. It is published below as a composite with Neugebauer's drawing (*MKT*: II Taf. 44); the complete tablet must have measured around 145 × 225 mm.[1]

Numbering of lines and problems runs consecutively between the fragments, including an estimated number of missing sections. Both the order of the problems and the solutions themselves are very clearly structured, which has allowed large parts of the missing text to be restored more or less accurately. The problems on the obverse and the first of the reverse deal with building brick walls and houses (Robson 1996*a*), using three different coefficients (§4.5). The second part of the text concerns 'Pythagorean' right triangles (Robson 1995*a*; fc.); the restoration of problems (xxii)–(xxv) is for the most part conjectural. Uniquely for an OB problem text, a colophon at the end of column V gives the scribe's name and some further information; column VI is blank.

Glosses have been written below the line of the main text in several places on the obverse (I 2, 3, 7, 8; II 6; III 3, 5, 10, 16, 22, 24, 32, 34).[2] They are represented in the translation by the phrases in parentheses. With one exception (I 7) these glosses enlarge on the numerical answer to a calculation by giving its Akkadian name, or logogram with an Akkadian phonetic complement. The gloss in line I 3 also shows that the number 7 30 is to be read as $7^1/_2$ sar_v. Most of the glosses are written in the gap between the numerical answer and the word *tammar* 'you will see', except in line III 10, where the gloss is inserted after the word *tammar*. In line I 7 the gloss *pānu* precedes the word it explains. All other published occurrences of *pānu* for IGI 'reciprocal' are from OB texts from the Eshnunna region, in which there is typically a much lower frequency of logograms (e.g. Haddad 104 (x): II 34, 37).

The gloss GAB.UŠ in line III 5 also occurs in III 24: GAB.UŠ.TA, gloss to *qà-qá-ra*; III 34: GAB.UŠ, gloss to *qà-qá-ra-tum*; IV 6: GAB.UŠ.TA. This is not a previously known mathematical logogram. 'Area' (*qaqqarum* or *eqlum*) is generally denoted logographically in OB maths by KI or A.ŠÀ It is unlikely that TA is a phonetic complement indicating the plural *qaqqarātum* as in É.GAR_8-*tum* for *igarātum* 'walls' in line II 6. In this case the form with TA glosses singular *qaqqarum* while *qaqqarātum* is glossed by GAB.UŠ alone (lines III 24, 34). In Standard Babylonian omen texts GAB-*ús-tu* is a writing for *pitrustu* 'a neutral or indecisive omen', which, however, is always written syllabically in the Old Babylonian period (Starr

[1] BM 96957 collated by the author in August 1996; VAT 6598 collated by Jeremy Black in April 1997. Farouk Al-Rawi made many useful suggestions on difficult readings in December 1995.

[2] Cf. Lieberman (1990: 332) on NA glosses: 'Variants are usually added in smaller script as supralinear "glosses". This practice already started in OB times, as is evident from the orthographic variant given in the oil omen *CT* 5 pls. 4–5 1, where all of the graphemes of *ḫa-pí* ^{pi}-⌈*ir*⌉ are written full-size on the line, as is usual with OB glosses.'

1975: 243 & n. 12). It is difficult to see how GAB.UŠ, if the signs are to be read like this, could refer to 'area', the meaning required by the context here. Another possibility might be to read the sign GAB as DUḪ, *pitrum* 'land which has not (yet) been built on', especially as this text is concerned with calculations in preparation for building. But this explanation leaves the signs UŠ and TA unaccounted for.

Column I

(i)[3] [É.GAR$_8$ 2 KÙŠ *ku*]-⸢*bu*⸣-*ru-ú* $2^1/_2$ NINDAN UŠ $1^1/_2$ NINDAN SUKUD
[SIG$_4$ EN.NAM ZA].E 2 KÙŠ ⸢SAG⸣ *a-na* $2^1/_2$ NINDAN UŠ *i-ši* 25$_{\text{qà-qá-rum}}$ *ta-mar*
[25] ⸢*a*⸣-*na* 18 SUKUD *i-ši* 7 30$_{7^1/_2\ \text{SAR SAḪAR-rum}}$ *ta-mar* 7 30 *a-na* 6
[IGI].GUB É.GAR$_8$ *i-ši* 45 *ta-mar* 45 SAR SIG$_4$ *ne-pé-šum*

[A wall.] The width is [2 cubits], the length $2^1/_2$ nindan, the height $1^1/_2$ nindan. [How many bricks? You]: multiply 2 cubits, the width, by $2^1/_2$ nindan, the length. You will see 0;25 (the area). Multiply [0;25⸣ by 18, the height. You will see 7;30 ($7^1/_2$ sar, the volume). Multiply 7;30 by 6, the coefficient of a wall. You will see 45. The bricks are 45 sar$_b$. The method.

(ii) ⸢É⸣.GAR$_8$ $2^1/_2$ NINDAN UŠ $1^1/_2$ NINDAN SUKUD 45 SAR SIG$_4$ *ku-bu-re-e*
[É].GAR$_8$-*ia* EN.NAM ZA.E 2 30 UŠ *a-na* 18 SUKUD *i-ši-ma*
[45 *ta*]-⸢*mar*⸣ *ki-il* $_{\text{pa-nu}}$IGI 6 IGI.GUB É.GAR$_8$ DU$_8$.A *a-na*
[45 SAR SIG$_4$ *i-ši*] 7 30$_{\text{SAḪAR-rum}}$ *ta-mar ki-il* IGI 45 *ša tu-ka-lu*!
[DU$_8$.A 1 20 *ta-mar* 1] ⸢20⸣ *a-na* 7 30$_{\text{SAḪAR}}$ *ša tu-ka-lu i-ši-ma*
[10 *ta-mar* 2 KÙŠ *ku-bu-re-e* É.GAR$_8$] ⸢*ne-pé-šum*⸣

A wall. The length is $2^1/_2$ nindan, the height $1^1/_2$ nindan, the bricks 45 sar$_b$. What is the width of my wall? You: multiply 2;30, the length, by 18, the height, and you will see [45]. Keep.[4] Take the reciprocal (the reciprocal) of 6, the coefficient of a wall; [multiply] by [45 sar$_b$, the bricks]. You will see 7;30 (the volume). Keep. [Take] the reciprocal of 45 that you are keeping. [You will see 0;01 20.] Multiply 0;01 20 by 7;30 (the volume) that you are keeping and [you will see 0;10. The width of the wall is 2 cubits.] The method.

(iii) Missing. Presumably on finding the bricks needed for the wall; cf. (xiv).

(iv) Missing. Presumably on finding the length and width given their sum; cf. (xv).

The last seven lines of the column might be restored as:

(v) [É.GAR$_8$ $1^1/_2$ NINDAN SUKUD 45 SAR SIG$_4$ UŠ *e-li ku-bu-re* É.GAR$_8$]
[2 20 NINDAN *i-te-er* UŠ *ù ku-bu-re* É.GAR$_8$-*ia* EN.NAM]

[3] For mathematical commentary on problems (i)–(xvi) see §4.5 and Robson (1996*a*).

[4] The more usual expressions for *kīl* 'keep' (I 7, 8, [6']; III 3, [16], 22, 24, 34) and *ša tukallu* 'that you are keeping' (I 8, 9; II 1; III 4, 17, 23, 26) are *rēška likīl* and *ša rēška ukallu*. These phrases have hitherto been translated as 'let your head hold (the number)' and '(the number) that your head is holding' (see *MKT*: II, 18; *MCT*: 167; Høyrup 1990*a*: 58). When VAT 6598 was published the longer version was restored, e.g. *rēš-‹ka li›-ki-il* in 'Obv. 18' (now III 34) (*MKT*: I 278). The first sign of the phrase, ÍB, should be read as *-tum*, attached to the preceding noun. However, von Soden [*AHw*: 503, *kullum* II (12b)] interprets the phrase as *rēšam kullum* 'let the number "hold your head" ', in conjunction with other non-mathematical occurrences. But in fact *rēšam kullum* 'to take care of someone (or something)' (Wilcke 1987: 83) is a completely different idiom: 'let the number take care of you(r head)' would be quite meaningless. This text (and 22 *ša tu-ka-al-lu,* VAT 6597 (iv): III 7) shows that in this context 'you' and 'your head' can be equated, vindicating the original translators' interpretation that 'you' or 'your head' hold the number while other calculations are performed.

[ZA.E IGI 6 IGI.GUB É.GAR$_8$ DU$_8$.A 10 *ta-mar* 10 *a-na*]
[45 SAR SIG$_4$ *i-ši* 7 30 *ta-mar* $7^1/_2$ SAR SAḪAR IGI 18 SUKUD DU$_8$.A]
[3 20 *ta-mar* 3 20 *a-na* 7 30 *i-ši* 25 *qà-qá-rum ta-mar*]
[*ki-il* $^1/_2$ 2 20 *ša* UŠ UGU *ku-bu-re* É.GAR$_8$ *i-te-rù ḫe-pé*]
[1 10 *ta-mar* 1 10 *a-di* 2 GAR.RA *šu-tam-ḫir*]

Column II

1 21 40 *ta-mar* 25 *ša tu-ka-lu a-na* 1 21 40
DAḪ.ḪA 1 46 40 *ta-mar* 1 46 40 EN.NAM ÍB.SI$_8$ 1 20 ÍB.SI$_8$
a-di 2 GAR.RA 1 10 *a-na* 1 20 DAḪ.ḪA 2 30 *ta-mar* 2 30 UŠ É.GAR$_8$
1 10 *i-na* 1 20 BA.ZI 10 ÍB.TAG$_4$ 2 KÙŠ *ku-bu-re* É.GAR$_8$ *ne-pé-šum*

[A wall. The height is $1^1/_2$ nindan, the bricks 45 sar$_b$. The length exceeds the width of the wall by 2;20 nindan. What are the length and width of my wall? You: take the reciprocal of 6, the coefficient of a wall. You will see 0;10. Multiply 0;10 by 45 sar$_b$, the bricks. You will see 7;30. $7^1/_2$ sar is the volume. Take the reciprocal of 18, the height. You will see 0;03 20. Multiply 0;03 20 by 7;30. You will see 0;25. The area is 0;25. Keep. Break in half 2;20, by which the length exceeds the width of the wall. You will see 1;10. Put down 1;10 twice. Square.] You will see 1;21 40. Add 0;25 that you are keeping to 1;21 40. You will see 1;46 40. What is the square root of 1;46 40? 1;20 is the square root. Put down twice. Add 1;10 to 1;20. You will see 2;30. The length of the wall is 2;30. Take 1;10 from 1;20. The remainder is 0;10. The width of the wall is 2 cubits. The method.

(vi) 5 SAR É *a-na* $2^1/_2$ NINDAN SUKUD SIG$_4$ EN.NAM *lu-ša-‹al›-bi-in* ZA.E
IGI.3.GÁL 5 SAR *le-qé* 1 40$_{\text{É.GAR}_8\text{-}tum}$ *ta-mar* 1 40 *a-na* $2^1/_2$ NINDAN SUKUD
i-ši 4 10 *ta-mar* $2^1/_2$ IKU SIG$_4$ *tu-ša-al-ba-ma* 5 SAR É
a-na $2^1/_2$ NINDAN SUKUD *ta-kà-mar ne-pé-šum*

The ‹area of a› house is 5 sar$_a$. For a height of $2^1/_2$ nindan how many bricks should I get made? You: take a third of 5 sar$_a$. You will see 1;40 (the walls). Multiply 1;40 by $2^1/_2$ nindan, the height. You will see 4;10. You will get $2^1/_2$ iku$_b$ bricks made and you will pile up 5 sar$_a$ house to a height of $2^1/_2$ nindan. The method.

(vii) *šum-ma* $2^1/_2$ IKU SIG$_4$ $2^1/_2$ NINDAN SUKUD É EN.NAM *e-pu-uš* ZA.E
⸢IGI $2^1/_2$ NINDAN SUKUD DU$_8$.A⸣ *a-na* 4 10 $2^1/_2$ IKU SIG$_4$ *i-ši* 1 40 *ta-mar*
[*ki-il* IGI 20 IGI.GUB É.DÙ].⸢A⸣ DU$_8$.A 3 *ta-mar* 3 *a-na*
[1 40 *ša tu-ka-lu i-ši* 5 *ta-mar*] ⸢5 SAR É⸣
[*ne-pé-šum*]

If the bricks are $2^1/_2$ iku$_b$, the height $2^1/_2$ nindan, how much house should I build? You: take the reciprocal of $2^1/_2$ nindan, the height. Multiply by 4;10, the $2^1/_2$ iku$_b$ bricks. You will see 1;40. [Keep.] Take [the reciprocal of 0;20, the coefficient of a built house]. You will see 3. [Multiply] 3 by [1;40 that you are keeping. You will see 5.] The ‹area of the› house is 5 sar. [The method.]

(viii)–(xi) The rest of the column is missing. It may have contained up to four problems of approximately the same length as (vi) and (vii), the first of which was almost certainly about finding the height of the house given the area and bricks.

Column III

(xii) É.GAR$_8$ SIG$_4$.AL.UR$_5$.RA 2 KÙŠ SAG 1 NINDAN SUKUD 9 SAR SIG$_4$.AL.UR$_5$.RA UŠ É.GAR$_8$-*ia* EN.NAM ZA.E IGI 2 15 IGI.GUB É.GAR$_8$ DU$_8$.A 26 40 *ta-mar* 26 40 *a-na* 9 SAR SIG$_4$.AL.UR$_5$.‹RA› *i-ši* 4 ⌈*ta-mar* SAḪAR-*rum ki-il*⌉[5] IGI 12 1 NINDAN SUKUD É.GAR$_8$ DU$_8$.A 5 *ta-mar* 5 *a-na* 4 *ša tu-ka-lu i-ši* 20 GAB.UŠ *ta-mar* IGI 10 SAG DU$_8$.A *a-na* 20 *ša ta-mu-rù i-ši* 2 *ta-mar* 2 NINDAN UŠ *ne-pé-šum*

A wall of baked bricks. The width is 2 cubits, the height 1 nindan, the baked bricks 9 sar$_b$. What is the length of my wall? You: take the reciprocal of 2;15, the coefficient of a wall. You will see 0;26 40. Multiply 0;26 40 by 9 sar$_b$, the baked bricks. You will see 4 (the volume). Keep. Take the reciprocal of 12, the 1 nindan height of the wall. You will see 0;05. Multiply 0;05 by 4 that you are keeping. You will see 0;20 (*the area*). Take the reciprocal of 0;10, the width; multiply by 0;20 that you saw. You will see 2. The length is 2 nindan. The method.

(xiii) 2 NINDAN UŠ É.GAR$_8$ SIG$_4$.AL.UR$_5$.RA 1 NINDAN SUKUD 9 SAR SIG$_4$.AL.UR$_5$.RA *ku-bu-re* É.GAR$_8$-*ia* EN.NAM ZA.E IGI 2 15 IGI.GUB É.GAR$_8$ DU$_8$.A 26 40 *ta-mar* 26 40 *a-na* 9 SAR SIG$_4$.AL.UR$_5$.RA *ša a-na* É.GAR$_8$-⌈*ka*⌉ ‹*ša*› *ta-at-bu-ku i-ši* 4 *ta-mar* SAḪAR-*rum* IGI 1 NINDAN SUKUD É.GAR$_8$-*ka* DU$_8$.A *a-na* 4 *ša ta-mu-rù i-ši* 20 *ta-mar* 20 $^1/_3$ SAR *qà-qá-ar* é.gar$_8$-*ka* IGI 2 NINDAN UŠ DU$_8$.A 30 *ta-mar* 30 ⌈*a*⌉-*na* 20 *i-ši* 10 *ta-mar* 2 KÙŠ *ku-bu-re-e* É.GAR$_8$ ⌈*ne*⌉-*pé-šum*

The length of a wall of baked bricks is 2 nindan, the height 1 nindan, the baked bricks 9 sar$_b$. What is the width of my wall? You: take the reciprocal of 2;15, the coefficient of a wall. You will see 0;26 40. Multiply 0;26 40 by 9 sar$_b$, the baked bricks ‹that› you spread out for your wall. You will see 4 (the volume). Take the reciprocal of 1 nindan, the height of your wall. Multiply by 4 that you saw. You will see 0;20. The area of your wall is $^1/_3$ sar. Take the reciprocal of 2 nindan, the length. You will see 0;30. Multiply 0;30 by 0;20. You will see 0;10. The width of the wall is 2 cubits. The method.

(xiv) 2 NINDAN UŠ É.GAR$_8$ 2 KÙŠ *ku-bu-‹re›-e* É.GAR$_8$ 9 SAR SIG$_4$.AL.UR$_5$.RA *a-na* É.GAR$_8$ ⌈*ša*⌉-*ak-na-at i-na* 9 SAR SIG$_4$.AL.UR$_5$.RA EN.NAM ⌈*li-li*⌉ ZA.E IGI 2 15 IGI.GUB [SIG$_4$].⌈AL⌉.UR$_5$.RA DU$_8$.⌈A 26⌉ 40 *ta-mar* 26 ⌈40 *a-na* 9⌉ [SAR SIG$_4$.AL].⌈UR$_5$.RA «*i-ši*»⌉ [*ša a-na*] ⌈É⌉.GAR$_8$ *ša-ak-na-at i-ši* 4 SAḪAR-[*rum*] *ta* ⌈*mar*⌉ [*ki-il* 2 NINDAN UŠ *a-na* 10] [*i-ši* 20 *ta-mar*] ⌈IGI 20 DU$_8$.A 3 *ta-mar* 3 *a-na* 4⌉ [*ša*] ⌈*tu-ka-lu i-ši*⌉ [12 *ta-mar* 1 NINDAN SUKUD] ⌈*i-na*⌉ 9 SAR SIG$_4$.AL.UR$_5$.‹RA› *tu-ša-qá*[6] *ne-pé-‹šum›*

The length of a wall is 2 nindan, the width of the wall 2 cubits. 9 sar$_b$ baked bricks are set for the wall. How high can it be[7] with 9 sar$_b$ baked bricks? You: take the reciprocal of 2;15, the coefficient of baked [bricks]. You will see 0;26 40. Multiply 0;26 40 by 9 [sar$_b$, the baked bricks that] are set [for] the wall. You will see 4 (the volume). [Keep. Multiply 2 nindan, the length, by 0;10. You will see 0;20.] Take the reciprocal of 0;20. You will see 3. Multiply 3 by 4 [that] you are keeping. [You will see 12. The height] that you will make high from 9 sar$_b$ baked bricks is [1 nindan]. The method.

---

[5] See collation at the end of this appendix.

[6] See collation at the end of this appendix.

[7] Reading *li-li* as *līli* (G-stem third person singular precative of *elûm* 'to be high').

(xv) [É.GAR$_8$ SIG$_4$.AL].UR$_5$.RA 1 NINDAN SUKUD É.GAR$_8$ 9 SAR SIG$_4$.AL.UR$_5$.RA UŠ *ù ku-bu-re*
[É.GAR$_8$ UL].GAR-*ma* 2 10 UŠ *ù ku-bu-re* É.GAR$_8$-*ia* EN.NAM
[ZA.E IGI 2 15] IGI.GUB SIG$_4$.AL.UR$_5$.RA DU$_8$.A 26 ˹40 *ta*˺-*mar* 26 40
[*a-na*] ˹9 SAR˺ SIG$_4$ AL.UR$_5$.RA *i-ši* 4 SAḪAR-*rum ta-mar* 4 SAḪAR SIG$_4$.AL.UR$_5$.‹RA›-*ka ki-il*
[IGI 12] 1 NINDAN SUKUD SIG$_4$.AL.UR$_5$.RA-*ka* DU$_8$.A 5 *ta-mar* 5 *a-na* 4 *ša tu-ka-lu*
[*i*]-˹*ši*˺ 20 *ta-mar* 20 *qà-qá-ra* GAB.UŠ.TA *ki-il* $^1/_2$ 2 10 [UL].˹GAR˺ UŠ *ù ku-bu-re* É.GAR$_8$
[*ḫe*]-*pé* 1 05 *ta-mar* 1 ˹05˺ *šu-tam-ḫir* 1 10 ˹25˺ [*ta-mar* 20 *qà-qá*]-*ra-tum*
[*ša tu*]-˹*ka-lu*˺ *i-na* 1 10 25 BA.ZI 50 25 [*ta*]-˹*mar*˺ 50 25 EN.NAM ÍB.SI$_8$
[55 ÍB.SI$_8$ 55] *a-na* 1 05 *ša* Ì.KÚ.KÚ DAḪ.ḪA 2 «10» *ta-mar* 2 NINDAN UŠ
[55 *i-na* 1 05 *ša tu*]-˹*uš*˺-*tam-ḫir* BA.ZI 10 ÍB.TAG$_4$ 2 KÙŠ *ku-bu-re ne-pé*-‹*šum*›

A wall of] baked [bricks]. The height of the wall is 1 nindan, the baked bricks 9 sar$_b$. The sum of the length and thickness [of the wall] is 2;10. What are the length and thickness of my wall? [You:] take the [reciprocal of 2;15], the coefficient of baked bricks. You will see 0;26 40. Multiply 0;26 40 [by] 9 sar$_b$, the baked bricks. You will see 4 (the volume). Keep 4, the volume of your baked bricks. Take [the reciprocal of 12], (or) 1 nindan, the height of your baked bricks. You will see 0;05. Multiply 0;05 by 4 that you are keeping. You will see 0;20. The area is 0;20 (the area?). Keep. Break in half 2;10, [the sum] of the length and thickness of the wall. You will see 1;05. Square 1;05. [You will see] 1;10 25. Subtract [0;20], the area [that] you are keeping, from 1;10 25. You will see 0;50 25. What is the square root of 0;50 25? [The square root is 0;55.] Add [0;55] to 1;05 that you squared. You will see 2. The length is 2 nindan. Subtract [0;55 from 1;05 that] you squared. The remainder is 0;10. The thickness is 2 cubits. The method.

(xvi) [É.GAR$_8$ SIG$_4$.AL.UR$_5$.RA] 9 ˹SAR˺ [SIG$_4$.AL.UR$_5$.RA] *at-bu-uk* UŠ *e-li*
[*ku-bu-re* É.GAR$_8$ 1 50 DIRIG 1] NINDAN *za*-˹X X UŠ!˺ É.GAR$_8$ *ù ku-bu-re* É.GAR$_8$-*ia*
[EN.NAM ZA.E IGI 2] 15 IGI.GUB É.GAR$_8$ ˹SIG$_4$˺.AL.UR$_5$.‹RA› DU$_8$.A 26 40 *ta-mar* 26 40
[*a-na* 9 SAR SIG$_4$.AL].UR$_5$.RA *i-ši* 4 SAḪAR-*rum ta-mar* SAḪAR SIG$_4$.AL.UR$_5$.RA 4 SAR
[IGI] 12 SUKUD É.GAR$_8$ DU$_8$.A 5 *ta-mar* 5 *a-na* 4 SAḪAR SIG$_4$.AL.UR$_5$.RA
i-ši 20 *ta-mar* 20 GAB.UŠ *qà-qá-ra ki-il* $^1/_2$ 1 ˹50 *ša*˺ UŠ UGU *ku-bu-re*
É.GAR$_8$ *i-te-rù ḫe-pé* 55 *ta-mar* [55 *a-di* 2] GAR.RA
55 *šu-tam-ḫir* 50 25 *ta-mar* 20 *qà-qá-ra a-na* 50 ‹2›5 DAḪ.ḪA
1 10 25 *ta-mar* 1 10 25 EN.NAM ÍB.SI$_8$ 1 05 [ÍB.SI$_8$] ˹*a-di*˺ 2 GAR.RA
[55 *ša*] *tu-uš-tam-ḫir a-na* 1 05 DAḪ.ḪA 2 ˹*ta*˺-[*mar* 2 NINDAN UŠ] É.GAR$_8$
[55 *i-na* 1 05 BA].˹ZI˺ 10 ÍB.TAG$_4$ 2 KÙŠ *ku-bu*-[*re* É.GAR$_8$]
[*ne-pé-šum*]

[A wall of baked brick.] I spread out 9 sar$_b$ of [baked] bricks. The length [exceeds the thickness of the wall] by [1;50]. The *height* is 1 nindan. [What are] the length of the wall and the thickness of my wall? [You:] take [the reciprocal of 2];15, the coefficient of a wall of baked bricks. You will see 0;26 40. Multiply 0;26 40 [by 9 sar$_b$], the baked [bricks]. You will see 4 (the volume). The volume of the baked bricks is 4 sar. Take [the reciprocal] of 12, the height of the wall. You will see 0;05. Multiply 0;05 by 4, the volume of baked bricks. You will see 0;20. Keep 0;20, the area. Break in half 1;50, by which the length exceeds the thickness of the wall. You will see 0;55. Put down [0;55 twice]. Square 0;55. You will see 0;50 25. Add 0;20, the area, to 0;50 25. You will see 1;10 25. What is the square root of 1;10 25? [The square root] is 1;05. Put (it) down twice. Add [0;55 that] you squared to 1;05. You will see 2. [The length] of the wall is [2 nindan. Subtract 0;55 from 1;05.] The remainder is 0;10. The thickness [of the wall] is 2 cubits. [The method.]

Column IV

(xvii) É.GAR$_8$ SIG$_4$.AL.UR$_5$.RA [2 NINDAN UŠ É.GAR$_8$ 2 KÙŠ]
a-na ša-ap-li-a-tim ka-ba-ar [1 KÙŠ *a-na e-le-nim ka-ba-ar*]
2 NINDAN SUKUD *ši-ip-ka-at* É.GAR$_8$*-ia* EN.NAM [*i-ša-pa-ak ù*]
i-na 1 KÙŠ EN.NAM *i-ku-ul* ZA.E 2 KÙŠ KI.KAL KI.TA [*ù* 1] ⸢KÙŠ⸣
ša a-na UGU *ip-ḫur* UL.GAR 15 *ta-mar* $^1/_2$ 15 *ḫe-pé*
⸢7⸣ 30 GAB.UŠ.TA *ta-mar* 7 30 *a-na* 2 NINDAN UŠ *i-ši* 15 *ta-mar*
[15 *a-na*] 24 SUKUD.«DA»[8] *i-ši* 6 *ta-mar* 6 *a-na*
[2 1]5 IGI.GUB SIG$_4$.AL.UR$_5$.‹RA› *i-ši* 13 30 *ta-mar* 13$^1/_2$ SAR SIG$_4$.AL.UR$_5$.RA
⸢$^1/_2$⸣ 13 30 *ḫe-pé* 6 45 *ta-mar* 6$^2/_3$ 5 GÍN
[*it*]*-ti* É.GAR$_8$*-ka ši-ip-ka-ti i-ša-pa-ka-ku* NIGIN$_2$.NA
⸢*ša*⸣ *i-na* 1 KÙŠ *i-ku-lu a-mur* 2 KÙŠ KI.KAL KI.TA *e-li*
5 *ša a-na e-le-nu* EN.NAM *i-ter* 5 *i-ter ki-il*
[*aš*]*-sum i-na* 1 KÙŠ *ú-ku-‹la›* EN.NAM *i-ku-ul* DUG$_4$.GA
[5 GAR.RA] IGI 5 DU$_8$.A 12 *ta-mar a-na* 2 NINDAN SUKUD
[*i-ši* 24 *ta-mar*] IGI 24 DU$_8$.A 2 30 *ta-mar* «2 30 *ta-mar*»
[2 30 *a-na*] 5 1 KÙŠ «X» *i-ši* 12 30 *ta-mar*
[12 30 *a-na* 6 *i-ši* 1 15] ⸢*ta*⸣*-mar i-na* 1 KÙŠ 1 ŠU.SI
⸢*ù* IGI.4.GÁL 1 ŠU.SI⸣ É.GAR$_8$ *ú-ku-la i-ku-ul ne-pé-šum*

A wall of baked bricks. [The length of the wall is 2 nindan.] It is [2 cubits] thick at the bottom; [it is 1 cubit thick at the top]. The height is 2 nindan. How much is the *heaping (?)* of my wall[9] [*heaped,* and] in 1 cubit how much does it slope? You: add 2 cubits, the lower thickness, [and 1 cubit], that *it tapers to* on the top. You will see 0;15. Break 0;15 in half. You will see 0;07 30 (the cross section?).[10] Multiply 0;07 30 by 2 nindan, the length. You will see 0;15. Multiply [0;15 by] 24, the height. You will see 6. Multiply 6 by [2;1]5, the coefficient of baked bricks. You will see 13;30. The baked bricks are 13$^1/_2$ sar$_b$. Break 13;30 in half. You will see 6;45. A *heaping* of 6$^2/_3$ ‹sar$_b$› 5 shekels will be heaped for you with your wall.[11] Return. See how much it slopes in 1 cubit. By how much does 1 cubit, the lower width, exceed 0;05 that is on the top? It exceeds by 0;05. Keep. Because they said 'what slope does it slope in 1 cubit?', [put down 0;05]. Take the reciprocal of 0;05. You will see 12. [Multiply] by 2 nindan, the height. [You will see 24.][12] Take the reciprocal of 24. You will see 0;02 30. Multiply [0;02 30 by] 0;05, (i.e.) 1 cubit, the ratio. You will see 0;00 12 30.[13] [Multiply 0;00 12 30 by 6 00.] You will see [1;15]. In 1 cubit the wall slopes a slope of 1 finger and a quarter of 1 finger. The method.

---

8 Erased by the scribe. (Collation, Jeremy Black, April 1997.)

9 Cf. 'What is the (volume of) the mound of earth (needed) for my wall?' (*CAD*: Š III 69).

10 i.e. the average width of the wall. This is then multiplied by the height, length and brick coefficient in the usual way to give the brick content of the wall.

11 The meaning of this procedure is unclear. It may be concerned with finding the bricks in a 1 nindan length of wall (*pace TMB*: 129 n. 6).

12 The height is converted from nindan into cubits using the linear conversion coefficient 0;05 (see §7.1). This operation is performed in response to the preceding question 'what slope does it slope in 1 *cubit* (i.e. not nindan)?'

13 i.e. (base – top)/height = slope, in nindan. This is converted into fingers with a linear conversion coefficient in the following step (§7.1).

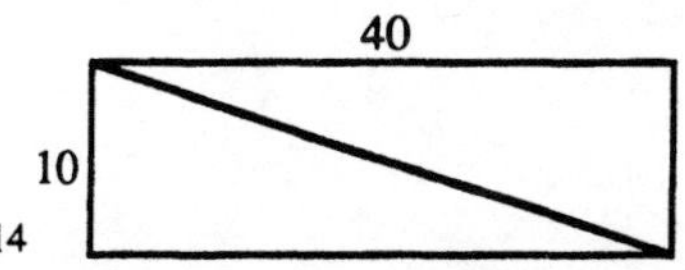

(xviii)[14]

KÁ $^1/_2$ ‹NINDAN› 2 KÙŠ SUKUD 2 KÙŠ DAGAL *ṣí-li-‹ip›-ta-šu* EN.NAM
ZA.E 10 DAGAL *šu-tam-ḫir* 1 40 *qà-qá-⌜ra ta-mar⌝*
IGI 40 KÙŠ SUKUD DU$_8$.A *a-na* 1 40 *qà-qá-ri i-ši*
2 30 *ta-mar* $^1/_2$ 2 30 *ḫe-pé* 1 15 *ta-mar* 1 15
[*a-na* 40 SUKUD DAḪ].ḪA 41 15 *ta-mar* 41 15
[*ṣí-li-ip-tum*] ⌜*ne*⌝*-pé-šum*

A gate, of height $^1/_2$ ‹nindan›, 2 cubits and breadth 2 cubits. What is its diagonal? You: square 0;10, the breadth. You will see 0;01 40, the base.[15] Take the reciprocal of 0;40 (cubits), the height; multiply by 0;01 40, the base. You will see 0;02 30. Break in half 0;02 30. You will see 0;01 15. [Add] 0;01 15 [to 0;40, the height]. You will see 0;41 15. [The diagonal] is 0;41 15. The method.

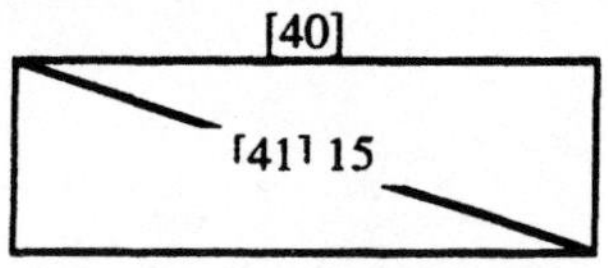

(xix)

šum-ma KÁ 40 KÙŠ SUKUD 41 15 *ṣí-li-ip-tum* DAGAL EN.NAM!
ZA.E 40 SUKUD *i-na* 41 15 *ṣí-li-ip-tum* BA.ZI
1 15 ÍB.TAG$_4$ 1 15 ⌜*a-na*⌝ 2 TAB.BA 2 30 *ta-mar*! 40 UŠ
a-na 2 30 A.RÁ ⌜*ša*⌝ *ta-mu-rù i-ši* 1 40 *ta-mar* EN.NAM ÍB.SI$_8$
10 ÍB.SI$_8$ 10 DAGAL *ne-pé-šum*

If the gate has height 0;40 (cubits) and diagonal 0;41 15, what is the breadth? You: take 0;40, the height, from 0;41 15, the diagonal.[16] 0;01 15 is the remainder. Double 0;01 15. You will see 0;02 30. Multiply 0;40, the length, by 0;02 30, the factor that you saw. You will see 0;01 40. What is the square root? 0;10 is the square root. The breadth is 0;10. The method.

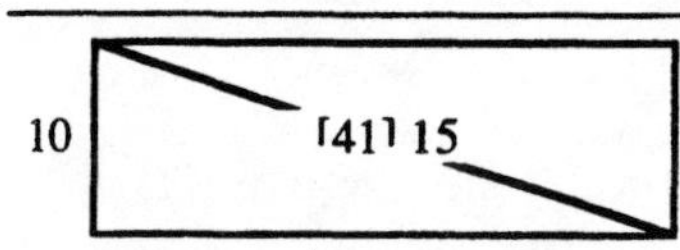

(xx)

2 KÙŠ DAGAL 41 15 *ṣí-li-ip-tum* SUKUD EN.NAM ⌜ZA⌝.E
NU
iš-te-‹en-ma›

The breadth is 2 cubits, the diagonal is 0;41 15. What is the height? You: no (solution).[17] «One.»[18]

---

[14] For mathematical commentary on problems (xviii)–(xxv), see Robson (1995*a*; forthcoming).

[15] When the breadth of the 'gate' is squared during the calculation, the result is sometimes called the 'base' *qaqqarum* (also (xxi)). There are no calculations left in which the height is squared, so that it is impossible to know whether there is an equivalent term for this result.

[16] In line IV 26 the text clearly shows *i-na ṣí-li-ip-tum* where one would expect *i-na ṣí-li-ip-tim*. Minor grammatical mistakes such as this are rather common in mathematical texts (e.g. *na-az-ba-al ar-hu-um* 'carriage of half-bricks', list G 15).

[17] Problem (xx) is stated, but no solution is given. Instead the sign NU — the logogram for negation — appears. This is the only occurrence of this phenomenon known so far. Because the primary function of an OB problem text is to convey the method of solution to a particular class of problems an insoluble problem serves no purpose.

Column V

(xxi) 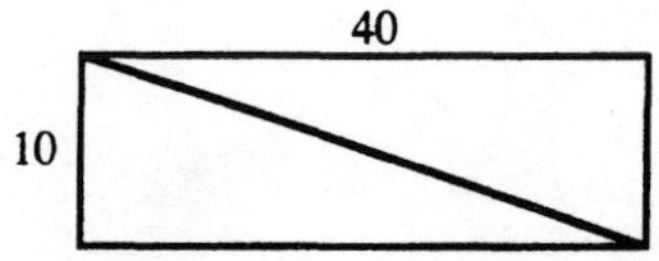

it-ta-šu[19]

2 KÙŠ DAGAL 40 KÙŠ SUKUD *ṣí-li-ip-ta-šu* EN.NAM ZA.E 10 SAG
šu-tam-ḫir 1 40 *ta-mar qà-qá-rum* 1 40 *a-na* 40 ⸢KÙŠ⸣ SUKUD *i-ši-ma*
1 06 40 *ta-mar a-na* ‹2› TAB.BA 2 13 20 *ta-mar a-na* 40 KÙŠ SUKUD
DAḪ.ḪA 42 13 20 *ṣí-li-ip-ta ta-mar ne-pé-šum*

Its diagram. The breadth is 2 cubits, the height 0;40 (cubits). What is its diagonal? You: square 0;10, the width. You will see 0;01 40. This is the base. Multiply 0;01 40 by 0;40 (cubits), the height, and you will see 0;01 06 40. Double it. You will see 0;02 13 20. Add it to 0;40 (cubits), the height. You will see 0;42 13 20, the diagonal. The method.

(xxii) 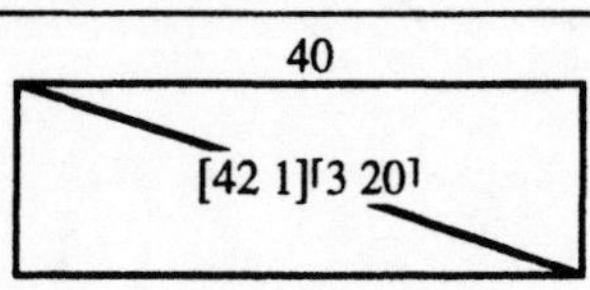

[20] [40 KÙŠ SUKUD 42 13 20 *ṣí-li-ip-tum* DAGAL EN.NAM ZA.E 40 SUKUD]
[*i-na* 42 13 20 BA.ZI 2 13 20 *ta-mar* 40 *a-na* 2 TAB.BA 1 20 *ta-mar* IGI DU$_8$.A]
[*a-na* 2 13 20 *i-ši-ma* 1 40 *ta-mar* EN.NAM ÍB.SI$_8$ 10 ÍB.SI$_8$ DAGAL *ne-pé-šum*][21]

[The height is 0;40 (cubits), the diagonal 0;42 13 20. What is the breadth? You: take 0;40, the height, from 0;42 13 20. You will see 0;02 13 20. Double 0;40. You will see 1;20. Take the reciprocal. Multiply it by 0; 02 13 20 and you will see 0;01 40. What is the square root? The square root is 0;1̣0, the breadth. The method.]

(xxiii)

[2 KÙŠ DAGAL 40 KUŠ SUKUD *ṣí-li-ip-ta-šu* EN.NAM ZA.E 10 DAGAL]
[*šu-tam-ḫir* 1 40 *ta-mar* 40 UŠ *šu-tam-ḫir* 26 40 *ta-mar a-na* 1 40 DAḪ.ḪA]
[28 20 *ta-mar* EN.NAM ÍB.SI$_8$ 41 13 51 48 ÍB.SI$_8$ *ṣí-li-ip-tum*] ⸢*ne-pé*⸣-‹*šum*›

[The breadth is 2 cubits, the height 0;40 (cubits). What is its diagonal? You: square 0;10, the width. You will see 0;01 40. Square 0;40, the length. You will see 0;26 40. Add it to 0;01 40.

Generally problems were carefully constructed so that they were soluble, with computationally simple answers (Høyrup 1990*a*: 330 n. 153). The size and spacing of the tablet suggest that the similarly insoluble problem after (xxii) was almost certainly omitted before V 8'. It might be reconstructed as:

[2 KÙŠ DAGAL 42 13 20 *ṣí-li-ip-tum* SUKUD EN.NAM ZA.E NU]

'The breadth is 2 cubits, the diagonal 0;42 13 20. What is the height? You: no (solution).'

Both problems are in fact soluble by modern algebraic techniques but would have been impossible using OB algorithmic methods: see Robson (1995*a*; fc.).

[18] The two final signs of column IV may be an abbreviation of *ištēnma*, or more likely an abandoned attempt to write that word (see below). However, there are no glosses of the sort found on the obverse of the tablet, which explain intermediate results in the calculations (see notes above).

[19] Written to the left of the diagram, perpendicular to the rest of the text.

[20] The hand-copy of VAT 6598 in *MKT*: II Taf. 44 does not show the sign '40' at the top of the illustration, but it is clearly visible on the photograph of the tablet (*MKT*: II Taf. 17).

[21] The next problem in the sequence, on finding the length of the rectangle given the width and diameter, was omitted from this tablet, presumably because it was considered insoluble. See Robson (1995*a*) and notes to the translation of problem (xx) below.

You will see 0;28 20. What is the square root? the square root is *0;41 13 51 48*. The diagonal.] The method.[22]

(xxiv)

[40 KÙŠ SUKUD 41 13 51 48 *ṣí-li-ip*]-⸢*tum* DAGAL⸣ EN.NAM ZA.E
[41 13 51 48 *ṣí-li-ip-ta šu-tam-ḫir* 28] 20 *ta-mar*
[40 SUKUD *šu-tam-ḫir* 26 40 *ta-mar* 26 40] ⸢*i-na*⸣ 28 20
[BA.ZI 1 40 *ta-mar* EN.NAM ÍB.SI$_8$ 10 ÍB.SI$_8$ DAGAL *ne-pé*]-*šum*

[The height is 0;40 (cubits), the diagonal 0;41 13 51 48.] What is the breadth? You: [square *0;41 13 51 48*.] You will see [0;28] 20. [Square 0;40, the height. You will see 0;26 40. Take 0;26 40] from 0;28 20. [You will see 0;01 40. What is the square root? The square root is 0;10, the breadth.] The method.

(xxv)

[2 KÙŠ DAGAL 41 13 51 48 *ṣí-li-ip-tum* SUKUD EN.NAM ZA.E]
[41 13 51 48 *ṣí-li-ip-ta*] ⸢*šu-tam-ḫir* 28 20 *šu-tam-ḫir-tim*!⸣
[10 DAGAL *šu-tam*]-⸢*ḫir*⸣ 1 40 *ta-mar* 1 40 *i-na* 28 20
[BA.ZI 2]⸢6⸣ [40 ÍB.TAG$_4$] EN.NAM ÍB.SI$_8$ 40 ÍB.SI$_8$ 40 SUKUD *ne-pé-‹šum›*

[The breadth is 2 cubits, the diagonal 0;41 13 51 48. What is the height? You:] square [*0;41 13 51 48*], the diagonal. 0;28 20 is the *squared number*.[23] [Square 0;10, the breadth.] You will see 0;01 40. [Take] 0;01 40 from 0;28 20. [0;26 40 is the remainder.] What is the square root? The square root is 0;40. The method.

---

iš-te-en-ma

BA RI (?) ŠE (?) RU (?) TI (?)
^{d}iškur-ma-an-sum
DUMU ^{m}30-*i-qí-ša-am*

One.[24]
Collated (?) ...[25]

[22] The word *ne-pe-šum* 'method' is twice abbreviated by omission of the final syllable, at the end of problems (xxiii) and (xxv). This phenomenon also occurs in other problems on this tablet, as well as in other similar texts (e.g. BM 85200 + VAT 6599, where *šu-tam-hir* 'square' and *ta-mar* 'you will see' regularly lose their final syllables; in BM 85210 some words are reduced to their initial signs only).

[23] The result of squaring the diagonal seems to be described as *šutamḫirtim* 'squared (number)'. This reading is very tentative, because there are no parallel attestations, and the word is in the genitive case where one would expect a nominative. The expected word *tammar* 'you will see' does not fit the traces here.

[24] This colophon does not conform to the general scheme of OB colophons laid out by Hunger (1968: 15). The word *ištēnma* just above the main body of the colophon, in between two horizontal rulings, might indicate that the tablet has been copied from an original and then collated: the text is correct, or 'one' with the original. (This reading was suggested by Dr Irving Finkel; this meaning for *ištēn* is not recorded in either *CAD* or *AHw*.) It seems that this note — and an abandoned attempt to write it at the bottom of column IV — was written a while after the rest of the text was copied and the clay was already slightly too hard, altering the shape of the signs slightly. If this interpretation is correct, this is a new type of evidence for a mathematical text being copied from an earlier original. For further proof that some OB mathematical texts were copied see §10.4.

[25] This first line of the colophon itself remains a mystery: if the first two signs are to be read *ba-ri* 'collated', which would be an unprecedentedly early occurrence (Hunger 1968: 15), the second half of the line, apparently ŠE RU TI — or just possibly SIG$_4$! $^2/_3$-*ti* 'two-thirds bricks' (cf. §4.3) — is still unclear. The line is most likely to be a description of the contents of the tablet, or the catch-line of the next in the series, as appears to be the case for the colophon of BM 85210 (cf. *MKT*: I 227 n. 9c). BM 85194, BM 85196 and BM 85200 + VAT 6599, also from Sippar,

Ishkur-mansum,[26]
son of Sin-iqisham.

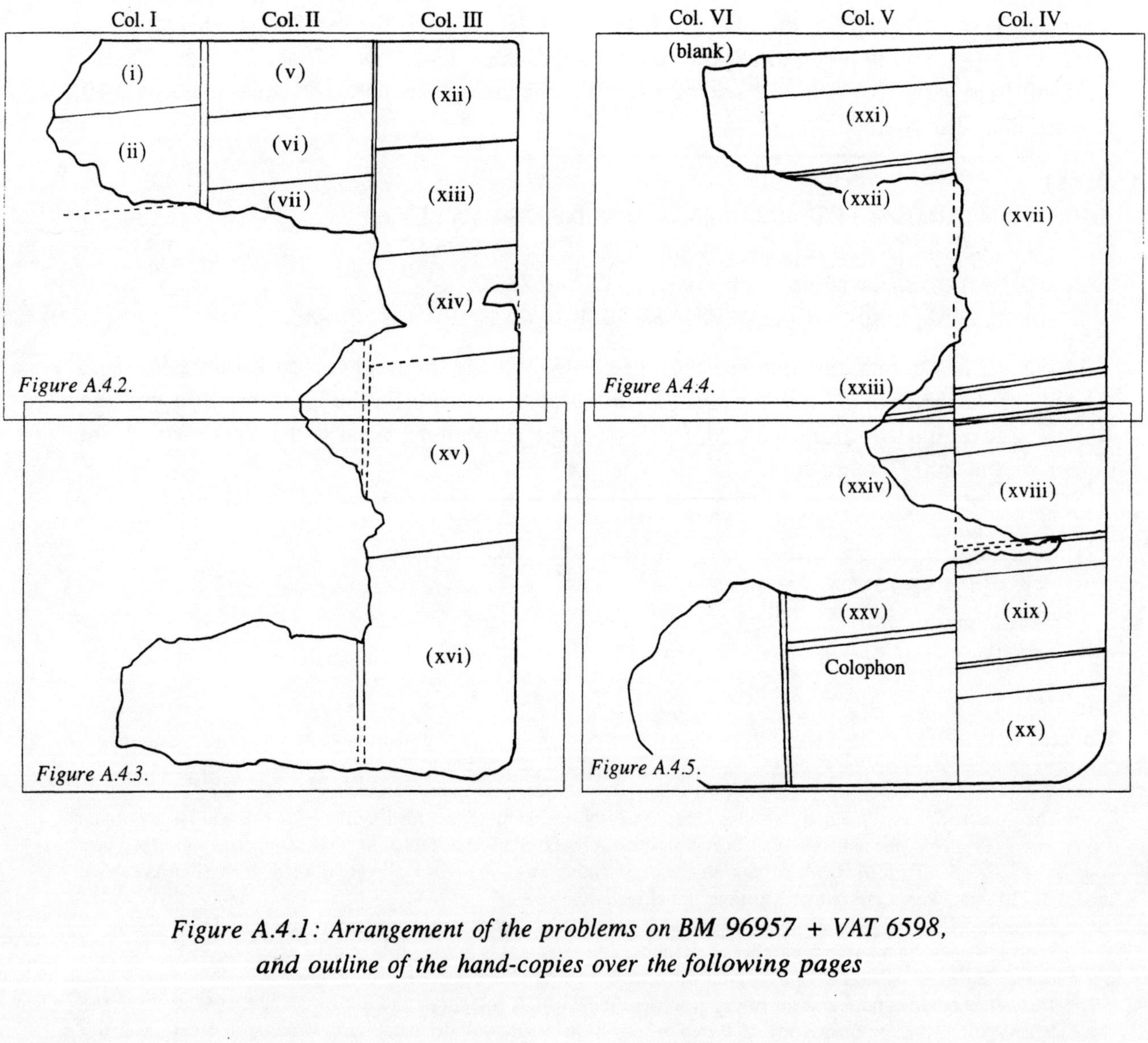

Figure A.4.1: Arrangement of the problems on BM 96957 + VAT 6598, and outline of the hand-copies over the following pages

have colophons consisting solely of the number of problems on the tablet. For more on OB mathematical colophons see Robson (fc.).

[26] This person cannot be identified unequivocally. There is no Ishkur-mansum, son of Sin-iqisham yet known from OB Sippar (cf. Ferwenda and Woestenburg 1992: ᵈIškur.mansum), although both names are widely attested in that city.

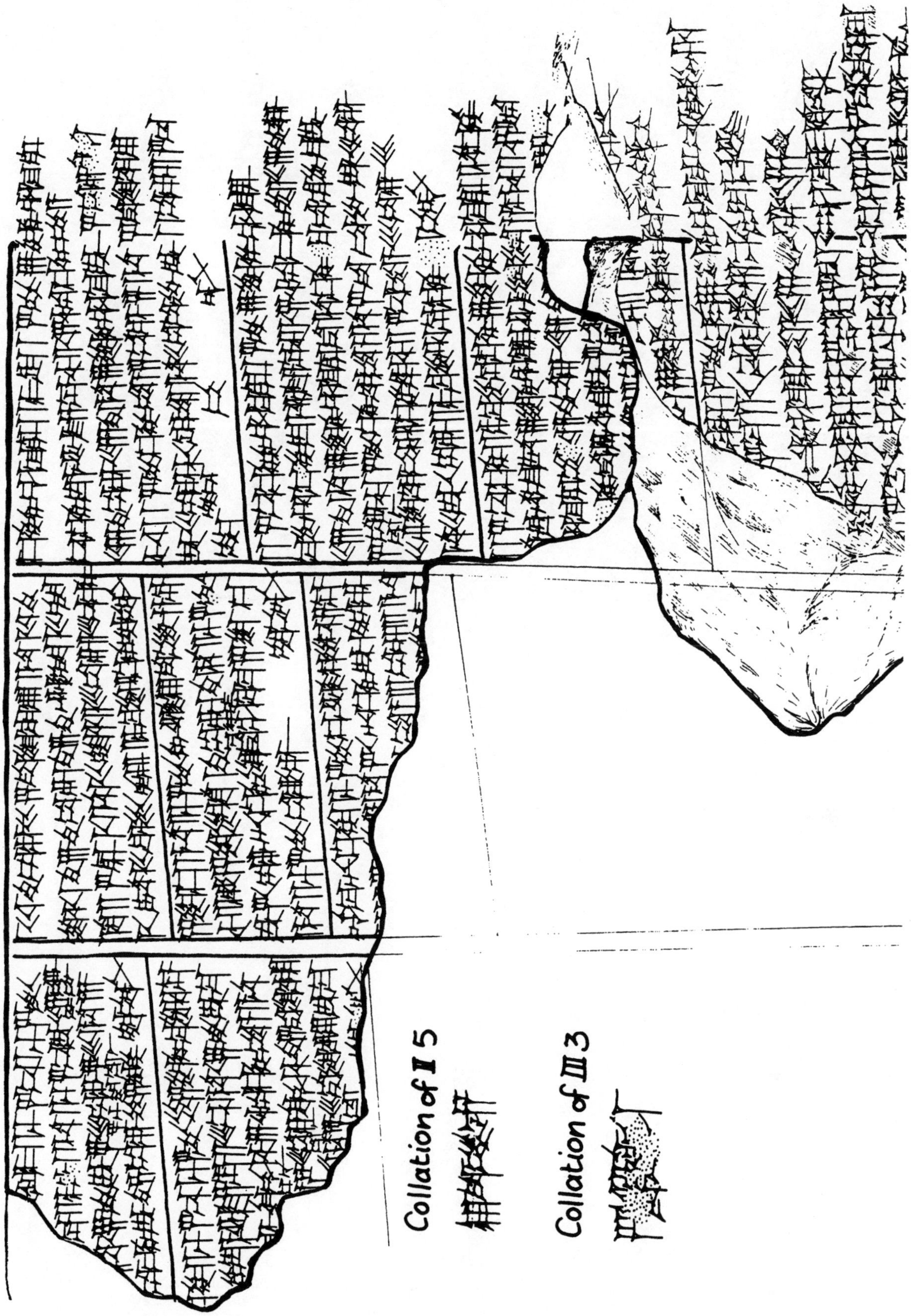

*Figure A.4.2: BM 96957 + VAT 6598 obverse top. Copy of BM 96957 (left) by Christopher Walker; copy of VAT 6598 (right) by Neugebauer (*MKT*: II Taf. 44)*

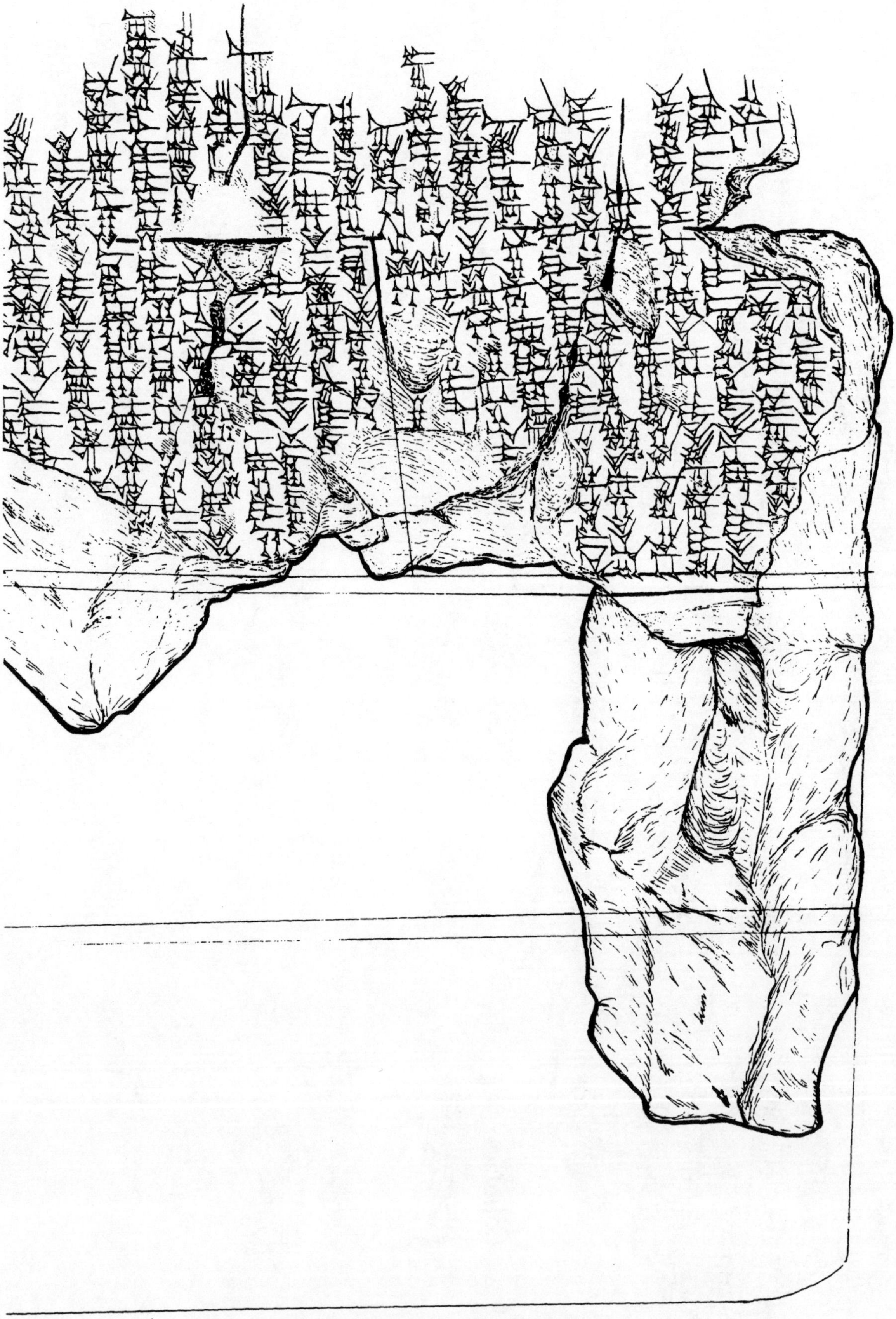

Figure A.4.3: BM 96957 + VAT 6598 obverse bottom.
Copy of VAT 6598 by Neugebauer (MKT: *II Taf. 44)*

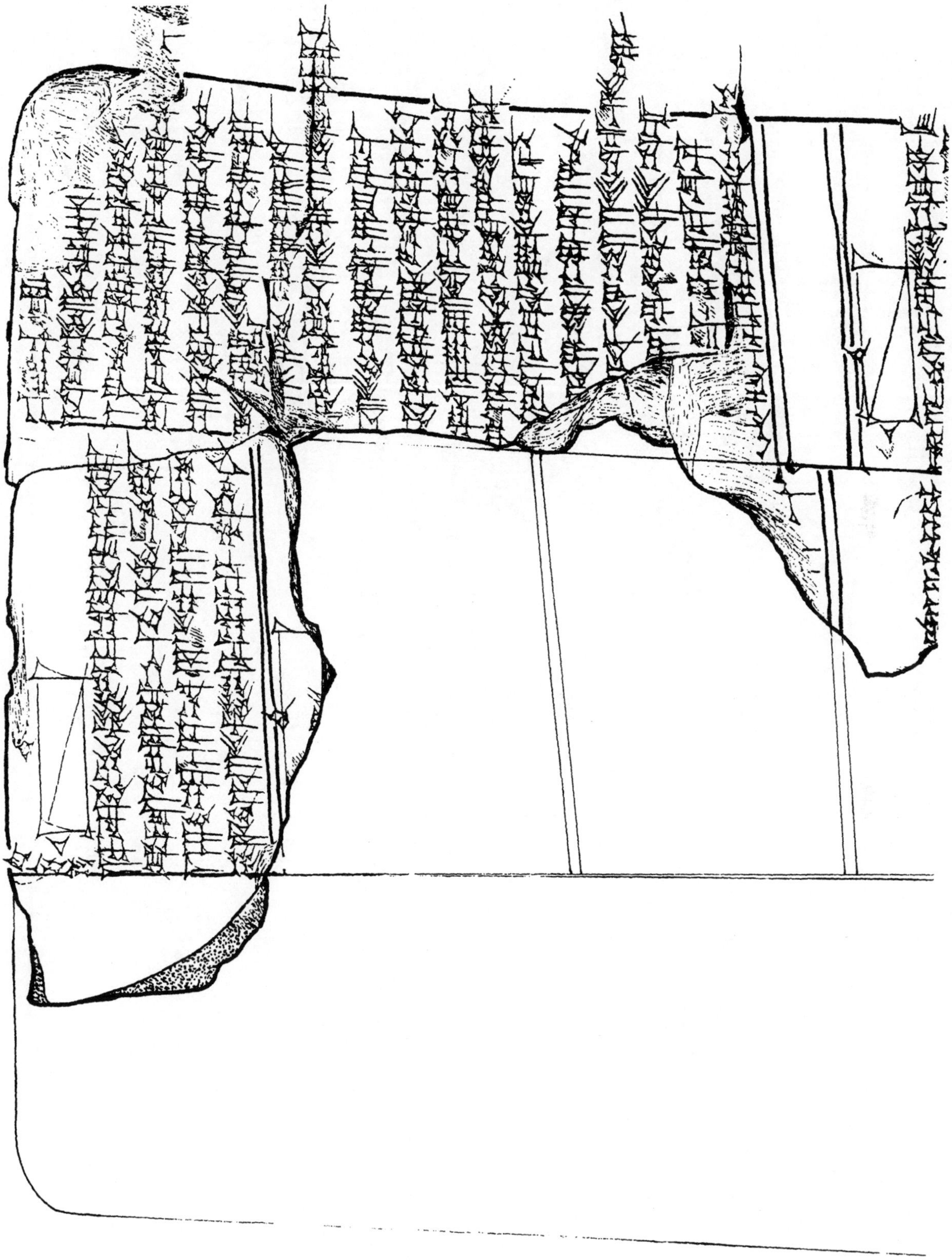

Figure A.4.4: BM 96957 + VAT 6598 reverse top.
Copy of VAT 6598 by Neugebauer (MKT*: II Taf. 44*)

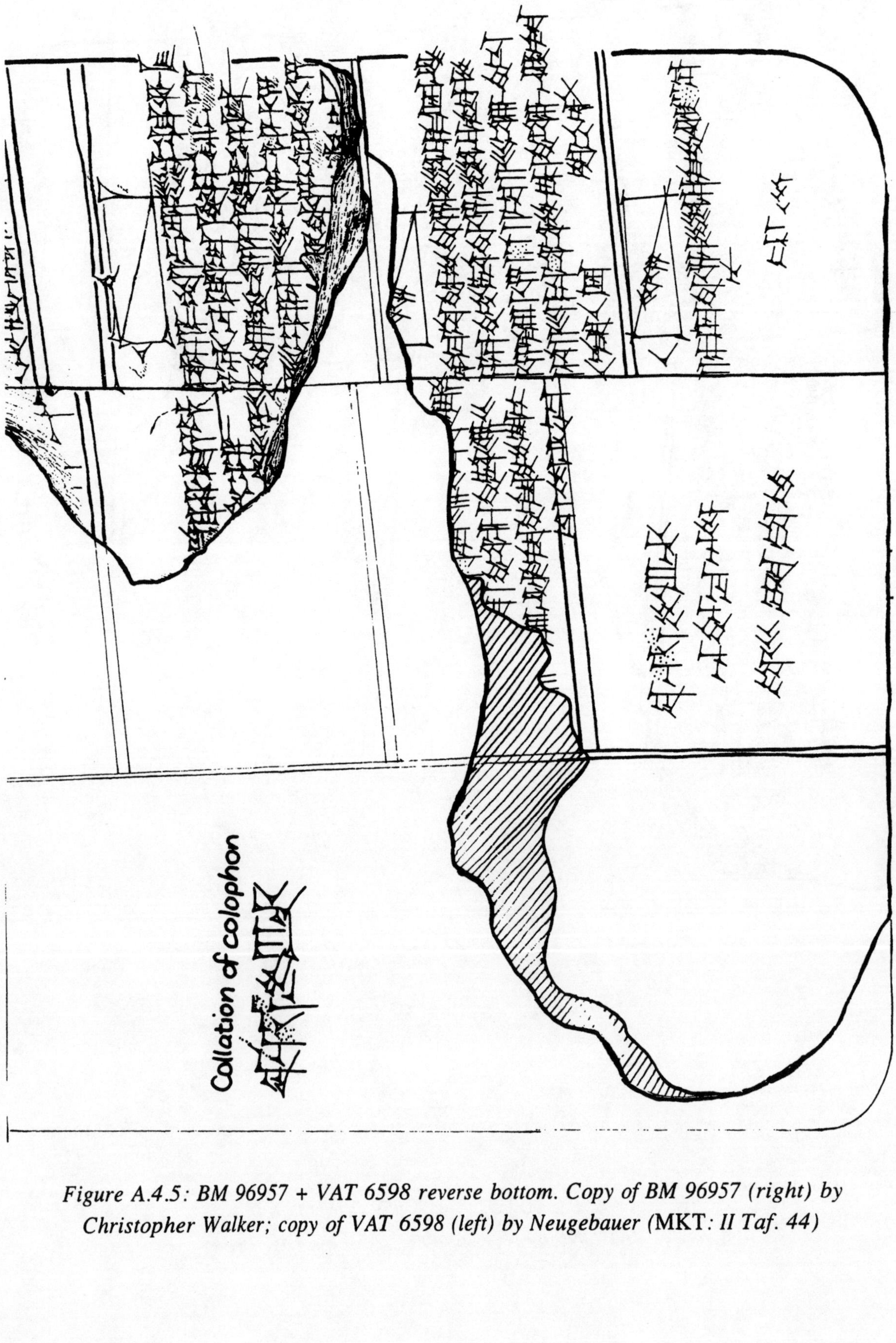

*Figure A.4.5: BM 96957 + VAT 6598 reverse bottom. Copy of BM 96957 (right) by Christopher Walker; copy of VAT 6598 (left) by Neugebauer (*MKT*: II Taf. 44)*

APPENDIX 5: MATHEMATICAL ROUGH WORK FROM OLD BABYLONIAN UR AND ELSEWHERE

While collating the school tablets from OB Ur published by Gadd and Kramer (1966), Dr Marie-Christine Ludwig discovered that 40 (about 20%) of the lenticular tablets containing Sumerian proverbs also have calculations on the back, which were not included in the original publication. (The remainder are blank on the reverse.) She very generously brought them to the author's attention, and they are provisionally published here.

This appendix also contains first editions of the mathematical rough work on six round tablets from OB Ur to be published by Shaffer (forthcoming),[1] and seven others from the Ashmolean Museum, Oxford, the British Museum, and the University Museum, Philadelphia. They are reproduced here with the kind permission of the Trustees of the three museums.

A.5.1 TABLETS FROM UR

The forty-six Ur tablets edited here were almost certainly found in secondary context, re-used as building rubble in 'No. 1, Broad Street' (§2.1.6). Their obverses almost all contain Sumerian proverbs; for transliterations, translations, and discussion see Alster (1997: I 309–28, II 466–73). The calculations fall into distinct groups, two or more of which may occur on the same tablet, as follows:

1. reciprocal pairs;
2. finding reciprocals: 'the Technique' (Sachs 1947: 222);
3. squared numbers;
4. series of multiplications;
5. tabular calculations;
6. fragments, scattered numbers, and unidentifiable calculations.

It is striking how consistent the layout is within each calculation group, implying that apprentice scribes were taught to lay out their calculations in a specific way. It is striking too that additive and subtractive operations are for the most part missing from these tablets — as they also appear to be from the published mathematical rough work listed in §1.3.3. Wherever possible, comparisons and contrasts are drawn with that published material.

Because of the damaged state of the tablets, the frequent erasures and rewritings, and the often haphazard arrangement of the signs, it has not always been possible to provide a satisfactory or unequivocal interpretation of these calculations. Nevertheless, this very preliminary study is offered here in the full expectation that it will enable others to derive a wealth of further information about the arithmetical practices of Old Babylonian Ur.

The tablets are listed by their *UET* 6 numbers in the following summary table, but are then dealt with according to the subject groupings 1–6 given above. All copies are the same size as the originals; their diameters have been recorded here to the nearest half-centimetre.

[1] *UET* 6/3 628 (U. 10144) was overlooked here. There is a proverb on the obverse, and the square of 2 24 in the same format as *UET* 6/2 211 on the reverse.

Table A.5.1: Mathematical rough work from Old Babylonian Ur

UET 6 number	Excavation number	Contents of obverse	Contents of reverse	Diameter
2 210	U. 17207, 38	Proverb: SP 1.128	6 Unidentified calculation	9.5 cm
2 211	U. 17207, 112	Proverb: SP 2.118	3 Squared number	9.5 cm
2 218	none	Proverb: SP 8 Sec. B 24	4.2 Multiplication: second series	—
2 219	none	Proverb: SP 8 Sec. B 28	6 Fragment of calculation	—
2 222	U. 17207, 6	Proverb: SP 5.107	3 Squared numbers	9.5 cm
2 233	U. 17207, 34	Proverb	4.1 Multiplications: first series	9.0 cm
2 236	U. 17207, 113	Proverb: SP 5.44	4.1 Multiplications: first series, and other multiplications	9.0 cm
2 247	U. 17207, 52	Proverb: "Minor SP Collections"	4.2 Multiplications: second series	9.0 cm
2 250	U. 17207, 30	Proverb: cf. SP 27.9	4.2 Multiplications: second series?	9.5 cm
2 254	none	Proverb	4.1 Multiplications: first series	8.5 cm
2 257	U. 17207, 9	Proverb	5 Tabulated calculations	9.5 cm
2 265	U. 17207, 145	Three proverbs: SP 1.6, SP 3.23, another	1 Reciprocal pairs?	10.5 cm
2 267	U. 17207, 21	Three proverbs: SP 2.37, another, 2.54; 1 reciprocal pair	6 Remains of calculations	10.5 cm
2 268	U. 17207, 88	Two proverbs: SP 2.38–39	6 Unidentified calculations	10.5 cm
2 269	none	Proverb: SP 2.49	4.2 Multiplications: second series; other calculations	9.5 cm
2 274	U. 17207, 3	Proverb	5 Tabulated calculation	9.0 cm
2 275	U. 17207, 5a	Proverb: SP 8 Sec. A 4	4.3 Multiplications; unfinished	9.0 cm
2 279	U. 17207, 18	Proverb	6 Unidentified calculation	9.0 cm
2 281	U. 17207, 25	Proverb: cf. SP 18.8	6 Numbers	9.5 cm
2 282	U. 17207, 26b	Proverb: SP 12 Sec. E 3	5 Tabulated calculation	9.0 cm
2 290	U. 17207, 55	Proverb: SP 2.39	4.2 Multiplications: second series	9.5 cm
2 291	U. 17207, 58	Two proverbs: SP 1.1–2, another	4.3 Multiplications and a reciprocal pair	10.5 cm
2 293	U. 17207, 62	Proverb: SP 2.134	4.1 Multiplications: first series	9.5 cm
2 295	none	Proverb	2 Finding reciprocals: 'the Technique'	8.5 cm
2 297	U. 17207, 71	Proverb	4.3 Multiplications; and a calculation, including a reciprocal	9.5 cm
2 298	U. 17207, 74	Proverb	4.1 Multiplication: first series	9.0 cm
2 300	U. 17207, 83	Two proverbs: SP 8 Sec. A 5–6	6 Numbers	9.0 cm
2 301	U. 17207, 85	Proverb: cf. SP 1.143	4.3 Multiplications; unfinished	8.5 cm
2 304	U. 17297, 94	2 proverbs	6 Numbers: reciprocal pairs?	9.5 cm
2 311	U. 17207, 119	Seven proverbs	4.3 Multiplications and a reciprocal pair	10.5 cm
2 313	U. 17207, 128	Proverb: SP 5.123	5 Tabulated calculation	9.0 cm
2 318	U. 17207, 143	Proverb	4.3 Fragment of a multiplication	10.0 cm
2 320	not given	Proverb: SP 1.9	6 Numbers	8.0 cm
2 321	not given	Proverb	2 Squared number	9.5 cm
2 322	not given	Two proverbs	5 Tabulated calculation	—
2 330	not given	Proverb: cf. SP 15 Sec. B 6	6 Numbers	8.0 cm
2 343	U. 17207, 93	Riddle	1 Reciprocal pairs	9.5 cm
2 367	U. 17207, 54	Two proverbs with Akk. glosses	6 Numbers	9.5 cm
2 371	U. 17207, 76	Four proverbs, one with Akk. glosses	1 Reciprocal pairs within an unidentified calculation	12.5 cm
2 387	U. 17207, 100	Proverb	1 Reciprocal pairs within an unruled multiplication-series	9.5 cm
3 120		Proverb	4.3 Multiplications; broken	9.5 cm
3 374		4.2 Multiplications: second series	Blank	—
3 378		4.3 Multiplication, reciprocal pair	Blank	—
3 385		5 Tabulated calculation	Blank	10.0 cm
3 452		4.2 Multiplications: second series	Blank	9.5 cm
3 466		Proverb	6 Numbers	—

A.5.1.1 Reciprocal pairs

Where space allows, pairs of mutual reciprocals are written next to each other (in horizontal alignment); they may be placed one on top of the other (in vertical alignment) otherwise. In the following discussions, the reciprocal relationship is symbolized by a tilde, and there is no attempt to mark sexagesimal place. For instance, $6 \sim 10$ should be understood to mean '6×60^{n} and 10×60^{-n} are mutually reciprocal'.

The five round tablets from Ur edited below contain reciprocal pairs, either alone or within unidentified calculations. There are also numerous reciprocal pairs on the other tablets, which may or may not form an integral part of some other calculation. Typically, many-place reciprocal pairs are written on the top of the tablet; two-place pairs are written near the bottom. The published tablets 2 N-T 115 (from Nippur) and FLP 1283 (from Uruk) also show reciprocal pairs.

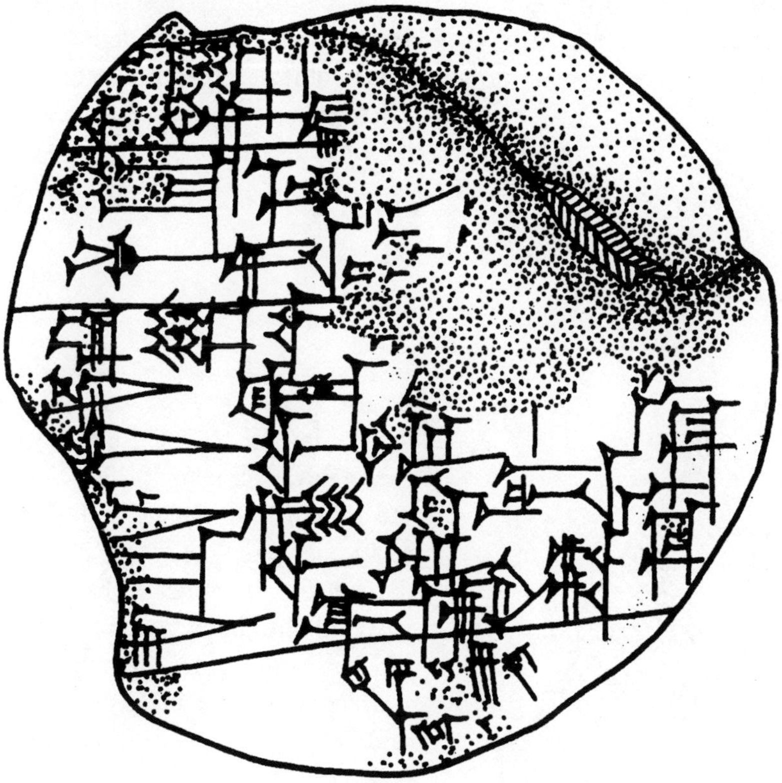

Figure A.5.1: UET *6/2 267 obverse*

The obverse of *UET* 6/2 267 contains several proverbs concerning scribes and scribal practice.[2] Beneath them, in a much smaller hand, is written a reciprocal pair from the standard table:

44 26 40
~ 1 2!1.

The 20 looks more like a 30 — perhaps the student has misremembered his reciprocal tables — and there is a stray Winkelhaken or 10 to the right.

[2] The proverbs read, with loose translations (cf. Alster 1997: I 311–12, II 365):

⸢dub⸣-sar-⸢me⸣-[en mu-ní-za nu-zu] **igi-ní-zu [zi-ga]**	You're a scribe and you don't even known your own name? You should slap yourself in the face!
ám-ma ⸢tuku?⸣ [...] **SUD UŠ X [...]**	
dub-sar pe-[el-lá] **lú KA×LÍ-⸢KA×LÍ-šè⸣** **nar pe-el-lá** **lú gi-di-da-sè** **⸢dam⸣-<gàr> pe-el-lá** **lú da-bi-RI+RI-šè**	A disgraced scribe becomes a *kalû*-priest; a disgraced *nar*-singer becomes a piper; a disgraced merchant becomes a ...

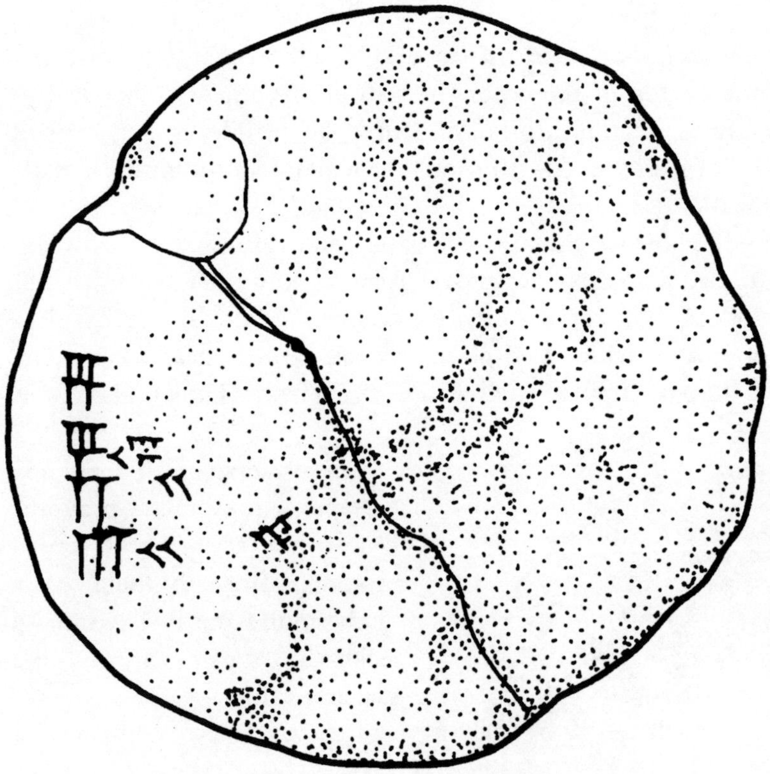

Figure A.5.2: UET *6/2 265 reverse*

Although the surface of *UET* 6/2 265 is now very damaged, enough remains of the numerals on the left-hand side to allow a tentative interpretation of them:

5 [~ 10]
4 ~ 15
2 ~ ⌜30⌝
3 ~ 20

The remains of a 40 are also discernible.

Of course, these signs may equally be the first digits of some other set of numbers. There are no traces of rulings on the tablet.

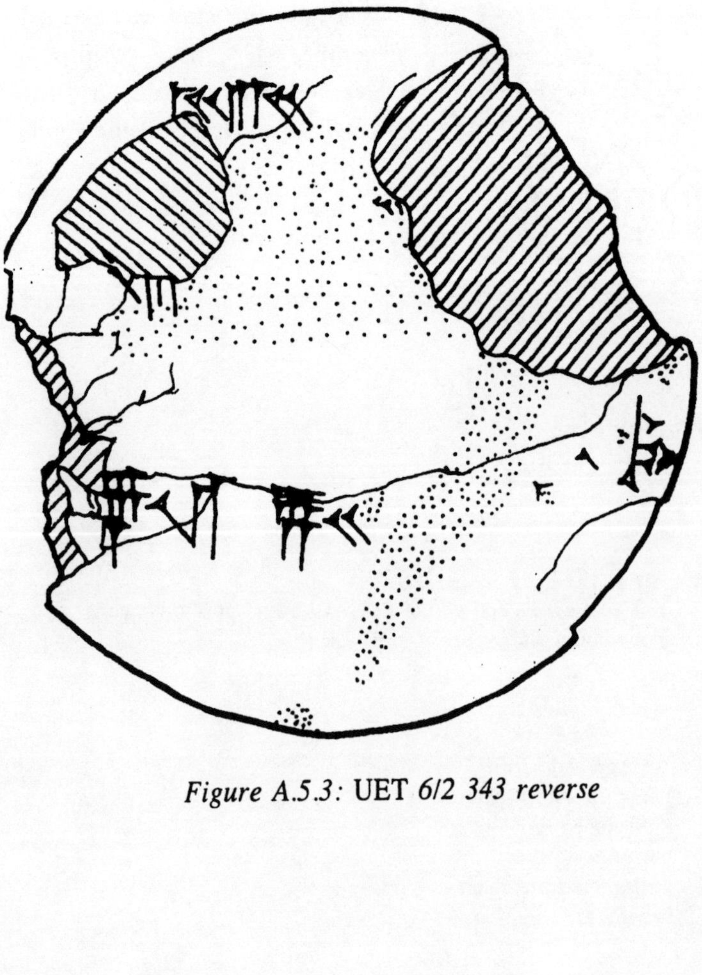

Figure A.5.3: UET *6/2 343 reverse*

All that remains of the calculation on *UET* 6/2 343 are the regular numeral 1 23 30 at the top of the tablet, and the reciprocal pair 7 12 ~ 8 20 at the bottom. The lower half of what might have been ⌜13⌝, ⌜56⌝ or some other number is visible below the break in the middle of the tablet. It is just possible that this is the remnants of 43 12, the reciprocal of 1 23 20.

There are several erasures, and a sign (KI or DI) which has run over from the end of a line of the proverb on the obverse of the tablet.

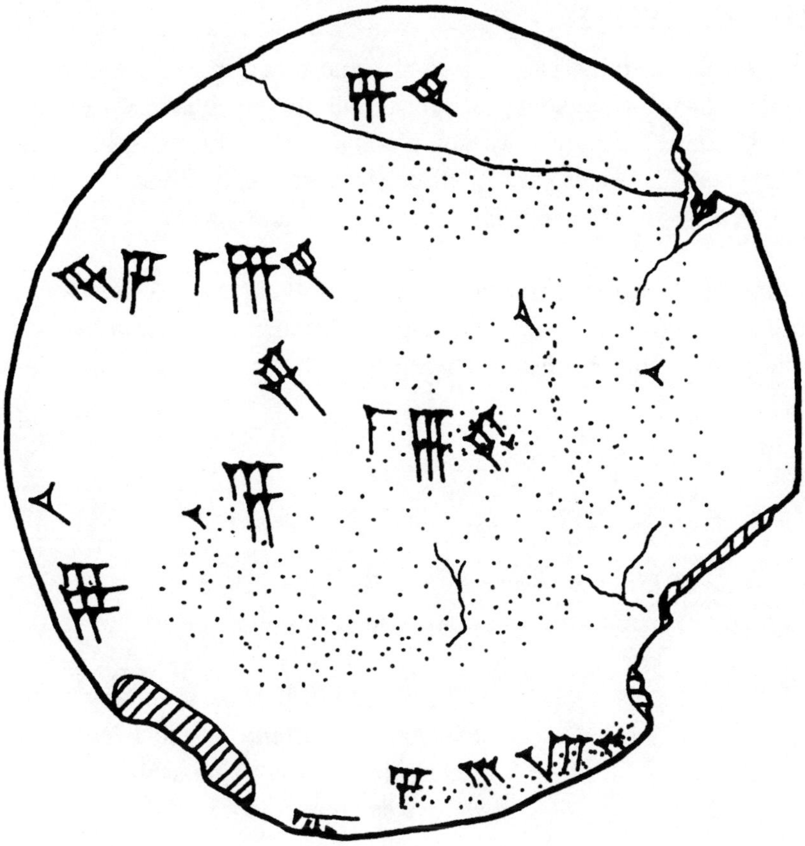

Figure A.5.4: UET *6/2 371 reverse*

The numerals on *UET* 6/2 371 do not seem to form a coherent calculation, but there are two recognisable reciprocal pairs. There are erasures and surface damage, in which the traces of Winkelhakens are visible. A sign from the proverb on the obverse is just discernible at the bottom of the tablet.

6 40

54 ~ 1 06 40
50
1 06 40
10 15
8

4 30 ~ 13 20

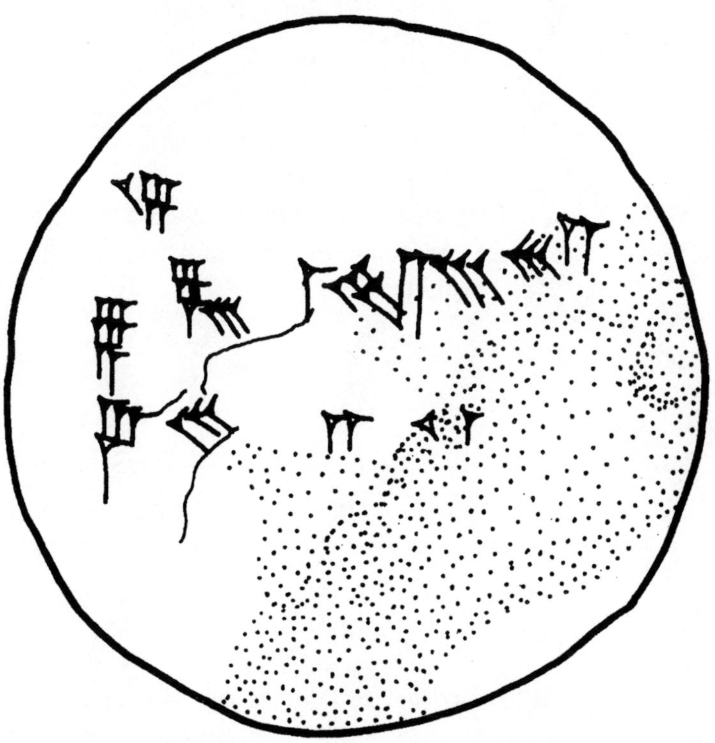

Figure A.5.5: UET *6/2 387 reverse*

There is an interesting set of relationships between the extant numbers on *UET* 6/2 387:

15
÷ 8 ~ 7 30
= 1 52 30 ~ 32
× 4 30 = 2 24

The visual arrangement is identical to that of the calculation on the right of *UET* 6/2 236 (§A.5.1.4). Unlike other multiplication series from Ur, the computation is formatted without vertical lines.

A.5.1.2 Finding reciprocals: 'the Technique'

The standard OB table of reciprocals contains pairs for all the one-place integers and for 1 04 and 1 21, the squares of 8 and 9 (thereby including reciprocal pair for all squares 1–9). It may also list the pairs for 1 12, 1 15, and 1 20 (thereby completing the set of two-place regular numbers up to 1 21) (cf. *MCT*: 11–12). Sachs (1947) discovered an OB method — which he named 'the Technique' — for finding the reciprocals of regular numbers which are not in the standard list.

The Technique can be summarized by modern means in the following way. A sexagesimally regular number c has to terminate with some one- or two-place integer in the standard list, a. Call the difference between them b, so that $c = b + a$. Therefore,

$$\begin{aligned}\frac{1}{c} &= \frac{1}{b+a}\\ &= \frac{1}{a}\cdot\frac{1}{1+b/a}\end{aligned}$$

There are no surviving instructions for this method, but five published tablets show the intermediate results from this type of procedure;[3] they are also found on one of the lenticular tablets from OB Ur.

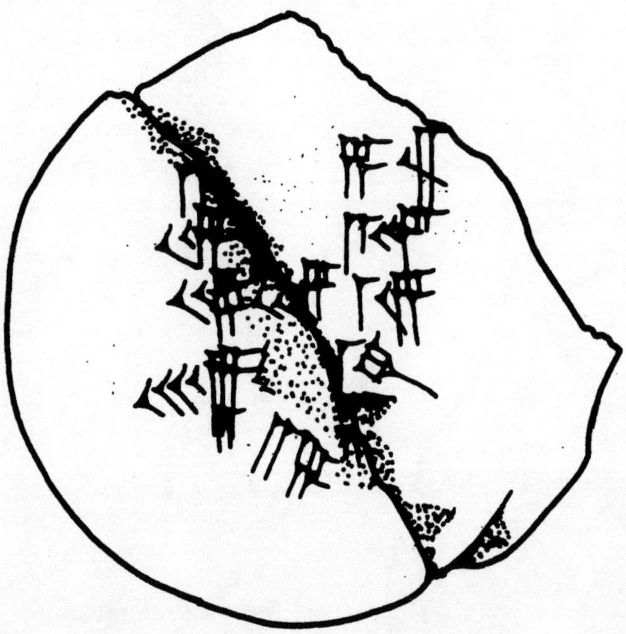

Figure A.5.6: UET *6/2 295 reverse*

UET 6/2 295 shows a very nice, if chaotically written, application of 'the Technique' to the regular number 2 05, where a is chosen to be 5:

2 05 ~ 12
25 (= 1 + 2 × 12) ~ 2 24
28 ⸢48!⸣ (= 12 × 2 24)

And then the same procedure is performed on the resulting reciprocal, presumably to check the arithmetic, with a = 48. (On the tablet, the 48 looks more like 45). As before, the number whose reciprocal is to be found is split into two widely spaced parts:

28 ⸢48!⸣ ~ 1 15
36! (= 1 + 28 × 1 15) ~ 1 40
2 05 (= 1 15 × 1 40)

What must be 36 in the fourth row looks more like 48 on the tablet. There are also traces of a 40 and a 5 along the central break which do not appear to belong to the calculation; perhaps they are erasures.

A.5.1.3 Squared numbers

Both tablets containing calculations on squares are laid out so that the number to be squared is written twice, in vertical alignment, as in a modern long multiplication. The answer is written underneath. There are no signs of intermediate calculations, or of horizontal or vertical rulings

[3] Those tablets are: N 3891 (Nippur), MLC 651, VAT 5457, YBC 1839, and YBC 10802 (all unprovenanced).

on the tablet. The numbers to be squared do not appear on standard tables of squares, which only include whole numbers from 1 to 60.

The same format appears on the square-shaped tablets CBS 3551 (with the word ÍB.SI$_8$ 'square' before the answer), HS 232, and N 3971 (§A.5.2) from Nippur.[4] On the unprovenanced YBC 7294 and YBC 10801, on the other hand, squarings are formatted in the same way as the multiplications from Ur, with the multiplicands vertically aligned to the left of a vertical line and their product to the right of it.

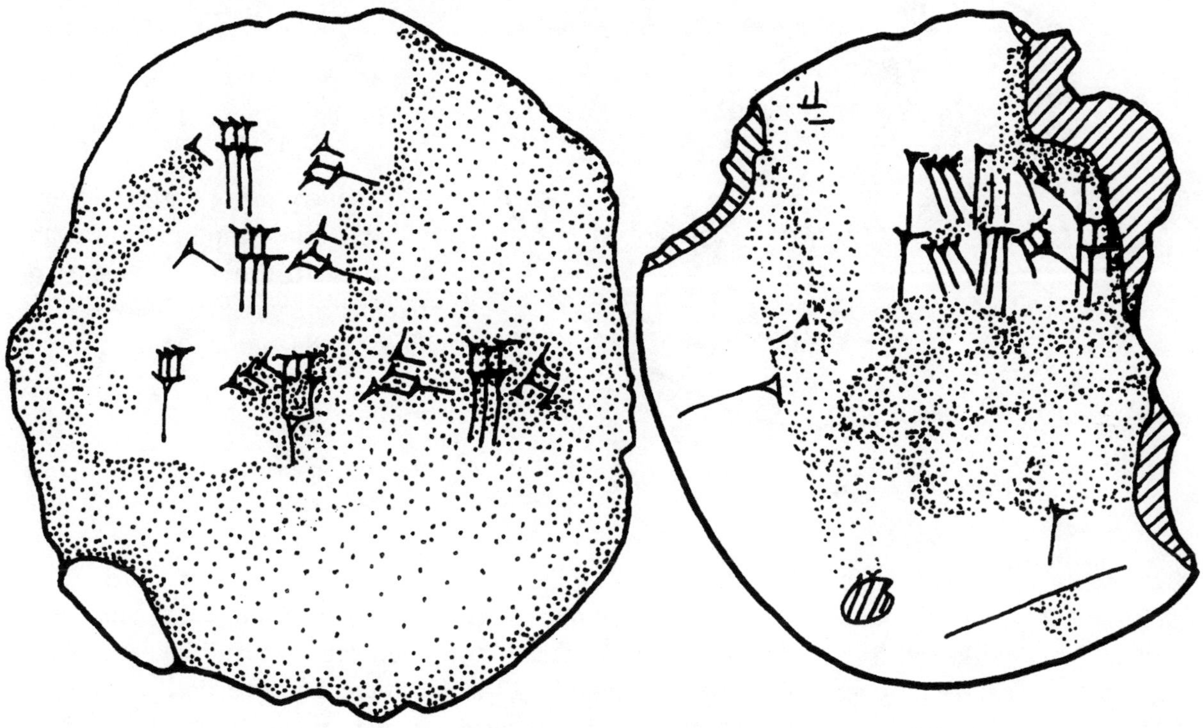

Figure A.5.7: UET *6/2 211 reverse and 321 reverse*

The surface of *UET* 6/2 211 is badly eroded and much has been erased from *UET* 6/2 321, but the calculations they contain can still be made out with ease. The answer has been deliberately removed from 321.

16 40	1 33 45
× 16 40	× 1 33 45
= 4 37 46 40	[= 2 26 29 03 45]

UET 6/2 222 is more complicated. It shows a repeated pattern of squares in a computation which comes down to:

$$(0;15^2 \times 17)^2 \times 16^2 = 17^2.$$

Because $15^2 \sim 16$ the relationship is trivially obvious, but is nevertheless appealing in its use of consecutive integers. This type of circular calculation — which seems to be an exercise in its own right, rather than the rough work for a problem text — is otherwise unknown.[5]

[4] See also these five members of the group tablets from Nippur discussed by Neugebauer and Sachs (1984: 246–8): CBS 11318, NBC 8082, NCBT 1913, UM 29-15-192, and 2N-T 30

[5] A simpler type of circular exercise, on an unpublished round tablet, is known from OB Sippar. The numbers 5, 6, 7 and 8 are written around the edge of the tablet. Between each pair of numbers is the product of those numbers, so that 30 is written between the 5 and 6, 42 between the 6 and 7, and 56 between the 7 and 8. 40, the product of 5 and 8, is missing. I thank Farouk Al-Rawi for sharing this information with me.

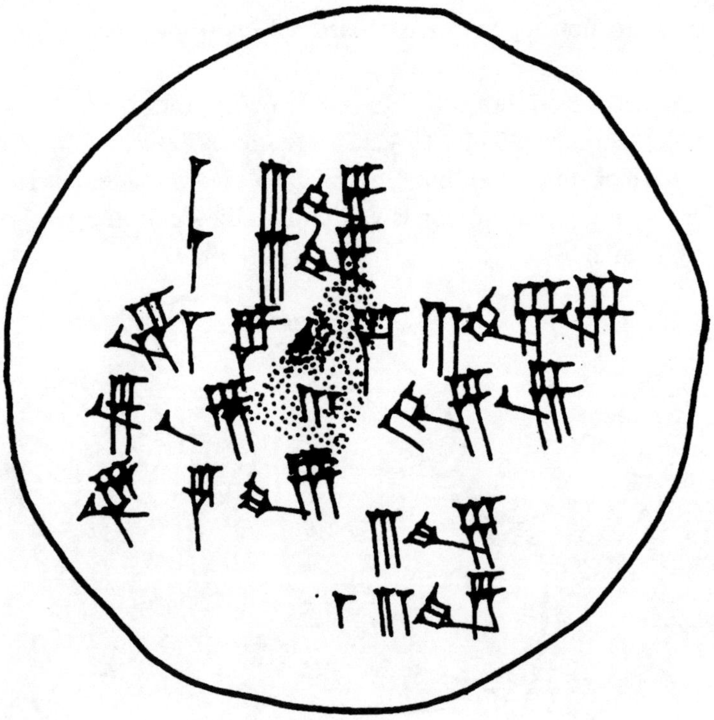

Figure A.5.8: UET *6/2 222 reverse*

The numbers are:

1 03 45
× 1 03 45
= 1 07 44 03 45
× 16
= 18 03 45
× 16
= 4 49 (= 17 × 17)
15
× 15
= 3 45
× 17
= 1 03 45

While the squarings are laid out vertically, other multiplications are recorded with the multiplicands in horizontal alignment. There are no rulings on the tablet.

A.5.1.4 Series of multiplications

The largest group of round tablets from Ur show several numbers written either side of a vertical line, on the left-hand side of the tablet. Those to the left are the multiplicands, and those to the right the products, of a series of multiplications which proceeds vertically down the tablet. The first two multiplicands are written to the left of the vertical line and their product is given to the right, placed vertically between them. This product is then multiplied by the number written to the left of the line underneath the first two. Its product is written underneath the first product on the right, aligned between the second and third multiplicands, etc. When a product has to be divided, the reciprocal of the divisor is written immediately to the right of it, although still on the left of the vertical line.[6] The product is then multiplied by the reciprocal. Typically there are three or four stages to the calculation, although there may be as many as six. There may be auxiliary, or apparently unrelated, calculations on the right-hand side of the tablet, sometimes separated from the other set by a second vertical line. Reciprocal pairs, multiplications and squarings are common on this right-hand side.

Two patterns of multiplications can be distinguished within the group as a whole. In one group of five tablets, the third operation is a division by 10 followed by a multiplication by 2 and, optionally, by some other multiplication. The third operation of the series is always a multiplication by 6 40 in the second series, which contains five, possibly seven tablets. These two sets of calculations are treated first, followed by the eight tablets with multiplication-series which cannot be grouped.

Apart from the squarings on YBC 7294 and YBC 10801 (§A.5.1.3), these are the only known calculations in this format.

[6] But in the first series immediately below the divisor 10 is always to the left of its reciprocal multiplicand 6.

First series: 10 and 2

Five tablets contain ruled series of calculations whose third and fourth stages are a division by 10 followed by a multiplication by 2 (or a division by 30). All but one are arranged around two vertical lines and have auxiliary calculations (or their remains) on the right of the tablet.

UET 6/2 233 shows the following series of multiplications:

5
× 2 = 10
× 3 = 30
÷ 10 ~ 6 = 3
× 2 = 6

The arrangement of the reciprocal pair in this and the other four calculations of this series, is the reverse of those found in the other multiplication tablets from Ur. Here the multiplicand 6 is written to the left of its reciprocal 10; elsewhere it is found to the right.

Nothing can be read of the erased numerals on the right of the tablet.

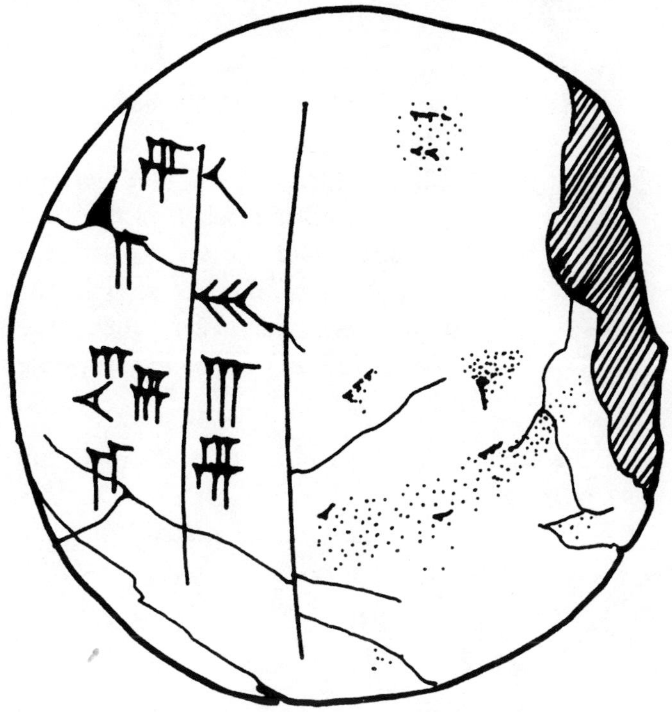

Figure A.5.9: UET *6/2 233 reverse*

The left of *UET* 6/2 236 reads thus:

50
× 30 = 25
× 3 = 1 15
÷ 10 (~ 6) = 7 30
÷ 30 (~ 2) = 15

Interestingly, the reciprocal pair 2 ~ 30 is reversed with respect to the 10 ~ 6: here the multiplicand is on the left.

The right hand reads:

30
× 3 = 1 30 ~ 40
× 10 = 6 40

(cf. UET 6/2 387, §A.5.1.1). These same multiplicands are also found in the same order on the left-hand multiplication (where, though, 10 is a divisor).

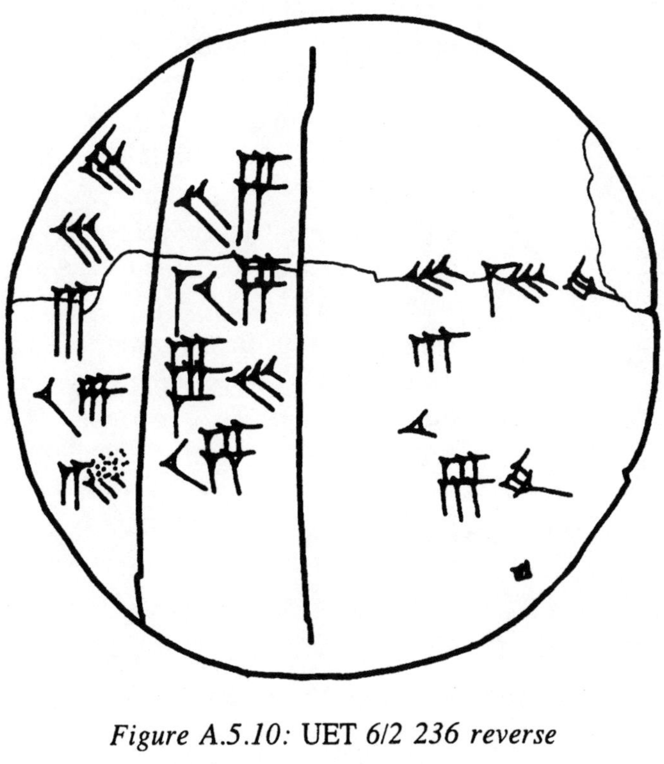

Figure A.5.10: UET *6/2 236 reverse*

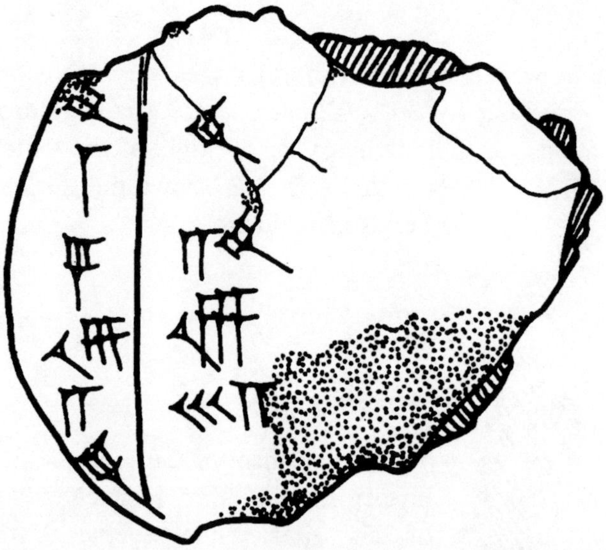

Figure A.5.11: UET *6/2 254 reverse*

Most of the numbers on *UET* 6/2 254 are easily explicable:

⌜40⌝
× 1 = 40
× 4 = 2 40
+ 10 (~ 6) = 16
× 2 = 32.

The role of the final 40, however, is difficult to explain. It *may* be a badly written 30 (~ 2; see below). If, on the other hand, it is the final multiplicand of the series (and the three others of this set stop at the multiplication by 2), then the product (22 24) ought to be visible below the 32.

There are no traces of any other numerals on the tablet; indeed, there is no central dividing line, which suggests that no further calculations were intended.

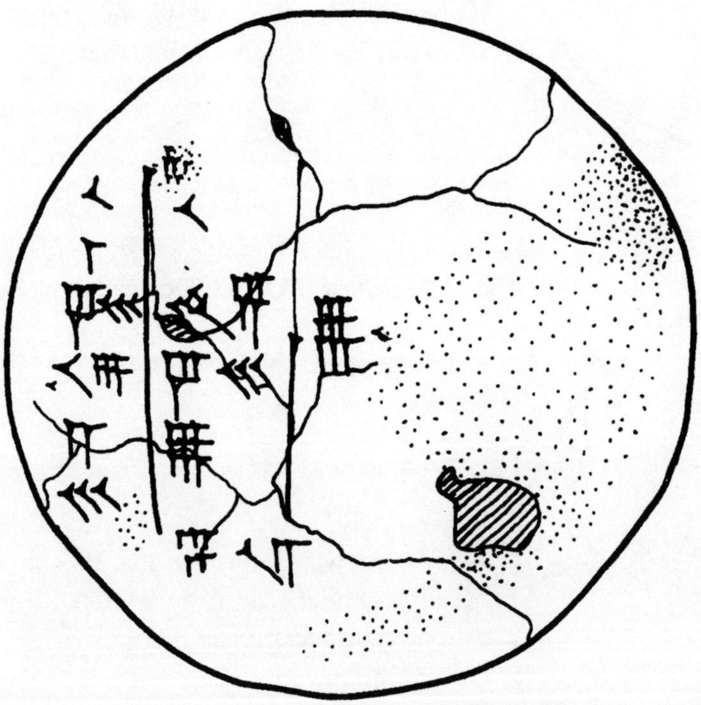

Figure A.5.12: UET *6/2 293 reverse*

The multiplication-series on *UET* 6/2 293 reads:

10
× 1 = 10
× 4 30 = 45
+ 10 ~ 6 = 4 30
+ 30 ~ 2 = 9

The reciprocal pair 2 ~ 30 is written in vertical alignment, even though there is enough room for a horizontal arrangement.

There is a further reciprocal pair, 5 ~ 12, below the products column — which might just conceivably be the conversion coefficient between nindan and cubits (§7.1). The number 9 (the final answer of the series) is also written at the centre of the tablet, perhaps marking the beginning of an extra calculation, now erased or eroded.

There is also a very small erased 8 at the top of the products column.

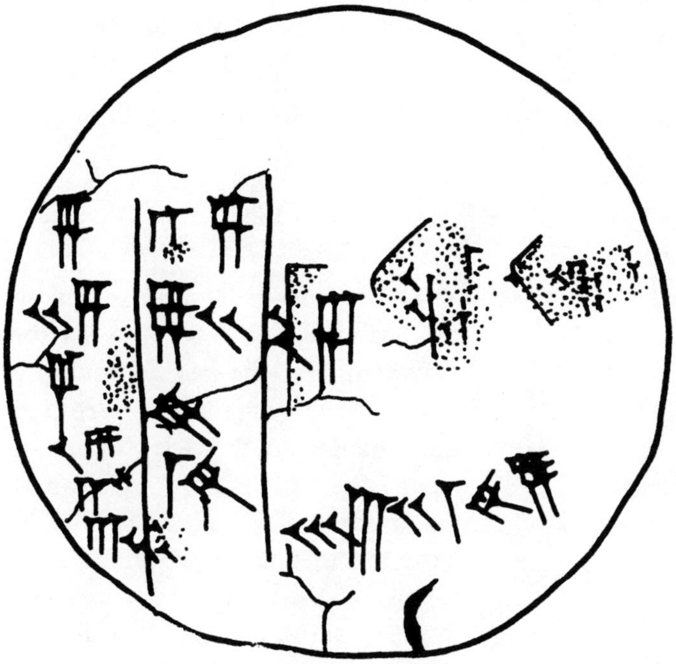

Figure A.5.13: UET *6/2 298 reverse*

The final multiplication-series of the set is on *UET* 6/2 298.

$$
\begin{array}{l}
\quad 5 \\
\times\ 25 = 2\ 05 \\
\times\ 4 = 8\ 20 \\
\div\ 10 \sim 6 = 50 \\
\times\ 2 \sim 30 = 1\ 40 \\
\div\ 3 \sim 20^{!} = 33\ 20 \sim 1\ 48
\end{array}
$$

There are also some partially erased numerals on the left of the central dividing line. The first may be 15 or 45; then what looks like 16 written twice in vertical alignment; then perhaps 18.

What might have been the purpose of these five computations? One plausible answer would be that they are the workings of students solving mathematical problems of the sort found in problem texts. Clearly, they all relate to the same type of problem: one which involves multiplying three variables together, dividing the answer by 10, then multiplying by 2. The fact that these last two operations involve constants suggests that we are dealing with coefficients. If so, then there is one problem type that would admirably fit this evidence: digging canals and other earthworks at a daily rate of 0;10 sar_v (§6.2–3). In that case, the first three variables would describe the dimensions of a rectangular prism (which might also explain the presence of the reciprocal pair 5 ~ 12 as helping to convert those measurements into the correct units, §7.1). The total volume is then divided by the daily volume of one worker to give the number of days needed to complete the project (cf. §6.1.3). The multiplication by 2 — or, rather, division by 30 — is used, finally, when the working time is better expressed in months (cf. BM 85196 (xiii), §5.2.2). This is, of course, a highly speculative, unprovable interpretation.

Second series: 6 40

A further five lenticular tablets from Ur contain a different series of multiplications, in which the third and final operation is always a multiplication by 6 40. All but one has a vertical line separating the multiplicands from their products, and only one — in which the computations have not been performed — has any other numbers on the right side of the tablet. There are two further tablets which might also belong to the series; they are discussed immediately afterwards.

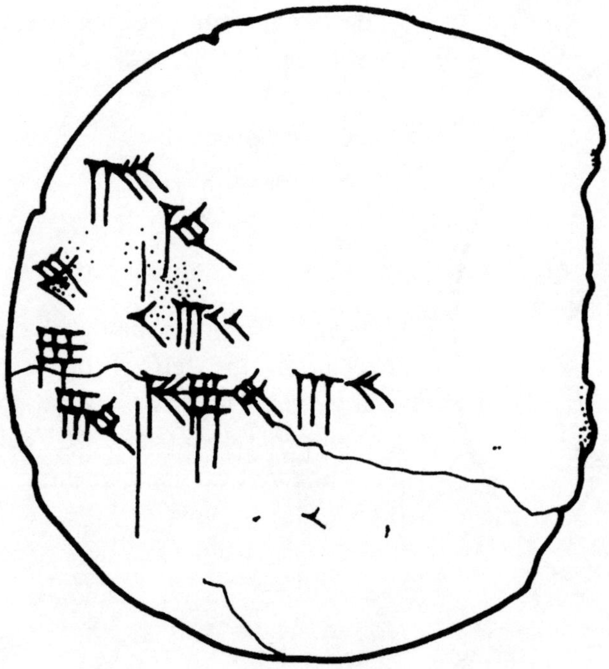

Figure A.5.14: UET *6/2 247 reverse*

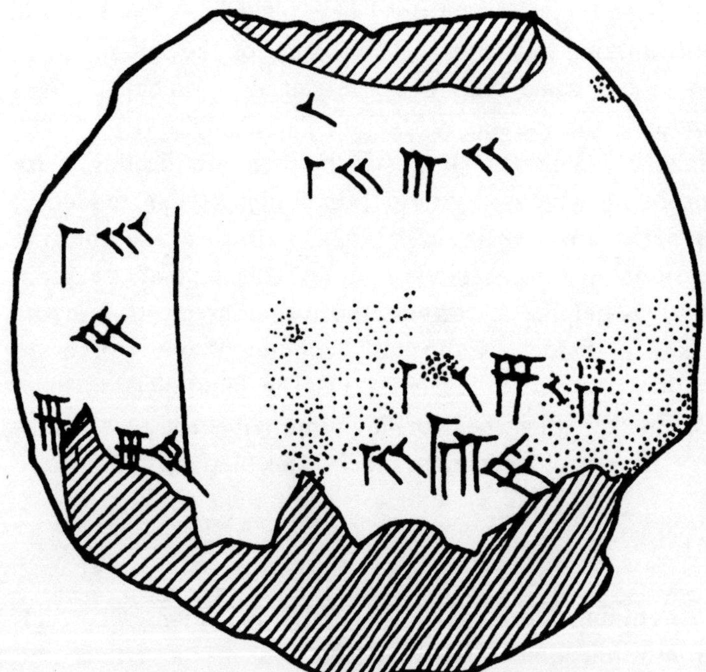

Figure A.5.15: UET *6/2 269 reverse*

The entire series of multiplications on *UET* 6/2 247 is preserved:

2 (~ 30)
× 50 = 1 40
× 8 = 13 20
× 6 40 = 1 28 53 20.

The first operation recorded is in fact a division: 50 ÷ 30. The divisor is written over the vertical dividing line, and its reciprocal is recorded immediately to the left, where it takes the place of a normal multiplicand. The pair are written so close together that it would be easy to mistake them for a single number. In many cases, however, the divisor can be distinguished from its mutual reciprocal by the much smaller, lighter script in which it is noted.

The multiplicands for a multiplication series have been laid out to the left of the vertical line on *UET* 6/2 269, but they seem to bear little relation to the numbers recorded on the right-hand side of the tablet. The completed calculation would have been:

1 30
× 50 = 1 15
× 6 = 7 30
× 6 40 = 50.

On the right we appear to have:

10
1 23 20

[...] 1! 15 ⌜15!⌝ [...]
1 24 5[1 10!!]

The restoration of this last number is *highly* speculative: there is no particular reason, in the author's opinion, to definitively prefer the famous approximation to √2 (§3.4) over other possible completions. Nevertheless, it remains a possibility.

However, there is another possible restoration, based on *UET* 6/2 385 (§A.5.1.6). If we have the same numbers on each, then the whole should read:

[54⌉
10
1 23 20
÷ 1 15 ~ ⌈48⌉
1 06!! 40!

In that case, though, it would be difficult to account for the last number, which is so unlike what would be expected here.

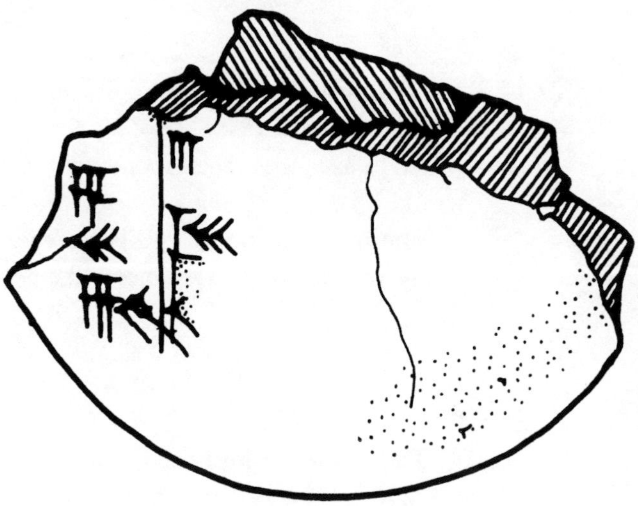

Figure A.5.16: UET *6/2 218 reverse*

The top of *UET* 6/2 218 has broken away, but the calculation can be fully restored:

[36]
× 5 = 3
× 30 = 1 30
× 6 40 = 10.

The missing 36 would have been immediately above the 5.

There is an erased 1 between the products 1 30 and 10, and a whole line of text has been removed from the bottom right of the tablet.

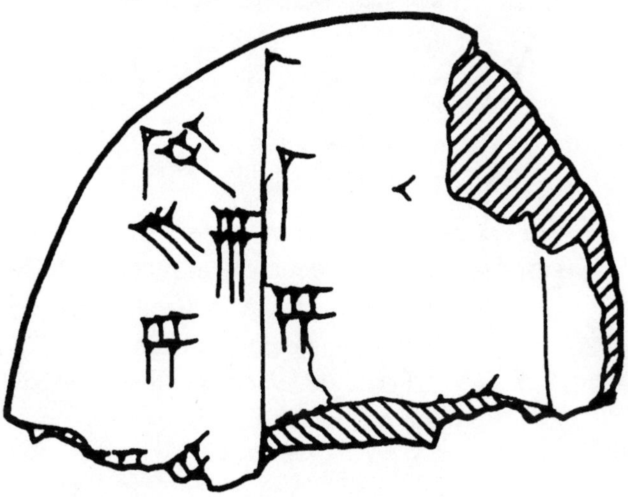

Figure A.5.17: UET *6/3 374: reverse*

Enough remains of the multiplication-series on *UET* 6/3 374 to allow it to be tentatively identified as one of the 6 40 group:

1 40
× 36 = 1
× 5 = 5
× ⌈6 40⌉ [= 33 20].

Of course, the vertical wedges visible above the break might equally represent any numeral from 3 to 9, and the Winkelhakens 40 or 50. Some traces of the final answer are visible to the right of the vertical line too.

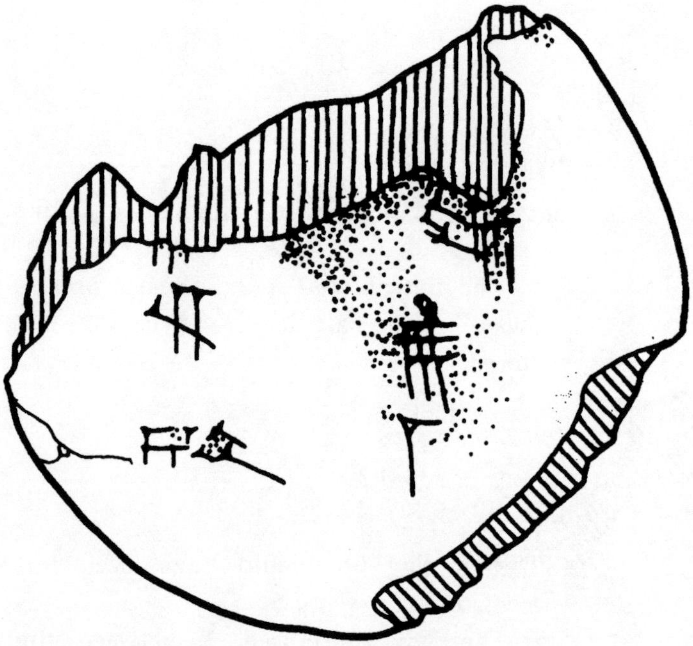

Figure A.5.18: UET *6/3 452 reverse*

Although no vertical dividing line is visible on *UET* 6/3 452, the remains of the calculation suggest that it too belongs to the 6 40 group of multiplication-series:

[X]
⌜× Y = 45⌝
× 12 = 9
× 6! 40 = 1

According to the remaining traces Y might be any one of 3, 6, or 9, in which case the corresponding X would be 15, 7 30, or 5.

The 6 of the final 6 40 is defectively written, and looks more like 5 than 6.

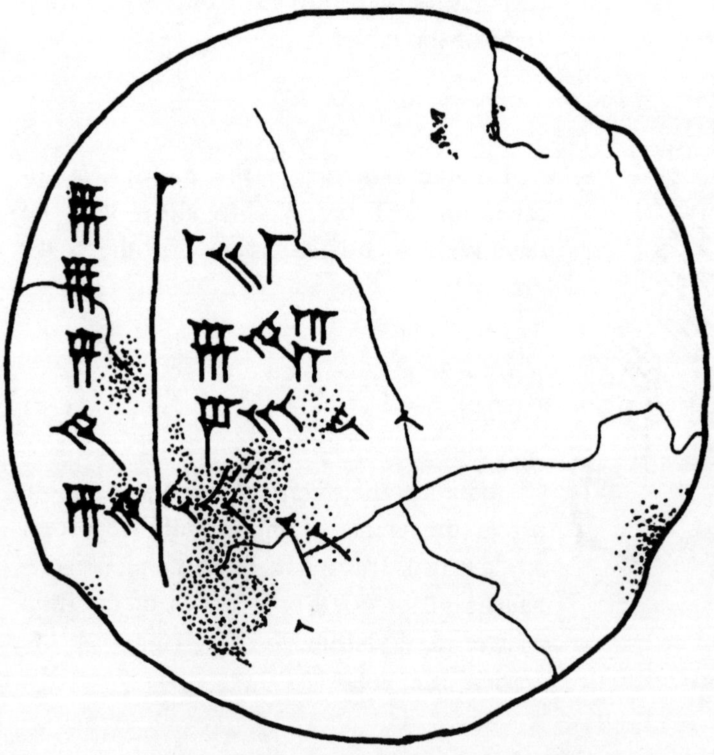

Figure A.5.19: UET *6/2 290 reverse*

The final multiplicand on *UET* 6/2 290 is also 6 40, although there is one more step to the procedure here than the others of the set:

9
× 9 = 1 21
× 5 = 6 45
× 40 = 4 30
× 6 40 = ⌜30⌝

It is not clear, then, whether it should be counted as one of this series or not: the last multiplicand may simply be coincidental.

There are various other erased numerals on the tablet.

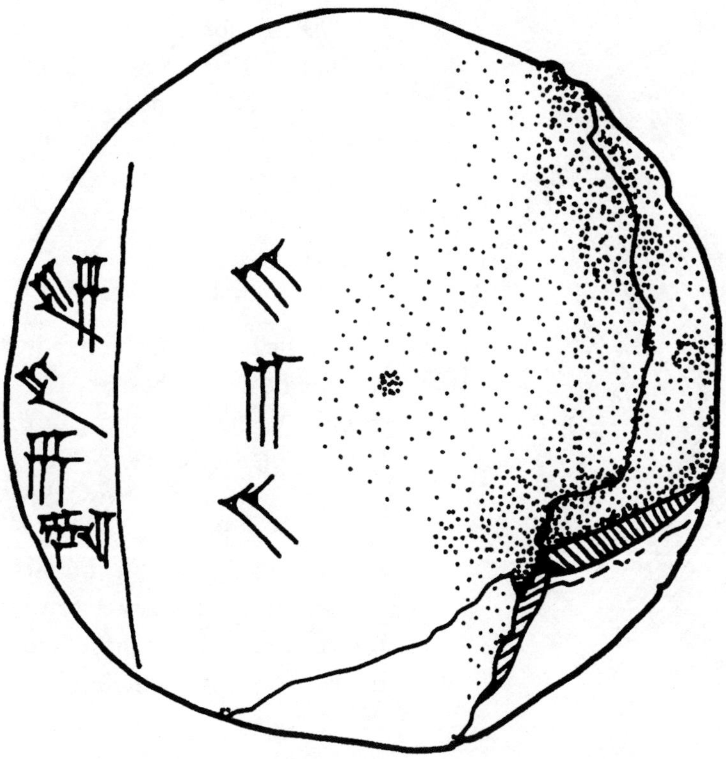

Figure A.5.20: UET *6/2 250 reverse*

The entire computation is preserved on *UET* 6/2 250, which contains an interesting error:

45
× 40 = 30
× 6 = 3
× 7 12 = 20.

The correct answer to this last operation is 3 × 7 12 = 21 36. As the copy shows, there is no final 1 36 hidden under damage or an erasure. To have obtained the 20 given on the tablet, the student must have multiplied 3 by 6 40 — the final multiplicand in the preceding six tablets here — instead of by 7 12.

There are several possible explanations for this error. Either the scribe meant to multiply by 7 12 but looked up the wrong table;[7] or he performed the correct calculation, having recorded the wrong multiplicand and failed to alter it.

Why might he have written down 7 12 instead of 6 40? The fact that this latter number occurs repeatedly throughout this group of multiplication-series, always in the same position, while the other numbers vary, suggests that it is a coefficient. There are several possible candidates for its identity, the most plausible in the author's opinion being the *našpakum* grain capacity coefficient 6 40 00, which converts sar$_v$ to sila (§7.2); 7 12 00 is also such a coefficient (§7.3). The first three multiplicands could quite conceivably be the measurements of some rectangular prism whose grain capacity was being sought. It is conceivable that the wrong coefficient was chosen, through misreading a coefficient list or through misjudgement.

Why, then, was the *correct* calculation performed despite the wrong choice of coefficient? Maybe the student noticed his error but failed to alter what was recorded on the tablet; or maybe the student simply cheated by writing what he already knew the numerical answer should be (these are often recorded on the problem text tablets) without bothering to perform the calculation. This interpretation is speculative and unprovable.

Miscellaneous multiplications

There are eight further tablets from OB Ur which contain multiplications. Three of them have not been completed, and all have evidence of other calculations on the right-hand side. All have a vertical line down the middle of the series of computations.

[7] Cf. Neugebauer's reconstructions (*MKT*: II) of the multiplication-series on CBS 19841 (Taf. 63), HS 208 (Taf. 65), Ist Ni 2739 (Taf. 67), all from Nippur, and. Ist O 4299 + Ist O 4654 (Taf. 68), from Kish, in which the tables for 6 40 and 7 12 are next to each other in adjacent columns.

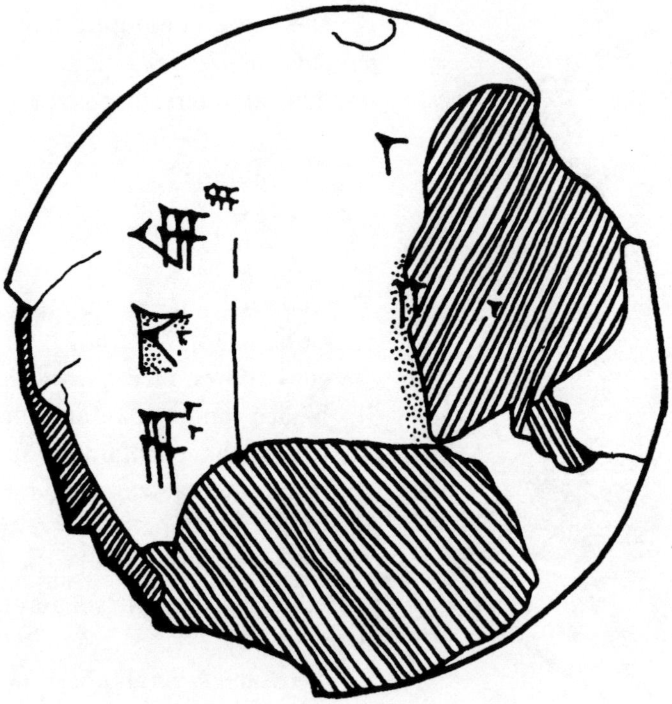

Figure A.5.21: UET *6/2 275 reverse*

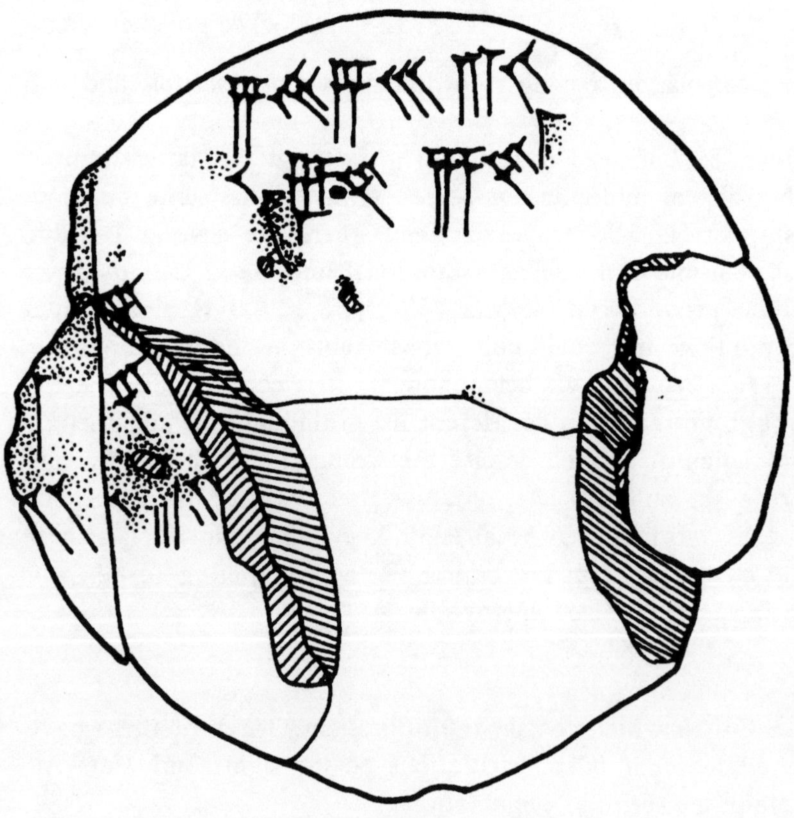

Figure A.5.22: UET *6/2 291 reverse*

The numbers stacked to the left of the vertical line on *UET* 6/2 275 appear to be laid out for a series of multiplications:

16
× 10 (= 2 40)
× 6 (= 16)

The 2 visible at the edge of the right-hand break just might be the beginning of the expected product, 2 40.

It is not clear how the small 6 at the top of the column and 1 on the upper portion of the tablet relate to the multiplication. There are traces of more numerals in the damage on the right of the tablet.

Although the multiplications on *UET* 6/2 291 are damaged, enough remains to allow most of the series to be restored:

[X]
[× Y] = 40
× 1 = 40
× 20 = ⌜13 20⌝.

At the top of the tablet is an erroneous reciprocal pair:

5 55 33 20
~ 10 07 46 40.

In fact 5 55 33 20 ~ 1 07 30; 10 07 46 40 is not a sexagesimally regular number. The student may have misremembered the pair, or miscalculated it.

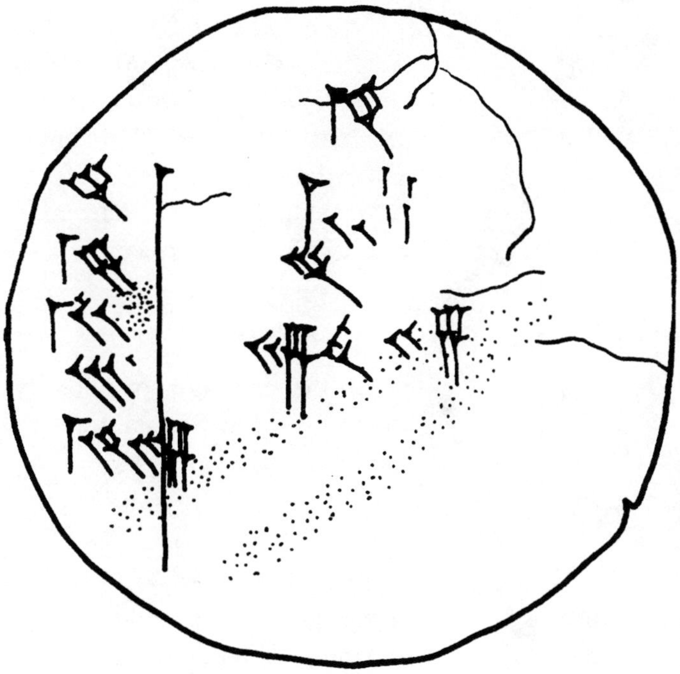

Figure A.5.23: UET 6/2 297 reverse

The left-hand side of *UET* 6/2 297 has been prepared for a series of multiplications which have not been carried out:

40
× 1 50 (= 1 13 20)
× 1 20 (= 1 37 46 40)
× 30 (= 48 53 20)
and
× 1 40 ~ 36 = 1 21 28 53 20
or
÷ 1 40 ~ 36 = 29 20.

Either of these last two operations is equally possible: the first is to be preferred because the multiplicand would be on the left of the reciprocal pair (the most common arrangement); and the second because it produces a two-place answer.

The numerals on the left — written over two erased lines — do not appear to form a coherent calculation. They are, from top to bottom, 1 40, 1 ⌜25!⌝, 40, 25 40, and 25.

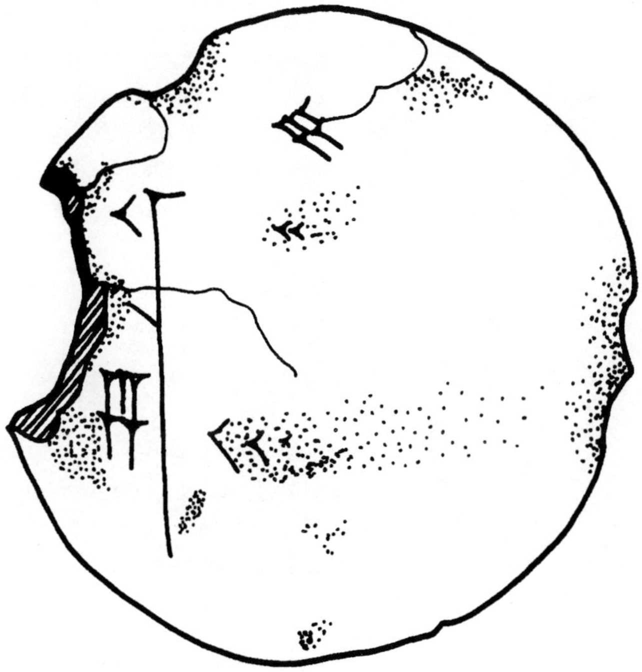

Figure A.5.24: UET *6/2 301 reverse*

UET 6/2 301 contains the preparations for another series of multiplications that was not completed:

10
⌜× 40!⌝ (= 6 40)
× 5 (= 33 20).

If this restoration is correct, then the traces under the large erasure towards the bottom of the tablet may be the first 30 of 33 20. On the other hand, the remains of the signs where one would expect the first product to be do not fit well with the required 6 40.

The 50 at the top centre does not appear to be related to the multiplications. There are also traces of other numerals.

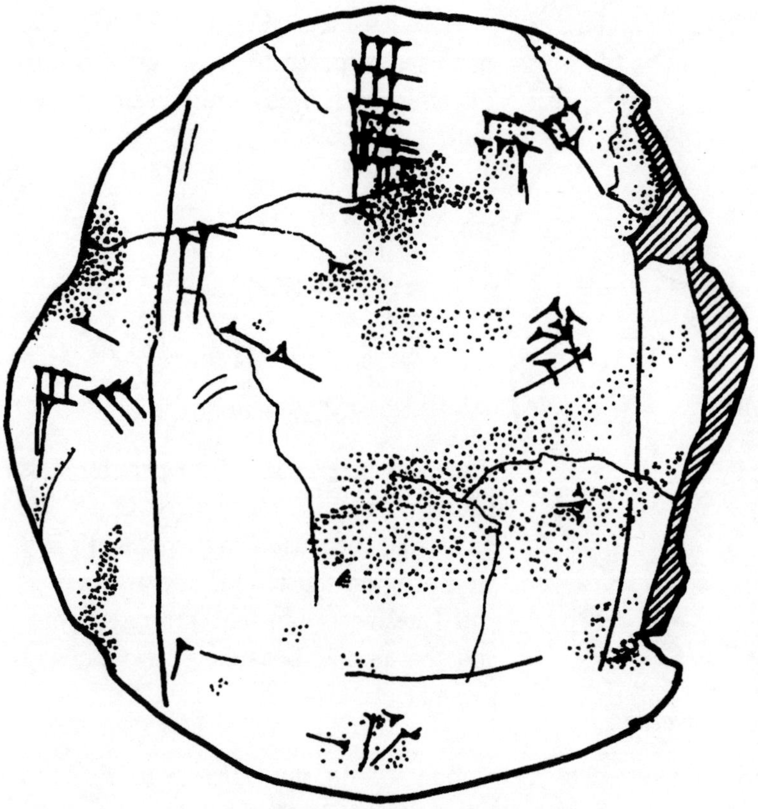

Figure A.5.25: UET 6/2 311 reverse

The reverse of *UET* 6/2 311 is a mess. There are many erasures, and plentiful traces of previous exercises. The single multiplication either side of the vertical line to the left may plausibly be restored as:

[26] ⌜40⌝
× 4 30 = 2.

At the top are two badly written 9s, in vertical alignment, and their reciprocal:

9
9 ~ 6 40.

There is another vertical line to the right of the tablet, and a horizontal line towards the lower edge, neither of which is obviously associated with any of the extant signs.

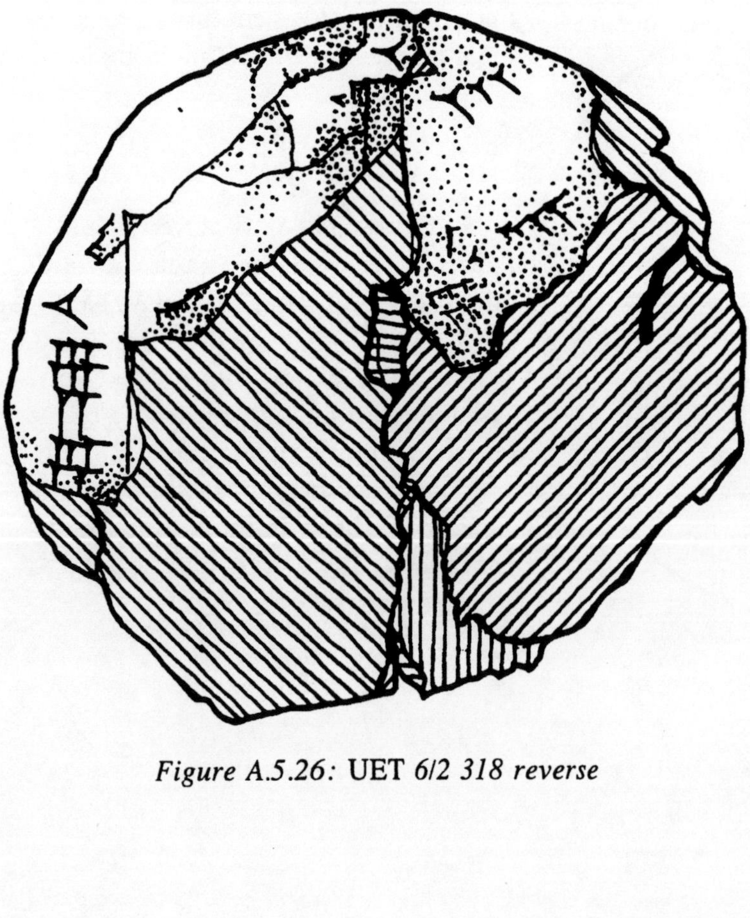

Figure A.5.26: UET 6/2 318 reverse

What little remains on the left-hand side of *UET* 6/2 318 can be identified as a multiplication-series because of the vertical dividing line:

10
× 8 = ⌜1 20!⌝
× 5 [= 6 40].

The many traces of numerals on the remaining surface, which are almost certainly unrelated to the multiplication, cannot be grouped into a coherent calculation of their own.

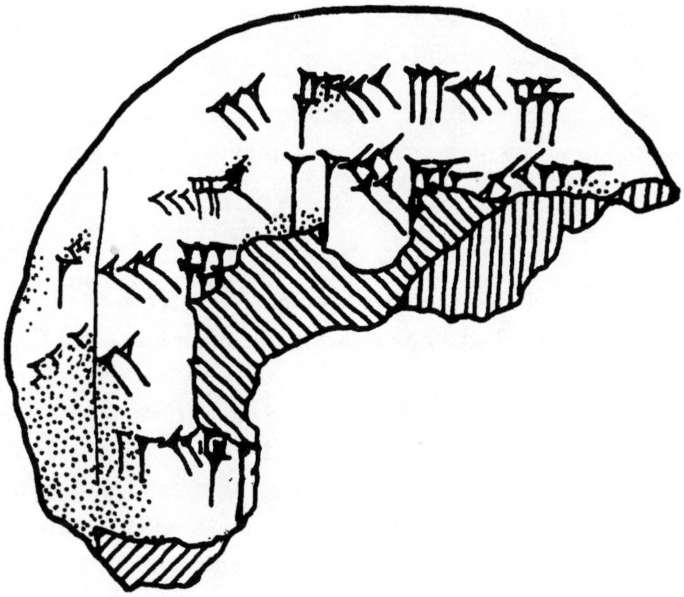

Figure A.5.27: UET *6/3 120 reverse*

The vertical line on the left of *UET* 6/3 120 indicates that the numerals around it form a series of multiplications. If so, the first multiplicand is written in small signs to the right of the line instead of the left:

36
× 1 = 36
× 40 = 2[4!]
[× 10] = 2 24.

One would expect the long numbers at the top of the tablet to be mutual reciprocals. However, instead of the anticipated pair 34 33 36 ~1 44 10, we apparently have:

34 33 36
(× 22 30 =) 12 5⌜7 36⌝.

There is a single multiplication — or, rather, division — on *UET* 6/3 378:

1
× 40 ~ 1 30 = 40.

At the top of the tablet is a 4-place, sexagasimally regular number which was presumably intended to be part of a reciprocal pair (cf. *UET* 6/2 291 and 311):

29 37 46 40
(~ 2 01 30).

There are several erasures on the tablet too.

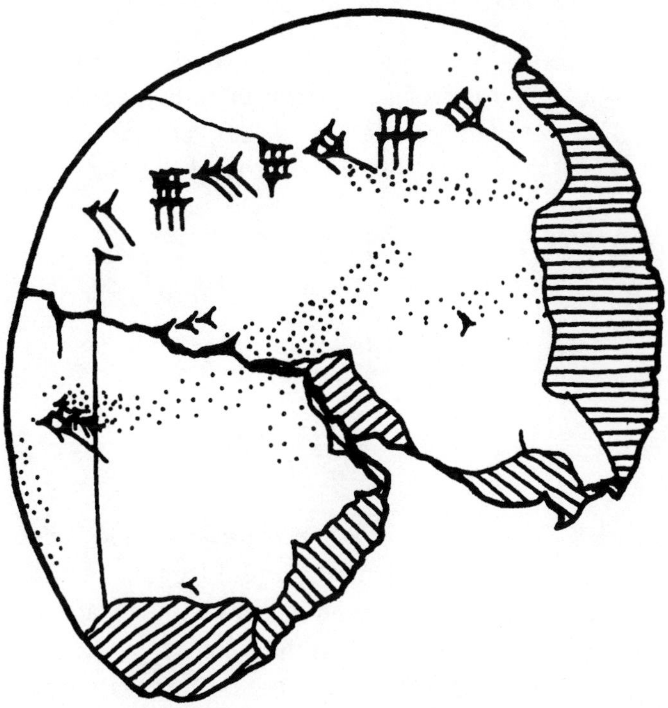

Figure A.5.28: UET 6/3 378 reverse

These multiplication-series as a whole, and *UET* 6/2 250 in particular, do allow some interesting if tentative conclusions to be drawn about the practice of mathematics in OB scribal education at Ur:

- first, that students were set problems to solve, and that the mathematics education was not restricted to learning arithmetical and metrological tables, and to copying out model solutions;
- second, that students were set problems of the same type but with different values for the variables — perhaps from the 'catalogue'-type lists (§1.3.2);

- third, that they were taught to lay out their calculations in standard formats;
- fourth, that students were prone to both calculation (especially place value) errors and mistakes through misreading coefficient lists and arithmetical tables;
- fifth, and most speculatively, that students knew the numerical results they were aiming for, and were not above fudging their calculations to fit.

Tabular calculations

Four lenticular tablets from Ur contain tabular arrangements of numbers, with vertical lines separating the columns. Nine similar are already published,[8] none of which contains exactly the same numerical relationships as these.

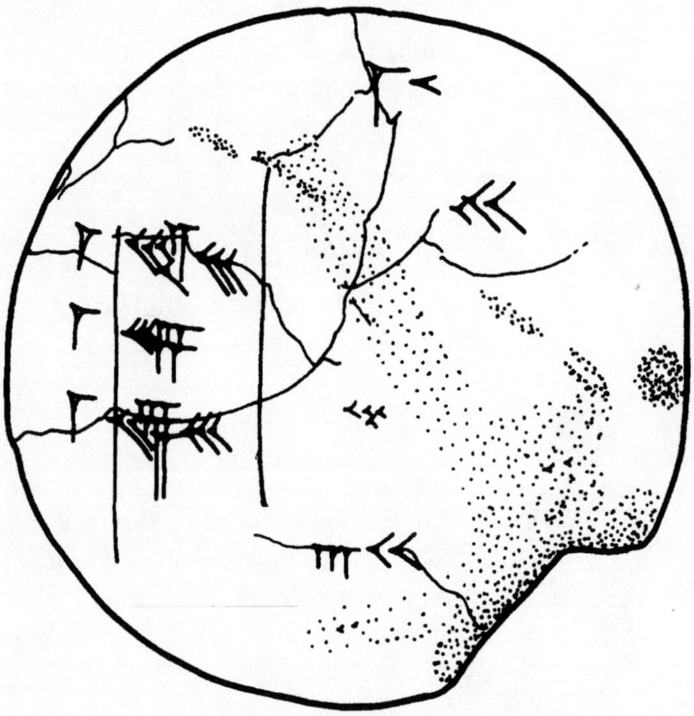

Figure A.5.29: UET *6/2 257 reverse*

The disposal of vertical lines on the left-hand side of *UET* 6/2 257 is reminiscent of a multiplication-series, but the numbers either side of them are horizontally aligned:

1	42 30
1	35
1	27! 30

If the numeral written as 28 should in fact be 27 then there is a constant difference of 7 30 between the figures in the right hand column.

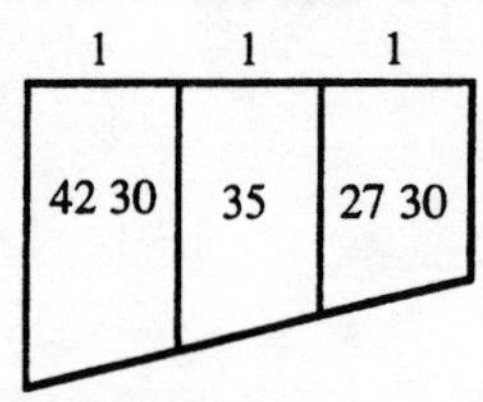

Figure A.5.30: Possible interpretation of UET *6/2 257*

The arrangement might represent, for instance, a trapezium whose non-parallel sides are trisected equally. The numbers of the first column would represent the equal lengths of the subdivisions (either as absolute values or ratios) and those in the second would give their areas. This is of course one of many possible interpretations of the numbers on the tablet.

The scattered numerals on the right of the tablet do not appear to be related. There is also a long erasure from the top left to the bottom right of the surface.

[8] YBC 7234, 7354, 7355, 7358, and 11127, which all share the same format; and PTS 247 and YBC 7235, 7353, and 11125. All are unprovenanced.

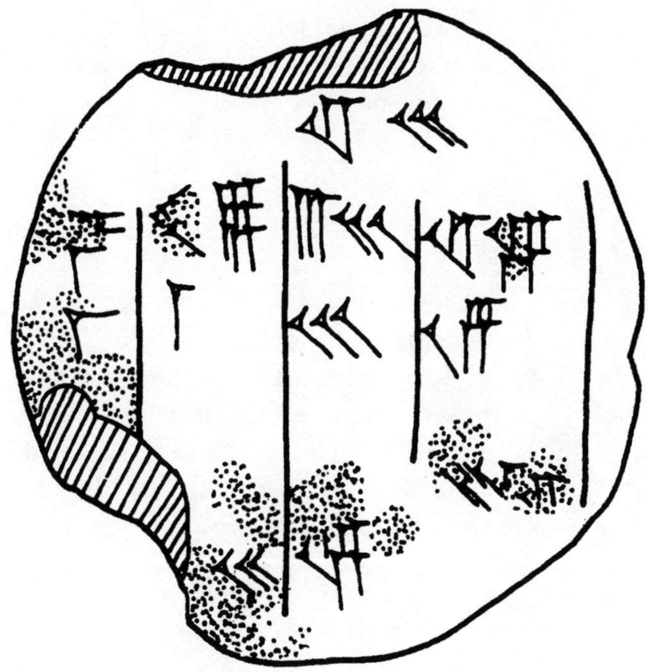

Figure A.5.31: UET 6/2 274 reverse

The structure of the table on *UET* 6/2 274 is rather more complex. Even though the bottom left has broken away there appears to be no text missing:

			12 30
7	49	3 30	12 15
1	1	30	15
	⌜50 ~ 1 12⌝		
	30	15	

The relationships in the table can be seen by replacing 7 with *a* and 1 with *b*:

	$(\frac{1}{2}a)^2+(\frac{1}{2}b)^2$		
a	a^2	$\frac{1}{2}a$	$(\frac{1}{2}a)^2$
b	b^2	$\frac{1}{2}b$	$(\frac{1}{2}\mathrm{b})^2$
	(a^2+b^2) ~ $1/(a^2+b^2)$		
	$\frac{1}{2}$	$(\frac{1}{2})^2$	

Its purpose is not clear, however.

Although the layout of *UET* 6/2 282 is highly reminiscent of 274, the numbers within the table do not seem to exhibit the same relationships. The damage in the middle of the tablet makes it difficult to determine its content exactly:

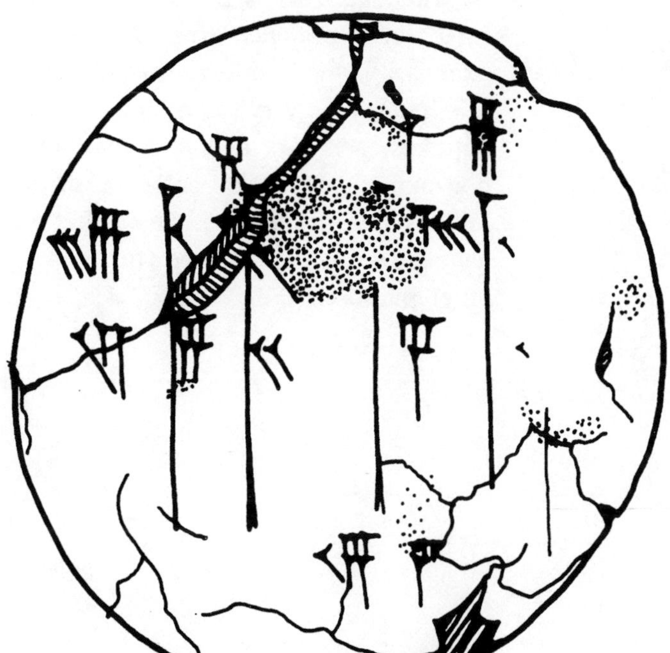

Figure A.5.32: UET 6/2 282 reverse

	5		1 06
36	10!	⌜40!⌝	⌜X⌝ 30
12	5	20	4
			15 ~ 4

The number-relationships in the two rows do not appear to be the same. If $12 = c$, then:

	$1/c$	1	$c/2$
$3c$	$2/c$?	$8/c$?	?
c	$1/c$	$4/c$	$4 = c/3$
	$3/c$ ~ $c/3$.		

Again, the table's function is unclear.

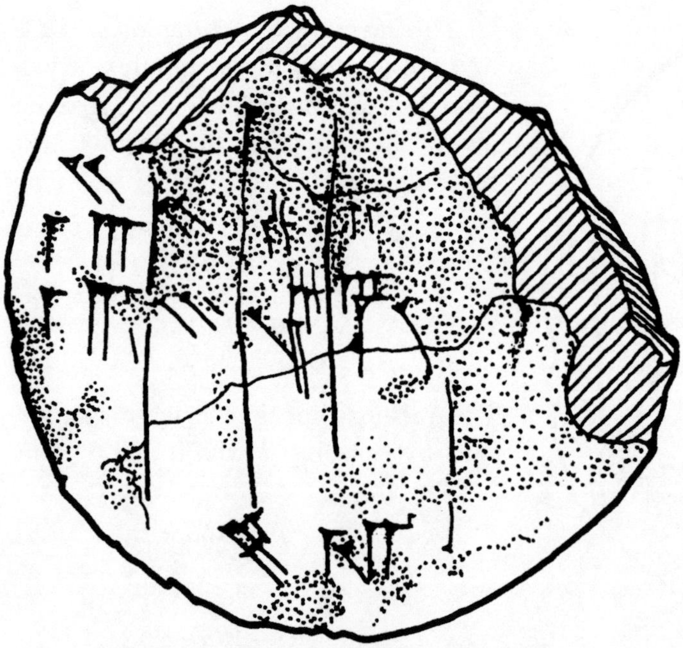

Figure A.5.33: UET *6/2 313 reverse*

The tabular format of *UET* 6/2 313 is clear, but most of the numerals are very damaged:

20	[		]
1 03	⌜20?	5?⌝	[ ]
1 03	⌜30⌝	18?	4 ⌜30?⌝
		50 ~ 1 12	

There may have been further rows towards the top of the tablet.

It is almost impossible to discern a pattern here, except that the table concludes with the sum of the second column and its reciprocal, as do the two preceding examples.

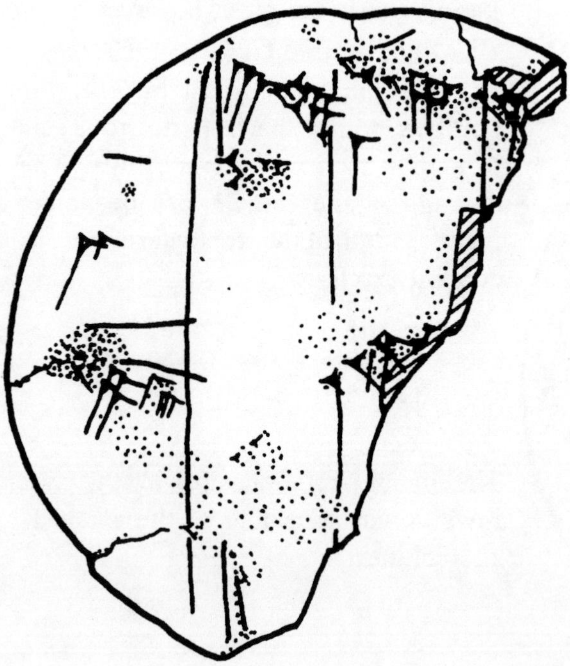

Figure A.5.34: UET *6/2 322 reverse*

Although *UET* 6/2 322 has been ruled in preparation for a tabular exercise, the only meaningful numbers, at the very top of the central columns, are 3 45 and 45. Scattered non-numerical signs and erasures lower down the tablet indicate that it had previously been used for some other purpose.

A.5.1.5 Fragments and miscellaneous calculations

Thirteen of the lenticular tablets from OB Ur cannot be assigned to any group of calculations. The first four presented here are visually very similar, with two columns of numbers but no vertical or horizontal rulings, reminiscent of the layout for 'the Technique'. Although the author cannot yet determine the nature of these calculations they are doubtless susceptible to interpretation. Two published tablets from Nippur, 2 N-T 174 and HS 231 also contain computations with no vertical rulings.

The remaining nine tablets are all very fragmentary or eroded; they are presented here with little or no comment.

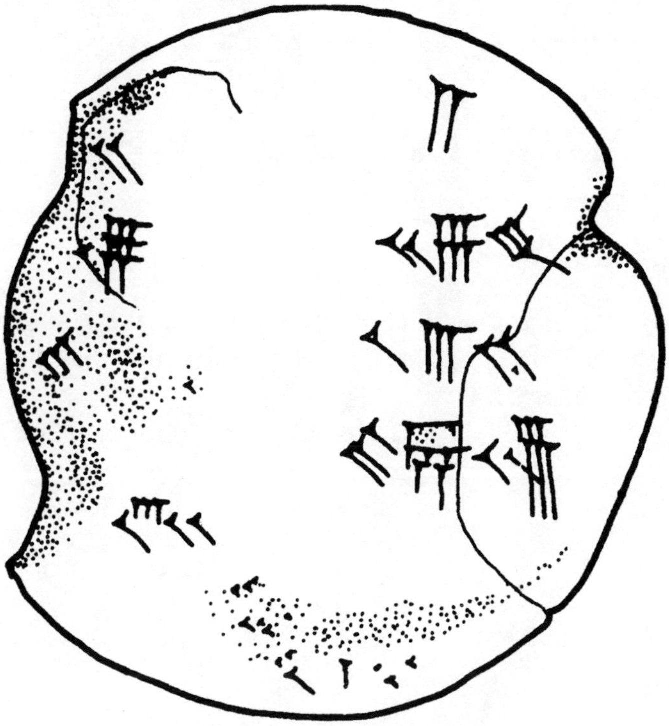

Figure A.5.35: UET *6/2 210 reverse*

The numbers on *UET* 6/2 210 read:

[...] ⌜50?⌝	2
[...] ⌜4?8⌝	26 40
⌜30?⌝ [...]	13 20
	38 26
13 20	

The only clear relationship between the numerals on the right is that:

26 40 = 2 × 13 20.

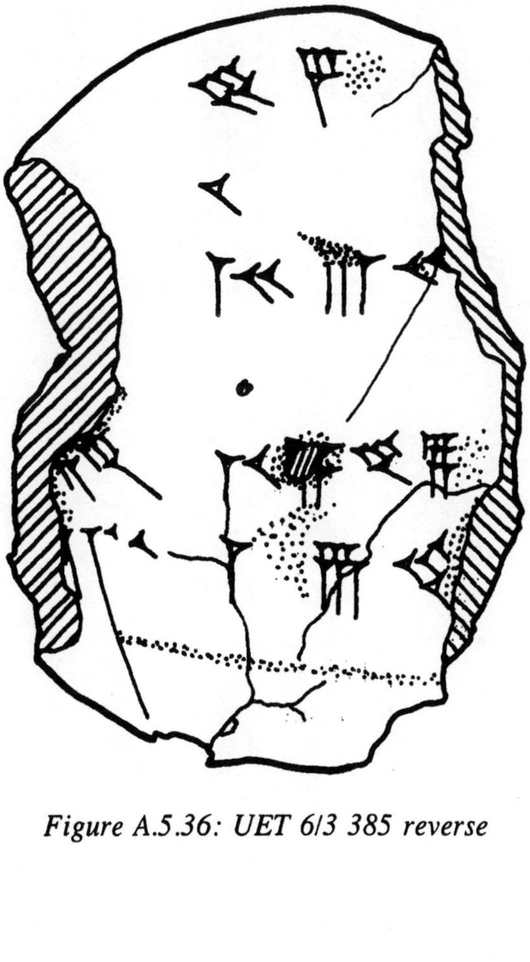

Figure A.5.36: UET 6/3 385 reverse

The figures on the central portion of *UET* 6/3 385 are all more or less undamaged. They read:

	54
	10
	1 23 20
⌜50?⌝ 40	1 15 ~ 48
	1 06 40

Is this a method for finding reciprocals? We have:

54 ~ 1 06 40

linking the numbers at the top and bottom of the tablet, and the 10 immediately beneath the 54 is the reciprocal of 6. This may be treated either as a factor of 54, in which case we might expect a procedure equivalent to:

$$\frac{1}{c}=\frac{1}{a}\times\frac{1}{b},$$

where $c = ab$; or as the difference between 1 00 and 54, in which case 'the Technique' could be adapted to give:

$$\frac{1}{c} = \frac{1}{b} \times \frac{1}{(\frac{1}{b} - 1)},$$

where $c = 1 - b$. But neither scenarios fit the following calculation, which is:

1 23 20
+ 1 15 ~ 48
= 1 06 40.

The significance of the numbers 1 23 20 and 1 15 is not at all clear; it may be that the key to interpretation lies in the damaged figures on the left of the tablet. The same procedure may be recorded on the right-hand side of *UET* 6/2 269 (§A.5.1.4).

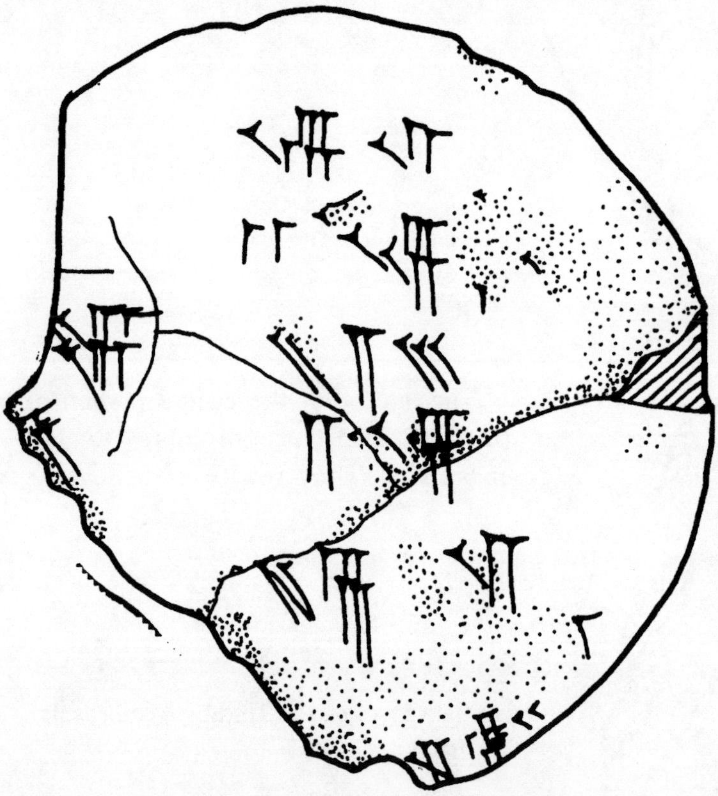

Figure A.5.37: UET *6/2 268 reverse*

The numbers on *UET* 6/2 268 are sometimes ambiguously written, but appear to be the following:

	16 12
⌈...⌉	2 40 15
[...] 56	22 30
[...] 50	2 1!5
	36 12

[7] 12 ~ 8 20

The only relationship discernible is that 22 30 is ten times 2 15.

The number in the second line may in fact be 2 55, and that in the penultimate line 26 instead of 36. The apparent 1 to the top right of the reciprocal pair is the remains of an erased line, which may or may not relate to the extant numbers.

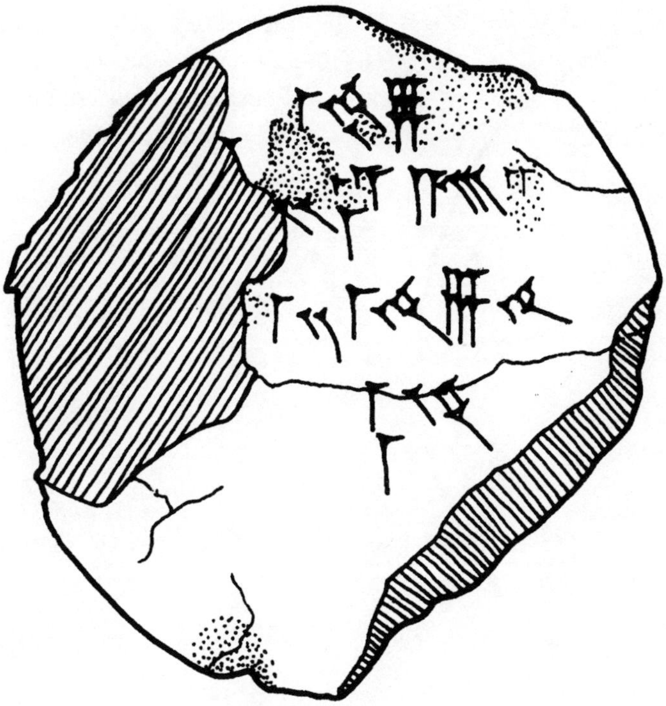

Figure A.5.38: UET *6/2 279 reverse*

Although the numerals remaining on *UET* 6/2 279 are for the most part clearly written, and the relationships between them identifiable, the nature of the calculation as a whole is not certain — or whether, indeed, it is a single calculation or several smaller ones.

1 58
24 ~ 2 30!
1 21(^{2} =) 1 46 40
1 40
1

There are also traces of wedges at the top left of the tablet, just below the 1 58, and the remains of an erased 2 after the 2 30 (which has a tiny extra wedge in the 30).

There are appear to be the remains of a multiplication-series or some tabular calculation to the left of *UET* 6/2 267; On the bottom right may be the remains of a squaring: 6 ⌜30⌝ × 6 ⌜30⌝ (= 42 15).

There are many erasures, and the final sign of an over-running line of the proverb on the obverse.

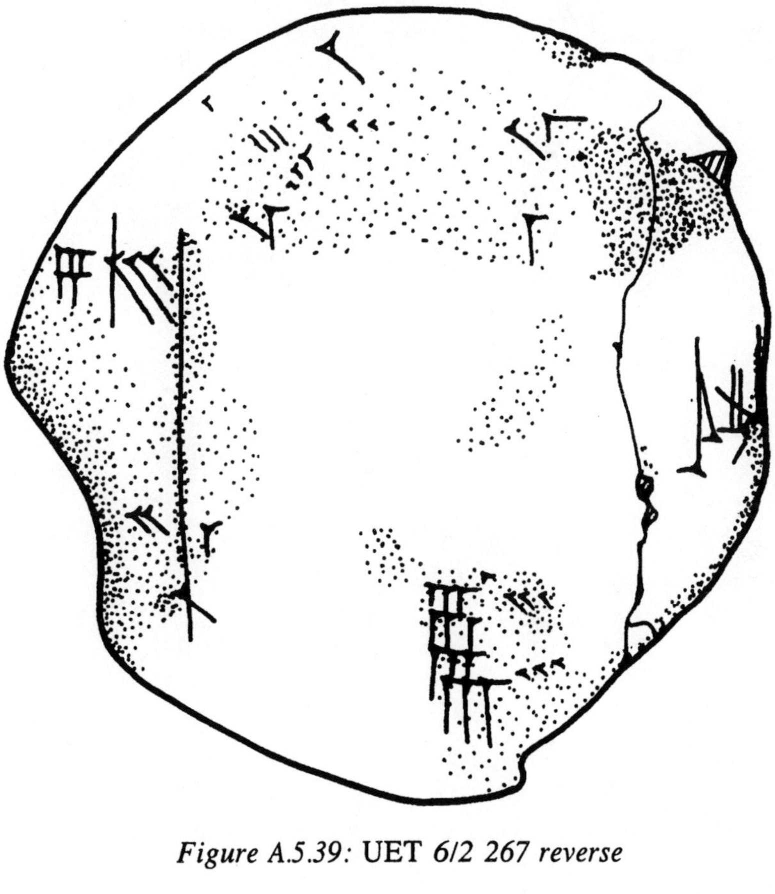

Figure A.5.39: UET *6/2 267 reverse*

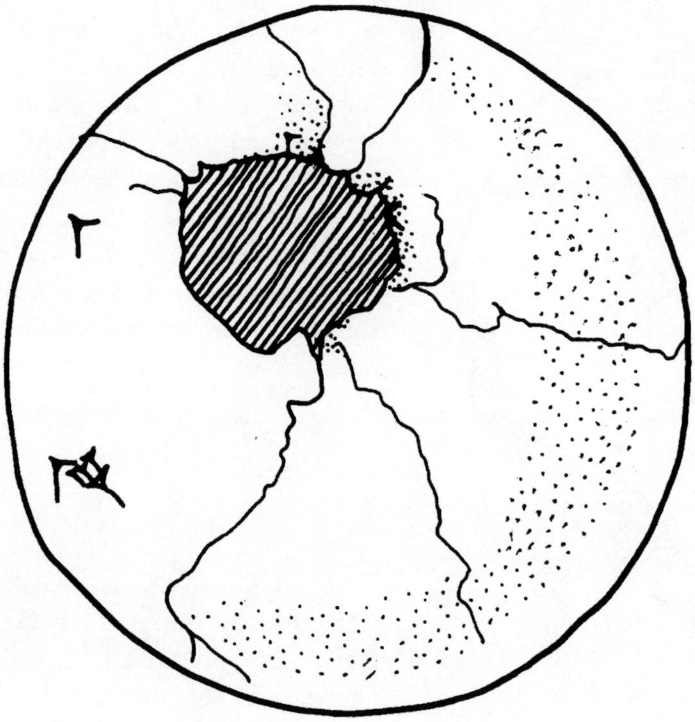

Figure A.5.40: UET 6/2 281 reverse

The only numerals visible on *UET* 6/2 281 are 1 and 1 40; there traces of further signs around the edges of the damaged area.

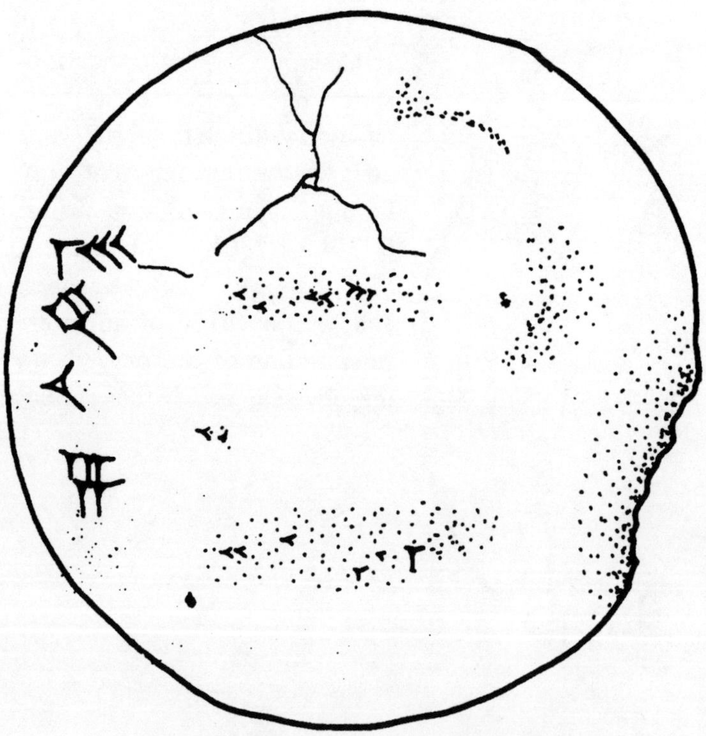

Figure A.5.41: UET *6/2 304 reverse*

Little can be said about the four numerals written on UET 6/2 304. They are written in vertical alignment, which may suggest that the scribe had started to write a multiplication series. If so, the calculation would have been:

1 40
× 30 [= 1]
× 10 [= 10]
× 5 [= 50].

However, there are no traces of a vertical line on the tablet, and the only other signs are the remnants of deliberate erasures.

1 30 and 40 are mutual reciprocals.

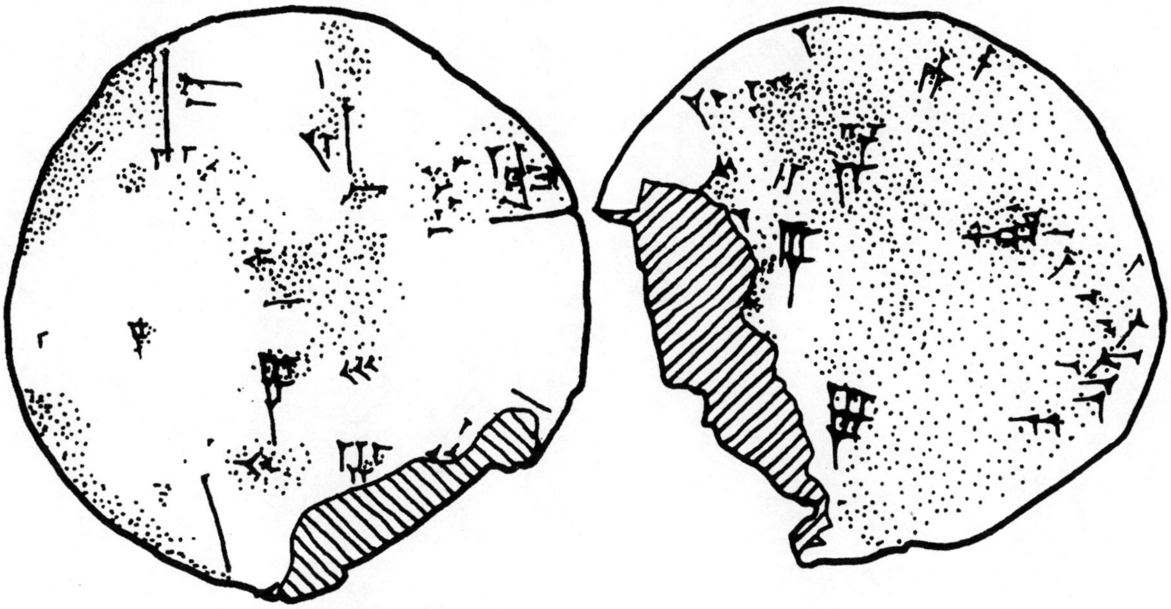

Figure A.5.42: UET 6/2 320 and 330 reverse.

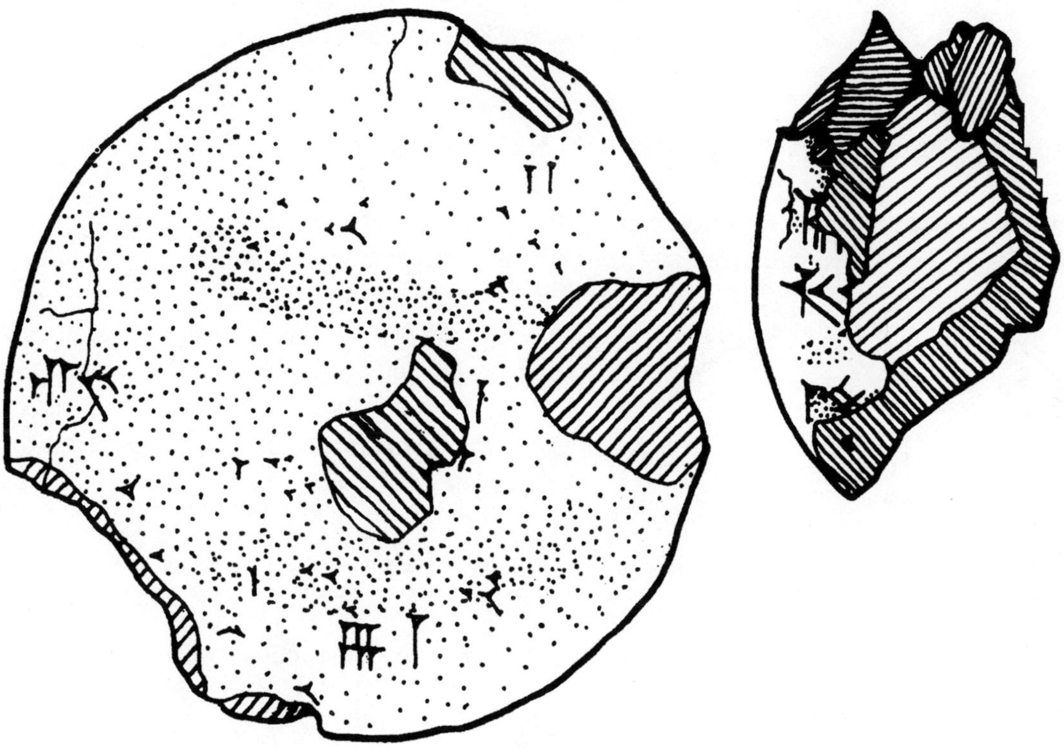

Figure A.5.43: UET 6/2 367 and 219 reverse

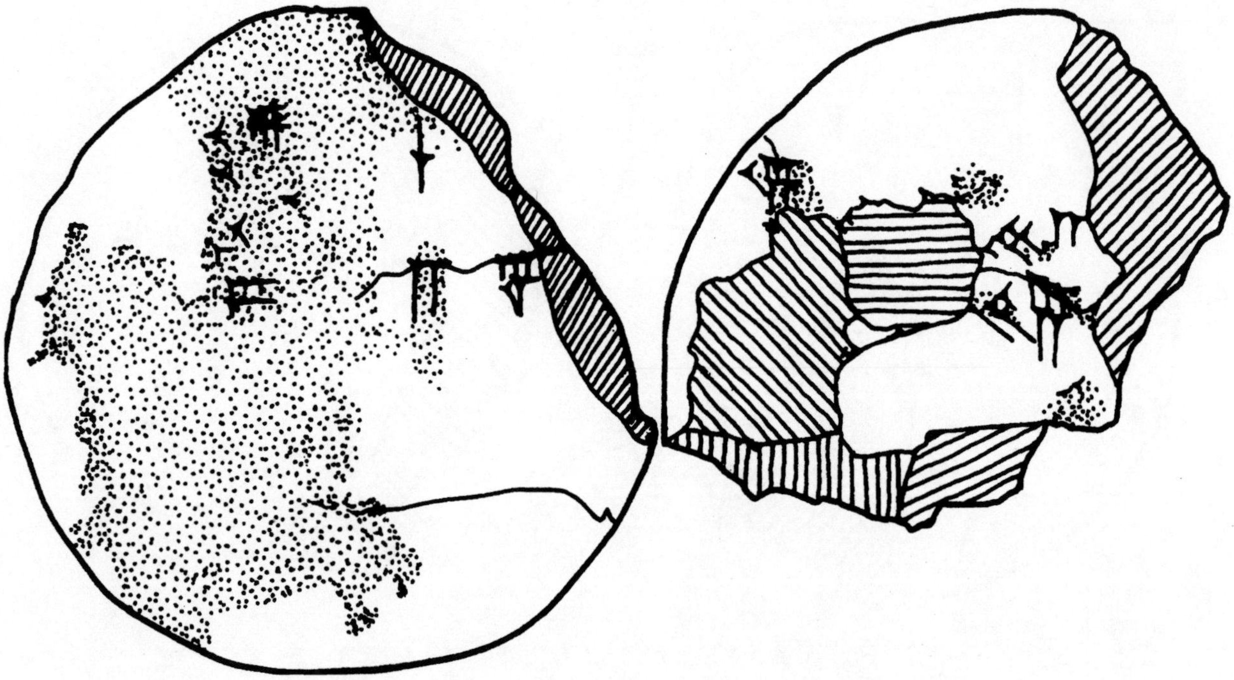

Figure A.5.44: UET 6/2 300 and UET 6/3 466 reverse

A.5.2 TABLETS FROM SIPPAR, NIPPUR, AND ELSEWHERE

The remaining tablets presented in this appendix do not form a coherent group in the way that those from Ur do; they are summarized and edited in no particular order (see Table A.5.2).

Figure A.5.45: BM 79937 obverse

BM 79937 is a very well-preserved, thick tablet, quite unlike the others the author has seen. It contains a simple division, laid out in a manner reminiscent of the calculation on YBC 7284 (cf. §4.4):

1 40
= 8 20
+ 12 ~ 5

The reverse of the tablet is blank.

Table A.5.2: Mathematical rough work from Sippar, Nippur and elsewhere

Museum no.	*Provenance*	*Contents of obverse*	*Contents of reverse*	*Shape of tablet*	*Measurements*
Ash 1922.168	Unknown	Diagram of trapezoid	Scattered numerals	Round	9.0 diameter
BM 79937	Sippar?	Multiplication	Blank	Round	6.5 diameter
BM 80762	Sippar?	Erased numerals	Multiplications	Round	6.5 diameter
BM 80765	Sippar?	Drawing of a circle	Part of a solution to a problem	Round	5.5 diameter
N 3971	Nippur	Calculation of square	Blank	Square	5.5 × 5.5
N 4281	Nippur	Multiplication table	Scattered numerals	Square	4.5* × 4.0*
UM 29-16-504	Nippur	Multiplication	Remains of calculation	Square	6.0 × 6.5

The lower half of the obverse of Ashmolean 1922.168 is covered in erasures, but the upper half contains a very fine diagram of a trapezium, with all the necessary measurements given.

The upper width, 6, and its reciprocal, 10, are at the top centre of the diagram. That width is divided into three unequal parts in an arithmetic progression: 1, 2, 3, written immediately underneath. The areas and lengths of the resulting sub-strips are given between and along the vertical lines. The length of the outer left edge, 15, and the area of the right-most strip, 26 15, are orientated in the same direction as the upper measurements; the others are all aligned to the vertical lines. The three 30s to the left and right of the diagram are coefficients:

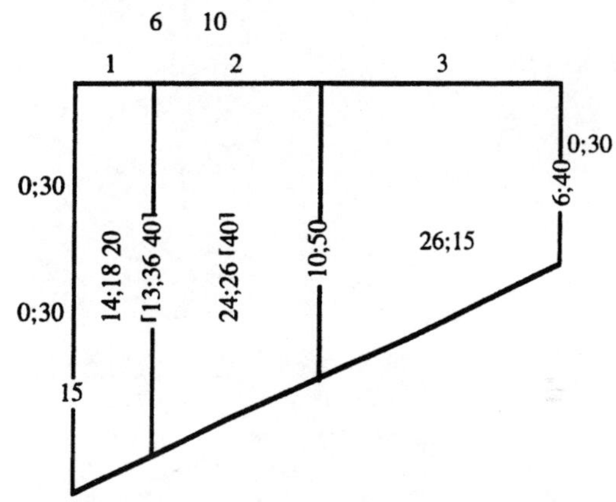

Figure A.5.46: The trapezium on Ashmolean 1922.168

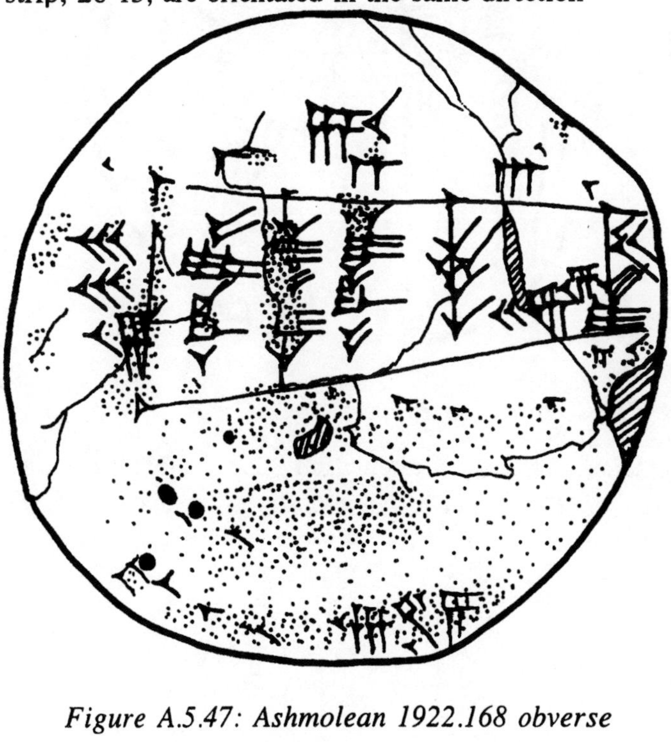

Figure A.5.47: Ashmolean 1922.168 obverse

Area = width × 0;30(length$_1$ + length$_2$)

14;18 20 = 1 × 0;30(15 + 13;36 40)
24;26 40 = 2 × 0;30(13;36 40 + 10;50)
26;15 = 3 × 0;30(10;50 + 6;40)

The total area, 1 05, is not mentioned. It is not clear which parameters were given and which the student had to calculate.

The numerals visible through the erasures on the lower half of the tablet may have been written in the course of the calculations.

Figure A.5.48: Ashmolean 1922.168 reverse

There are many scattered numbers covering the reverse of Ashmolean 1922.168, which may or may not relate to the diagram of the trapezium on the obverse.

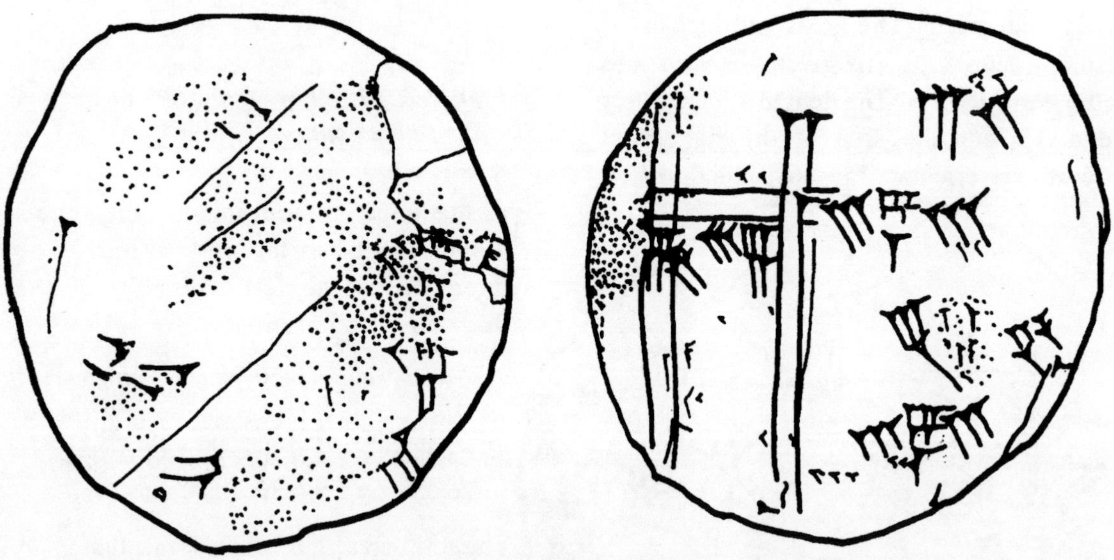

Figure A.5.49: BM 80762 obverse and reverse

One face of BM 80762 contains only the remains of an erased calculation. The double lines on the other side are reminiscent of house plans (cf. §9.4), but it is not clear how, if at all, they relate to the numerals within and to the left of them:

		3 20	
[...]5 50	35	37 30	
		40?	50
		37 30	

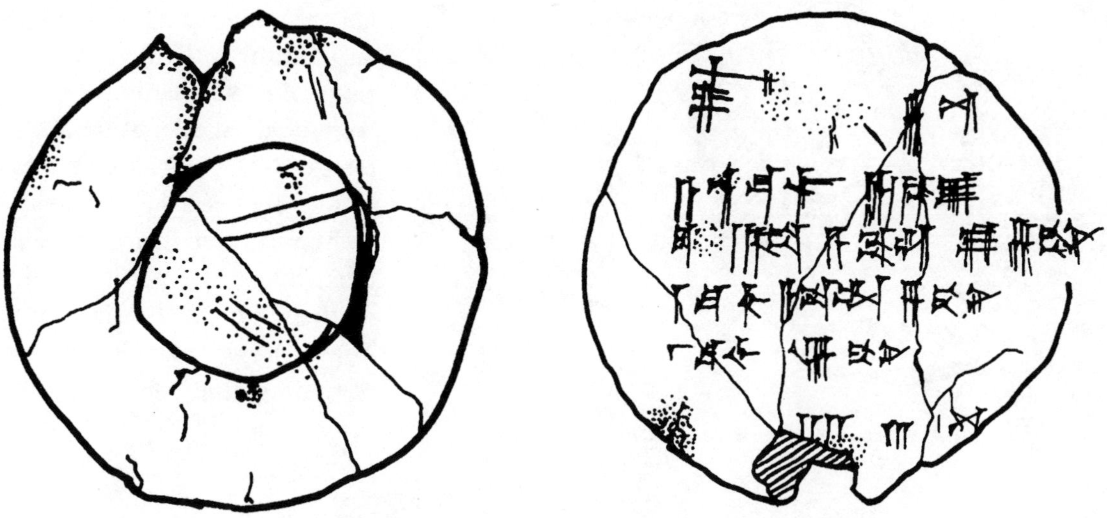

Figure A.5.50: BM 80765 obverse and reverse

On one side of BM 80765 is a roughly drawn picture of a circle, with the semi-erased traces of a previous diagram still visible. The other side contains an extract from the solution to a mathematical problem, which appears to be unrelated to the sketched circle. The signs are written with a damaged, split reed and are therefore difficult to decipher in places.

É?? … … 6 SÌLA	House?? 6 litres
a-na šu-ši 5 *a-na* É	for sixty 5 to the house
ša a-ši-ta a-ša-al É 5 GA.NI	of … I ask(ed) house 5 …
1 *šu-ši a-ša-al* 5 GA.NI	1 sixty I asked 5 …
1 *šu-ši* 18 GA.NI	1 sixty 18 …
⸢9?⸣ 3 SÌLA	9?? 3 litres

The sign transliterated as É might equally be GÍN, or some other similar sign; the sequence written here as GA.NI is probably a single sign and must be a unit of measure, always coming as it does after numerals. The word *a-ša-al* might be either the first person singular past or present of *šâlum* 'to ask' (*ašāl* or *ašâl*), or the absolute *ašal* of the length measure *ašlum* (= 10 nindan, *c*.60 m). The sequence *a-ši-ta* is apparently meaningless — as, for the moment, is the passage as a whole.

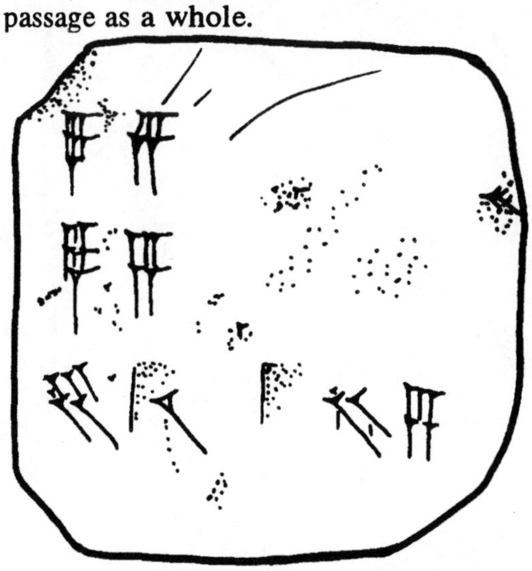

Figure A.5.51: N 3971 obverse

The front of N 3971 shows a very neatly laid out computation of a square. There are several erasures, near the answer in particular. They may be the traces of an earlier calculation, the remains of intermediate results, or an erroneous answer.

7 05
× 7 05
= 50 10 25

The reverse of the tablet is anepigraphic.

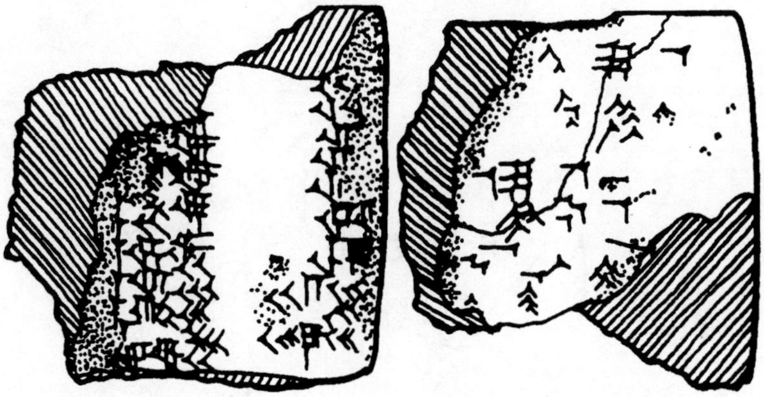

Figure A.5.52: N 4281 obverse and reverse

On the obverse of N 4281 is an untidily written extract from the standard series of reciprocal and multiplication tables. No vertical ruling is visible between the columns.

The reverse contains scattered numerals, perpendicular to their expected orientation. They may have helped in compiling the tables on the obverse but now mean little.

[...			]
[30]	⌜2⌝	[	]
[32	1] ⌜52⌝ 30	[14]	10 ⌜30⌝
[36	1] ⌜40⌝	⌜15⌝	11 ⌜15⌝
[40	1 ⌜30⌝	16	12
[45]	1 20	17	12 [45]
[48]	1 15!	18	⌜13⌝ [30]
[50]	1 12!	19	14 ⌜15⌝
[54]	1 ⌜06 40⌝	20	15
[1]	1	30	22 30
[1 04]	56 15	40	30
[1 21	46 26 40]	50	3!7 30

If the tablet followed the usual order of tables, starting with reciprocals and then moving on to the 50 and 45 times tables, this fragment must be roughly two-thirds of its original width and a third of its length.

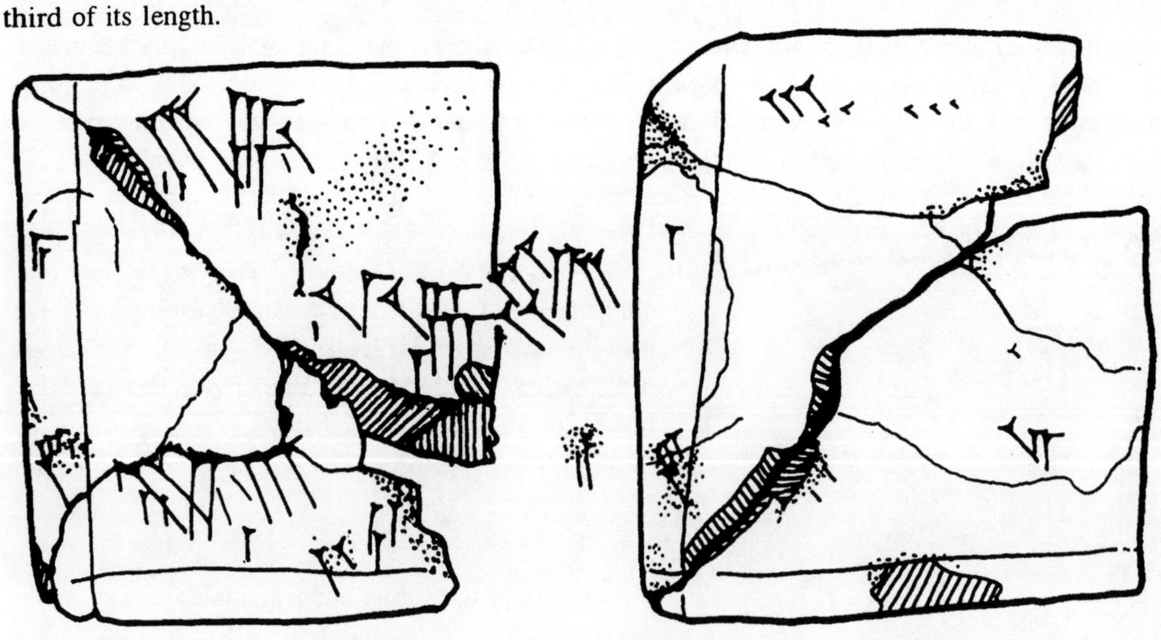

Figure A.5.53: UM 29-16-504 obverse and reverse

The obverse and reverse of UM 29-16-504 both show the same rulings close to the left and lower edges, with the numerals 1 and 40 inscribed between the left edge and the vertical line. However, neither lines nor numerals appear to relate to the computation on the obverse;

perhaps they are left over from some previous calculation, odd traces of which are still visible. Only scattered numerals remain of whatever was recorded on the reverse of the tablet.

The multiplication on the obverse is laid out in a similar manner to those from Ur, with the multiplicands in vertical alignment and the product to their right, centred vertically between them. However, there is no vertical ruling to separate them:

30 05	
	= 11 16 52 30!
× 22 30	

There is a 10 missing from the final place of the answer.

One would normally expect the number 30 05 to be written with a clear physical space between the 30 and the 5; the fact that there is no gap here leads one to wonder whether the calculation was not meant to have been 35 × 22 30 (= 13 07 30) instead. Place value errors are common enough in OB arithmetic; see §5.3 for further examples.

APPENDIX 6: STATISTICAL ANALYSIS OF BAKED BRICKS IN BRITISH COLLECTIONS

A.6.1 INTRODUCTION

Some years ago Powell signalled the necessity of, and difficulties inherent in, a systematic examination of the sizes of Mesopotamian baked bricks:

> The material does not lend itself well to statistical analysis, because every archaeologist (or brick-measurer) operates according to slightly different principles, which creates a significant margin of error. [...] Ideally, one should have many careful measurements of datable (i.e., stamped) bricks, but this is presently unavailable, and perhaps unattainable, given the costs involved. Thus, there are practical as well as theoretical obstacles to using bricks for metrological evidence. Nevertheless, their importance is considerable, because they are the only surviving artefacts for which textual evidence attests that they incorporate norms of length, area, volume, capacity, and weight — a rather remarkable combination in the history of pre-modern metrology. (Powell 1982*a*: 116–17)

These remarks are taken into account in the following short study of the baked, stamped bricks from Mesopotamia now in British collections. Its aim is to compare bricks recovered from excavations with the theoretical, standardized brick sizes attested in OB mathematics (Chapter 4). Its starting point was the measurements of bricks made by Christopher Walker, which he very generously made available to the author. Some additional bricks in the Ashmolean Museum were also examined, with the kind assistance of Roger Moorey. The length, width and thickness of the bricks were all measured to the nearest 0.5 cm,[1] and classified according to the inscriptions of the rulers found on them (Walker 1981).

Table A.6.1: Number of baked, stamped bricks in the survey

Period	*CBI nos.*	*L, W, T*	*L, W*	*L, T*	*W, T*	*Length*	*Width*	*Thickness*	*Total*
Pre-Ur III	1–11	9	1	1	3	0	3	5	22
Ur III	12–27	86	5	21	2	0	0	43	157
Isin dynasty	28–35	25	2	3	3	0	0	17	50
Larsa dynasty	36–46	12	3	8	5	0	0	36	64
Later Old Babylonian	47–54	5	3	1	0	0	1	13	23
Kassite	55–71	21	2	5	2	0	0	21	51
Middle Babylonian	72–88	11	1	3	4	0	0	6	25
Neo-/Late Babylonian	89–114	129	4	6	15	0	0	89	243
Old Assyrian	118–20	3	1	2	0	0	0	1	7
Middle Assyrian	121–41	6	5	5	1	0	0	37	54
Neo-Assyrian	142–90	65	11	22	8	2	1	84	193
Total		372	38	77	43	2	5	352	889

[1] The length is defined here as the longest side of the brick, with the width shorter than or equal to the length. Where one of these sides is broken, if the complete side is longer than the broken one this has been called the length, and if shorter the width.

Table A.6.1 gives the numbers of bricks for which some or all of the length L, width W and thickness T were recoverable. It shows widely differing, and often very small, numbers of bricks from each period. The results of this study should thus be treated with caution. In particular, it only concerns *baked* bricks and in general does not say anything about the sizes of unbaked bricks used in Mesopotamia at any period. Further, this study can only show to what extent brick sizes were standardized, and reveal some of the sizes that were used; it cannot reveal which of the OB types were *not* used at any time.

For each period the sizes of bricks for which length and width are known are plotted on a scatter diagram like the one below. Each point on the graph is marked with a cross representing a margin of error of 1 cm. The grey circles on the chart each represent one or more of the sizes of OB mathematical brick types (§4.1), assuming that 1 cubit = 50 cm. In acknowledgement of the uncertainty of this equivalence, each circle represents a spread of 1.5 cm. Crosses on the diagram which overlap one of these circles can reasonably be said to represent bricks of the standard size represented by that circle. The sample size and the brick types attested are given to the right of the chart.

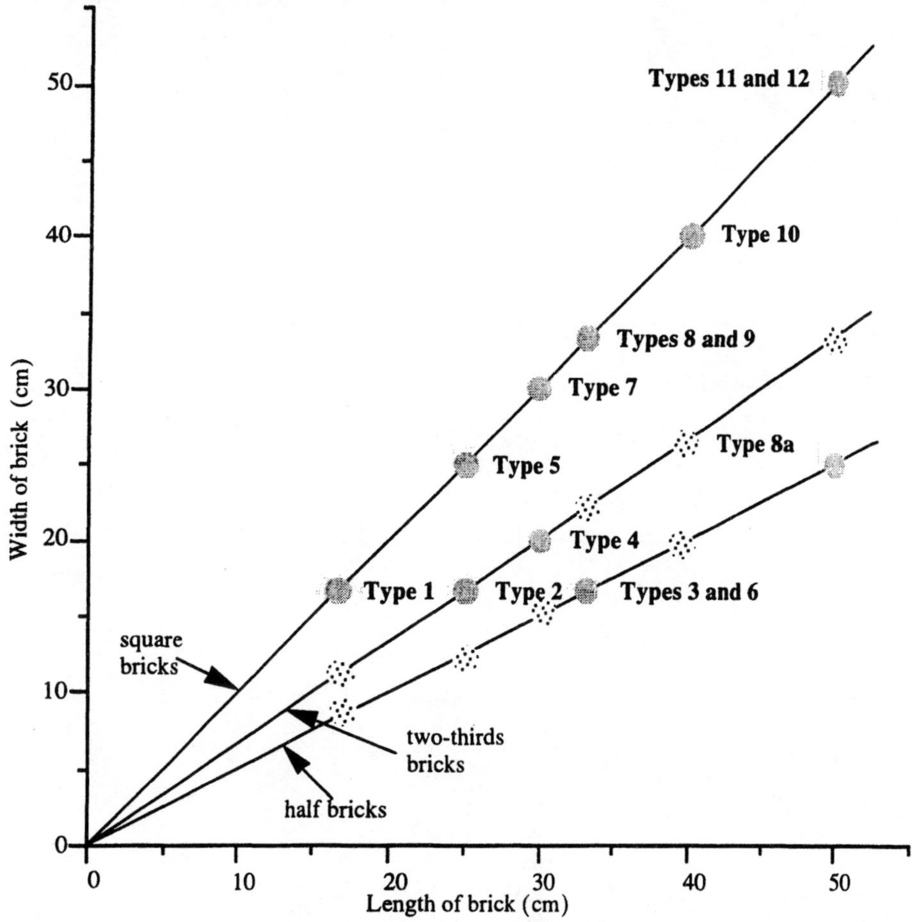

Figure A.6.1: Sizes of bricks in Old Babylonian mathematics

Below each scatter diagram is a set of descriptive statistics — to 3 significant figures — for each of the length, width and thickness of the bricks in the sample. The sample size varies for each dimension, as the bricks are broken in different ways. Brief comments are made underneath each table. For an analysis and comparison of the results see §10.4.1.

A.6.2 PRE-UR III BAKED BRICKS

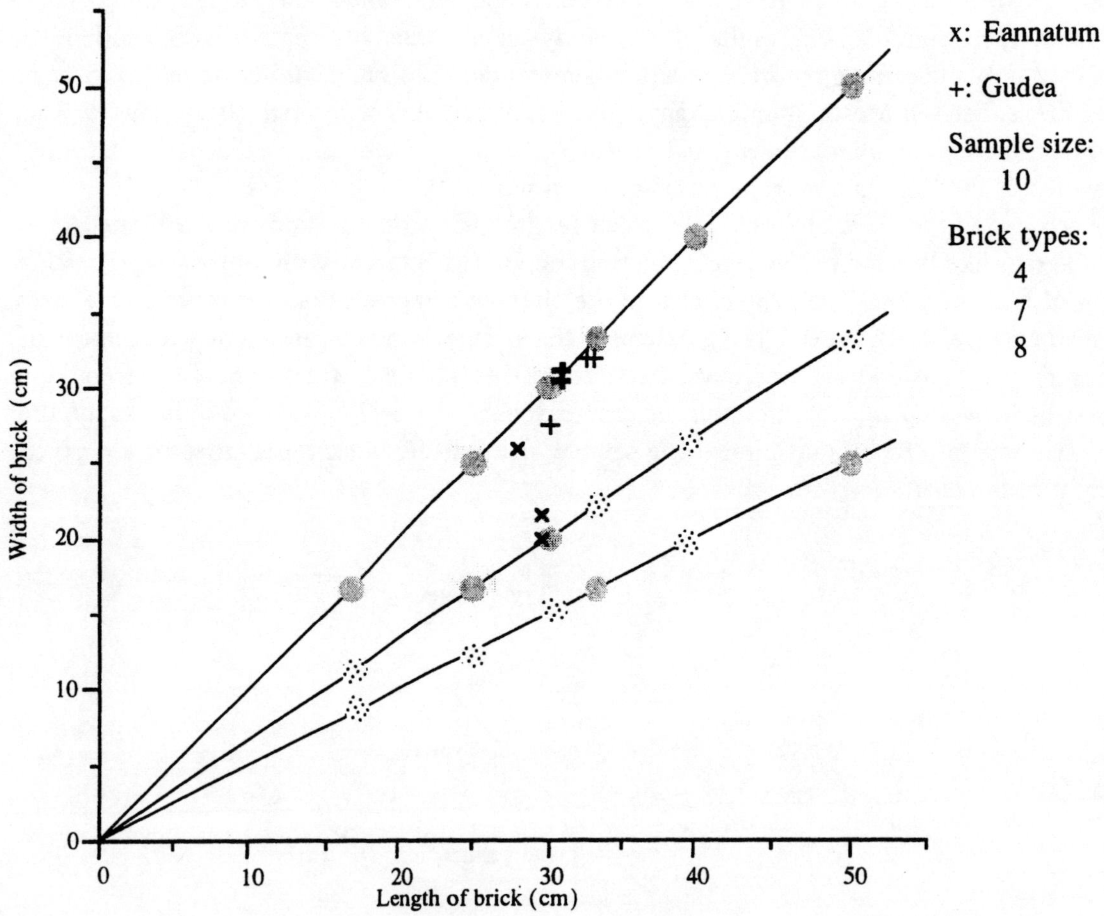

Figure A.6.2: Pre-Ur III baked bricks

Descriptive statistics are given only for the Gudea dynasty, because there are not enough ED bricks for a statistically significant sample. The Gudea bricks correspond roughly to OB square types 7 and 8; the associated two-thirds brick type 4 is attested for ED. The bricks are rather thinner than the OB mathematical ones.

Table A.6.2: Pre-Ur III baked bricks

Statistic	*Length*	*Width*	*Thickness*
Mean	31.2	30.4	6.71
Median	31.0	30.8	6.75
Mode	31.0	30.5	7.0
Standard deviation	0.810	1.52	0.865
Variance	0.656	2.32	0.748
Range	2.5	4.5	2.5
Minimum	30.0	27.5	5.5
Maximum	32.5	32.0	8.0
Count	7	8	12

A..6.3 UR III BAKED BRICKS

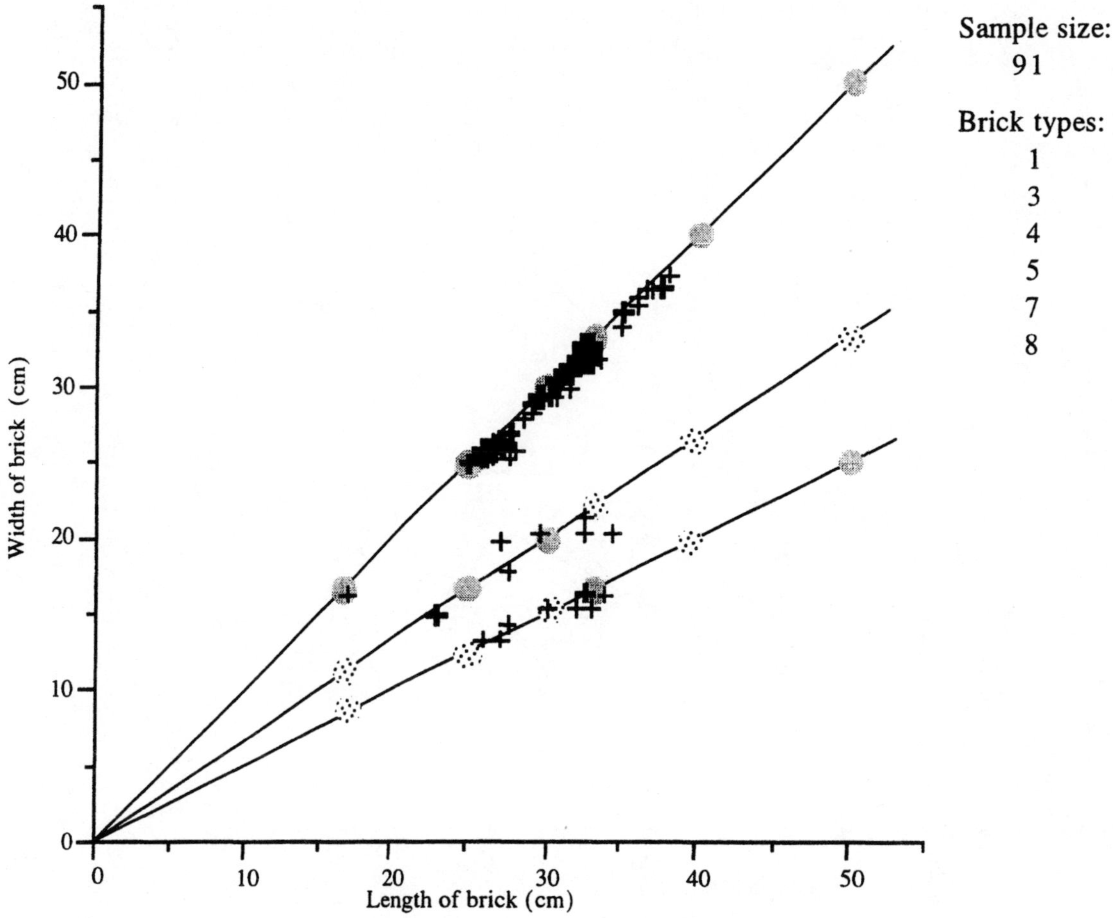

Figure A.6.3: Ur III baked bricks

The large majority of these Ur III baked bricks are square, of length between 25 and 38 cm. There are three clusters of sizes, two of which correspond roughly to the OB mathematical brick types 5, 7 and 8. The smaller square bricks are also cut into two-thirds and half-bricks. The bricks are all around 7 cm — just under 5 fingers — thick.

Table A.6.3: Ur III baked bricks

Statistic	*Length*	*Width*	*Thickness*
Mean	30.5	27.3	7.13
Median	31.0	29.5	7.0
Mode	32.0	32.0	7.0
Standard deviation	3.65	6.45	1.08
Variance	13.3	41.5	1.17
Range	21.0	24.0	5.0
Minimum	17.0	13.5	5.0
Maximum	38.0	37.5	10.0
Count	112	93	152

A.6.4 ISIN DYNASTY BAKED BRICKS

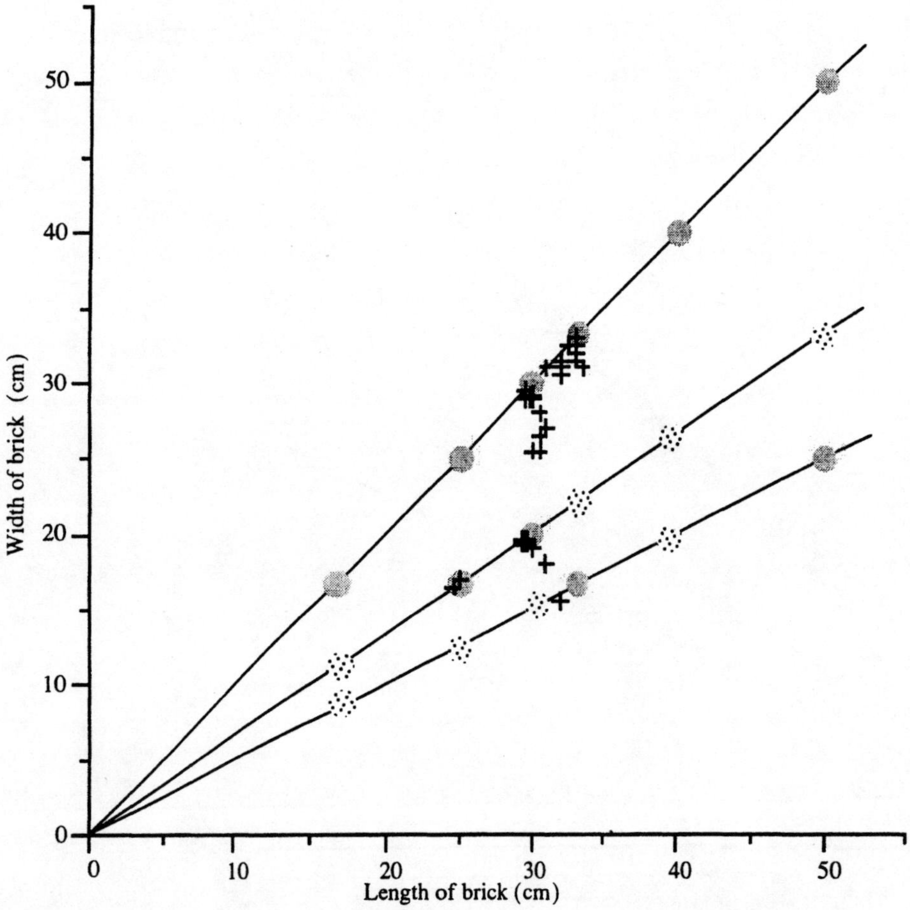

Sample size:
27

Brick types:
2
4
7
8

Figure A.6.4: Isin dynasty baked bricks

The bricks in this sample cluster closely around the OB mathematical brick types, including Types 2 and 8, the most commonly attested in the school texts. Although their mean thickness is rather less than 8 cm (5 fingers), the mode — the thickness of the greatest frequency — is exactly 8 cm.

Table A.6.4: Isin dynasty baked bricks

Statistic	*Length*	*Width*	*Thickness*
Mean	29.9	26.1	7.41
Median	30.8	29.3	7.5
Mode	32.0	31.0	8.0
Standard deviation	3.62	6.24	0.938
Variance	13.1	38.9	0.879
Range	18.0	18.5	4.5
Minimum	15.5	14.5	5.5
Maximum	33.5	33.0	10.0
Count	30	30	48

A.6.5 LARSA DYNASTY BAKED BRICKS

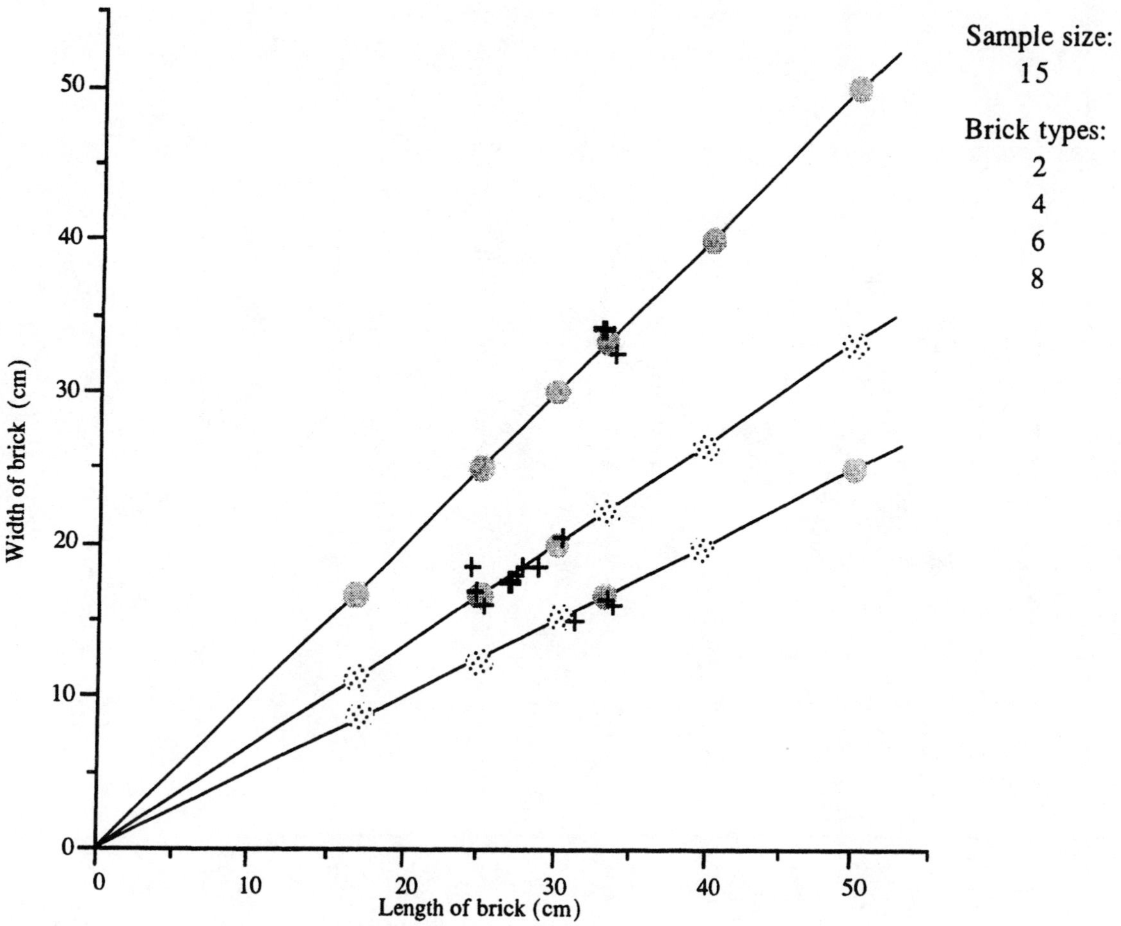

Figure A.6.5: Larsa dynasty baked bricks

Although only one size of square brick is attested — which exactly matches OB type 8 — but the existence of the two-thirds types 2 and 4 suggests that the corresponding square bricks 5 and 7 were used as well. The mean, median and mode of the thickness are all approximately 8 cm, or 5 fingers.

Table A.6.5: Larsa dynasty baked bricks

Statistic	*Length*	*Width*	*Thickness*
Mean	30.0	20.5	7.88
Median	31.0	17.5	8.0
Mode	34.0	17.0	8.0
Standard deviation	4.23	6.74	0.853
Variance	17.9	45.4	0.728
Range	17.5	19.0	4.5
Minimum	18.0	15.0	5.0
Maximum	35.5	34.0	9.5
Count	23	20	61

A.6.6 LATER OLD BABYLONIAN BAKED BRICKS

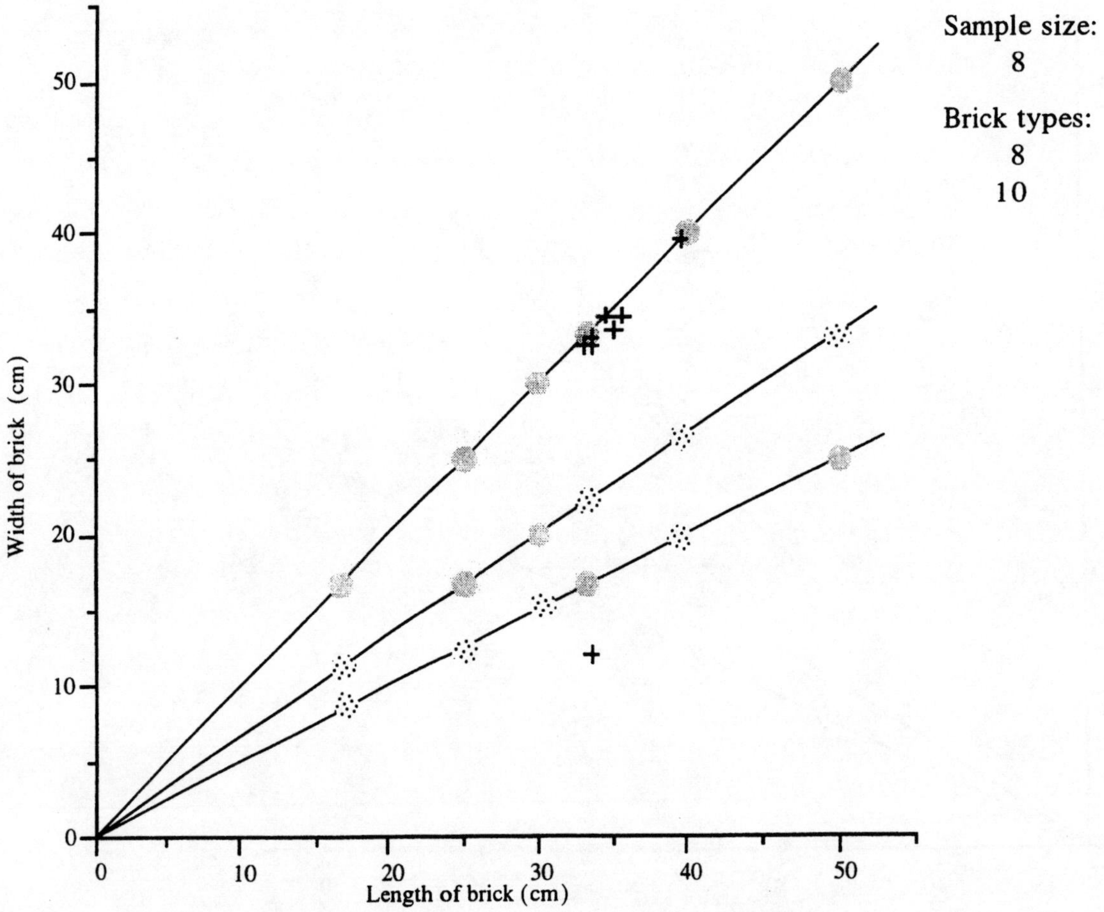

Figure A.6.6: Later Old Babylonian baked bricks

The small sample size may explain why only two OB brick sizes are attested here — albeit very clearly. The single brick which is not roughly square is a third of a Type 8 brick; this suggests that two-thirds of Type 8 bricks were used too. Type 10 bricks attested here are not found in the samples from the earlier periods. The average thickness is again 5 fingers, that of the OB mathematical bricks.

Table A.6.6: Later Old Babylonian baked bricks

Statistic	*Length*	*Width*	*Thickness*
Mean	32.7	29.8	7.76
Median	33.5	33.0	8.0
Mode	33.5	34.5	8.0
Standard deviation	6.23	9.25	1.22
Variance	38.8	85.5	1.48
Range	22.5	27.5	5.0
Minimum	17.0	12.0	5.0
Maximum	39.5	39.5	10.0
Count	9	9	19

A.6.7 KASSITE PERIOD BAKED BRICKS

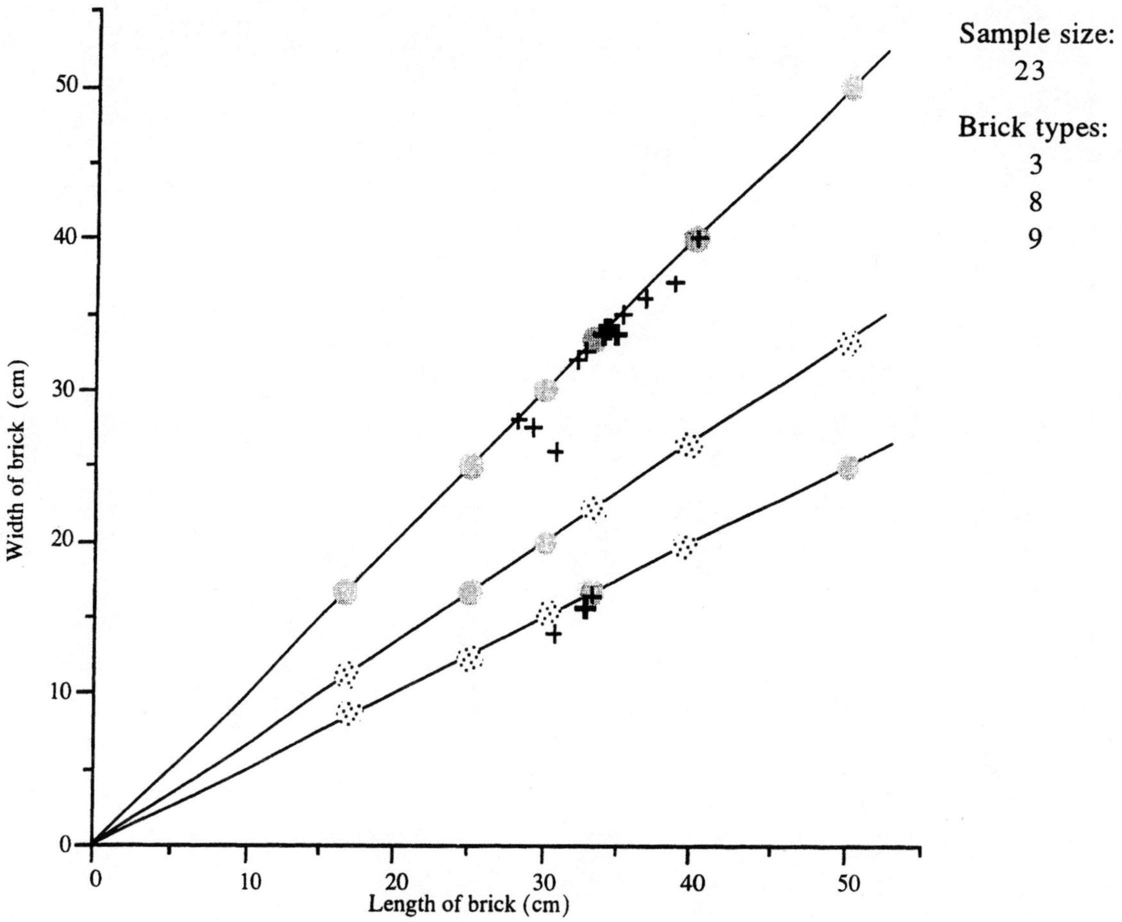

Figure A.6.7: Kassite period baked bricks

Again, most of the bricks in the sample are square, but — apart from a large cluster around Type 8 — they do not seem to be made in standardized sizes. The half bricks are approximately Type 3, half the width of Type 8 bricks. The bricks are on average thicker than those of earlier periods.

Table A.6.7: Kassite period baked bricks

Statistic	*Length*	*Width*	*Thickness*
Mean	31.8	27.3	8.35
Median	32.5	32.0	8.5
Mode	33.5	14.0	8.0
Standard deviation	5.03	9.10	1.22
Variance	25.4	82.8	1.49
Range	29.0	29.0	6.5
Minimum	11.0	11.0	5.0
Maximum	40.0	40.0	11.5
Count	28	25	49

A.6.8 MIDDLE BABYLONIAN BAKED BRICKS

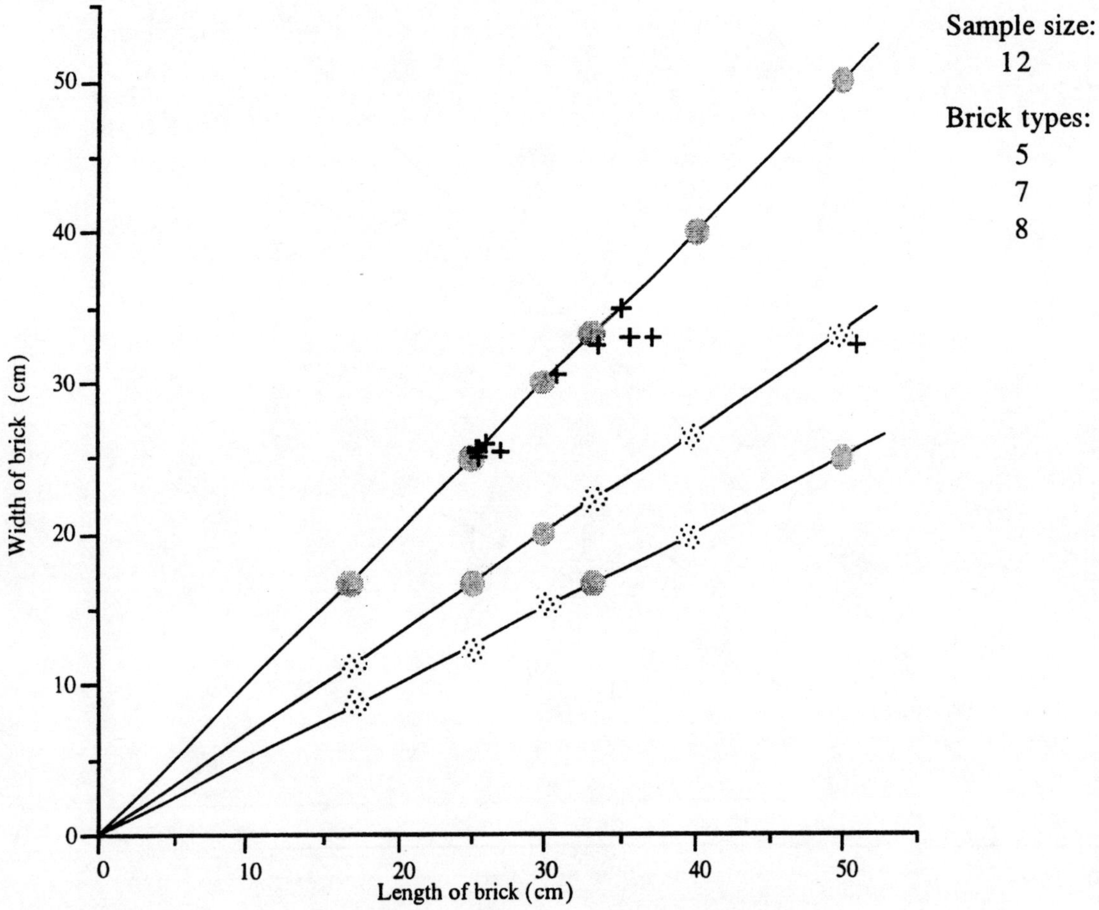

Figure A.6.8: Middle Babylonian baked bricks

The scatter diagram of complete MB bricks gives a rather different picture from the statistics on all complete measurements. The scatter diagram reveals square bricks of rather shorter length than in earlier periods, but the averages of the whole sample fall between Types 7 and 8.

Table A.6.8: Middle Babylonian baked bricks

Statistic	*Length*	*Width*	*Thickness*
Mean	32.0	30.7	7.21
Median	30.5	32.5	7.5
Mode	25.5	25.5	7.5
Standard deviation	7.25	4.63	0.845
Variance	52.5	21.4	0.714
Range	25.5	13.0	3.5
Minimum	25.5	25.0	5.0
Maximum	51.0	38.0	8.5
Count	15	13	21

A.6.9 NEO-BABYLONIAN AND LATE BABYLONIAN BAKED BRICKS

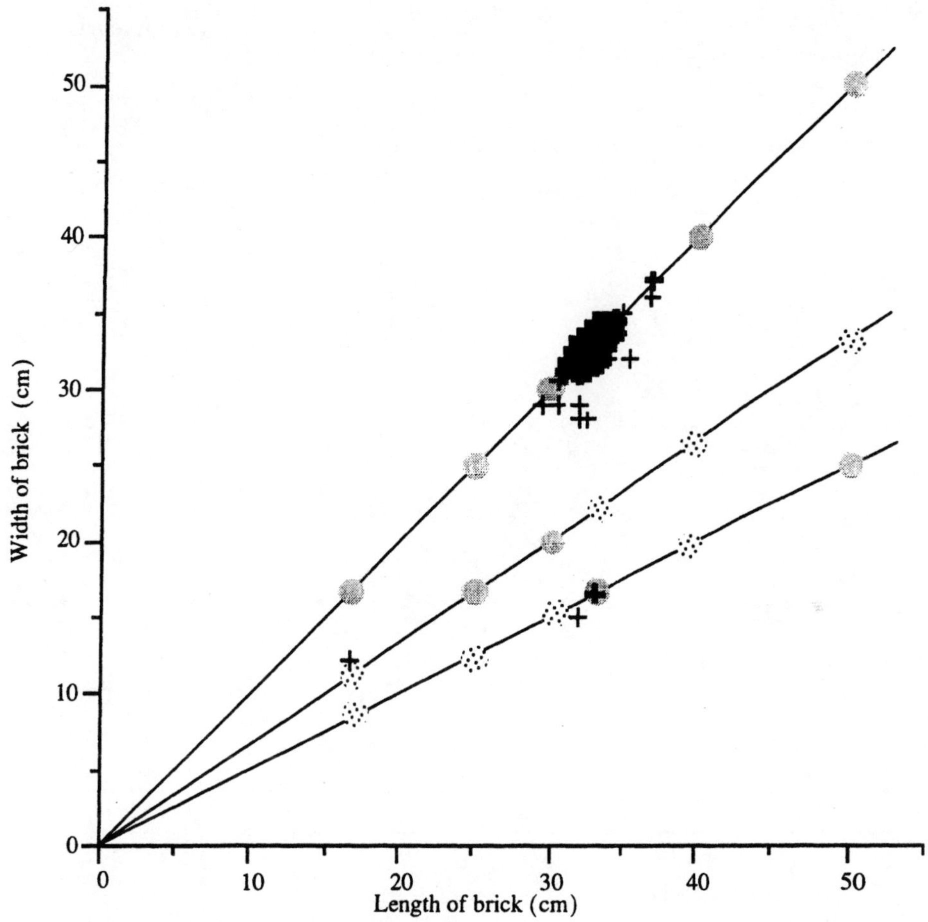

Sample size: 133

Brick types: 3 7 8 9

Figure A.6.9: Neo-Babylonian and Late Babylonian baked bricks

These baked bricks are, almost without exception, the same size as OB Type 8 bricks, or their corresponding half-bricks, Type 3. Some of them are rather thicker, and so could be considered as Type 9s.[2]

Table A.6.9: Neo-Babylonian and Late Babylonian baked bricks

Statistic	*Length*	*Width*	*Thickness*
Mean	32.7	31.8	7.70
Median	33.0	32.5	8.0
Mode	33.0	33.0	8.0
Standard deviation	2.25	3.37	0.947
Variance	5.06	11.4	0.897
Range	36.5	22.0	5.5
Minimum	11.0	15.0	4.5
Maximum	37.5	37.0	10.0
Count	144	135	295

[2] See George (1995: 178 nn 34, 37) for a brief survey of the sizes of brick excavated from the late millennium layers of Esagil, Marduk's temple in Babylon.

A.6.10 OLD ASSYRIAN AND MIDDLE ASSYRIAN BAKED BRICKS

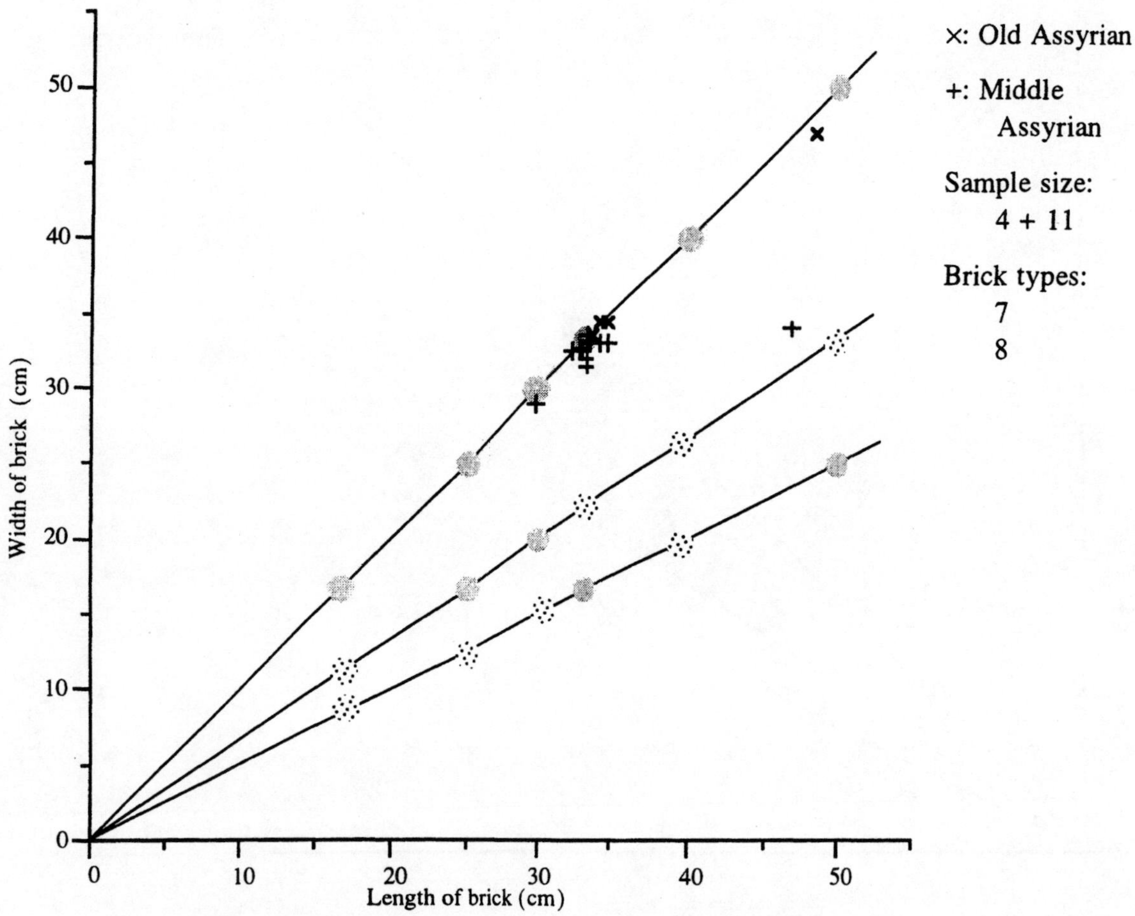

Figure A.6.10: Old Assyrian and Middle Assyrian baked bricks

The Old Assyrian bricks in this sample are all square, but noticeably thinner than the OB mathematical bricks. The Middle Assyrian sample includes a two-thirds brick as well as the square ones. They are slightly thicker than the Old Assyrian ones but still rather thinner than the OB mathematical brick sizes. The samples are too small to determine whether their distributions are significantly different.

Table A.6.10: Old Assyrian and Middle Assyrian baked bricks

	Old Assyrian			*Middle Assyrian*		
Statistic	*Length*	*Width*	*Thickness*	*Length*	*Width*	*Thickness*
Mean	31.9	37.4	5.0	34.7	31.0	5.92
Median	34.3	34.5	5.0	33.5	32.8	6.0
Mode	–	34.5	4.0	33.5	33.0	6.0
Standard deviation	10.9	6.43	1.0	4.56	5.21	0.868
Variance	118	41.4	1.0	20.8	27.1	0.754
Range	29.0	13.5	2.5	17.0	19.0	4.0
Minimum	19.5	33.5	4.0	30.0	15.0	5.0
Maximum	48.5	47.0	6.5	47.0	34.0	9.0
Count	6	4	6	16	12	49

A.6.11 NEO-ASSYRIAN BAKED BRICKS

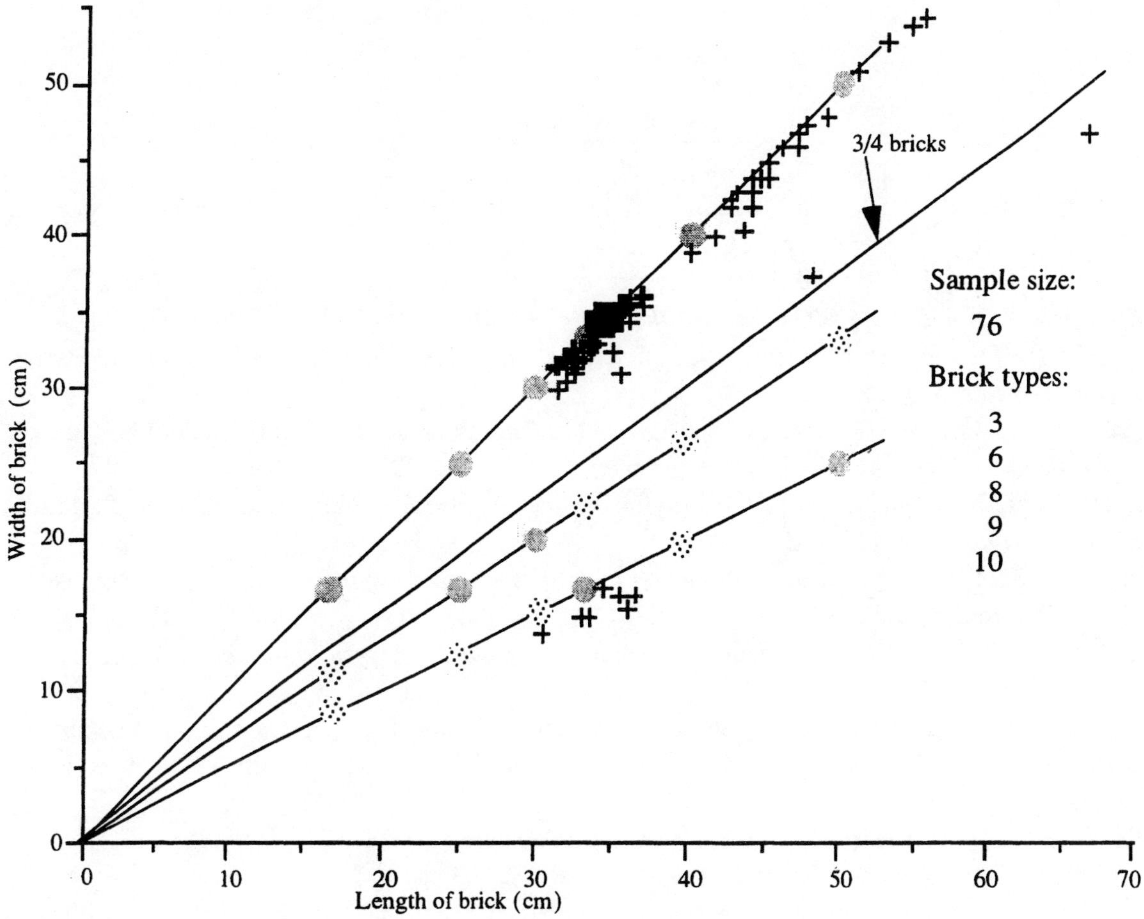

Figure A.6.11: Neo-Assyrian baked bricks

The scatter diagram shows that Neo-Assyrian baked bricks were not made in closely defined sizes. Most baked bricks are square, measuring 30–7 cm or 40–55 cm in length. Smaller square bricks are roughly halved, but two-thirds bricks are not attested. Three-quarter bricks may have been used instead. In general, Neo-Assyrian baked bricks are both larger and thicker than those of other periods.

Table A.6.11: Neo-Assyrian baked bricks

Statistic	*Length*	*Width*	*Thickness*
Mean	38.0	33.6	9.58
Median	35.5	34.5	10.0
Mode	34.5	34.5	11.0
Standard deviation	6.93	10.2	1.82
Variance	48.0	104	3.32
Range	36.0	43.5	8.0
Minimum	30.5	11.0	5.0
Maximum	66.5	54.5	13.0
Count	77	86	179

REFERENCES

Adamson, P B. (1990). 'Some anatomical and pathological terms in Akkadian'. *Revue d'Assyriologie et d'Archéologie Orientale*, 84: 27–32.

Allotte de la Fuÿe, F.-M. (1907). 'La mesure des volumes dans les textes archaïques de la Chaldée'. *Revue d'Assyriologie et d'Archéologie Orientale*, 6: 75–8.

—— (1915). *Documents présargoniques, fascicule supplémentaire (DP 468 à 669)*. Paris: Librairie Ernest Leroux.

Al-Rawi, F. N. H., and George, A. R. (1991–2). 'Enūma Anu Enlil XIV and other early astronomical tables'. *Archiv für Orientforschung*, 38–9: 52–73.

—— and Roaf, M. (1984). 'Ten Old Babylonian mathematical problems from Tell Haddad, Himrin'. *Sumer*, 43: 175–218.

Alster, B. (1997). *Proverbs of ancient Sumer: the world's oldest proverb collections*. Bethesda, MD: CDL Press.

Archi, A. (1989). 'Tables de comptes eblaïtes'. *Revue d'Assyriologie et d'Archéologie Orientale*, 83: 1–6.

Arnaud, D. (1994). *Texte aus Larsa: die epigraphischen Funde der 1. Kampagne in Senkereh-Larsa 1933* (Berliner Beiträge zum Vorderen Orient. Texte, 3). Berlin: Dietrich Reimer Verlag.

Baqir, T. (1946). 'Tell Harmal: a preliminary report'. *Sumer*, 2: 22–30.

—— (1948). 'Excavations at Harmal'. *Sumer*, 4: 137–9.

—— (1950*a*). 'An important mathematical problem text from Tell Harmal (on a Euclidean theorem)'. *Sumer*, 6: 39–54.

—— (1950*b*). 'Another important mathematical text from Tell Harmal'. *Sumer*, 6: 130–48.

—— (1951). 'Some more mathematical texts from Tell Harmal'. *Sumer*, 7: 28–45.

—— (1959). *Tell Harmal*. Baghdad: Directorate General of Antiquities.

—— (1962). 'Tell Dhiba'i: new mathematical texts'. *Sumer*, 18: 11–14.

—— and Mustafa, M. A. (1945). 'Iraq government sounding at Dêr'. *Sumer*, 1/2: 37–54.

Beckmann, P. (1971). *A history of π (pi)*. New York, NY: St. Martin's Press.

Birot, M. (1960). *Textes administratifs de la salle 5 du palais* (Archives royales de Mari (transcription et traduction), 9). Paris: Imprimerie Nationale.

Black, J. A. (1980). 'Review of Cooper (1978)'. *Archiv für Orientforschung*, 27: 154–9.

—— (1988). 'Review of Civil *et al.* (1986)'. *Wiener Zeitschrift für die Kunde des Morgenlandes*, 78: 271–4.

—— (1991). *Sumerian grammar in Babylonian theory* (2nd edn, Studia Pohl: Series Maior, 12). Rome: Biblical Institute Press.

Borger, R (1967–75). *Handbuch der Keilschriftliteratur*, I–III. Berlin: Walter de Gruyter & Co.

Boyer, C. (1989). *The history of mathematics* (2nd edn, rev. U. C. Merzbach). New York, NY: Wiley.

Brack-Bernsen, L., and Schmidt, O. (1990). 'Bisectable trapezia in Babylonian mathematics'. *Centaurus*, 33: 1–38.

Braidwood, R. J., and Howe, B. (1960). *Prehistoric investigations in Iraqi Kurdistan* (Studies in Ancient Oriental Civilization, 31). Chicago, IL: University of Chicago Press.

Brinkman, J. A. (1977). 'Mesopotamian chronology of the historical period', in Oppenheim (1977), 335–48.

Britton, J. P. (1991–3). 'A table of 4th powers and related texts from Seleucid Babylon'. *Journal of Cuneiform Studies*, 43–5: 71–87.

Bruins, E. M. (1953). 'Revision of the mathematical texts from Tell Harmal'. *Sumer*, 9: 241–53.

—— (1954). 'Some mathematical texts'. *Sumer*, 10: 55–61.

—— (1962). 'Interpretation of cuneiform mathematics'. *Physis*, 4: 277–341.

—— (1971). 'Computation in the Old Babylonian period'. *Janus*, 58: 222–67.

—— (1986–90). 'Ptolemaic and Islamic trigonometry. The problem of the Qibla'. *Janus*, 73: 125–48.

—— and Rutten, M. (1961). *Textes mathématiques de Suse* (Mémoires de la Mission Archéologique en Iran, 34). Paris: Librairie Orientaliste Paul Geuthner.

Burton, D. M. (1995). *Burton's history of mathematics: an introduction* (3rd edn). Dubuque, IA: William C Brown.

Cagni, L. (1980). *Briefe aus dem Iraq Museum* (Altbabylonische Briefe in Umschrift und Übersetzung, 8). Leiden: E. J. Brill.

Calvet, Y. (1994). 'Les grandes résidences paléo-babyloniennes de Larsa', in H. Gasche, M. Tanret, C. Janssen, and A. Degraeve (eds.), *Cinquante-deux reflexions sur le Proche-Orient ancien offertes en hommage à Léon De Meyer* (Mesopotamian History and Environment Occasional Publications, 2). Leuven: Peeters, 215–28.

Carter, E., and Stolper, M. W. (1984). *Elam: surveys of political history and archaeology* (University of California Publications: Near Eastern Studies, 25). Berkeley, CA: University of California Press.

Castellino, G. R. (1972). *Two Shulgi hymns (BC)* (Studi Semitici, 42). Rome: Istituto di Studi del Vicino Oriente.

Cavigneaux, A. (1980–3). 'Lexikalische Listen', in D. O. Edzard (ed.), *Reallexikon der Assyriologie und vorderasiatischen Archäologie*, 6. Berlin/New York, NY: Walter de Gruyter, 609–41.

—— (1996). *Uruk. Altbabylonische Texte aus dem Planquadrat Pe XVI–4/5 nach Kopien von Adam Falkenstein* (Ausgrabungen in Uruk-Warka Endberichte, 23). Mainz am Rhein: Verlag Philipp von Zabern.

Charpin, D. (1986). *Le clergé d'Ur au siècle d'Hammurabi (XIXe–XVIIIe siècles av. J.-C.)* (Hautes Études Orientales, 22). Paris: Librairie Droz.

—— (1993). 'Données nouvelles sur la poliorcétique à l'époque paléo-babylonienne. 1) Le plan d'une fortification (M.288)'. *MARI, Annales de Recherches Interdisciplinaires*, 7: 193–7.

Chiera, E. (1916–19). *Lists of personal names from the temple school of Nippur*, I–III (Publications of the Babylonian Section, 11). Philadelphia, PA: University Museum.

Çig, M., and Kizilyay, H. (1965). *Neusumerische Rechts- und Verwaltungsurkunden aus Nippur*, I (Türk Tarih Kurum Yayinlarindan, 6/7). Ankara: Türk Tarih Kurumu Başimevi.

Civil, M. (1964). 'A hymn to the beer goddesss and a drinking song', in R. D. Biggs and J. A. Brinkman (eds.), *Studies presented to A. Leo Oppenheim*. Chicago, IL: University of Chicago Press, 67–89.

Civil, M. (1969). *The series lú = ša and related texts* (Materials for the Sumerian Lexicon, 12). Rome: Biblical Institute Press.

—— (1976). 'Lexicography', in S. J. Lieberman (ed.), *Sumerological studies in honor of Thorkild Jacobsen on his 70th birthday* (Assyriological Studies, 20). Chicago, IL: University of Chicago Press, 123–58.

—— (1979). *Ea A* = nâqu, *Aa A* = nâqu, *with their forerunners and related texts* (Materials for the Sumerian Lexicon, 14). Rome: Biblical Institute Press.

—— (1980). 'Les limites de l'information textuelle', in M. Barrelet (ed.), *L'archéologie de l'Iraq du début de l'époque néolithique à 333 avant notre ère: perspectives et limites de l'interprétation anthropologique des documents* (Colloques Internationaux du Centre National de la Recherche Scientifique, 580). Paris: Éditions du Centre National de la Recherche Scientifique, 225–32.

—— (1985). 'Sur les "livres d'écolier" à l'époque paléo-babylonienne', in J.-M. Durand and J.-R. Kupper (eds.), *Miscellanea Babylonica: mélanges offerts à Maurice Birot*. Paris: Éditions Recherche sur les Civilisations, 67–78.

—— (1987). 'Ur III bureaucracy: quantitative aspects', in McG. Gibson and R. D. Biggs (eds.), *The organization of power: aspects of bureaucracy in the ancient Near East* (Studies in Ancient Oriental Civilization, 46). Chicago, IL: Oriental Institute of the University of Chicago, 43–53.

—— (1994). *The Farmer's Instructions. A Sumerian agricultural manual* (Aula Orientalis Supplementa, 5). Barcelona: Editorial AUSA.

——Gurney, O. R., and Kennedy, D. A. (1986). *The sag-tablet; lexical texts in the Ashmolean Museum; Middle Babylonian grammatical texts; miscellaneous texts* (Materials for the Sumerian Lexicon, Supplementary Series, 1). Rome: Biblical Institute Press.

Clay, A. T. (1915). *Miscellaneous inscriptions in the Yale Babylonian Collection* (Yale Oriental Series, Babylonian Texts, 1). New Haven, CT: Yale University Press.

Collon, D. (1980–3). 'Leier. B. Archäologisch', in D. O. Edzard (ed.), *Reallexikon der Assyriologie und vorderasiatischen Archäologie*, 6. Berlin/New York, NY: Walter de Gruyter, 576–82.

Cooper, J. S. (1978). *The return of Ninurta to Nippur: an-gim dím-ma* (Analecta Orientalia, 52). Rome: Biblical Institute Press.

Damerow, P., and Englund, R. K. (1987). 'Die Zahlzeichensysteme der archäischen Texte aus Uruk', in M. W. Green and H. J. Nissen (eds.), *Zeichenliste der archäischen Texte aus Uruk* (Archäische Texte aus Uruk, 2). Berlin: Gebr. Mann Verlag, 117–66.

Deakin, M. (1996*a*). 'More on Babylon'. *Function*, 20: 128–34.

—— (1996*b*). 'Yet more on Babylon'. *Function*, 20: 163–5.

Delaporte, L. (1911). 'Document mathématique de l'époque des rois d'Our'. *Revue d'Assyriologie et d'Archéologie Orientale*, 8: 131–3.

De Meyer, L. (ed.) (1978). *Tell ed-Dēr*, 2. Leuven: Peeters.

Donald, T. (1962). 'A Sumerian plan in the John Rylands Library'. *Journal of Semitic Studies*, 7: 184–90.

Dossin, G. (1972). ' "*adaššum*" et "*kirḫum*" dans les textes de Mari'. *Revue d'Assyriologie et d'Archéologie Orientale*, 66: 111–30.

Duchesne-Guillemin, M. (1984). *A Hurrian musical score from Ugarit: the discovery of Mesopotamian music* (Sources from the Ancient Near East, 2/2). Malibu: Undena Publications.

Dunham, S. S. (1980). 'A study of ancient Mesopotamian foundations'. Ph.D. thesis, Columbia University.

—— (1982). 'Bricks for the temples of Šara and Ninurra'. *Revue d'Assyriologie et d'Archéologie Orientale*, 76: 27–41.

Durand, J.-M. (1979). 'Textes de collections diverses'. *Revue d'Assyriologie et d'Archéologie Orientale*, 73: 23–35.

—— (1983). 'Relectures d'*ARMT* VIII, II: *ARMT* VIII, 89 et le travail du métal à Mari'. *MARI, Annales de Recherches Interdisciplinaires*, 2: 123–39.

—— (1987). 'Questions de chiffres'. *MARI, Annales de Recherches Interdisciplinaires*, 5: 605–10.

Edzard, D. O. (1962). 'Texts and fragments'. *Journal of Cuneiform Studies*, 16: 78–81.

—— (1969). 'Eine altsumerische Rechentafel (OIP 14, 70)', in W. Röllig (ed.), *lišān mitḫurti, Festschrift Wolfram Freiherr von Soden* (Alter Orient und Altes Testament, 1). Kevelaer/Neukirchen-Vluyn: Verlag Butzon und Bercker/Neukirchener Verlag des Erziehungsvereins, 101–4.

—— (1970). *Altbabylonische Rechts- und Wirtschaftsurkunden aus Tell ed-Dēr im Iraq Museum, Baghdad* (Bayerische Akademie der Wissenschaften, philosophisch-historische Klasse, Abhandlungen. Neue Folge, 72). Munich: Verlag der Bayerischen Akademie der Wissenschaften.

—— (1971). *Cuneiform texts. Altbabylonische Rechts- und Wirtschaftsurkunden aus Tell ed-Dēr bei Sippar* (Texts in the Iraq Museum, 7). Wiesbaden: Otto Harrassowitz.

Englund, R. K. (1988). 'Administrative timekeeping in ancient Mesopotamia'. *Journal of the Economic and Social History of the Orient*, 31: 121–85.

—— (1990). *Organisation und Verwaltung der Ur III-Fischerei* (Berliner Beiträge zum vorderen Orient, 10). Berlin: Dietrich Reimer Verlag.

—— (1991). 'Hard work — where will it get you? Labor management in Ur III Mesopotamia'. *Journal of Near Eastern Studies*, 50: 255–80.

Eves, H. W. (1983). *An introduction to the history of mathematics* (5th edn). New York, NY: Saunders.

Faivre, X. (1995). 'Le recyclage des tablettes cunéiformes'. *Revue d'Assyriologie et d'Archéologie Orientale*, 89: 57–66.

Falkowitz, R. S. (1983–4). 'Round Old Babylonian school tablets from Nippur'. *Archiv für Orientforschung*, 29–30: 18–45.

Farber, G. (1989). 'al-tar im Edubba: notwendige Arbeitsgänge beim Bau eines Schulhauses', in H. Behrens, D. Loding, and M. T. Roth (eds.), *DUMU-E$_2$-DUB-BA-A. Studies in honor of Åke W. Sjöberg* (Occasional Publications of the Samuel Noah Kramer Fund, 11). Philadelphia, PA: University Museum, 137–47.

Ferwerda, G. T., and Woestenburg, E. (1992). 'Index personal names Old Babylonian "Sippar" '. Unpublished manuscript.

Figulla, H. H., and Martin, W. J. (1953). *Letters and documents of the Old-Babylonian period* (Ur Excavations. Texts, 5). London/Philadelphia, PA: Joint Expeditions of the British Museum and of the University Museum, University of Pennsylvania, Philadelphia, to Mesopotamia.

Foster, B. R. (1982). 'Education of a bureaucrat in Sargonic Sumer'. *Archív Orientální*, 50: 238–41.

Fowler, D. H. (1992). '*Dýnamis*, *mitḫartum*, and *square*'. *Historia Mathematica*, 19: 418–19.

Fowler, D. H. and Robson, E. (forthcoming). 'Square root approximations in Old Babylonian mathematics: YBC 7289 in context'. To appear in *Historia Mathematica.*

Foxvog, D. A. (1986). 'A summary of non-sealed labor assignments from Umma'. *Acta Sumerologica,* 8: 59–75.

Frayne, D. (1983). 'Review of Klein (1981)'. *Bibliotheca Orientalis,* 40: 92–101.

Friberg, J. (1978). 'The third millennium roots of Babylonian mathematics 1: a method for the decipherment, through mathematical and metrological analysis, of proto-Sumerian and proto-Elamite semi-pictographic inscriptions'. Department of Mathematics, Chalmers University of Technology, Preprint 1978–9, Gothenburg.

—— (1981). 'Methods and traditions of Babylonian mathematics, II: an Old Babylonian catalogue text with equations for squares and circles'. *Journal of Cuneiform Studies,* 33: 57–64.

—— (1982). 'A survey of publications on Sumero-Akkadian mathematics, metrology and related matters (1854–1982)'. Department of Mathematics, Chalmers University of Technology, Preprint 1982–17, Gothenburg.

—— (1983). 'On the big 6-place tables of reciprocals and squares from Seleucid Babylon and Uruk and their Old Babylonian and Sumerian predecessors'. *Sumer,* 42: 81–7.

—— (1986). 'The early roots of Babylonian mathematics, III. Three remarkable texts from ancient Ebla'. *Vicino Oriente* 6: 3–25.

—— (1987–90). 'Mathematik', in D. O. Edzard (ed.), *Reallexikon der Assyriologie und vorderasiatischen Archäologie,* 7. Berlin/New York, NY: Walter de Gruyter, 531–85.

—— (1993). 'On the structure of cuneiform metrological table texts from the –1st millennium', in H. D. Galter (ed.), *Die Rolle der Astronomie in Kulturen Mesopotamiens* (Grazer Morgenländische Studien, 3). Graz: RM Druck- und Verlagsgesellschaft, 383–405.

—— (1996*a*). 'Bricks and mud in metro-mathematical cuneiform texts'. Max-Planck-Institut für Wissenschaftsgeschichte Preprint 32, Berlin.

—— (1996*b*). 'Pyramids and cones in ancient mathematical texts. New hints of a common tradition'. *Proceedings of the Cultural History of Mathematics,* 6: 80–95.

——Hunger, H., and Al-Rawi, F. (1990). ' "Seed and reeds", a metro-mathematical topic text from Late Babylonian Uruk'. *Baghdader Mitteilungen,* 21: 483–557.

Gadd, C. J. (1922). 'Forms and colours'. *Revue d'Assyriologie et d'Archéologie Orientale,* 19: 149–59.

—— and Kramer, S. N. (1966). *Literary and religious texts: second part* (Ur Excavations. Texts, 6/2). London/Philadelphia, PA: Joint Expeditions of the British Museum and of the University Museum, University of Pennsylvania, Philadelphia, to Mesopotamia.

Gelb, I. J. (1967). 'Growth of a herd of cattle in ten years'. *Journal of Cuneiform Studies,* 21: 64–9.

—— (1970). *Sargonic texts in the Ashmolean Museum, Oxford* (Materials for the Assyrian Dictionary, 5). Chicago, IL: University of Chicago Press.

Genouillac, H. de (1922). *Textes économiques d'Oumma de l'époque d'Our* (Textes Cunéiformes, 5). Paris: Librairie Orientaliste Paul Geuthner.

—— (1924–5). *Premières recherches archéologiques à Kich. Mission d'Henri de Genouillac, 1911–1912, I–II* (Fouilles françaises d'El-Akhymer). Paris: Librairie ancienne Édouard Champion.

George, A. R. (1991). 'Review of Hunger and Pingree (1989)'. *Zeitschrift für Assyriologie und vorderasiatische Archäologie,* 81: 301–6.

—— (1993). 'Ninurta-pāqidāt's dog-bite, and notes on other comic tales'. *Iraq,* 55: 63–75.

—— (1995). 'The bricks of E-sagil'. *Iraq*, 57: 173–97.

Goetze, A. (1945). 'The Akkadian dialects of the Old-Babylonian mathematical texts', in *MCT*, 146–51.

—— (1946). 'Number idioms in Old Babylonian'. *Journal of Near Eastern Studies*, 5: 185–202.

—— (1948*a*). 'Review of *MCT*'. *Journal of Cuneiform Studies*, 2: 33–7.

—— (1948*b*). 'Umma texts concerning reed mats'. *Journal of Cuneiform Studies*, 2: 165–202.

—— (1951). 'A mathematical compendium from Tell Harmal'. *Sumer*, 7: 126–55.

—— (1957). 'Old Babylonian documents from Sippar in the collection of the Catholic University of America'. *Journal of Cuneiform Studies*, 11: 15–40.

—— (1962). 'Two Ur-dynasty tablets dealing with labor'. *Journal of Cuneiform Studies*, 16: 13–16.

Gordon, E. I. (1959). *Sumerian proverbs: glimpses of everyday life in ancient Mesopotamia* (Museum Monographs, 19). Philadelphia, PA: University Museum.

Grégoire, J.-P. (1979). 'Quelques nouveaux documents cunéiformes provenant de l'ancienne collection Allotte de la Fuÿe'. *Oriens Antiquus*, 18: 227–39.

—— (1981). *Inscriptions et archives administratives cunéiformes (I^{e} partie)* (Materiali per il Vocabolario Neosumerico, 10). Rome: Multigrafica Editrice.

Groneberg, B. (1980). *Die Orts- und Gewässernamen der altbabylonischen Zeit* (Répertoire Géographique des Textes Cunéiformes, 3). Wiesbaden: Dr. Ludwig Reichert.

Gundlach, K.-B., and von Soden, W. (1963). 'Einige altbabylonische Texte sur Lösung "quadratischer Gleichungen" '. *Abhändlungen aus den dem Mathematisches Seminar der Universität Hamburg*, 26: 248–63.

Gurney, O. R. (1973). *Middle Babylonian legal documents and other texts* (Ur Excavations. Texts, 7). London/Philadelphia, PA: Joint Expeditions of the British Museum and of the University Museum, University of Pennsylvania, Philadelphia, to Mesopotamia.

—— (1994). 'Babylonian music again'. *Iraq*, 56: 101–6.

Hackman, G. G. (1937). *Temple documents of the Third Dynasty of Ur from Umma* (Babylonian Inscriptions in the Collection of James B. Nies, 5). New Haven, CT: Yale University Press.

Hallo, W. W. (1966). 'The coronation of Ur-Nammu'. *Journal of Cuneiform Studies*, 20: 133–41.

—— and Simpson, W. K. (1971). *The ancient Near East: a history*. New York, NY: Harcourt Brace Jovanovich.

Harper, P. O., Aruz, J., and Tallon, F. (1992). *The royal city of Susa. Ancient Near Eastern treasures in the Louvre*. New York, NY: Metropolitan Museum of Art.

Heath, T. (1981). *The history of Greek mathematics*, I–II (reprint). New York, NY: Dover Publishers, Inc.

Hecker, K. (1990). *Rückläufiges Wörterbuch des Akkadischen* (SANTAG, 1). Wiesbaden: Otto Harrassowitz.

Heimpel, W. (1987). 'Proportional guidelines in ancient Near Eastern art. Appendix B: Gudea's fated brick'. *Journal of Near Eastern Studies*, 46: 205–11.

Heinrich, E., and Seidl, U. (1967). 'Grundrißzeichnungen aus dem Alten Orient'. *Mitteilungen der Deutschen Orientgesellschaft*, 98: 24–45.

Herrero, P., and Glassner, J.-J. (1996). 'Haft Tépé: choix de textes, IV'. *Iranica Antiqua*, 31: 51–82.

Hilprecht, H. V. (1906). *Mathematical, metrological and chronological tablets from the temple library of Nippur* (Babylonian Expedition of the University of Pennsylvania. Series A: Cuneiform Texts, 20/1). Philadelphia, PA: University of Pennsylvania.

Horowitz, W. (1993). 'The reverse of the Neo-Assyrian planisphere CT 33, 11', in H. D. Galter (ed.), *Die Rolle der Astronomie in Kulturen Mesopotamiens* (Grazer Morgenländische Studien, 3). Graz: RM Druck- und Verlagsgesellschaft, 149–59.

Høyrup, J. (1982). 'Investigations of an early Sumerian division problem, c. 2500 BC'. *Historia Mathematica*, 9: 19–36.

—— (1985). 'Varieties of mathematical discourse in pre-modern socio-cultural contexts: Mesopotamia, Greece, and the Latin Middle Ages'. *Science and Society*, 49: 4–41.

—— (1990*a*). 'Algebra and naive geometry. An investigation of some basic aspects of Old Babylonian mathematical thought'. *Altorientalische Forschungen*, 17: 27–69, 262–354.

—— (1990*b*). '*Dýnamis*, the Babylonians, and *Theaetetus* 147c7–148d7'. *Historia Mathematica*, 17: 201–22; 18: 89.

—— (1992). 'Mathematics, algebra, and geometry', in D. N. Freedman (ed.), *The Anchor Bible Dictionary*, IV. New York, NY: Doubleday, 602–12.

—— (1993*a*). 'On subtractive operations, subtractive numbers, and purportedly negative numbers in Old Babylonian mathematics'. *Zeitschrift für Assyriologie und vorderasiatische Archäologie*, 83: 42–60.

—— (1993*b*). ' "Remarkable numbers" in Old Babylonian mathematical texts: a note on the psychology of numbers'. *Journal of Near Eastern Studies*, 52: 281–6.

—— (1994). 'Babylonian mathematics', in I. Grattan-Guinness (ed.), *Companion encyclopedia of the history and philosophy of the mathematical sciences*. London: Routledge, 21–9.

—— (1996*a*). 'Changing trends in the historiography of Mesopotamian mathematics: an insider's view'. *History of Science*, 34: 1–32.

—— (1996*b*). 'The finer structure of the Old Babylonian mathematical corpus: elements of classification, with some results'. Filosofi og Videnskagsteori pa Roskilde Universitetscenter, Preprint 1996 no. 4, Roskilde.

Hruška, B. (1988). 'Die Bewässerungsanlagen in den altsumerischen Königsinschriften von Lagaš'. *Bulletin on Sumerian Agriculture*, 4: 61–72.

Hunger, H. (1968). *Babylonische und assyrische Kolophone* (Alter Orient und Altes Testament, 2). Kevelaer/Neukirchen-Vluyn: Verlag Butzon & Berker/Neukirchener Verlag des Erziehungsvereins.

—— and Pingree, D. (1989). *MUL.APIN. An astronomical compendium in cuneiform* (Archiv für Orientforschung, Beiheft 24). Horn, Austria: Verlag Ferdinand Becker und Söhne Gesellschaft.

Hunt, R. C. (1988). 'Hydraulic management in southern Mesopotamia in Sumerian times: some observations'. *Bulletin on Sumerian Agriculture*, 4: 189–206.

Jestin, R. (1937). *Tablettes sumériennes de Shuruppak conservées au Musée de Stamboul*. Paris: E. de Boccard.

Jeyes, U. (1989). *Old Babylonian extispicy: omen texts in the British Museum* (Uitgaven van het Nederlands Historisch-Archaeologisch Instituut te Istanbul, 64), Leiden/Istanbul: Nederlands Historisch-Archaeologisch Instituut.

Joannès, F. (1993–7). 'Metalle und Metallurgie. A. I. In Mesopotamien', in D. O. Edzard (ed.), *Reallexikon der Assyriologie und vorderasiatischen Archäologie*, 8. Berlin/New York, NY: Walter de Gruyter, 96–112.

Jones, T. B., and Snyder, J. W. (1961). *Sumerian economic texts from the third Ur dynasty: a catalogue and discussion of documents from various collections*. Minneapolis: University of Minnesota Press.

Joseph, G. G. (1992). *The crest of the peacock: non-European roots of mathematics*. Harmondsworth: Penguin.

Jursa, M., and Radner, K. (1995–6). 'Keilschrifttexte aus Jerusalem'. *Archiv für Orientforschung*, 42–3: 89–108.

Keiser, C. E. (1914). *Cuneiform bullae of the third millennium BC* (Babylonian Records in the Library of J. Pierpont Morgan, 3). New Haven, CT: Yale University Press.

—— (1919). *Selected temple documents of the Ur dynasty* (Yale Oriental Series, Babylonian Texts, 4). New Haven, CT: Yale University Press.

Kepinski-Lecomte, C. (ed.) (1992). *Haradum, I: une ville nouvelle sur le Moyen-Euphrate (XVIIIe–XVIIe siècles av. J.-C,)*. Paris: Éditions Recherche sur les Civilisations.

Killick, R., and Black, J. A. (1985). 'Excavations in Iraq, 1983–84'. *Iraq*, 47: 215–39.

Kilmer, A. D. (1960). 'Two new lists of key numbers for mathematical operations'. *Orientalia*, 29: 273–308.

—— (1980–3). 'Leier. A. Philologisch', in D. O. Edzard (ed.), *Reallexikon der Assyriologie und vorderasiatischen Archäologie*, 6. Berlin/New York, NY: Walter de Gruyter, 571–6.

—— (1985). 'Additions and corrections to Akkadian lists of mathematical coefficients (*igigubbû*)'. Unpublished manuscript.

—— (1990). 'Sumerian and Akkadian names for designs and geometrical shapes', in A. C. Gunter (ed.), *Investigating artistic environments in the ancient Near East*. New York, NY: Smithsonian Institution, 83–91.

—— (1992). 'Musical practice at Nippur', in M. de J. Ellis (ed.), *Nippur at the centennial: papers read at the 35^e RAI, Philadelphia 1988* (Occasional Publications of the Samuel Noah Kramer Fund, 14). Philadelphia, PA: University Museum, 101–12.

King, L. W. (1900). *Cuneiform texts from Babylonian tablets, &c., in the British Museum*, IX. London: Trustees of the British Museum.

Klein, J. (1981). *Three Šulgi hymns: Sumerian royal humns glorifying King Šulgi of Ur*. Ramat-Gan: Bar-Ilan Universtiy Press.

—— (1986). 'On writing monumental inscriptions in the Ur III scribal curriculum'. *Revue d'Assyriologie et d'Archéologie Orientale*, 80: 1–7.

Kline, M. (1972). *Mathematical thought from ancient to modern times*. New York, NY: Oxford University Press.

Knuth, D. E. (1972). 'Ancient Babylonian algorithms'. *Communications for the Association of Computing Machinery*, 15: 671–7.

—— (1976). 'Reciprocal table of Inakabit-Anu incorrectly rendered'. *Communications for the Association of Computing Machinery*, 19: 108.

Kraus, F. R. (1959–62). 'Briefschreibübungen im altbabylonischen Schulunterricht'. *Jaarbericht van het Voor-Asiatisch-Egyptisch Genootschap Ex Oriente Lux*, 16: 16–39.

—— (1973). *Vom mesopotamischen Menschen der altbabylonischen Zeit und seiner Welt. Eine Reihe Vorlesungen*. Amsterdam: North-Holland Publishing Company.

—— (1975). 'Akkadische Wörter und Ausdrücke IX. — *ḫabātum*'. *Revue d'Assyriologie et d'Archéologie Orientale*, 69: 31–40.

Krecher, J. (1993–7). 'Miete. A.I. Fāra-Zeit bis Ur III. Mobiliarmiete (Personen, Schiffe, Tiere)', in D. O. Edzard (ed.), *Reallexikon der Assyriologie und vorderasiatischen Archäologie*, 8. Berlin/New York, NY: Walter de Gruyter, 156–62.

Krispijn, T. J. H. (1990). 'Beiträge zur altorientalischen Musikforschung, 1. Shulgi und die Musik'. *Akkadica*, 70: 1–27.

Krušina-Černý, L. J. (1957). 'Two Prague collections of the Sumerian tablets of the Third Dynasty of Ur'. *Archív Orientální*, 25: 547–62.

Kümmel, H. M. (1970). 'Zur Stimmung der babylonischen Harfe'. *Orientalia*, 39: 252–63.

Kupper, J.-R. (1953). *Lettres* (Archives royales de Mari, 6). Paris: Librairie Orientaliste Paul Geuthner.

—— (1954). *Correspondance de Baḫdi-Lim, préfet du palais de Mari* (Archives royales de Mari (transcription et traduction), 6). Paris: Imprimerie Nationale.

Lackenbacher, S. (1982). 'Un texte vieux-babylonien sur la finition des textiles'. *Syria*, 59: 129–49.

Lafont, B. (1980). 'Un nouveau texte d'Ur III sur l'irrigation'. *Revue d'Assyriologie et d'Archéologie Orientale*, 74: 29–42.

Lawergren, B., and Gurney, O. R. (1987). 'Sound holes and geometrical figures: clues to the terminology of ancient Mesopotamian harps'. *Iraq*, 49: 37–52.

Leemans, W. F., and Bruins, E. M. (1951). 'Un texte vieux-babylonien concernant des cercles concentriques', in *Compte rendu de la 2e Rencontre Assyriologique Internationale*. Paris: Imprimerie Nationale, 31–5.

Legrain, L. (1935). 'Quelques textes anciens, II'. *Revue d'Assyriologie et d'Archéologie Orientale*, 32: 128–9.

—— (1937–47). *Business documents of the Third Dynasty of Ur*, I–II (Ur Excavations. Texts, 3). London/Philadelphia, PA: Joint Expeditions of the British Museum and of the University Museum, University of Pennsylvania, Philadelphia, to Mesopotamia.

Leichty, E. (1986). *Tablets from Sippar*, I (Catalogue of the Babylonian Tablets in the British Museum, 6). London: British Museum Publications.

Lerberghe, K. van (1986). *Old Babylonian legal and administrative texts from Philadelphia* (Orientalia Lovaniensia Analecta, 21). Leuven: Peeters.

—— and Voet, G. (1991). *Sippar-Amnānum: the Ur-Utu archive, I.* (Mesopotamian History and Environment, III/1). Ghent: University of Ghent.

Lewy, H. (1947). 'Marginal notes on a recent volume of Babylonian mathematical texts'. *Journal of the American Oriental Society*, 67: 305–20.

—— (1949). 'Studies in Assyro-Babylonian mathematics and metrology'. *Orientalia*, 18: 40–67, 137–70.

Lieberman, S. J. (1990). 'Canonical and official cuneiform texts: towards an understanding of Ashurbanipal's personal tablet collection', in T. Abusch, J. Huehnergard, and P. Steinkeller (eds.), *Lingering over words* (Harvard Semitic Studies, 37). Atlanta, GA: Scholars' Press, 305–36.

Limet, H. (1973). *Étude de documents de la période d'Agadé appartenant à l'Université de Liège* (Bibliothèque de la Faculté de Philosophie et Lettres de l'Université de Liège, 206). Paris: Société d'Éditions "Les Belles Lettres".

Livingstone, A. (1986). *Mystical and mythological explanatory works of Assyrian and Babylonian scholars*. Oxford: Clarendon Press.

McCown, D. E., Haines, R. C., and Hansen, D. P. (1967). *Nippur, I: temple of Enlil, scribal quarter and soundings* (Oriental Institute Publications, 78). Chicago: University of Chicago Press.

Maeda, T. (1984). 'Work concerning irrigation canals in pre-Sargonic Lagash'. *Acta Sumerologica*, 6: 33–53.

Maekawa, K. (1990). 'Cultivation methods in the Ur III period'. *Bulletin on Sumerian Agriculture*, 5: 115–45.

Margueron, J. (1986). 'Quelques remarques concernant les archives retrouvées dans le palais de Mari', in K. R. Veenhof (ed.), *Cuneiform archives and libraries: papers read at the 30e Rencontre Assyriologique Internationale*, Leiden, 4–8 July 1983 (Publications de l'Institut historique-archéologique néerlandais de Stamboul, 57). Leiden/Istanbul: Nederlands Historisch-Archaeologisch Instituut, 141–52.

Meer, P. E. van der (1935). *Textes scolaires de Suse* (Mémoires de la Mission Archéologique de Perse, 27). Paris: Librairie Ernest Leroux.

Michalowski, P. (1983). 'Review of Cagni (1980)'. *Journal of Cuneiform Studies*, 35: 221–8.

—— (1987). 'Charisma and control: on continuity and change in early Mesopotamian bureaucratic systems', in McG. Gibson and R. D. Biggs (eds.), *The organization of power: aspects of bureaucracy in the ancient Near East* (Studies in Ancient Oriental Civilization, 46). Chicago, IL: Oriental Institute of the University of Chicago, 55–68.

Michel, C. (1989). 'Un problème mathématique relatif à la métallurgie'. *Nouvelles Assyriologiques Brèves et Utilitaires*, 1989/4: 82–3, no. 110.

Mieroop, M. van de (1992*a*). 'Old Babylonian Ur: portrait of an ancient Mesopotamian city'. *Journal of the Ancient Near Eastern Society*, 21: 119–30.

—— (1992*b*). 'Reed in the Old Babylonian texts from Ur'. *Bulletin on Sumerian Agriculture*, 6: 147–53.

Miller, R., McEwen, E., and Bergman, C. (1986). 'Experimental approaches to ancient Near Eastern archery'. *World Archaeology*, 18: 178–95.

Moorey, P. R. S. (1994). *Ancient Mesopotamian materials and industries: the archaeological evidence*. Oxford: Clarendon Press.

Muhly, J. D. (1993–7). 'Metalle. B. Archäologisch', in D. O. Edzard (ed.), *Reallexikon der Assyriologie und vorderasiatischen Archäologie*, 8. Berlin/New York, NY: Walter de Gruyter, 119–36.

Muroi, K. (1986). 'An interpretation of Babylonian mathematical text VAT 8522 vs I'. *Kagakusi Kenkyu*, 25: 261–6.

—— (1992*a*). 'Small canal problems of Babylonian mathematics'. *Historia Scientiarum*, 1/3: 173–80.

—— (1992*b*). 'Re-examination of Susa mathematical text no. 3: alleged value $\pi \approx 3^1/_8$'. *Historia Scientiarum*, 2/1: 45–9.

—— (1994). 'Re-examination of the first problem of the Susa mathematical text no. 9'. *Historia Scientiarum*, 3/3: 231–3.

Nemet-Nejat, K. R. (1982). *Late Babylonian field plans in the British Museum* (Studia Pohl Series Maior, 11). Rome; Biblical Institute Press.

—— (1988). 'Cuneiform mathematical texts as training for scribal professions', in E. Leichty, M. DeJ. Ellis, and P. Gerardi (eds.), *A scientific humanist: studies in memory of Abraham Sachs* (Occasional Publications of the Samuel Noah Kramer Fund, 9). Philadelphia, PA: University Museum, 285–300.

—— (1993). *Cuneiform mathematical texts as a reflection of everyday life in Mesopotamia* (American Oriental Series, 75). New Haven, CT: American Oriental Society.

—— (1995). 'Systems for learning mathematics in Mesopotamian scribal schools'. *Journal of Near Eastern Studies*, 54: 241–60.

Neugebauer, O. (1928). 'Zur Geschichte des Pythagoräischen Lehrsatzes'. *Nachrichten von der Gesellschaft der Wissenschaften zu Göttingen, mathematisch-physikalische Klasse*, 45–8.

Neugebauer, O. (1931–2). 'Über die Approximation irrationaler Quadratwurzeln in der babylonischen Mathematik'. *Archiv für Orientforschung*, 7: 90–9.

—— (1934). *Vorgriechische Mathematik* (Vorlesungen über Geschichte der antiken mathematischen Wissenschaften, 1). Berlin: Springer Verlag.

—— (1935–7). *Mathematische Keilschrift-Texte*, I–III (Quellen und Studien zur Geschichte der Mathematik, Astronomie und Physik, A3). Berlin: Verlag von Julius Springer.

—— (1952). *The exact sciences in antiquity*. Princeton, NJ: Princeton University Press.

—— and Sachs, A. (1945). *Mathematical cuneiform texts* (American Oriental Series, 29). New Haven, CT: American Oriental Society.

—— —— (1984). 'Mathematical and metrological texts'. *Journal of Cuneiform Studies*, 36: 243–51.

—— and Struve, W. (1929). 'Über die Geometrie des Kreises in Babylonien'. *Quellen und Studien zur Geschichte der Mathematik, Astronomie und Physik*, B1: 81–92.

Nikolskii, M. V. (1915). *Dokumenty chozjajstvennoj otčetnosti drevnej Chaldei iz sobranija N. P. Lichačeva, Čast' 2: epocha dinastii Agade i epocha dinastii Ura* (Drevnosti Vostočnyja, 5). Moscow.

Nissen, H. (1976–80). 'Kanal(isation). B. Archäologisch', in D. O. Edzard (ed.), *Reallexikon der Assyriologie und vorderasiatischen Archäologie*, 5. Berlin/New York, NY: Walter de Gruyter, 365–8.

—— Damerow, P., and Englund, R. K. (1990). *Frühe Schrift und Techniken der Wirtschaftsverwaltung im alten Vorderen Orient: Informationsspeicherung und -verarbeitung vor 5000 Jahren*. Berlin: Verlag Franzbecker.

Oates, D. (1990). 'Innovations in mud-brick: decorative and structural techniques in ancient Mesopotamia'. *World Archaeology*, 21: 388–406.

Oberhuber, K. (1960). *Sumerische und akkadische Keilschriftdenkmäler des Archäologischen Museums zu Florenz* (Innsbrucker Beiträge zur Kulturwissenschaft, 8). Innsbruck: Sprachwissenchaftliche Institut der Universität Innsbruck.

Oppenheim, A. L. (1947). 'Review of *MCT*'. *Journal of Near Eastern Studies*, 6: 126–8.

—— (1948). *Catalogue of the cuneiform tablets of the Wilberforce Eames Babylonian Collection in the New York Public Library. Tablets of the time of the Third Dynasty of Ur* (American Oriental Series, 32). New Haven, CT: American Oriental Society.

—— (1977). *Ancient Mesopotamia: portrait of a dead civilization* (2nd edn, rev. E. Reiner). Chicago, IL: University of Chicago Press.

—— *et al.* (eds.) (1956–). *The Assyrian Dictionary of the Oriental Institute of the University of Chicago*. Chicago, IL: Oriental Institute of the University of Chicago.

Owen, D. I. (1972). 'Neo-Sumerian texts from American collections, I'. *Journal of Cuneiform Studies*, 24: 137–73.

—— (1982). *Neo-Sumerian archival texts primarily from Nippur in the University Museum, the Oriental Institute and the Iraq Museum*. Winona Lake, IN: Eisenbrauns.

—— (1991). *Neo-Sumerian texts from American collections* (Materiali per il Vocabolario Neosumerico, 15). Rome: Multigrafia Editrice.

Parrot, A. (1948). *Tello. Vingt campagnes de fouilles (1877–1933)*. Paris: Éditions Albin Michel.

—— (1958). *Mission archéologique de Mari, II: Le palais*. Paris: Librairie Orientaliste Paul Geuthner.

Pemberton, W., Postgate, J. N., and Smyth, R. F. (1988). 'Canals and bunds, ancient and modern'. *Bulletin on Sumerian Agriculture*, 4: 207–21.

Pohl, A. (1935). *Vorsargonische und sargonische Wirtschaftstexte* (Texte und Materialen der Frau Professor Hilprecht Collection, 5). Leipzig: J. C. Hinrichs'sche Buchhandlung.

Pomponio, F. (1981). 'Tabella di concordanze di misure per aridi: n. 74', in G. Pettinato (ed.), *Testi lessicali monolingui della bibliotheca L. 2769* (Materiali Epigrafici di Ebla, 3). Naples, Istituto Universitario Orientale di Napoli, 270–1.

Postgate, J. N. (1974). 'Review of Salonen (1972)'. *Journal of the Royal Asiatic Society*: 52–3.

—— (1992). *Early Mesopotamia: society and economy at the dawn of history.* London/New York, NY: Routledge.

Powell, M. A. (1971). 'Sumerian numeration and metrology' (unpublished Ph.D. thesis), Minneapolis: University of Minnesota.

—— (1976*a*). 'The antecedents of Old Babylonian place notation and the early history of Babylonian mathematics'. *Historia Mathematica*, 3: 417–39.

—— (1976*b*). 'Two notes on metrological mathematics in the Sargonic period'. *Revue d'Assyriologie et d'Archéologie Orientale*, 70: 97–102.

—— (1981). 'Collations to Nikolskii (1915)'. *Acta Sumerologica*, 3: 125–47.

—— (1982*a*). 'Metrological notes on the Esagila tablet and related matters. Appendix II: bricks as evidence for metrology'. *Zeitschrift für Assyriologie und vorderasiatische Archäologie*, 72: 116–23.

—— (1982*b*). 'Merodach-Baladan at Dur-Jakin: a note on the defense of Babylonian cities'. *Journal of Cuneiform Studies*, 34: 59–61.

—— (1984). 'Late Babylonian surface mensuration: a contribution to the history of Babylonian agriculture and arithmetic'. *Archiv für Orientforschung*, 31: 32–66.

—— (1987–90). 'Maße und Gewichte', in D. O. Edzard (ed.), *Reallexikon der Assyriologie und vorderasiatischen Archäologie*, 7. Berlin/New York, NY: Walter de Gruyter, 457–530.

—— (1988). 'Evidence for agriculture and waterworks in Babylonian mathematical texts'. *Bulletin on Sumerian Agriculture*, 4: 161–72.

Price, D. J. de Solla (1964). 'The Babylonian "Pythagorean triangle" tablet'. *Centaurus*, 10: 1–13.

Reiner, E. (1970). '*ana nalban*'. *Archiv für Orientforschung*, 23: 89–91.

Renger, J. (1990). 'Rivers, water courses and irrigation ditches and other matters concerning irrigation based on Old Babylonian sources (2000–1600 BC)'. *Bulletin on Sumerian Agriculture*, 5: 31–46.

Ritter, J. (1990). 'Les practiques de la raison en Mésopotamie', in J.-F. Mattéi (ed.), *La naissance de la raison en Grèce (Actes du Congrès de Nice, mai 1987)*. Paris: Presses Universitaires de France, 99–110.

Roaf, M. (1990). *Cultural atlas of Mesopotamia and the ancient Near East.* Oxford: Facts on File.

Robson, E. (1995*a*). 'More on Pythagoras Theorem in ancient Mesopotamia'. *British Society for the History of Mathematics Newsletter*, 28: 34–6.

—— (1995*b*). 'Review of Nemet-Nejat (1993)'. *Bibliotheca Orientalis*, 52: 424–32.

—— (1996*a*). 'Building with bricks and mortar: quantity surveying in the Ur III and Old Babylonian periods', in K. R. Veenhof (ed.), *Houses and households in ancient Mesopotamia: papers read at the 40e Rencontre Assyriologique Internationale, Leiden, July 5–8, 1993* (Publications de l'Institut historique et archéologique Néerlandais de Stamboul, 78). Leiden/Istanbul: Nederlands Historisch-Archaeologisch Instituut, 181–90.

Robson, E. (1996*b*). 'From Uruk to Babylon: 4500 years of Mesopotamian mathematics', in M. J. Lagarto, A. Vieira, and E. Veloso (eds.), *História e educação matemática. Proceedings: ICME-8 satellite meeting of the International Study Group on the Relations between History and Pedagogy of Mathematics, 24–30 Julho 1996, Braga, Portugal,* I. Braga: HEM Braga 96, 35–44.

—— (forthcoming). 'Three Old Babylonian methods for dealing with "Pythagorean" triangles'. To appear in *Journal of Cuneiform Studies.*

Rochberg-Halton, F. (1983). 'Stellar distances in early Babylonian astronomy: a new perspective on the Hilprecht text (HS 229)'. *Journal of Near Eastern Studies,* 42: 209–17.

Roth, M. T. (1979). 'Scholastic tradition and Mesopotamian law: a study of FLP 1287, a prism in the collection of the Free Library of Philadelphia' (unpublished Ph.D. thesis). Philadelphia, PA: University of Pennsylvania.

—— (1995). *Law collections from Mesopotamia and Asia Minor* (Writings from the Ancient World, 6). Atlanta, GA: Scholars Press.

Rouault, O., and Masetti-Rouault, M. G. (1993). *L'Eufrate e il tempo. Le civiltà del medio Eufrate e della Gezira siriana.* Milan: Electra.

Sachs, A. J. (1947). 'Babylonian mathematical texts, I. Reciprocals of regular sexagesimal numbers'. *Journal of Cuneiform Studies,* 1: 219–40.

—— (1952). 'Babylonian mathematical texts, II–III. Approximations of reciprocals of irregular numbers in an Old-Babylonian text; the problem of finding the cube root of a number'. *Journal of Cuneiform Studies,* 6: 151–6.

—— (1955). *Late Babylonian astronomical and related texts copied by T. G. Pinches and J. N. Strassmaier.* Providence, RI: Brown University Press.

Saggs, H. W. F. (1960). 'A Babylonian geometrical text'. *Revue d'Assyriologie et d'Archéologie Orientale,* 54: 131–46.

Salonen, A. (1972). *Die Ziegeleien im Alten Mesopotamien* (Annales Academie Scientiarum Fennicae, 171). Helsinki: Academia Scientiarum Fennica.

Sarzec, E. de (1884–1912). *Découvertes en Chaldée,* I–II (ed. L. Heuzey). Paris: Ernest Leroux.

Scheil, V. (1915). 'Le calcul des volumes dans un cas particulier à l'époque d'Ur'. *Revue d'Assyriologie et d'Archéologie Orientale,* 12: 161–72.

Schmandt-Besserat, D. (1977). 'An archaic recording system and the origin of writing'. *Syro-Mesopotamian Studies,* 1: 31–70.

—— (1978). 'The earliest precursor of writing'. *Scientific American,* 238/6: 38–47.

—— (1979). 'An archaic recording system in the Uruk-Jemdet Nasr period'. *American Journal of Archaeology,* 83: 19–48.

—— (1992). *Before writing, I: from counting to cuneiform.* Austin, TX: University of Texas Press.

Schneider, N. (1930). 'Die Geschäftsurkunden aus Drehem und Djoḫa in den Staatlichen Museen (VAT) zu Berlin'. *Orientalia,* 47–9.

Shaffer, A. (forthcoming), *Literary and religious texts: third part* (Ur Excavations. Texts, 6/3). London/Philadelphia, PA: Joint Expeditions of the British Museum and of the University Museum, University of Pennsylvania, Philadelphia, to Mesopotamia.

Sigrist, M. (1983). *Textes économiques néo-sumériens de l'Université de Syracuse.* Paris: Éditions Recherche sur les Civilisations.

—— (forthcoming). *Andrews University Cuneiform Texts, VI.* Berrien Springs, MI: Andrews University Press.

——and Gomi, T. (1991). *The comprehensive catalogue of published Ur III tablets*. Bethesda, MD: CDL Press.

—— Owen, D. I., and Young, G. D. (1984). *The John Frederick Lewis Collection, part II* (Materiali per il Vocabolario Neosumerico, 13). Rome: Multigrafia Editrice.

Sjöberg, Å. W. (1975). 'Der Examenstext A'. *Zeitschrift für Assyriologie und vorderasiatische Archäologie*, 64: 137–76.

—— (1976). 'The Old Babylonian eduba', in S. J. Lieberman (ed.), *Sumerological studies in honor of Thorkild Jacobsen on his 70th birthday* (Assyriological Studies, 20). Chicago, IL: University of Chicago Press, 159–80.

——*et al.* (eds.) (1984–). *The Sumerian dictionary of the University Museum of the University of Pennsylvania*. Philadelphia, PA: University Museum.

Smeur, A. J. E. M. (1969–70). 'On the value equivalent to π in ancient mathematical texts. A new interpretation'. *Archive for the History of Exact Sciences*, 6: 249–70.

Snell, D. C. (1982). *Ledgers and prices: early Mesopotamian merchant accounts* (Yale Near Eastern Researches, 8). New Haven, CT: Yale University Press.

Soden, W. von (1952*a*). 'Zu den mathematischen Aufgabentexten vom Tell Harmal'. *Sumer*, 8: 49–56.

—— (1952*b*). *Grundriss der akkadischen Grammatik* (Analecta Orientalia, 33). Rome: Pontifical Biblical Institute.

—— (1959–81). *Akkadisches Handwörterbuch*, I–III. Wiesbaden: Otto Harrassowitz.

—— (1964). 'Review of *TMS*'. *Bibliotheca Orientalis*, 21: 44–50.

—— (1977). 'Mathematische Konstantenlisten als Zeugnisse für Arbeitsnormen in Babylonien', in W. Voigt (ed.), *XX Deutscher Orientalistentag, vom 3. bis 8. Oktober 1977 in Erlangen* (Zeitschrift der Deutschen Morgenländischen Gesellschaft, Supplement 4). Wiesbaden: Franz Steiner Verlag, 107–9.

—— (1994). *The ancient Orient: an introduction to the study of the ancient Near East* (trans. D. G. Schley). Grand Rapids, MI: Wm. B. Eerdmans.

Soubeyran, D. (1984). 'Textes mathématiques de Mari'. *Revue d'Assyriologie et d'Archéologie Orientale*, 78: 19–48.

Spycket, A. (1983). ' "Louez-le sur la harpe et la lyre" '. *Anatolian Studies*, 33: 39–49.

Starr, I. (1975). 'Notes on some technical terms in extispicy'. *Journal of Cuneiform Studies*, 27: 241–7.

Steele, F. R. (1951). 'Writing and history: the new tablets from Nippur'. *University Museum Bulletin*, 16/2: 21–7.

Steinkeller, P. (1987). 'The administrative and economic organization of the Ur III state: the core and the periphery', in McG. Gibson and R. D. Biggs (eds.), *The organization of power: aspects of bureaucracy in the ancient Near East* (Studies in Ancient Oriental Civilization, 46). Chicago, IL: Oriental Institute of the University of Chicago, 19–42.

——(1988). 'Notes on the irrigation system in third millennium southern Babylonia'. *Bulletin on Sumerian Agriculture*, 4: 73–92.

Stève, M.-J., Gasche, H., De Meyer, L., and Amiet, P. (1980). 'La Susiane du deuxième millénaire: à propos d'une interprétation des fouilles de Suse'. *Iranica Antiqua*, 15: 49–154.

Stol, M. (1976–80). 'Kanal(isation). A. Philologisch', in D. O. Edzard (ed.), *Reallexikon der Assyriologie und vorderasiatischen Archäologie*, 5. Berlin/New York, NY: Walter de Gruyter, 355–65.

Stol, M. (1993–7). 'Miete. B.I. Altbabylonisch', in D. O. Edzard (ed.), *Reallexikon der Assyriologie und vorderasiatischen Archäologie*, 8. Berlin/New York, NY: Walter de Gruyter, 162–74.

—— (1994). 'Constant factors in Old Babylonian texts on ploughing with the *inītum*', in H. Gasche, M. Tanret, C. Janssen, and A. Degraeve (eds.), *Cinquante-deux reflexions sur le Proche-Orient ancien offertes en hommage à Léon De Meyer* (Mesopotamian History and Environment Occasional Publications, 2). Leuven: Peeters, 229–35.

Stone, E. (1981). 'Texts, architecture and ethnographic analogy: patterns of residence in Old Babylonian Nippur'. *Iraq*, 43: 19–33.

Tanret, M. (1982). 'Les tablettes "scholaires" découvertes à Tell ed-Dēr'. *Akkadica*, 27: 46–9.

—— (1986). 'Fragments de tablettes pour des fragments d'histoire', in L. De Meyer, H. Gasche, and F. Vallat (eds.), *Fragmenta historiae Elamicae: mélanges offerts à M. J. Stève*. Paris: Éditions Recherche sur les Civilisations, 139–50.

Thomsen, M.-L. (1984). *The Sumerian language: an introduction to its history and grammatical structure* (Mesopotamia, 10). Copenhagen: Akademisk Vorlag.

Thureau-Dangin, F. (1903). *Recueil de tablettes chaldéennes*. Paris: Ernest Leroux.

—— (1932). 'Notes assyriologiques, LXX. — Carré et demi-cercle'. *Revue d'Assyriologie et d'Archéologie Orientale*, 29: 136–9.

—— (1935). 'La mesure des volumes d'après une tablette inédite du British Museum'. *Revue d'Assyriologie et d'Archéologie Orientale*, 32: 1–28.

—— (1936). 'Textes mathématiques babyloniens'. *Revue d'Assyriologie et d'Archéologie Orientale*, 33: 65–84.

—— (1938). *Textes mathématiques babyloniens* (Ex Oriente Lux, 1). Leiden: E. J. Brill.

Vaiman, A. A. (1959). 'O geometriceskoi figure absamikku klinopisnich matematičeskich tekstov (On the geometric figure absamikku in cuneiform mathematical texts)'. *Vestnik Drevnej Istorii*, 67: 91–4.

—— (1960). *Über die sumerisch-babylonische angewandte Mathematik* (XXV Internationaler Orientalisten Kongress, Vorträge der Delegation des UdSSR). Moscow: Verlag für Orientalische Literatur.

—— (1961). *Shumero-vavilonskaya matematika III–I tysyaceletiya do n. e.* (Sumero-Babylonian mathematics of the III–I millennia BC). Moscow.

—— (1962). 'In alpim, tallum i pirkum v klinopisnich matematicheskich tekstich (*īn alpim, tallum* and *pirkum* in cuneiform mathematical texts)'. *Drevnej Mir* (Struve volume), 214–17.

—— (1963). 'Istolkvanie geometričeskich postoyannich iz suzskogo klinopisnogo spiska I (Suzi) (Interpretation of the geometric constants in the cuneiform tablet I from Susa)'. *Vestnik Drevnej Istorii*, 83: 75–86.

—— (1976). 'Issledovanie po sumero-vavilonskoi metrologij (Studies on Sumero-Babylonian metrology)'. *Drevnej Vostok*, 2: 37–66.

—— (1989). 'Protosumerische Maß- und Zählsysteme'. *Baghdader Mitteilungen*, 20: 114–20.

Vanstiphout, H. L. J. (1978). 'Lipit-Eštar's praise in the edubba'. *Journal of Cuneiform Studies*, 30: 33–51.

—— (1979). 'How did they learn Sumerian?' *Journal of Cuneiform Studies*, 31: 118–26.

Veldhuis, N. (1997). 'Elementary education at Nippur: the lists of trees and wooden objects'. Unpublished doctoral thesis, University of Groningen.

Vogel, K. (1959). *Vorgriechische Mathematik II: Die Mathematik der Babylonier* (Mathematische Studienhefte, 2). Hanover: Hermann Schroedel Verlag KG.

Volk, K. (1996). 'Methoden altmesopotamischer Erziehung nach Quellen der altbabylonischen Zeit'. *Saeculum*, 47: 178–216.

Waerden, B. L. van der (1954). *Science Awakening*, I (trans. A. Dresden). Groningen: P. Noordhoff Ltd.

—— (1983). *Geometry and algebra in ancient civilizations*. Berlin: Springer-Verlag.

Waetzoldt, H. (1970–1). 'Zwei unveröffentliche Ur-III-Texte über die Herstellung von Tongefässen'. *Die Welt des Orients*, 6: 7–41.

—— (1972). *Untersuchungen zur neusumerischen Textilindustrie* (Studi Economici e Tecnologici, 1). Rome: Centro per le Antichità e la Storia dell'Arte del Vicino Oriente.

—— (1978). 'Miscellanea neo-Sumerica, VII'. *Oriens Antiquus*, 17: 35–59.

—— (1990). 'Zu den Bewässerungseinrichtungen in der Provinz Umma'. *Bulletin on Sumerian Agriculture*, 5: 1–29.

—— (1992). ' "Rohr" und dessen Verwendungsweisen anhand der neusumerischen Texte aus Umma'. *Bulletin on Sumerian Agriculture*, 6: 125–46.

—— and Bachmann, H. G. (1984). 'Zinn- und Arsenbronzen in den Texten aus Ebla und aus dem Mesopotamien des 3. Jahrtausends'. *Oriens Antiquus*, 23: 1–18.

Walker, C. B. F. (1981). *Cuneiform brick inscriptions in the British Museum, the Ashmolean Museum, Oxford, the City of Birmingham Museums and Art Gallery, the City of Bristol Museum and Art Gallery*. London: British Museum Publications.

—— (1987). *Cuneiform* (Reading the Past). London: British Museum Publications.

—— (1995). 'Mesopotamian chronology', in D. Collon, *Ancient Near Eastern art*. London: British Museum Press, 230–8.

Walters, S. D. (1970). *Water for Larsa: an Old Babylonian archive dealing with irrigation*. New Haven, CT: Yale University Press.

Waschow, H. (1936–7). 'Review of *MKT*: I–II'. *Archiv für Orientforschung*, 11: 245–7.

—— (1937–9). 'Review of *MKT*: III'. *Archiv für Orientforschung*, 12: 277.

Weidner, E. F. (1916). 'Die Berechnung rechtwinkliger Dreiecke bei den Akkadern um 2000 v. Chr'. *Orientalistische Literaturzeitung*, 19: 257–63.

—— (1927). 'Eine Beschreibung des Sternhimmels aus Assur'. *Archiv für Orientforschung*, 4: 73–85.

—— (1957–8). 'In aller Kürze'. *Archiv für Orientforschung*, 18: 393–4.

Weiher, E. von (1993). *Uruk: spätbabylonische Texte aus dem Planquadrat U 18* (Ausgrabungen in Uruk-Warka Endberichte, 12). Mainz am Rhein: Verlag Philipp von Zabern.

West, M. L. (1993–4). 'The Babylonian musical notation and the Hurrian melodic texts'. *Music and Letters*, 75: 161–79.

Westenholz, A. (1974–7). 'Old Akkadian school texts: some goals of Sargonic scribal education'. *Archiv für Orientforschung*, 25: 95–110.

Whiting, R. M. (1984). 'More evidence for sexagesimal calculations in the third millennium BC'. *Zeitschrift für Assyriologie und vorderasiatische Archäologie*, 74: 59–66.

Wilcke, C. (1987). 'A riding tooth: metaphor, metonymy and synecdoche, quick and frozen in everyday language', in M. Mindlin, M. J. Geller, and J. E. Wansborough (eds.), *Figurative language in the ancient Near East*. London: School of Oriental and African Studies, 77–102.

Woolley, C. L. (1931). 'Excavations at Ur, 1930–1'. *Antiquaries Journal*, 11: 343–81.

Woolley, C. L. (1939). *The ziggurat and its surroundings* (Ur Excavations, 5). London/Philadelphia, PA: Joint Expeditions of the British Museum and of the University Museum, University of Pennsylvania, Philadelphia, to Mesopotamia.

—— and Mallowan, M. (1976). *The Old Babylonian period* (ed. T. C. Mitchell) (Ur Excavations, 7). London/Philadelphia, PA: Joint Expeditions of the British Museum and of the University Museum, University of Pennsylvania, Philadelphia, to Mesopotamia.

Yildiz, F., and Gomi, T. (1988). *Die Puzriš-Dagan-Texte der Istanbuler archäologischen Museen, Teil II: nr. 726–1379* (Freiburger Altorientalische Studien, 16). Wiesbaden: Franz Steiner Verlag.

—— —— (1993). *Die Umma-Texte aus den archäologischen Museen zu Istanbul, Band III (nr. 1601–2300)*. Bethesda, MD: CDL Press.

Young, G. D. (1972). 'Utu and justice: a new Sumerian proverb'. *Journal of Cuneiform Studies*, 24: 132.

INDEX OF TEXTS DISCUSSED

The index of texts discussed lists all cuneiform texts mentioned in this study. It also gives the text type (e.g. problem text, house plan, etc.), date (Ur III, OB, LB, etc.) and provenance, if known. Under each index entry is a short list of previous studies of that text, except for those mathematical texts listed in Nemet-Nejat's bibliography (1993: 251–90). In these instances, only the main place of publication is referred to (*MKT*, *MCT*, *TMS*, or other), with Nemet-Nejat's bibliography entry and additions and corrections to it. Entries in **bold** type refer to transliterations and translations of a text. References to the OB coefficient lists are not given for §§3-8, as they are mentioned throughout those chapters.

The second index, §I.2, lists the numerical values of all coefficients discussed in this work, with references to the sections in which they are discussed. It includes those found explicitly and implicitly in all mathematical texts, not just the coefficient lists. Only those from list K and from B II 18-33 have been excluded, because they are not discussed here and their numerical values are all damaged or unclear. Entries in *italic* type refer to implicit coefficients; those followed by an asterisk (*) refer to coefficient-like information in the statements of problems. The numbers are given in transliteration only, that is, with no indication of absolute value; those in bold type are sexagesimally irregular. The index entries are then sorted by text and line number, with the coefficients given before other types of mathematical text. Coefficient list entries with more than one numerical value are listed under each of those values. The transliteration of the coefficient lists, §A.1, also serves as an index, ordered by text and line number.

All Akkadian and Sumerian mathematical terms discussed in the study are also indexed, in §I.3. Where appropriate, both forms of each word are given. The index is in alphabetical order; translations are not included.

INDEX OF NUMERICAL VALUES OF COEFFICIENTS

[1] Scribal error for 12.

[2] Scribal error for 1 36.
[3] Scribal error for 2 15.
[4] Scribal error for 2 42.
[5] Scribal error for 4 10 37 30.

6 Scribal error for 3.

7 Scribal error for 14 24.

[8] Scribal error for 5 12 30.

[9] Scribal error for 6.

[10] Scribal error for 12 39 22 30.

[11] Scribal error for 12 39 22 30.

[12] Scribal error for 26 40?

INDEX OF AKKADIAN AND SUMERIAN WORDS DISCUSSED